Planets
and
Their
Atmospheres
Origin
and
Evolution

This is Volume 33 in

INTERNATIONAL GEOPHYSICS SERIES

A series of monographs and textbooks

Edited by WILLIAM L. DONN

A complete list of the books in this series is available from the Publisher upon request.

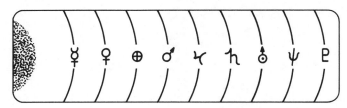

Planets
and
Their
Atmospheres
Origin
and
Evolution

JOHN S. LEWIS

Department of Planetary Sciences
The University of Arizona
Tucson, Arizona

RONALD G. PRINN

Department of Earth, Atmospheric and Planetary Sciences
Massachusetts Institute of Technology
Cambridge, Massachusetts

1984   ACADEMIC PRESS, INC.

(Harcourt Brace Jovanovich, Publishers)

Orlando   San Diego   San Francisco   New York   London
Toronto   Montreal   Sydney   Tokyo   São Paulo

Cover design for paperback edition by ALMA ORENSTEIN

ACADEMIC PRESS, INC.
Orlando, Florida 32887

United Kingdom Edition published by
ACADEMIC PRESS, INC. (LONDON) LTD.
24/28 Oval Road, London NW1 7DX

Library of Congress Cataloging in Publication Data

Lewis, John S.
    Planets and their atmospheres: origin and evolution.

    (International geophysics series)
    Includes bibliographical references and index.
    1. Planetology. I. Prinn, Ronald G. II. Title.
QB981.L398 1983   521'.54   83-10001
ISBN 0-12-446580-3
ISBN 0-12-446582-X(pbk)

PRINTED IN THE UNITED STATES OF AMERICA

84 85 86 87   9 8 7 6 5 4 3 2 1

Contents

3

Evolutionary Processes

4

The Atmospheres of the Planets

5

Conclusions

Preface

The past 20 years have seen an explosive growth in our knowledge of the solar system. Both Earth-based observational techniques (radio, radar, and high-resolution infrared spectroscopy) and spacecraft missions (reaching Mercury, Venus, the Moon, Mars, Jupiter, and Saturn, and en route to Uranus and Neptune) have provided vast stores of data quantitatively far superior to any prior information and often qualitatively different from any observations resulting from remote sensing. At present, there is a pause in the pace of the spacecraft missions to the planets. Our store of planetary data has been extensively studied, and new observational discoveries have become much less frequent. This is an ideal time for a summary of our present understanding of planetary atmospheres, including the planetary-scale and solar-system-scale processes that give rise to atmospheres and oceans, that cause them to evolve, and that deplete them.

This book is intended to be appropriate for the upper-division undergraduate or graduate student in any of the "neighbor sciences" germane to the study of planetary evolution and the behavior of planetary atmospheres (geology, astronomy, chemistry, physics, meteorology). Heavily referenced to provide access to the research literature by professional scientists in these disciplines, it, moreover, leads the reader through references back to fundamental texts in these "neighbor sciences" to help fill possible information gaps in one or more of these subjects. The book is based on courses taught by the authors at M.I.T. in the Departments of Earth and Planetary Sciences, Meteorology, and Chemistry during the years 1968–1982. It presents a 1982–1983 perspective on a wide variety of research problems that, in the past, have evolved and changed from year to year. The publication of this book at this time does not mean that we think this evolution of understanding is at an end. Rather, we have found this period of slower expansion of knowledge to be the most opportune time for a review of the past 20 years.

The literature searches on most areas covered in the book end at various points in 1982. A few important items from early 1983 have been incorporated. A severe limitation has been imposed by the diffuseness of the subject of planetary evolution. We have tried to paint a sufficiently detailed background of information from geology, meteorology, aeronomy, meteoritics, geochemistry, and cosmogony to put the evolutionary aspects of the story into a recognizable planetary context. In the case of Earth, we have selected from a vast literature so as to discriminate in favor of evidence bearing on the origin, long-term evolution, and stability of the atmosphere. We touch only lightly on such important and fascinating subjects as short-term chemical and climatic stability and the details of present geochemical cycles of the elements, which are reviewed well in other easily accessible recent sources.

While we are hesitant to reach absolute conclusions about exactly how planets acquire, regulate, and lose their atmospheres, we nonetheless believe that the range of competing theories has been greatly narrowed by information derived from the flood of new data. At the same time, new issues and fruitful new research areas have emerged. Issues such as the effects of asteroidal and cometary bombardment on Earth's biosphere have come to the fore and have generated tremendous ongoing interest. We fully expect that future discoveries will show 1983 to have been the time of a temporary lull in the rate of our acquisition of knowledge proceeding toward the still-far-off great and final synthesis.

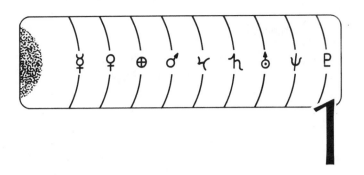

Introduction

As residents of Earth, we naturally are curious about the origin, evolution, and stability of our planetary environment. We find ourselves resident on a planet that seems, from a perhaps pardonably parochial viewpoint, to be ideally suited for life. We see all about us temperatures within the liquid range of water, and indeed liquid water covers nearly three-quarters of our planet. An atmosphere rich in oxygen, and containing significant amounts of carbon dioxide, provides for basic needs of both animals and plants. Solar ultraviolet (UV) radiation capable of destroying complex organic molecules is stopped by a stratospheric layer containing sufficient ozone to protect the Earth's surface. It would appear that our planetary environment was designed to be congenial to life! Could liquid water, abundant oxygen, an ozone layer, etc. possibly be the normal result of planetary evolution? If so, might not nearly all planets serve as abodes for some form of Earthlike life? These questions are virtually impossible to answer from our limited experience on this single planet. We must turn elsewhere to gain perspective and overcome our petty terrestrial preconceptions.

With even a cursory examination of the other planets of our Solar System, however, we find our wonder and astonishment growing faster than our comprehension. Earth's near-twin, Venus, is found to be a red-hot, parched desert with a crushing atmosphere of carbon dioxide and traces of corrosive and noxious gases. Mars is a frozen, windblown desert bearing tantalizing evidence of ancient river systems. Its atmosphere, though CO_2-rich like that of Venus, is 10^4 times lower in pressure. The Jovian planets present us with a situation far removed from that of the terrestrial planets: Each planet is largely atmosphere, and these atmospheres are remarkably similar in elemental composition to the Sun. Saturn's largest satellite, Titan, has a nitrogen–methane atmosphere far more massive than the carbon dioxide envelope of Mars. Jupiter's inner Galilean satellite, Io, is embedded in a diffuse cloud of atomic hydrogen, sodium, potassium, and sulfur ions! A few of these basic facts about the abundances of known volatiles on nine planets, Io, and Titan are summarized in Figure 1.1.

Surely our first impression must be one of utter chaos. What systematics can we discover or invent to bring order to this chaos? Just what do these incredibly diverse atmospheres have in common? In what ways are they similar, and why do they differ? Which of the properties which we take so readily for granted on Earth are truly universal, and which are highly idiosyncratic? Indeed, how can we treat such complex and diverse phenomena?

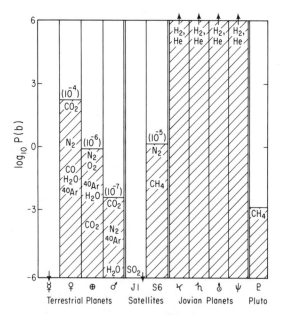

FIGURE 1.1. **Abundances of major volatiles as mass fractions (in parentheses) and as partial pressures in bars. To first approximation, the Solar System divides into the terrestrial planets, with thin atmospheres rich in CO_2, H_2O, N_2, ^{40}Ar, and other oxidized and acidic species, and the outer planets, rich in hydrogen and helium, with important traces of CH_4, NH_3, H_2O, and other reduced compounds. The volatile-element inventories of the terrestrial planets make up roughly 0.1% of the planet's mass, whereas the outer planets are almost entirely composed of volatiles. Detailed compositional data are given in Section 4; this graph is purely schematic. The symbols below the abscissa denote Mercury, Venus, Earth, Mars, Io, Titan, Jupiter, Saturn, Uranus, Neptune, and Pluto.**

1.1. ON MODELING COMPLEX PROCESSES

The approach of science is to disassemble the impossibly complex into manageable (but intentionally oversimplified) pieces. These pieces are then tested, studied, and manipulated until they are well understood. Then they are reassembled to, at best, produce a working model which reproduces the observed behavior of the natural system.

Idealized, simplified principles of structure and function are crystallized in the minds of scientists as *models*. It is very important to understand that models are intended to have practical use, but that they often do *not* reflect all that nature offers. Models are not to be categorized as "true" or "false": This is an elementary misunderstanding. Instead, models are categorizable as "useful" or "useless" with respect to their ability *to explain a certain selected range of facts* and *to predict the outcome of measurements not yet carried out.* Thus, two apparently contradictory models may both be in use to explain different related but incompletely overlapping sets of facts. The function of the creative scientist is to invent and combine models in such a way as to reconcile different successful but fragmental descriptions within a unified framework; to see the real unity under the superficial diversity.

In pursuit of this ultimate goal, models are constructed on several levels, differing in the way in which they depend on time. The simplest of these is the class of models that are invariant with time: These are called *equilibrium* models. In equilibrium models, time does not appear as a variable; all state functions are constant; there are no gradients or fluxes; the entropy is a maximum and the Gibbs free energy is a minimum. Natural systems that are perturbed slightly from equilibrium tend to approach equilibrium spontaneously, and thus this totally nonevolutionary approach provides a very valuable description of the background or ultimate state upon which all activity is superimposed.

The second class of models, often wrongly confused with the first, is that in which *local* conditions do not change with time, in that certain state functions are constant as a result of some generalized flow through the system. Thus, the local conditions lie within a *gradient* in some external field. (There are *no gradients* in an equilibrium model and thus *no time derivatives* either.) This second class of models, in which local conditions are maintained by a flow through the system, is called the *steady state*. In such a system, entropy is produced. The surface temperature of a rock orbiting at a fixed distance from a star is constant because the absorption and emission rates of radiation (both *time* derivatives) are equal. Thus, we can speak of the *steady-state* (*not* equilibrium) temperature of the rock. An object in space can have an *equilibrium* temperature only if the temperature everywhere in the Universe is the same. One may speak loosely of the *interior* of the rock being in chemical equilibrium if one remembers that the temperature governing the equilibrium is in fact established by an external physical steady state.

The third level of model involves state functions that vary cyclically, in such a way as to repeat perfectly their prior performance *ad infinitum*. Examples would be the local surface temperature on a rotating airless planet, or the mean temperature of the rock mentioned above if it were in an eccentric rather than circular orbit. Each state function varies over a range between certain limits, and does so in a cyclic manner. These cycles are generally analyzed in terms of the rates of competing processes, which are self-regulated in a stable way.

The second (steady-state) and third (cyclic) modes of behavior lie within the scope of nonequilibrium thermodynamics. It can be shown theoretically that sufficiently complex self-regulatory networks of processes may have several different configurations (different cyclic modes) depending on the initial static and dynamic conditions, and that catastrophic changes from one cyclic (or steady-state) mode to another may occur for certain critical values of the boundary conditions, even if the latter are varied slowly and smoothly.

Highly complex disequilibrium systems in which energy flow maintains cyclic processes are termed *dissipative structures*. These systems structure themselves in such a way that the rate of production of entropy (disorder) is minimized. Living organisms are examples of dissipative structures exhibiting minimum production of entropy.

At the fourth level of modeling, we recognize that the previous two types often are based on the *mathematical* assumption of constancy of certain physical parameters. For example, the rock in circular orbit must change slightly in temperature over long periods of time because the luminosities of evolving stars are not constant. These *evolutionary* or *secular* changes make it impossible in reality for any cycle ever to repeat perfectly. Nonequilibrium thermodynamics warns us that radically differing modes of operation may succeed one another with catastrophic abruptness as these boundary conditions slowly change. The different individual configurations may each be treatable to excellent precision by a steady-state of cyclic model, but the critical *transitions* between states can be explained and predicted only by a fully *evolutionary* model.

Smoothly evolving systems, or those persisting in a limited number of such cycles, may be said to reflect the *uniformitarian* principle. Discontinuous changes in the nature of the structure or behavior of evolving systems, even those resulting from smooth variation of boundary conditions, illustrate the principle of *catastrophism*. Neither principle is "true" or "false." It is in fact now very clear that the overwhelming majority of the modern history of a planet can be well described by a strictly uniformitarian evolutionary view, whereas extremely rare but violent catastrophic events, most often of a stochastic nature, can cause major changes in climatic and compositional regulatory cycles. These catastrophic interventions are of greatest significance on Earth, where the extreme complexity of biospheric processes admits of vast numbers of possible structures for transporting energy and matter.

Thus, we are led naturally to a need to understand the laws governing evolutionary changes. Such laws must be framed in universal terms: They must be equally valid everywhere and at any time; they must comprehend catastrophism as well as uniformitarianism. On the abstract level of principles, we have reached complete understanding of the changing system only when we can explain it in terms of nonchanging principles. Only then can we rationalize rigorously our use of the less complex levels of modeling and delineate precisely their limits of usefulness. Figure 1.2 summarizes schematically the behavior of each of these types of models.

It must be kept in mind that it is the synthesis of all of these approaches which we seek: We must see the background equilibrium state, the energy flows, the gradients, the thermodynamic forces, the regulatory force balances, the cyclic excursions, the evolutionary changes, and the timeless principles underlying change all at once, or we do not understand completely.

But in order to develop an understanding of the atmospheres of the planets we must not only attempt to decipher the natural laws governing evolution, but also collect a vast body of factual constraints on the application of these laws. Both initial and boundary conditions must eventually be clarified. The gathering of factual information to constrain these theories of evolution is a vast task, and any source of useful information is welcome. It is clear that many specialities must contribute to this synthesis.

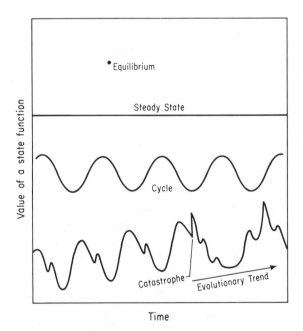

FIGURE 1.2. **Time dependence of several classes of models. Equilibrium models are time-independent; steady-state models are established to maintain certain state variables constant by equating the rates of opposing processes in the presence of gradients and net energy flow. Complex systems may adapt to such boundary conditions by establishing cyclic processes to transport the required fluxes with minimum production of entropy. Evolutionary processes show a steady drift in properties, usually modulated by a variety of cycles, and often display catastrophic functional or morphological changes in transit.**

1.2. INTERPLAY WITH OTHER AREAS
OF INVESTIGATION

Our sources of information about atmospheres impose certain nonphilosophical constraints on the conduct of this effort. We, of course, have an immensely detailed body of data on Earth's present atmospheric and hydrospheric composition, structure, and motions. Historical data, deteriorating markedly in both quantity and quality with increasing age, are available from geological studies of ancient sediments, and direct observations of crustal motions and heat flow are available. The overwhelming preponderance of human knowledge of atmospheres concerns Earth. But what aspects of the composition, distribution, and history of Earth's volatiles are faithful reflections of universal trends, and what aspects are highly specific to Earth? The very low terrestrial abundance of neon, which is an important component of primitive solar gases, has been used to argue against any simple derivation of Earth's atmosphere from a solar-composition progenitor. But is the neon abundance on Venus or Mars high or low? Do we have any good a priori reason to expect it to be similar to that on Earth?

We know that Venus is so hot (above the critical temperature for H_2O) that H_2O exists only as water vapor; that Earth's surface is dominated by liquid water; that water ice is widely stable on Mars. But how much H_2O is there in each planet? How much of the primordial water content was released into the atomsphere, hydrosphere, and crust during the thermal evolution of the planet? How much water has been destroyed by solar ultraviolet (UV) light, and how much hydrogen has been irreversibly lost to space? What sources of later injection of water may have been important? How do these processes depend on the mass of the planet and its distance from the Sun?

From our vantage point on Earth, we can study in detail many of the processes that act to maintain the stability of the biosphere. However, the immense complexity of the system makes this very difficult. If the planetary sciences were a laboratory discipline, we would synthesize a large number of planets with varying initial compositions and boundary conditions and explore the entire range of possibilities experimentally. But when one considers the consequences of carrying out such experiments with the planet we live on, the prospect of learning anything useful suddenly seems very remote: If we were to cause a change in, let us say, Earth's reflectivity, and thus cause Earth to change modes and enter an ice age, the cost of the experiment would surely seem prohibitive. The additional necessity of filing an environmental impact statement before each adjustment would surely be the last straw. And yet, entirely inadvertently, industrial and agricultural perturbations produced by mankind are in many cases causing serious disturbances of natural regulatory cycles. The scale of man's activities is now so large that his wastes have become a factor in global balances. The consequences of our failure to understand the way our atmosphere functions are unacceptable—yet we must gain this knowledge in only a few years.

The only obvious source of data to test the generality and completeness of terrestrial theories without waiting for and observing ecological or climatological disasters is to observe these same processes at work in other natural systems where nature has provided other sets of boundary conditions; that is, on other planets.

It is not that other planets are necessarily simpler than our own (although life on Earth is a very substantial complicating factor). Rather, the various planets are complex in *different ways.* Mechanisms that are hopelessly obscured by complexity on Earth may be quite clear on Mars, and vice versa. Examination of atmospheric balances and cycles on Venus, where several parameters (such as atmospheric composition and mass) are very different, whereas others (such as solar heating rate and escape velocity) are very similar, gives us a better and more immediate test of our theories than mere observation of processes on Earth. Thus, we see the crucial importance of comparative parallel studies of the atmospheres of Earth, Venus, Mars, Jupiter, and other planets and satellites. The general, universal laws that govern all these atmospheres are the same, but any particular aspect of these laws is not necessarily most easily understood from the study of Earth.

The dissipative structures encountered in, for example, the atmospheric circulation regimes of the planets, are extremely diverse. By comparative study of the atmospheres of Earth and its sister planets we may hope to develop a body of meteorological theory which spans these regimes without ever having to witness the catastrophic conversion of one structure to another on any one planet.

Finally, it is essential that the reader begin this book with the clear realization that, although its focus is the origin and evolution of atmospheres, it is wholly impossible to divorce this subject matter from the general realm of planetology. We must understand the principles governing condensation and stability of preplanetary minerals in the primitive solar nebula. We must see how these solids accreted to form planets in order to assess the bulk composition, oxidation state, and volatile content of the newly formed planets. We must be able to account for the major heat sources leading to the heating and melting of planetary bodies. We must understand the geochemistry of the melting, differentiation, and crystallization processes which give rise to cores, mantles, and crusts. We must know how the volatile elements migrate during geochemical differentiation, how the composition of surface magmatic gases evolves with time, and how the oxidation state of gases and minerals varies with depth and with time. We must allow for the photochemical destruction of complex molecules and the escape of light atomic and molecular species such as H, H_2, and He from the upper atmosphere, with a consequent change of oxidation state of the atmosphere and of crustal sediments. We must also know the rates of injection of exogenic material by meteorite and comet infall, and by capture of solar wind gases. Finally, we must understand the intricate network of gas–crust reactions and geological, biological, and meteorological cycles that maintain the incredibly diverse range of environments on planetary surfaces.

The purpose of this book is to describe the present state of our knowledge of each of these aspects of planetary atmospheres, and to indicate the essential features and areas of usefulness of present theoretical models. It is not and cannot be a review of all relevant considerations from every science. Much of the background literature for the subjects treated in this book lies well outside the conventional boundaries of planetary science, or is presented in reference to Earth with little or no useful mention of the other planets. Most readers will be familiar with almost all the background we assume; however, so wide-ranging is the scope of the planetary sciences that few readers will possess *all* the necessary background. To alleviate this problem, we cite many reviews and source books throughout the text. However, some readers may prefer or require an even more general introduction to some areas. We here provide a short list of such general references.

For a valuable introduction to the theory of chemical equilibrium, we suggest *The Principles of Chemical Equilibrium*, by K. G. Denbigh (Cambridge: Cambridge University Press, 1957). A broad perspective on the study of our planet is given in *Earth*, by F. Press and R. Siever (San Francisco: W. H. Freeman, 1974). The nomenclature and principles of geochemistry are well given in *Introduction to Geochemistry*, 2nd ed., by K. B. Krauskopf (New York: McGraw-Hill, 1979). The regulatory cycles of the volatile elements on Earth are discussed by J.C.G. Walker in *Evolution of the Atmosphere* (New York: Macmillan, 1977), who, however, interprets planetary history in the context of a genetic model we regard as very unlikely. The subject of atmospheric chemistry is introduced by P. M. Banks and G. Kockarts in *Aeronomy* (New York: Academic Press, 1973). Numerous excellent introductions to physics and physical chemistry are available.

1.3. CONDENSATION AND ACCRETION

The first touchstone in our search for unifying principles must be in our fundamental cosmogenic concept of the derivation of the Sun, its planets, satellites, asteroids, and comets from a primordial gas and dust cloud called the solar nebula. This gas and dust cloud, though originally very diffuse and chemically and physically homogeneous, appears to have passed through a dense, collapsed state in which there were strong temperature and pressure gradients outward from the center. Thus, the condensed materials may have varied enormously in composition with distance from the center due to the stability of condensates against evaporation. Those species with very low vapor pressures were everywhere condensed, whereas highly volatile substances were condensed only at very great distances from the center. The accretion of these solids into large bodies must then occur with some degree of radial (compositional) mixing. Thus, conditions of origin profoundly affect the resulting compositions of the planets which accrete from these materials.

All such condensation–accretion scenarios are proposed within a particular framework of knowledge and assumptions regarding the physicochemical nature of the solar nebula and the early lives of stars. It is widely (but not universally) accepted that stars form from moderately dense nebulae comprising gases and dust with overall elemental abundances essentially identical to those in the Sun and in other normal ("Main Sequence") hydrogen-burning stars. Temperatures in such a nebula are in the range from a few tens of degrees absolute to a few thousand degrees. The temperature is believed to fall off rapidly with heliocentric distance and with distance from the midplane due to the large thermal opacity. The nebula is a sufficiently good insulator to permit temperature gradients large enough to drive convection over large regions of its interior. The properties of a particular solar nebula model by Cameron and Pine (1973) are detailed in Figures 1.3–1.5. This model is characterized by a steady state in which the nebular mass is constant and the heat flux through the nebula is constant. More recent and more realistic models by Cameron (1978) are based on the assumption of a steady-state configuration in which there is a constant flux of mass into the "accretion disk," further modified by consideration of the time-dependent phenomena attendant on turning this mass flux on and, finally, off. The *peak* temperatures found in this model for the regions of condensation of preplanetary solids are strikingly similar to those illustrated in Figure 1.2, and the pressure profiles are sufficiently similar so as to have no large effect on the chemical behavior of the system.

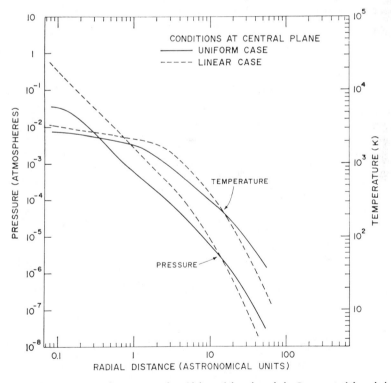

FIGURE 1.3. **Temperatures and pressures at the midplane of the solar nebula. Over most of the nebula inside Jupiter's orbit the conditions are adiabatic or nearly so. The flattening of the temperature profile at and above 2000 K is due to the absorption of heat by the thermal dissociation of H_2. The only important opacity source in the inner solar system is grains of metallic iron. (After Cameron and Pine, 1973.)**

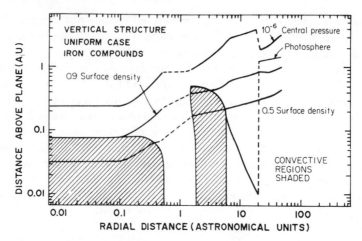

FIGURE 1.4. Vertical structure of the nebula. Most of the nebular mass at any heliocentric distance is the thick, relatively isothermal and isobaric layer bracketing the midplane, where the acceleration due to the gravitational force of the gas is small. Far from the midplane, g is large, the temperature and pressure gradients are steep, and little mass is present. Solids condensed in this cooler region will sediment rapidly toward the midplane and will experience the higher temperatures there. (After Cameron and Pine, 1973.)

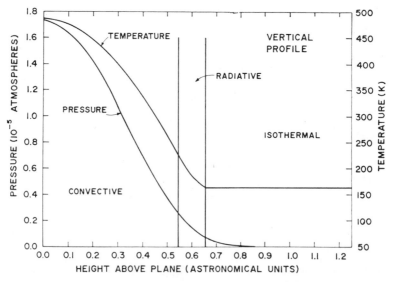

FIGURE 1.5. Vertical temperature and pressure profiles of the nebula. In the dense, opaque gas near the midplane, heat transport is by convection. At lower pressures, the temperature gradient is smaller due to radiative cooling. Yet farther out, the small amount of remaining gas is quite transparent and nearly isothermal. (After Cameron and Pine, 1973.)

During this early nebular stage, those solid particles which are chemically stable at any distance from the center of the nebula will accrete to form larger solid bodies. Electrostatic, magnetostatic, and surface adhesion forces may allow individual mineral grains to adhere *selectively* to one another.

Over a few thousand years, large particles in the nebula have ample opportunity to settle "down" onto the midplane of the nebula under the influence of the gravitational field of the gas–dust mixture, and thus build up a thin central layer of enhanced dustiness. Goldreich and Ward (1973) have shown that gravitational instabilities in this dense dust layer can, if the layer is dense enough, lead to rapid formation of a host of sub-lunar-sized (asteroidal) bodies in low-eccentricity orbits. The time scale estimated for the duration of the nebular phase, about 10^5 years from nebular formation until ignition of the Sun, is adequate for the formation of such bodies.

Estimates of the lifetime of the nebula against various possible instabilities are of course dependent on the nebula model, and ideally should be based on a fully time-dependent evolutionary model with reasonable initial and boundary conditions. The closest available approach to this ideal is the accretion disk model already mentioned (Cameron, 1978), in which a nebular lifetime of $\sim 10^5$ years is suggested. The outer part of the nebula may be subject to powerful instabilities which give rise directly (DeCampli, 1978) or indirectly (Perri and Cameron, 1974) to gaseous giant protoplanets.

Theoretical and observational studies of the early collapse-phase evolution of stars have been thoroughly reviewed by Larson (1974) and by Woodward (1978). It is generally accepted on the basis of these studies that the newly formed Sun will be superluminous by about a factor of three until the gravitational potential energy of contraction has been given off. During this period of $\sim 10^7$ years the Sun would have had roughly the same surface temperature it has today, but, because of its larger radius and surface area, it would have been more luminous with a lower surface gravity. During this phase, the Sun would expel mass as a dense "solar wind," a radially expanding proton/electron plasma many times denser than the present solar wind, but with similar particle speeds. Many young superluminous stars, which are recognized by their being slightly too luminous for their color relative to stable Main Sequence (MS) stars, are found associated with dense interstellar clouds of gas and dust and with extremely massive and luminous ($L \simeq 10^4$ $L_\odot$) MS stars. Since the lifetime of such spendthrift MS stars are short ($< 10^8$ years), they must be young—and thus the non-MS stars found in the same clusters must be young also. Those young stars which are superluminous relative to the MS are called T Tauri stars after the prototypical example. A number of such stars have been observed to be surrounded by moderately dense optically thin envelopes with large radial velocities. The stars are thus shedding (or gaining) mass at a rate sufficient to change the mass of the star by tens of percent over the duration of this "T-Tauri phase" in its evolution, preceding the star's arrival upon the MS (Kuhi, 1964). According to both the observed mass fluxes and theoretical studies of stellar evolution, a star which ends up on the MS with one solar mass ($M = 1\ M_\odot$) may shed ~25% of its mass over its T-Tauri phase, which lasts several times 10^7 years. Alternatively (Ulich, 1976), the T-Tauri phenomenon may be due to *infall* of a comparable mass of material. Stars in the T-Tauri phase have typical luminosities of two or three times their ultimate value when established on the MS.

In the 1960s, there was much discussion of the possibility of a much shorter-duration earlier phase in pre-MS evolution, in which the luminosity could briefly attain 10^4 or even 10^6 times the ultimate MS luminosity of the star. This putative extremely luminous state, called the Hayashi phase, has never been observed, and is believed by many astrophysicists to reflect artificial and unrealistic mathematical and physical assumptions in the model, such as assuming zero rotational angular momentum for the collapsing protostar. The interested reader is referred to the detailed treatment in the review paper by Larson (1974), and the review of early stellar evolution by Woodward (1978).

The ignition of the Sun, which occurred when the central conditions became adequate to begin and contain hydrogen fusion reactions, probably occurred within ~10^4–10^5 years of the formation of the solar nebula. The solar nebula was then rapidly dissipated by infall into the Sun or expulsion from the Solar System by the new T-Tauri phase solar wind. Chemical reactions between gases and solids were quenched rapidly. All unaccreted dust particles were at that time swept along with the gas and thus lost. Thus, in order for planets to form, accretion of mineral grains into sizeable chunks must have occurred in $<10^5$ years. The minimum size necessary to allow these chunks to survive dissipation of the nebula is ~10 m, but of course depends on distance from the Sun.

Goldreich and Ward's treatment of accretion suggests that bodies of sizes ~10 m to ~1000 km will dominate the solar system at the time of dispersal of the nebula.

Safronov, in an extended series of articles, has investigated the accretion of such a host of sub-lunar-sized bodies into planets (see, especially, Safronov, 1972). He has found a typical accretion time scale of ~10^8 years in the inner solar system. Weidenschilling (1974, 1976) has independently treated this problem and arrived at remarkably similar conclusions, with accretion times of a few times 10^7 to ~10^9 years for the inner planets, and Cox and Lewis (1980), in a more sophisticated treatment, again find ~10^8 years.

There must be significant radial mixing of materials during accretion (Wetherill, 1975; Hartmann, 1976), with consequent blurring of the compositional differences frozen in at the time of dissipation of the nebular gases. Thus, each planet that eventually accretes will necessarily be inhomogeneous in *origin*, although not necessarily very inhomogeneous in *composition*: high- and low-temperature components may be well-mixed throughout the planet, as in most classes of meteorites. Unfortunately, no unified condensation–accretion model for the planets has yet appeared in the literature, and qualitative assertions unsupported by tractable models and quantitative predictions are rife. Studies of the application of radial mixing models of Wetherill, Hartmann, and of Cox *et al.* (1978) to the accretion of the terrestrial planets are now available (Cox and Lewis, 1980), and the first combined condensation–accretion model has just been completed by Barshay and Lewis (1981).

It might seem that our understanding of condensation and accretion would be remarkably simplified by considering meteorites. After all, most meteorites date from the very earliest days of the Solar System, and most show little or no evidence of any alteration of their composition or structure since that time. Many chondrites, when dated by the radioactive decay of ^{232}Th, ^{235}U, and ^{238}U by α and β^- emission, and by decay of ^{40}K to ^{40}Ar and ^{40}Ca by β^+ and β^- emission, give closely concordant ages of 4.5×10^9 years. It is often assumed that this was the time of separation of meteoritic solids from the solar nebula gases. All meteorites whose ages deviate from this figure are younger, and most of these show obvious signs of severe mechanical shock and reheating. The effect of heating is, of course, to drive off a portion (or all) of the accumulated radiogenic gases ^{4}He and ^{40}Ar, making the meteorite appear younger than it really is.

Most meteorites contain angular fragmented mineral grains, ranging from very fine submicron dust to millimeter-sized particles, including spherical or spheroidal blebs of glass, which may be partly devitrified and crystallized into the common meteoritic minerals. The spheroidal objects are called "chondrules," and those meteorites which contain them are called "*chondrites*." The dominant minerals in most chondrites are olivine, $(Mg, Fe)_2SiO_4$; pyroxenes, $(Mg, Fe, Ca)SiO_3$; troilite, FeS; metal, (Fe, Ni); and plagioclase $(NaAlSi_3O_8–CaAl_2Si_2O_8$ solid solutions). The bulk compositions of meteorites reflect closely the observed composition of the Sun. Neighboring grains may have, however, very different Fe:Mg ratios. All textural evidence suggests that the chondrites are a low-temperature mixture of materials with different thermochemical histories. Some chondrites have been "cooked" at a sufficiently high temperature to approach closely the state of internal chemical equilibrium, but this temperature seems to have been far below the melting temperature. It is common to regard the chondrites as polymict breccias (multicomponent, multisource fragmental mixtures) which have been metamorphosed (heated and recrystallized) to varying degrees.

A small proportion of the ~3000 known meteorites, about 20% of the total, have been so severely heated that melting and magmatic differentiation have occurred. All traces of chondrules have been erased, internal chemical equilibrium has been closely approached, and the texture is of igneous origin. Meteorites of this type, called *achondrites*, are subdivided according to their bulk composition and mineralogy. Meteorites composed dominantly of iron–nickel alloy crystals are called *irons*, and those with igneous textures which contain comparable amounts of metal and silicates are called *stony-irons*. Almost all irons, stony-irons, and achondrites appear to be the results of melting and density-dependent separation in a gravitational field: They are thus of igneous, not metamorphic, origin.

Because of their properties, it is customary to refer to chondrites as "primitive" meteorites. This is taken by some authors to mean that they originated at the earliest point in time; by others to mean that they are closest to the Sun in elemental composition; and by others to suggest that they are the parent material from which the other (igneous) meteorite types and the present terrestrial planets were derived. These conflicting definitions and assumptions became even more sharply focused when the variations between different types of chondrites are considered in detail (Section 2.3).

The various classes of chondrites are usually taken to represent material accreted under different conditions of temperature and pressure, perhaps in the simplest case reflecting conditions at different heliocentric distances in the nebula.

Other authors consider that the entire nebula underwent cooling from very high to very low temperatures, so that the composition of solid grains in the nebula at any given heliocentric distance changed with time in such a way so as to generate progressively more oxidized and more volatile-rich material. This material is pictured as accumulating on large bodies in the sequence of condensation during cooling. In this view, the most refractory minerals are most primitive in the sense that they are slightly older, but the carbonaceous, volatile-rich last condensate is most primitive in the sense that it most accurately reflects the composition of the Sun.

It was formerly believed by some meteoriticists that these *carbonaceous chondrites* were the parent material of other classes of meteorites. However, careful isotopic analyses of meteorites have shown that this is not possible with any combination of simple known processes (see Section 2.3).

Of great interest to us in our attempt to understand the origin of planetary atmospheres is the quantity and composition of volatiles in each class of meteorites. For the moment, it suffices to remark that the *carbonaceous chondrites* are the most volatile-rich meteorites.

Anders (1976) and Anders *et al.* (1976) have shown that one may generally account for the abundances of trace volatile elements in the chondritic meteorites by picturing these objects as mixtures of two components. One, wholly volatile-free, is a relatively coarse-grained high-temperature mineral assemblage such as that which dominates the ordinary (noncarbonaceous) chondrites. The other, a very fine-grained volatile-rich mixture of low-temperature minerals, is closely similar to the dominant portion of the carbonaceous chondrites. These two components are then assembled at a particular "formation temperature" to make the observed groups of chondrites.

In practice, this scheme is found to be rather less elegant than might be hoped. The compositions of both the high- and low-temperature components are found to be variable, not constant. No relationships between the pairs of components needed to make a given meteorite class have been proposed.

The crucial central question of *why* meteorite compositional groups are so sharply defined and not continuous seems to be answerable only by asserting that each class represents one or two parent bodies, and that this "quantization" of groups is solely due to the nondemocratic nature of the sampling of parent bodies by the Earth. If, as almost all authors have done, we were to regard the asteroid belt as the source of ordinary chondrites, then we would have to assert that the meteorites falling on Earth must originate on those few asteroids and comets which have particularly favorable dynamical mechanisms to supply them. Asteroids in Earth-crossing orbits are an obvious choice (although such orbits are so short-lived that we must ask where *they* in turn came from).

Recent astronomical studies of asteroids have permitted the determination of the spectral reflectivities of over 100 asteroids and of the approximate bulk densities of the three asteroids that are large enough for mass determinations (Chapman, 1976). The results are remarkable in that the asteroid belt is found to be dominated throughout by carbonaceous chondrite material, whereas ordinary chondrite material may be totally absent from the asteroid belt. The only asteroids whose spectral reflectivities are compatible with ordinary chondrite composition are Earth-crossers. It is also interesting that a number of asteroids resemble stony-irons, irons, or achondrites in composition. The large asteroid 4 Vesta was found by McCord and co-workers (1970) to have the reflection spectrum of a basaltic achondrite. Several of the largest asteroids have very dark surfaces (albedos near 0.025), and density data on 1 Ceres and 3 Pallas suggest that they are also much less dense than ordinary chondrites. There are no clear compositional trends with size: The primitive material throughout the belt seems to have been carbonaceous.

The accretion of terrestrial planets from primitive (early) solids found in the nebula may sample distant regions of the Solar System to a small but important degree. Earth's oceans make up only $\sim 0.03\%$ of the mass of the Earth, but carbonaceous chondrites contain up to 10% water by weight: Thus, only 0.3% by weight of typical asteroidal material incorporated into the early Earth could provide all the water now in the oceans! It is clearly important to know how broad a region, and how wide a range of formation temperatures, is sampled by each planet as it forms. Several simple models for this sampling function are sketched in Figure 1.6. The first simply asks what material is condensed at the exact heliocentric distance at which the planet condenses and accretes, and wholly neglects radial motion of solids. The second allows each planet to accrete all the solid material that is closer to it than to the next planet, with perfect efficiency. The third model is similar, but divides the source zones at the points of equal gravitational influence instead. The fourth allows for radial diffusion of solids, and approximates a Gaussian sampling efficiency, with very small, but finite, amounts of material collected from distant parts of the solar system. Reality is more complex than any of these cases, and we can expect the sampling efficiency curves to change shape as the mass distribution and orbits of solid bodies evolves. However, we should bear in mind the general shape of these accretion efficiency curves as a guide to realistic behavior. It is worth mentioning that the first unified (although still oversimplified) condensation–accretion modeling for the planets has only recently been undertaken (Barshay and Lewis, 1981).

But it is not sufficient merely to trace the variation in condensation and accretion conditions with distance from the center of the nebula: We also must unravel the

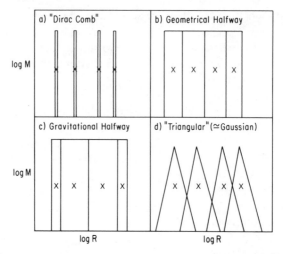

FIGURE 1.6. **Sampling efficiencies for accreting planets as a function of heliocentric distance.** (a) The simplest and least realistic model, and the easiest to quantify, considers the material of a planet to originate exclusively at a single *P, T* point and a single heliocentric distance. Other relatively simple sampling scenarios entail dividing the solar system into sharply bounded "feeding zones," either (b) by the purely geometrical criterion of requiring each particle to be accreted by the planet with the closest orbit, or (c) by assigning the boundaries between zones on the basis of the magnitudes of the gravitational effects of the planets. The latter, which apparently improves on the former, introduces the added difficulty that the relative masses must then be known as a function of time. In none of these cases could a planet accrete any material of distant origin, either from much closer to the Sun or much farther from it. (d) A sampling probability curve that dies exponentially on both sides, thus permitting almost all mass to be derived from the local "feeding zone," while a small amount of very distant condensate is also swept up. Note that the inner wing of the curve is of relatively small import, since it can only enrich elements which are already abundant in the planet. The outer side, however, can add materials that otherwise would be wholly absent from the planet: highly volatile low-temperature condensates. This trace of material could profoundly affect the volatile budget of the planet.

immensely complex sequence of changes and regulatory mechanisms that gave rise to and maintain current conditions on the planets. The next two chapters are devoted to general discussions of the origin of planetary atmospheres and of regulatory processes.

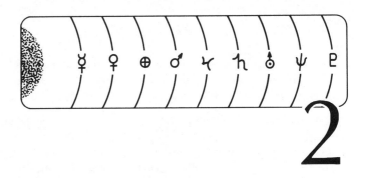

Retention of Volatiles
by Planets

The raw material of the Solar System, the solar nebula, was a mixture of gases and dust with very nearly the same overall elemental composition as the Sun. In the nebula, about 98% of the total mass was a mixture of hydrogen and helium, a little under 2% was ice-forming materials (H_2O, NH_3, CH_4, etc.), and about 0.5% was rock-forming solids (SiO_2, MgO, Fe, FeO, FeS, etc.). The present best estimates of the compositions of Jupiter and Saturn are close to these solar proportions.

On the other hand, the "solid" bodies in the Solar System, especially the terrestrial planets and the lunar-sized satellites, have atmospheric masses ranging from $\sim 10^{-4}$ of the total planetary mass downward (see Figure 1.1). The dominant constituents of these atmospheres include N_2, CO_2, O_2, radiogenic ^{40}Ar from ^{40}K decay, H_2O, SO_2, and CH_4. Neon, which is very abundant in the Sun (nearly equal to the nitrogen abundance), is very rare on Earth. There are no obvious compositional similarities between these atmospheres and solar material; further, the atmospheric masses of the terrestrial planets are 10^7–10^{10} times smaller relative to the rocky matter than in the Sun. How is it that these bodies have atmospheres, and how can they deviate so markedly in composition from the Sun?

Several models have been proposed and explored to varying degrees in an effort to explain the presence and properties of these atmospheres or of their components.

We shall discuss six rather diverse major models that span the full range of speculation: Two are based on gravitational capture of solar nebula gases, and four are
based on the observed and theoretically predicted compositions of meteorites.

1. At one extreme, gaseous protoplanets with solar elemental composition have
been suggested. Gravitational instability in the nebular gas–dust mixture would be a
possible source of such bodies. Gravitational settling of condensates to the centers of
mass of these protoplanets occurs. As each protoplanet cools, the core, mantle, and
crust would be deposited sequentially beneath an enormously massive hydrogen-
rich atmosphere, which would then have to be dispersed with almost perfect
efficiency.

2. In a closely related concept, condensate accretion by electrostatic, magnetic
and surface adhesion forces, and eventually by gravitation, would give rise to
growing solid planetesimals embedded in the gaseous nebula. Sufficiently large
solid planetesimals could later gravitationally capture portions of the nebular gas to
produce protoplanets somewhat akin to those described above on model 1.

3. The class of meteorites most closely related to the composition of the Sun,
the Carbonaceous Chondrites Type C1 (or CI after its prototype, the Ivuna
meteorite), has been suggested as the parent material of the terrestrial planets. Since
C1 material is considerably more volatile-rich than the terrestrial planets, and since
the terrestrial planets vary greatly in composition and density, extensive loss of
volatiles from all planets (and evaporation of silicates from Mercury) is required.

4. Formation of the Earth, and perhaps the other terrestrial planets, from ordinary chondritic material has also been suggested. The class of meteorites most
similar in bulk composition to the Earth, the high-iron (H-group) chondrites, also
has the advantage of a closely similar oxidation state and volatile content.

5. Accretion of a largely volatile-free Earth, comparable to achondritic material,
followed by addition of a trace of very volatile-rich (C1 or cometary) later infall, has
been suggested. The resulting layered planet may closely resemble the planetary
"core" of condensed matter formed in Model 1 above.

6. Finally, it has been proposed that the bulk compositions, oxidation states,
densities, and perhaps the volatile contents of the terrestrial planets were determined by the temperature at which preplanetary material last equilibrated with the
solar nebula. Here the main variable is the dependence of accretion temperature on
heliocentric distance, and the oxidation state, density, and volatile content are
strongly and uniquely interrelated through their dependence on temperature.

Models 1 and 2 partake of many common features, and differ significantly only
in the mechanism suggested for triggering accretion. In Model 1, one must understand the criteria for gravitational collapse in a rotating gaseous medium in order to
show the feasibility (or infeasibility) of the process. Unfortunately, this problem has
not been adequately studied. Even the physical parameters describing the structure
and dynamics of the unperturbed nebula are subject to wide-ranging uncertainties.
Nonetheless, such models must certainly be taken seriously, since not only stars in
multiple systems but also the Jovian planets of the Solar System, which have

compositions closely similar to that of the Sun, must be formed by one or the other of these mechanisms. But the most difficult task is not to explain the atmospheres of the outer planets: the real problem is to explain how the atmospheres of the terrestrial planets, which deviate very widely from one another and from solar composition, can be derived from a nebular parent material that was both quite uniform and solar in composition.

The remainder of this chapter will deal with the assumptions and results of each of these models. We include in Section 2.3 a summary of the compositions of chondritic meteorites for use in our discussion of Models 3–6.

2.1. SOLAR-COMPOSITION PROTOPLANETS

One early idea for the origin of the terrestrial planets was the suggestion that gravitational instability of the nebular gas gave rise to "Gazkugeln" of solar composition, with very high central temperatures and pressures. Jeans (1928) provided a simple criterion that must be met by a parcel of gas before it may collapse: that the gravitational potential energy of the (spherical) parcel exceed its internal thermal energy. If G is the universal gravitational constant, M the mass of gas within radius r of a point in a gas cloud, k the molecular gas constant, and T the absolute temperature, then collapse can occur within r only if

$$GM/r \geq kT. \tag{2.1}$$

Since $M = 4\pi r^3 \bar{\rho}/3$, where $\bar{\rho}$ is the mean gas density within r, this criterion may be restated as

$$T \leq \frac{4}{3} \frac{\pi \bar{\rho}}{k} Gr^2. \tag{2.2}$$

Unfortunately, this criterion is valid only for nonrotating gas clouds. Its applicability to a rotating nebula with a highly nonspherical density distribution is clearly limited. Once such gas balls were formed, the central temperature quickly became too high for condensates to be present, but radiative cooling from their outer surfaces allowed condensation of materials in those spherical shells just cool enough for saturation. Because of the strong radial temperature gradient, these gas balls could be convective, with such rapid overturn that each rising gas parcel would expand as a closed system. This means that the pressure–volume work done during expansion is exactly balanced by the decrease in internal thermal energy of the gas. This kind of process, with no exchange of heat with the surroundings, is called an *adiabatic* process. Knowledge of a single pressure–temperature point anywhere in the gas ball permits calculation of the entire radial profile of temperature versus pressure.

Eucken, writing in 1944, examined the condensation chemistry of such a gas ball at high temperatures (rock-forming mineral condensation) and sketched out the expected sequence of condensation of minerals during cooling of the gas ball. These calculations, although, of course, very imprecise at the very high pressures in the center of a gas ball, nonetheless showed the feasibility of initial condensation of metallic iron–nickel alloy, followed after further cooling by condensation of magnesium silicates. In the gravitational field of the gas ball, each condensate grain or droplet would, of course, rapidly sediment inward, destined either to reach a layer hot enough to vaporize it or to accrete into a core at the center. Thus, cooling of the gas ball leads to sequential deposition of metal, magnesium silicate, and finally volatile-rich condensate layers, in a crude imitation of the core–mantle–crust structure of the Earth.

Temperature, pressure, and density models of these gas balls have since been constructed to greater precision by Ostic (1965). The treatment by Eucken of condensation in an adiabatic, solar-composition atmosphere has been extended by Lewis (1969a,b) and by Barshay and Lewis (1976), for the higher-pressure case of the atmospheres of the Jovian planets, to a very large number of gaseous and condensed species. Such calculations can be done without large nonideality corrections.

The chemistry of sequential condensation in such a system, with rapid isolation of condensates from the gas explicitly assumed, has been investigated in detail for both high-pressure adiabats (the atmospheres of the Jovian planets: Lewis, 1969a,b; Weidenschilling and Lewis, 1973; Barshay and Lewis, 1978; Fegley and Lewis, 1979) and for low-pressure adiabats (the solar nebula: Lewis, 1972a, 1973b, 1974a; Fegley and Lewis, 1980).

The thermochemical problems presented by Eucken's model are of a particularly severe nature. Not only is condensation hypothesized to be occurring at very high temperatures, but also in a gas of such high density that it is very strongly nonideal. Real-gas equations of state for H_2 and He are of dubious accuracy above 10 kbar, and the fugacity coefficients of minor (condensible) species dissolved in the dense H_2–He fluid are completely unknown. There exists a real possibility that *all* solids and liquids may dissolve in the hot fluid phase at very high densities, and hence that no rocky core may form.

To the best of our present understanding, the sequence of condensates in Eucken's model (and hence the sequence of layers deposited from the center of the protoplanetary core outward) is as follows:

(1) Refractory oxides (minerals rich in CaO, Al_2O_3, Ti_2O_3, rare earth oxides, MgO, etc.)
(2) Metallic iron–nickel–silicon alloy
(3) Mg_2SiO_4 and $MgSiO_3$
(4) Silica (SiO_2)
(5) Volatile element halides and sulfides (Na_2S, NaF, K_2S, etc.)
(6) Ammonium salts ($NH_4H_2PO_4$, NH_4Cl, NH_4Br, etc.)
(7) Water (H_2O, dissolved NH_3, etc.)
(8) Ammonium hydrosulfide (NH_4SH)
(9) Ammonia (NH_3)
(10) Methane and argon (CH_4, Ar).

(11) Hydrogen and neon (H_2, Ne) might condense next, and finally

(12) Helium (He), but only at impossibly low temperatures.

If the temperatures in the gaseous envelop never fall low enough for condensation of the *Mth* entry in this list, then all species M, $M + 1$, etc. with *lower* condensation temperatures will remain well-mixed in the gaseous envelope of the protoplanet.

The most fundamental consequence of this model with respect to the origins of atmospheres is that each planet must contain a solar complement of all volatiles from inception. There then exists the embarrassing necessity of discovering a mechanism that removes this primitive atmosphere almost without a trace, since retention of even one-millionth of the primordial gas envelope would leave amounts of many volatiles that are far larger than their present abundances on the terrestrial planets.

Furthermore, it is very unclear whether the planet left behind after dissipation of such a gas sphere would bear any semblance to the present Earth or to any other planet. Certainly one might expect an Earth devoid of oxidized iron, possibly with the majority of its silicon reduced to elemental Si. The mineralogy would differ substantially from that of any known planet, satellite, asteroid, or meteorite (Fegley and Lewis, 1980).

In both this and the following model, the composition of the primitive atmosphere is taken to be that of the Sun. Cameron (1980) has recently reviewed these abundances, and an extract of abundance data for the most volatile elements is given in Table 2.1.

The isotopic compositions of selected volatile elements in the Sun are given in Table 2.2, also from Cameron (1980).

An interesting and particularly extreme version of this class of hypotheses was that of Kuhn and Rittmann (1941), who argued that the *present* Earth has a dense solid-hydrogen interior, and hence has very nearly solar composition even today. Modern knowledge of the state equation of hydrogen at high pressures permits the secure conclusion that the density of Earth's core is 15–20 times too high to be hydrogen (DeMarcus, 1958; Smoluchowski, 1967a; Hubbard, 1968, 1969, 1970; Zapolski and Salpeter, 1969).

The most recent development of the idea of gas instabilities giving rise to giant gaseous protoplanets is that of Cameron (1978). Basing his discussion on the concept of a dynamic, viscously coupled accretion disk as introduced by Lynden-Bell and Pringle (1974), Cameron assumes a period of steady infall of interstellar cloud material into the disk. He follows the evolution of the thermodynamic state variables in the accretion disk both during infall and after the (abrupt) cessation of infall, during which mass is lost from the outer surface of the disk, for a suite of models spanning five sets of assumed input parameters. A general prediction of this model is that axisymmetric instabilities will repeatedly generate annular structures. These gas and dust rings can then undergo collapse so that one or more gaseous giant protoplanets will form at the location of each of the rings.

TABLE 2.1
Abundances of the Most Common Volatile Elements in the Sun[a]

Element	Abundance (Si abundance = 10^6)
H	2.66×10^{10}
He	1.80×10^9
O	1.84×10^7
C	1.11×10^7
Ne	2.60×10^6
N	2.31×10^6
S	$5.0 \ \times 10^5$
Ar	1.06×10^5
P	6.50×10^3
Cl	4.74×10^3
F	7.80×10^2
Ge	1.17×10^2
Se	67
Kr	41.3
Br	9.2
Te	6.5
As	6.2
Xe	5.84
I	1.27
Sb	2.1×10^{-1}
Hg	2.1×10^{-1}

[a] After Cameron (1980).

The background run of pressure and temperature in the unperturbed dynamic accretion disk is remarkably similar to that deduced for the static models of Cameron and Pine (1973). Perhaps even more striking is the discovery (first made from comparison of numerical results of several accretion disk evolution runs) that the $P–T$ profiles of these disks are virtually identical. Furthermore, these $P–T$ profiles are essentially perfectly adiabatic throughout the region of planetary formation. Cameron's calculated evolutionary histories for the gas density at the points of formation of each of the planets are shown in Figure 2.1. The temperature–density relations for two radically different cases of disk evolution are compared with an adiabat in Figure 2.2.

It is, however, difficult to relate such protoplanets to the observed properties of the terrestrial planets. The central pressure and temperature of a protoplanet would be extremely high, so that the chemical criticisms of Eucken's model given above will apply here as well. Also, tidal stripping of the gaseous envelopes from forming terrestrial protoplanets would be very rapid closer to the Sun than Jupiter (De-Campli and Cameron, 1979). Even if a terrestrial gaseous giant protoplanet could form, other problems remain: DeCampli (1978) has shown that the body would rapidly collapse to the point where no known mechanism could dissipate the light

TABLE 2.2
Isotopic Compositions of Selected Volatile Elements in the Sun[a,b]

Element and isotope	Isotopic abundance
^{1}H	2.66×10^{10}
D(^{2}H)	$4.4 \ \times 10^5$
^{3}He	$3.2 \ \times 10^5$
^{4}He	1.80×10^9
^{16}O	1.84×10^7
^{17}O	$6.9 \ \times 10^3$
^{18}O	3.75×10^4
^{12}C	1.11×10^7
^{13}C	1.23×10^5
^{14}N	2.31×10^6
^{15}N	$8.5 \ \times 10^3$
^{20}Ne	2.31×10^6
^{21}Ne	$7.0 \ \times 10^3$
^{22}Ne	2.82×10^5
^{32}S	4.75×10^5
^{33}S	$3.8 \ \times 10^3$
^{34}S	2.11×10^4
^{36}S	68
^{36}Ar	8.93×10^4
^{38}Ar	1.67×10^4
^{40}Ar[c]	$2 \ \ \times 10^{-2}$
^{35}Cl	3.58×10^3
^{37}Cl	1.16×10^3
^{78}Kr	1.46×10^{-1}
^{80}Kr	$8.4 \ \times 10^{-1}$
^{82}Kr	4.77
^{83}Kr	4.77
^{84}Kr	23.5
^{86}Kr	7.17
^{79}Br	4.65
^{81}Br	4.55
^{124}Xe	$7.4 \ \times 10^{-3}$
^{126}Xe	$6.7 \ \times 10^{-3}$
^{128}Xe	1.27×10^{-1}
^{129}Xe	1.61
^{130}Xe	$2.5 \ \times 10^{-1}$
^{131}Xe	1.25
^{132}Xe	1.51
^{134}Xe	$5.9 \ \times 10^{-1}$
^{136}Xe	$4.9 \ \times 10^{-1}$

[a]After Cameron (1980).

[b]The elements F, P, and I have only one stable isotope and do not appear here.

[c]Begemann *et al.* (1976).

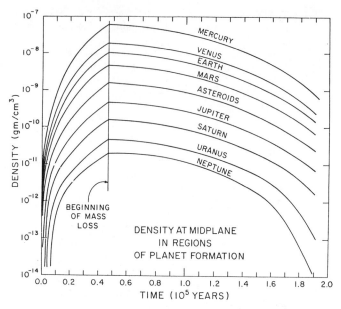

FIGURE 2.1. **Variation with time of the density in the regions of planet formation. After the results of Cameron (1978) for an accretion disk of one solar mass. Accretion if terminated at 0.47 × 10⁵ yr.**

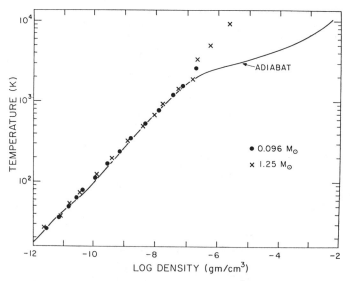

FIGURE 2.2. **Comparison of temperatures and densities in an accretion disk with an adiabatic relationship during nebular accretion for two different values of the central (protosolar) mass: (●) 0.096$M_\odot$; (×) 1.25$M_\odot$. (After Cameron 1978.)**

gases. Thus, any predominantly solid (terrestrial) planet which we try to form by sedimentation of solids within such a protoplanet (Slattery, 1978) would retain the envelope forever, and remain as a Jovian type planet today.

Furthermore, it is possible that this mechanism was not the dominant one even in the outer solar system. There the presence of large masses of solids may permit the growth of bodies with appreciable gravitational fields to be so rapid that gravitational gas capture by these bodies may be more important than gasdynamic instabilities of the nebula. This gravitational capture process will be discussed in the following section.

2.2. SOLAR GAS CAPTURE ON GRAVITATING BODIES

The possibility of gas capture by solid protoplanets embedded in the solar nebula has been mentioned by numerous authors over the past 20 years. We shall here examine the thermodynamics of gas capture for bodies ranging in size from the Earth down to the largest asteroids. Two extreme cases, isothermal and adiabatic gas capture, will be discussed. The presence of planetary bodies within the solar nebula is an integral part of some theories of the origin of the planets, most especially some of the models due to Cameron and co-workers.

We shall assume, following Cameron and Pine (1973) and Cameron (1978), an adiabatic background structure for the nebula. Into this medium we shall introduce massive bodies, as proposed by Cameron (1973), having the masses of the present terrestrial planets and satellites. Finally, we shall briefly consider the question of gravitational collapse of Jovian-size planets onto even more massive cores, which was the subject of articles by Perri and Cameron (1974) and by Mizumo *et al.* (1978). For ease of comparison we shall adopt the same initial adiabat used by the latter authors, which is based on deductions of the pressure–temperature regions within which the several classes of ordinary and carbonaceous chondrites were formed.

2.2.1. Isothermal Capture

We shall calculate the magnitude of the enhancement of local gas pressure produced by introducing a gravitating body into a uniform medium of constant temperature. Tidal distortion of the gaseous envelope by the Sun is neglected, and spherical symmetry is assumed. The captured atmosphere is assumed to remain isothermal at the local nebular temperature; that is, heating due to release of gravitational potential energy by the captured gas is ignored.

In this case, the only acceleration experienced by a mass element of gas is that due to the gravitational attraction of the solid planet:

$$g(r) = GM/r^2, \tag{2.3}$$

where G is the gravitational constant, M is the mass of the planet, r is the distance from the center of the planet, and g is the acceleration due to gravity.

The gaseous envelope is assumed to be in hydrostatic equilibrium:

$$dP = -\rho g(r)\, dr, \tag{2.4}$$

where ρ is the local gas density and P is the gas pressure.

For all cases of quantitative interest to us the gas behaves ideally:

$$Pv = RT, \tag{2.5}$$

where R is the gas constant, T is the Kelvin temperature, and v is the molar volume of the gas. The density of the gas is the molecular weight, μ, divided by the volume per mole, v:

$$\rho = \mu/v = P\mu/RT, \tag{2.6}$$

where T is the (constant) atmospheric temperature. Note that μ may be a function of T and P because of chemical processes in the gas such as hydrogen molecule dissociation, but is essentially constant below ~ 1500 K.

Substituting for g and ρ in Eq. (2.4), we have

$$dP = \frac{GM\mu P}{r^2\, RT_\infty}\, dr, \tag{2.7}$$

whence

$$\int_{P_r}^{P_\infty} \frac{dP}{P} = -\frac{G\mu M}{RT} \int_r^\infty \frac{dr}{r^2}, \tag{2.8}$$

where P_∞ is essentially the background nebular pressure. This equation is easily integrated for the surface pressure P_s:

$$P_s = P_\infty \exp\left(\frac{G\mu M}{RT_\infty r_s}\right). \tag{2.9}$$

This equation is valid only for constant μ; otherwise Eq. (2.7) must be integrated numerically. Using the planetary data in Table 2.3, the results given in Figure 2.3 are calculated.

For Venus and Earth, the mass of captured gas is a major contributor to the mass of the planet, and the neglect of the atmospheric mass in Eq. (2.3) leads to a substantial underestimate of the surface temperature and pressure.

Under these circumstances, our neglect of the atmospheric mass in Eq. (2.3) becomes a significant source of error, and we must instead write

$$dP = -\frac{\rho GM(r)}{r^2}\, dr, \tag{2.10}$$

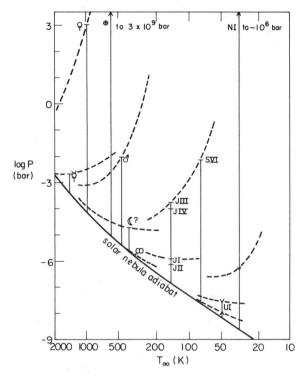

FIGURE 2.3. **Isothermal gas capture from the solar nebula.** The consequences of injecting certain planets and satellites into a smooth model nebula are portrayed for the case in which the gravitational capture of nebular gas is isothermal. Earth would capture enough gas to become another Jupiter. However, the capture process releases vast amounts of energy, and this isothermal case is not realistic for most cases of interest to us. Errors in assigning P_∞ and T_∞ points along the adiabat for each body will influence their calculated surface pressures in the manner illustrated by the dashed lines. From the left, the symbols denote Mercury, Venus, Earth, Mars, the Moon, Ceres (1), Jupiter's satellites JI (Io), JII (Europa), JIII (Ganymede), and JIV (Callisto), Saturn's satellite Titan (SVI), Uranus' satellite Titania (UI), and Neptune's satellite Triton (NI).

where the incremental contribution of each gas shell of thickness dr to the total mass is just

$$dM(r) = 4\pi r^2 \rho \, dr. \tag{2.11}$$

The use of the ideal gas equation also becomes invalid because of the large gas densities encountered. The worst problem, however, is our neglect of the heating of the captured gas by liberation of gravitational potential energy. Rather than attempting to develop further this isothermal model, we shall instead turn to more realistic capture scenarios.

2.2.2. Adiabatic Capture

There are both theoretical and observational reasons to doubt that isothermal gas capture ever occurred on the terrestrial planets. First, as we have seen in the

TABLE 2.3

Primordial Surface Atmospheric Pressure, P_s, for Isothermal Capture[a] for P_∞ and T_∞ and Observed Densities and Radii for M and r_s[b]

Object	Density (g cm^{-3})	r_s (km)	T_∞ (K)	ln (P_s/P_∞)	P_s/P_∞
Mercury	5.47	2434	1400	1.8	6
Venus	5.24	6052	900	16.5	1.5×10^7
Earth	5.51	6370	600	28.9	3.6×10^{12}
Mars	3.90	3390	450	7.7	2.3×10^3
Moon	3.32	1735	370?	2.1?	8.?
Ceres (1)	1.80	540	~300	0.136	1
Vesta (4)	2.45	285	~300	0.051	1
JI (Io)	3.4	1820	370?	2.3?	10.?
JII (Europa)	3.0	1550	300?	1.8?	6.?
JIII (Ganymede)	1.8	2700	~180	5.7	1.4×10^3
JIV (Callisto)	1.6	2500	~140	5.2	7.4×10^2
SVI (Titan)	2.1	2500	~ 80	12.7	3.3×10^5
UI (Titania)	2?	500–1000	~ 50	0.4–1.7	1.5–5.5
NI (Triton)	2?	1100–2600	~ 35	5.4–30	$2 \times 10^2 - 10^{13}$

[a]Using the nebular model of Cameron and Pine (1973).

[b]The ease of capture of nebular gas goes in the order: (Jupiter, Saturn, Uranus, Neptune) > (Earth, Venus) > (Titan, Triton?) > (Mars, Ganymede, Callisto, Triton?) > (Io, Europa, Mercury, Moon, Tethys, Rhea, Iapetus, Titania) > (Ceres, Pallas, Vesta).

previous section in Figure 2.3, Earth would have irreversibly turned into a Jovian planet, which it plainly did not. Second, it is very hard to imagine any mechanism that can accrete terrestrial planets and capture massive atmospheres in the very short period of ~10^4 years while leaving the materials involved even approximately isothermal. Rapid accretion of the earth would yield a surface temperature well in excess of 10^4 K, and it would be a reasonable approximation to assume an adiabatic, convective atmosphere that is rapidly mixed by accretion heating, with the mixing time shorter than the radiative cooling time. Accordingly, we shall now examine the surface conditions of bodies that capture adiabatic atmospheres. We shall splice the adiabat for each atmosphere onto the adiabat for the solar nebula, since accretion heating might keep the adiabat in the planetary protoatmosphere as high as possible. We would therefore also anticipate rapid mixing with nebular gases due to bulk turbulence.

In order to calculate the surface atmospheric pressure and temperature on a massive planet embedded in the nebula for the case of adiabatic, rather than isothermal, structure, we shall again assume ideal gas behavior and hydrostatic equilibrium. Further, we shall require that the expansion process take place at constant entropy. From the first law of thermodynamics, we equate the expansive ($P\, dv$) work done by a parcel of gas to the decrease in its internal thermal energy:

$$P\, dv = -C_v\, dT, \qquad (2.12)$$

where C_v is the heat capacity at constant volume. Since the gas is ideal ($Pv = RT$), then in differential form,

$$P\,dv + v\,dP = R\,dT \tag{2.13}$$

and

$$C_v + R = C_p, \tag{2.14}$$

the heat capacity at constant pressure, whence

$$dP = -\frac{\rho GM}{r^2}\,dr = \frac{C_p}{v}\,dT, \tag{2.15}$$

or

$$-\frac{\mu GM}{r^2}\,dr = C_p\,dT. \tag{2.16}$$

Integrating,

$$T_r = T_\infty - \frac{\mu GM}{C_p}\int_r^\infty \frac{dr}{r^2}, \tag{2.17}$$

which, at the surface of the planet, gives

$$T_s = T_\infty + \frac{\mu GM}{C_p r_s}. \tag{2.18}$$

We define the last term in (2.18) as ΔT_{ad}, the adiabatic temperature rise produced by reversible compression of a parcel of nebular gas from the nebular pressure, P_∞, to the surface pressure, P_s. This term is a strong function of the properties of the planet, but only weakly related to the local nebular condition (through μ and C_p).

The surface temperature and pressure for adiabatic atmospheric capture are related to the nebular conditions (P_∞, T_∞) by integrating Eq. (2.15):

$$P_s/P_\infty = (T_s/T_\infty)^{\gamma/(\gamma-1)} \tag{2.19}$$

where γ is the ratio of specific heats, C_p/C_v.

The surface temperatures calculated in Table 2.4 for all bodies except Venus and the Earth are below ~1700 K. At higher temperatures, the expression (2.12) for the first law must be modified to take into account the absorption of heat by thermal dissociation of molecular hydrogen, the molecular weight μ must be allowed to vary, and (2.16) must then be integrated numerically. The appropriate expression for the first law is now

$$C_p\,dT + 0.5(\Delta H^\circ_{H_2})dn_H = -\mu g\,dr, \tag{2.20}$$

where $\Delta H^\circ_{H_2}$ is the heat of dissociation of H_2 and n_H is the mole fraction of atomic hydrogen. The results of these calculations are displayed in Figure 2.4. It is now clear that adiabatic gas capture provides only a negligible amount of captured gas.

TABLE 2.4
Primordial Atmospheric Surface Pressures, P_s, for Adiabatic Capture from the Nebula[a]

Object	ΔT_{ad}	T_s (K)	P_∞ (bar)	P_s (bar)
Mercury	315	1715	4×10^{-4}	1×10^{-3}
Venus	1900	2800	7×10^{-5}	9×10^{-2}
Earth	2200	2800	7×10^{-5}	1×10^{-1}
Mars	1000	1450	4×10^{-6}	2×10^{-4}
Moon	240	610	2×10^{-6}	1×10^{-5}
Ceres	8	308	1×10^{-6}	1×10^{-6}
J III (Ganymede)	270	450	1×10^{-7}	2×10^{-6}
S VI (Titan)	360	440	2×10^{-8}	4×10^{-6}
U I (Titania)	<30	<80	7×10^{-9}	2×10^{-8}

[a]Data sources are given in Table 2.3. The effect of thermal dissociation of H_2 has been included in calculating the Venus and Earth adiabats since very high surface temperatures are predicted.

For example, the residual neon, nitrogen, and argon contents from a 10^{-1} bar solar-composition protoatmosphere would be 2×10^{-5}, 1×10^{-5}, and 7.3×10^{-7} bar, respectively. This is 10^5 times less than the terrestrial nitrogen inventory, but overestimates the amount of solar rare gases on Earth: The neon abundance on Earth is ~40 times less than the amount provided here.

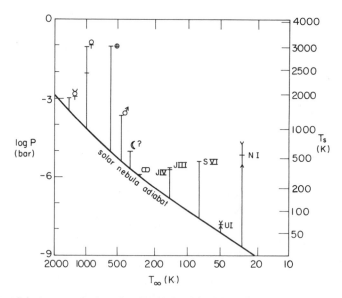

FIGURE 2.4. Adiabatic gas capture from the solar nebula. The surface temperatures of the bodies may be read from the right-hand scale, and the surface pressures from the left-hand scale. This model assumes perfect opacity of the gas, and probably somewhat exaggerates the surface temperature, while leading to somewhat low estimates of the surface pressure. Such a process would, for example, leave Earth with vastly less water, carbon, nitrogen, etc., than it actually has, while simultaneously providing far too much neon. Symbols are explained in the caption to Figure 2.3.

2.2.3. Radiative–Convective Captured Atmospheres

Whether the actual gas capture process in such a situation would be nearly isothermal or nearly adiabatic depends directly on the opacity of the gas to thermal radiation. If dust grains can radiate heat freely over distances large compared to the size of the captured atmosphere, then compressive heating of the captured gas will be rapidly offset by radiative cooling from embedded grains. At low pressures, the opacity of the atmosphere is due to metallic grains and to water vapor. At higher pressures, above 0.1–1 bar, opacity due to molecular hydrogen becomes very important. Thus, a more plausible scenario for gas capture might be to allow compression to occur nearly isothermally until pressures of 0.1–1 bar are reached, then nearly adiabatically thereafter, and to consider more carefully the effect of opacity due to absorption by H_2.

The pure rotational fundamental band of H_2, centered at 17 μm, is formally forbidden by reason of the mirror-image symmetry of the H_2 molecule. However, this symmetry is momentarily disturbed during intermolecular collisions, thus allowing the H_2 molecule to absorb thermal infrared radiation. The strength of absorption depends on the number of molecules engaged in collisions, and hence varies as P^2. This high opacity will permit the gas to maintain a closely adiabatic profile at densities near 0.1 amagat or higher. [The density of a gas at standard temperature and pressure is defined as 1 amagat (agt) equivalent to 2.68×10^{19} molecules cm^{-3}.] The opacity due to H_2 is, of course, strongest at 17 μm, and hence hydrogen is most effective for temperature near 170 K, at which the peak of the Planck function is at 17 μm, but the disturbance of the rotational energy levels during (relatively) very energetic thermal collisions is sufficient to provide strong absorption to wavelengths as short as 2 or 3 μm. We can easily demonstrate the importance of hydrogen opacity by a very approximate calculation.

Taking the thermal IR absorption coefficient for pressure-induced absorption by H_2 to be of order $\alpha = 0.1$ km^{-1} agt^{-2}, the optical thickness of a slab of gas d km thick will be

$$\tau = \alpha \left(\frac{273}{T} \right)^2 P^2 d. \tag{2.21}$$

Our criterion for opacity being important in the captured gas is simply that $\tau \geq 1$ over a distance d comparable to one scale height of the gas:

$$d \simeq H = \frac{RT}{\mu g} = \frac{1}{\alpha P^2} \left(\frac{T}{273} \right)^2 \tag{2.22}$$

or, from Eq. (2.3),

$$\frac{R}{\mu g} = \frac{(104 \times 10^{-4})T}{P^2} = \frac{Rr^2}{\mu GM}. \tag{2.23}$$

This is satisfied near Earth ($T \simeq 600$ K) at about 3.0 Earth radii, at a pressure of ~0.15 bar. Closer to Earth, the compression of the gas will be nearly adiabatic. At the surface of the Earth, $T = 2070$ K and $p = 20$ bar. The conditions estimated in the same way for Venus are $T = 1900$ K and $P = 0.5$ bar, with the adiabatic region extending out to two Venus radii. At higher temperatures H_2 is extensively dissociated and much less opaque.

Since we have here considered only hydrogen opacity, and since this is the irreducible minimum opacity, we must conclude that all reasonable gas-capture models should lie between the adiabatic and the radiative–convective cases discussed above. Thus, if Venus and Earth had been present in the solar nebula, they would have captured atmospheres with pressures in the range 10^{-1}–$10^{+1.3}$ bar at their surfaces. Our conclusions regarding the paucity of neon and the heavy rare gases on Earth argue strongly against both such mechanisms. The relative rarity of Ne and inert gases on Venus and Mars likewise lead us to conclude that no evidence for capture of an appreciably massive solar-composition protoatmosphere can be found in the inner solar system. The available evidence does not, however, suffice to rule out such a capture process categorically.

A much more detailed treatment of opacity (Hayashi *et al.*, 1979) permits a more accurate estimate of the surface temperature and pressure on Venus- and Earth-sized bodies embedded in the nebula. Their model for gas capture on Earth assumed a very low nebular temperature of only 225 K, following Kusaka *et al.* (1970), and hence has a much smaller scale height and much higher surface pressure than we expect. Figure 2.5 contains the details of their opacity model incorporating H_2, H_2O, and dust, as well as a comparison of the P, T profiles connecting the nebular conditions to the surface conditions of Earth and Venus.

We thus conclude that, if the terrestrial planets were present in the solar nebula, only tiny traces of heavy gases could have been deposited on them as escape residues from captured solar-composition protoatmospheres. However, even this trace must have been removed from Earth with nearly perfect efficiency. As we shall discuss in detail in Section 4.3, the rare gases on Earth show a far greater resemblance to the

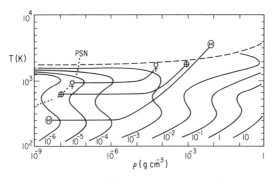

FIGURE 2.5. **Total opacity for solar material.** The capacity K (cm² g⁻¹) due to H_2O, H_2, and dust is contoured as a function of the gas density and temperature. The opacities are from Trafton (1967) and from Cameron and Pine (1973), and the graph is after Hayashi *et al.* (1979). The light dashed line marks the temperature above which both H_2 and H_2O become highly dissociated. The Venus and Earth P, T profiles are marked with their respective symbols; the lowest profile, marked H, is the Hayashi, *et al.* profile for the Earth using the very cold nebular model of Kusaka *et al.* (1970). The dotted line is a low solar nebula adiabat.

"planetary" rare gas component of meteorites than to solar rare gases (Wasson, 1969; Fanale, 1971b).

On the other hand, it seems clear that any *massive* body present in the nebula is certain to be accompanied by a *temporary*, low-pressure solar protoatmosphere. This atmosphere may readily blow off upon dissipation of the nebular gases, since, as $P_\infty \rightarrow 0$, adiabatic structure requires $P_S \rightarrow 0$ also. Every parcel of an adiabatic atmosphere extending to infinity has a thermal energy content sufficient for escape. Nonetheless, a temporary atmosphere is capable of maintaining high surface temperature as long as the nebula is still present. Thus, accretion by growing protoplanets in the nebula must take place onto a surface substantially warmer than the nebula. This circumstance is of particular relevance to the inhomogeneous accretion model for planetary origin (Turekian and Clark, 1969), in which newly formed condensates accrete rapidly and sequentially onto the surfaces of large bodies as the nebula cools. This model will be discussed in Section 2.2.5.

The capture of gas onto much more massive solid cores has been suggested as a means of forming the Jovian planets by Perri and Cameron (1974), who found that core masses of about 60 Earth masses were needed in order to produce runaway collapse of nebular gases. Since they assumed that the collapse process was adiabatic, they clearly must have overestimated the core mass needed to effect collapse. Mizuno *et al.* (1978) have shown that a more detailed radiative–convective treatment of the sort described above will permit collapse for a Jupiter core mass as small as 15 Earth masses, or a Saturn core as small as $6 M_\oplus$. Mizuno *et al.* further conclude that this model rules out the formation of a Jovian planet about Earth even for the cold nebular model (Kusaka *et al.*, 1970) that they employ. They find that a core mass of 7 to about 30 Earth masses would be required to bring about capture of enough gas so that hydrostatic equilibrium becomes impossible and collapse occurs.

In all of these models we have tacitly assumed quasistatic embedding of the protoplanet or planetary core in the nebula. However, a massive body will be little affected by the pressure gradient force which helps support the nebular gas, and it will therefore follow a nearly Keplerian orbit. The gas, which feels the solar gravitational force partially offset by the pressure gradient force, will, of course, travel about the Sun at less than the Keplerian speed. Relative velocities of as much as 1 km sec^{-1} are attainable. Miki (1980) has considered the steady-state flow of nebular gas past an embedded massive body. Curiously, however, he has neglected the differential velocity between the planet and the nebula, and has assumed Keplerian motion for the gas (that is, no radial pressure gradient in the Sun's gravity field). These assumptions generate a radial shear in the gas motion due to the dependences of orbital speed on heliocentric distance. Net motion of the planet relative to the gas is generated post hoc by later introducing finite eccentricity in the orbit of the planet. Miki points out that stirring of the gas by the planets could lead to radial mixing over distances comparable to the spacings of the planets after about 2×10^8 years. We should, however, recall that there are various instabilities that can destroy the nebula on a time scale of only 10^5 years.

Although the available work does not answer our questions about how an accreting planet might interact with a realistic nebula, we must view the question as very important. Up to now, with our static view of planet embedding in the nebula, we have simply assumed that the outer, low-opacity envelope will be radiatively controlled and nearly isothermal, while the inner, denser layers of captured gas will, because of their high opacity, build up a steep temperature gradient and become convectively controlled. The prospect of embedding this delicate model in a Mach 0.1–1 wind is somewhat daunting. If the outer envelope is replaced by new nebular gas on a time scale short compared to the radiative cooling time, then adiabatic structure will be forced on the entire atmosphere, and we can scarcely regard it as captured: This gas would flow off in instant response to any lowering of the nebular pressure. The dynamic pressure due to the gas flow is about 1 mbar (10^3 dyn cm^{-2}) at Earth's orbit, and the flow-through time at about 6 Earth radii is about 4×10^4 sec. With an effective opacity of about 3×10^6 cm^2 g^{-1}, a 600-K gas will radiate about 20 erg g^{-1} sec^{-1}. The amount of energy that can be radiated as the gas passes near Earth is then about 10^6 erg g^{-1}, which is completely negligible. If Mach 1 is in fact reached, a bow shock will be set up and the interaction will be made even more complex by radiation from the shock. We conclude that proper modeling of the interaction of embedded planets with the nebula is so difficult a task that it should not be undertaken unless concrete evidence for a captured solar-composition gas component can be found in the atmosphere of a terrestrial planet.

2.2.4. Related Capture Mechanisms

Ozima and Nakazawa (1979), in attempting to explain the strong mass-dependent fractionation observed in the planetary rare gases relative to solar abundances, have proposed that these rare gases were initially captured by asteroid-sized preplanetary bodies. During capture, gravitational sorting of rare gases by atomic weight was assumed to occur by diffusion. They claim that the elemental and isotopic fractionation of the rare gases can be explained in a consistent manner. However, it turns out that a body of radius 9×10^7 cm (10^{25} gm) is required to explain the elemental fractionation, while the body must simultaneously have a mass of about 1.2×10^{24} gm to explain the isotopic fractionation of xenon. The authors fail to mention that the predicted isotopic fractionation for krypton from this mechanism is far larger than the observed effect and opposite in sign. Furthermore, diffusive separation of heavy rare gases cannot possibly occur in a turbulent gas in viscous flow, which is the flow regime for an asteroid embedded in the nebula.

An alternative explanation for the rare gas fractionations has been proposed by Izakov (1979), who discusses ion capture from the solar wind. A distinctive feature of this model is that the solar nebula is assumed to have had highly nonsolar composition. The "planetary" or fractionated rare gas abundance patterns are simply assumed to have arisen in the nebula. Extremely long interaction times ($>10^8$ years) are required for the gas and solid bodies.

A model involving nongravitational capture of rare gases by a preaccretion solar wind implantation mechanism has recently been suggested by Wetherill (1980). We will discuss this model in Section 4.2.2. Other solar wind capture models are generally based on the idea of the solar wind impinging on fully accreted planetary bodies. These models are introduced in conjunction with the concept of solar wind sweeping of light planetary gases in Section 3.7.2.

2.2.5. Chemical Effects

As an example of the chemical consequences of the presence of a temporary captured atmosphere on the accretion process, let us consider accretion of material by a lunar-sized object embedded in the nebula. We shall start with the relatively simple case of an object like Ganymede accreting gas adiabatically in a medium of $P_\infty = 10^{-6}$ bar: ΔT_{ad} is ~270 K. T_s is therefore ~400 K, and P_s is ~2×10^{-5} bar. At this temperature all ices are vaporized, and even serpentine, $(Mg, Fe)_3Si_2O_5(OH)_4$, is decomposed into olivine, $(Mg, Fe)_2SiO_4$, pyroxenes, $(Mg, Fe)SiO_3$, and water vapor. The effective radiating layer in the atmosphere would be either solid NH_3 clouds near 150 K or the tops of ice-crystal clouds near 250 K, if these clouds were sufficiently dense. However, the gas density is so low that direct radiative cooling of the surface will occur.

It can easily be shown that the full adiabatic temperature rise need *not* occur in order to get very effective evaporation of infalling material. Further, it can be shown that, for common materials of interest to us, adiabatic compression always leads to evaporation, not condensation. For simplicity of treatment, let us consider a condensible pure substance i of molar enthalpy of vaporization λ_i. By the Clausius–Clapeyron equation, the vapor pressure P_i of the substance varies with the natural thermodynamic temperature function, $-1/T$, as

$$\frac{\delta \ln P_i}{\delta (-1/T)} = \frac{\lambda_i}{R} \quad . \tag{2.24}$$

Along an adiabat, the total gas pressure P varies with temperature as

$$d \ln P = \frac{C_p}{R} \, d \ln T, \tag{2.25}$$

or, equivalently,

$$\frac{d \ln P}{d (-1/T)} = \frac{C_p T}{R} \quad . \tag{2.26}$$

For condensation to occur during adiabatic compression, the slope of the vapor pressure curve *must* be less than the slope of the adiabat, i.e.,

$$\lambda_i < C_p T. \tag{2.27}$$

Ingersoll (1969) has shown that this condition is met for all real substances, contrary to the suggestion by Brunt (1934): Since, for condensation from an adiabatic solar gas, each substance saturates at a characteristic temperature T_i, we may compare the λ_i for various condensates to the corresponding product $C_p T_i$, to see whether condensation during adiabatic compression is possible. Invariably, the calculated values of $C_p T_i$ are far smaller than the λ_i; hence condensation will *not* take place during adiabatic compression. If the compression is isothermal, then, of course, condensation will eventually occur at any temperature below T_c, the critical temperature: But if even as little as 5% of the P–V work of compression is stored as heat in the gas (i.e., the atmosphere is strongly subadiabatic), the temperature rise during compression will be steep enough to prevent condensation of any of the cosmically abundant volatiles (H_2O, NH_3, CH_4, etc.). The plain expectation is that, if a fresh condensate from the nebula is being deposited on the surface of a gravitating body as condensation occurs in the nebula, the dominant thermodynamic effect will be for this substance to be evaporated from the surface of the body and reinjected into the nebula as a gas. The energy supplied by the gas to evaporate the condensates will cause the temperature gradient to be slightly shallower than the dry adiabatic gradient near the surface. Figure 2.6 illustrates this idea. Thus, in a nebula, bodies with small gravitational fields will possibly grow more rapidly than lunar-sized objects. If the gravity is large enough to develop a surface temperature near 2000 K, then all incoming material will be vaporized if the P–T profile is adiabatic. An accretion-energy heat input would force the atmosphere toward adiabatic structure, and the short time scale for motions in such an atmosphere (which is, after all,

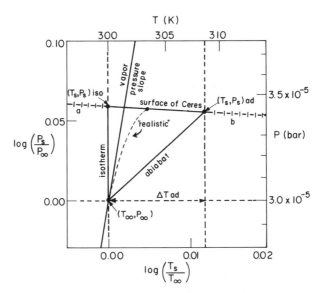

FIGURE 2.6. Comparison of slopes of an adiabat and a typical vapor pressure curve. The trivial example shown here, Ceres, with a ΔT_{ad} of only 8 K, illustrates qualitatively an effect that becomes quantitatively important only for lunar-sized bodies, where ΔT_{ad} becomes comparable to T_∞, the nearby unperturbed nebular temperature. From Table 2.4 it may be seen that this effect would be important for the Moon and very important for the Galilean satellites if they were present in the solar nebula. For common materials in a solar-composition gas, the slope of the log P versus log T vapor pressure curve is ~20 times that of the adiabat. Note also the relationship between the surface pressure for isothermal gas capture and for adiabatic capture.

continuous with the nebula and moving ~ 1 km sec^{-1} through it) may permit little net accretion of solid matter. It is important to realize that planetary accretion while the nebula is still present (i.e., on a time scale of $\sim 10^4$ years) *requires* that any infalling debris be a recent condensate, that the condensates be at a temperature only very slightly lower than their saturation temperature in the nebula, and that the accretion energy released per gram be very large (see the treatment of accretion energy in Section 4.7). Interestingly, it seems that, under these circumstances, a planet may refuse to capture such condensates by vaporizing them (via adiabatic compression) while they are still at great distances from the surface. We have seen in the above discussion that even a markedly subadiabatic atmosphere can vaporize infalling solids in this manner if the solids are only marginally saturated in the nebula.

This problem may be to some degree avoided if the infalling solids are at a temperature far below their condensation temperature and if the rate of accretion is too low to form massive bodies while the nebula is still present. This means that the time scale for accretion must be much longer than the time scale for cooling of the nebula, that is, probably $>10^5$ years. It also suggests that accretion of bodies larger than Mars may not occur until after the dissipation of the nebula. During the short lifetime of the nebula, adiabatic compression and accretion heating may provide a kind of leveling influence on the mass distribution of objects in the inner solar system. In the outer solar system, where a substantial portion of the condensed material was ice at a temperature not far below its condensation temperature, accretion to larger bodies may have been similarly discouraged. Hayashi *et al.* (1979) have used similar arguments to claim that massive gas capture will cause extensive melting of the planet.

In any case, we can conclude that, if massive solid bodies had been present in the solar nebula, they would have unavoidably become endowed with temporary primordial atmospheres. These atmospheres would very likely have been closely adiabatic in structure for several reasons, as discussed above. As one direct consequence, we are led to infer that rapid accretion of freshly condensed nebular material by bodies larger than the Moon may become almost impossible due to vaporization and convective dispersal of infalling condensates. Second, we conclude that massive (>10 bar) solar protoatmospheres derived from capture of nebular gases by the terrestrial planets, asteroids, or satellites of the outer planets are probably impossible. Third, we expect that, as soon as the external pressure is relieved through dissipation of nebular gases by the T-Tauri phase of the Sun, all protoatmospheres will at once flow off in a hydrodynamic escape mode, driven by their internal thermal energy, on a time scale far too short for molecular separation by diffusion. Thus, if such protoatmospheres ever existed, they were of low mass and no remnant of them could survive to the present. We are also led to question the hypothesis that rapid accretion of planets occurred in the solar nebula. We may now see the importance of a critical discussion of the available evidence for and against rapid accretion and strong accretional heating. This discussion will be presented in Section 4.3.

The remaining models to be considered are all closely related to the compositions of chondritic meteorites. For that reason, we insert here a summary of the properties of the chondrites, with emphasis on the chemistry and isotopic composition of the volatile elements.

2.3. COMPOSITIONS OF CHONDRITIC METEORITES

Meteorites are free samples of extraterrestrial material that have fallen on the surface of a planet. About 3000 meteorites are known, mostly ranging in mass from a few grams to about 30 tons. Larger objects do fall to Earth, but they generally reach the ground with such high speeds that they are destroyed, and even vaporized, by impact. The majority of all meteorites contain the major rock-forming elements Fe, Mg, Si, S, Ca, Al, Na, Ni, Cr, Mn, P, K, etc., in very nearly the same relative proportions found in the Sun, with fractionations seldom larger than ±10 or 20% from the mean. The abundances of these elements are given in Table 2.5. These meteorites have textures that clearly show that they are not of igneous origin, sometimes with mineral grains of incompatible compositions adjacent to one another, and often with tiny droplets of quenched glass, ranging from about 100 μm to about 1 mm in diameter. (These spheroidal and often glassy particles are called *chondrules*, and the meteorites that bear them are termed *chondrites*, as we mentioned earlier.)

A few percent of all the meteorites observed to fall on Earth are clearly of igneous origin. Their textures attest to formation by crystallization from a melt, and they are generally substantially fractionated relative to solar abundances. Because melting in

TABLE 2.5
Cosmic Abundances of the Major Rock-Forming Elements[a]

Element	Atomic abundance $(Si = 10^6)$
Mg	1.060×10^6
Si	1.000×10^6
Fe	0.900×10^6
S	0.500×10^6
Al	0.085×10^6
Ca	0.063×10^6
Na	0.060×10^6
Ni	0.048×10^6
Cr	0.013×10^6
Mn	0.009×10^6
All others combined	0.022×10^6

[a]After Cameron (1980).

a gravitational field permits density-dependent fractionation of the component materials, we would expect the differentiates formed from heating a solar-proportion mixture of rock-forming minerals to include individuals with compositions ranging from virtually pure metal to virtually metal-free silicates. Such is indeed the case. The igneous meteorites are divided into three major categories according to their overall composition: Objects with 90% or more metal are called *irons*, those containing less than about 2% metal are called *achondrites*, and those containing roughly equal proportions of metal and silicates are called *stony-irons*.

Because of the primitive, undifferentiated nature of the chondritic meteorites, they are often taken as the composition standards for discussions of the chemistry of solar system bodies. The interpretation of these compositions is, however, complicated by the fact that chondrites span a considerable range of oxidation state, metal content, and content of volatile elements. We shall first familiarize ourselves with the nomenclature of the most important minerals in chondrites, then explore the classification of the chondrites through chemical, mineralogical, and petrological considerations. Finally, we shall summarize what is known about the elemental abundances and isotopic compositions of the volatile elements in chondrites for our eventual use in constructing planetary models.

2.3.1. Major Element Mineralogy

The mineral assemblage found in any chondrite class is strongly influenced by the degree of oxidation of iron and the content of volatiles, especially water. In a highly reduced assemblage, iron is found as troilite and kamacite-rich metal (see Table 2.6 for nomenclatural guidance), and the FeO content is negligible. This means that the olivine and pyroxene components are dominated by forsterite and enstatite, respectively. Because Fe, Mg, and Si have nearly equal abundances, and because they are far more abundant than any other silicate-forming (lithophile) elements, the highly reduced mineral assemblage contains roughly one divalent cation per silicon. Thus, the stoichiometry of the system demands that $MgSiO_3$ (enstatite) be by far the most abundant silicate. Modest oxidation of the system, with partial conversion of metal to FeO, will have several mineralogical consequences. First, the less easily oxidized Ni will remain in the metal, so that the taenite : kamacite ratio will increase. Second, the FeO will be accommodated in the silicates to provide olivine and pyroxene solid solutions with appreciable fayalite and ferrosilite concentrations. Third, the ratio of the abundance of divalent cations to that of silicon will increase, so that the overall stoichiometry of the silicates will shift from pyroxene toward olivine. These phenomena were once thought to account strictly for all the chemical differences between the major chondrite types. Now, however, the fractionations between the rock-forming elements are well known, and we know that the picture is quite a bit more complicated than these relationships, called Prior's Rules, assume. Still, Prior's Rules work quite well in interpreting the range of compositions found *within* each group of chondrites.

Highly oxidized chondritic material will suffer total loss of metal (which by this

TABLE 2.6
Major Minerals in Chondrites

Olivine solid solution series

Mg_2SiO_4————Fe_2SiO_4
(forsterite)   (fayalite)

Pyroxenes
Clinopyroxene: $CaSiO_3$ (wollastonite)
Orthopyroxenes: $CaMgSi_2O_6$ (diopside)
$MgSiO_3$————$FeSiO_3$
(enstatite)   (ferrosilite)

Sulfides

FeS (troilite)
$(Fe,Ni)_8S_9$ (pentlandite)

Metal

Fe, Ni with <6% Ni (α-iron; kamacite)
Fe, Ni with >6%, Ni (γ-iron; taenite)

Feldspars
K-spar: $KAlSi_3O_8$ (orthoclase)
Plagioclase: $NaAlSi_3O_8$————$CaAl_2Si_2O_8$
(albite)   (anorthite)

Phyllosilicates

$(Mg, Fe)_3Si_2O_5(OH)_4$ (serpentine)
$(Mg, Fe, Al)_6(Si, Al)_4O_{10}(OH)_8$ (chlorite)

Carbides, phosphides, etc.

Fe_3C (cohenite)
$(Ni, Fe)_3P$ (schreibersite)
C (graphite, carbynes, diamond)
S (sulfur)

Oxysalts

$Ca_5(PO_4)_3(F, Cl, OH)$ (apatite)
$CaSO_4·H_2O$ (gypsum)
$CaMg (CO_3)_2$ (dolomite)
$CaCO_3$ (calcite)

Oxides

Fe_3O_4 (magnetite)

point is extremely Ni-rich taenite), followed by the appearance of magnetite. The next step is to oxidize sulfides and carbon, leading to production of the Fe-poor sulfide pentlandite, then elemental sulfur, then sulfates and carbonates. Similarly, phosphides give way to phosphates with increasing oxidation. As we shall see in Section 2.7, oxidation is a consequence of *cooling* in a system with solar elemental abundances. Thus, the most highly oxidized mineral assemblages coexist with silicates, rich in oxidized iron, that have taken on water in alteration reactions. These hydroxyl silicates are very fine-grained and very difficult to characterize: They are phyllosilicates with layer-lattice structure similar to serpentine or chlorite, with high magnesium contents. These same low-temperature meteorites are rich in organic matter, and show strong evidence of having been formed in, or thoroughly altered by, liquid water.

This brief introduction to the major-element mineralogy of chondrites will now enable us to discuss the chemical classification of chondrite groups.

2.3.2. Chemical–Mineralogical Classification of Chondrites

Urey and Craig (1953) first demonstrated that reliable chemical analyses of chondrites could resolve several distinct and nonoverlapping chemical groups. For reasons of easy recognition of the major chemical differences between these groups, they are usually portrayed on a diagram of total reduced iron (Fe metal plus FeS) versus total oxidized iron (as FeO). Such a diagram, simplified from that given by Bild and Wasson (1977), is shown in Figure 2.7. As we have seen, the most highly reduced chondrites are dominated by metal, troilite, and enstatite. The members of this class are called the E (enstatite) chondrites. Four other groups are resolved on the basis of total iron content and oxidation state. These include the progressively more oxidized and less iron-rich H (high-iron), L (low-iron), and the much less common LL (very low iron) groups. The most highly oxidized and volatile-rich chondrites belong to the C (carbonaceous) group, of which some contain sulfides and up to a few percent of metal, whereas others have iron fully oxidized to ferrous and ferric silicates plus magnetite. The H, L, and LL chondrites are often collectively called the *ordinary* chondrites.

Various suggestions have been made regarding the relationships between the compositions of the chondrite classes and the terrestrial planets (see, for example, Tandon and Wasson, 1968; Miyashiro, 1968; Lewis, 1974a; Shimizu, 1979). We shall examine the predictions of these models later in our treatment of the planets, whereas astronomical evidence regarding the mineralogy of the asteroids is described in Section 4.7.

Not all the compositional differences between these meteorite classes can be deduced from the information in Figure 2.7. For this reason, representative bulk

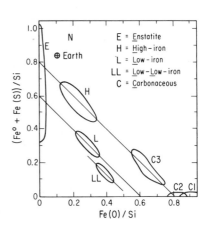

FIGURE 2.7. **Urey–Craig chemical classification of chondrites.** The total iron:silicon ratios 0.8 and 0.6 define the H and L groups; C1 chondrites average about 0.9, the densest E chondrites reach over 1.0, and Earth is about 0.96. The dispersion of each group is described by Prior's Rules (see text). N denotes the Netschaëvo chondrite described by Bild and Wasson (1977).

chemical analyses of members of the E, H, L, C3, and C2 groups from Mason (1962) are given in Table 2.7.

Van Schmus and Wood (1967) have used petrological criteria, including the degree of recrystallization of matrix material, the state of devitrification of chondrules, diffusive blurring of chondrule rims, etc., to rank the members of each of these groups in a sequence of "metamorphic grades." It is by no means clear that we are in fact seeing evidence of metamorphism in the usual sense, involving reheating from a colder early state. It may be simply that the different petrologic grades were accreted and stored at somewhat different temperatures or with different cooling rates. Whatever the significance of these grades, however, they are of only limited utility to us in our present discussion. It suffices to say that the meteorites preserving the most delicate temperature-sensitive structural and chemical features, the C1 and C2 chondrites, are the only meteorites known of petrologic grades 1 and 2. The ordinary chondrites are mostly of grades 4, 5, and 6, with only a few individuals of grade 3.

TABLE 2.7
Analyses of Chondrites Belonging to Different Groups[a]

Component	Weight percent in members of types:				
	E(5,6)	H(4,5,6)	L(4,5,6)	C3	C2
Fe	23.70	15.15	6.27	4.02	—
FeS	8.09	6.11	5.89	5.12	3.66[b]
FeO	0.23	10.21	15.44	24.34	27.34
SiO_2	38.47	36.75	39.93	33.82	28.61
MgO	21.63	23.47	24.71	23.57	19.46
Ni	1.78	1.88	1.34	1.43	—
Al_2O_3	1.58	1.91	2.10	2.68	2.15
CaO	1.03	2.41	1.70	2.17	1.66
Na_2O	0.69	0.76	0.87	0.47	0.51
Cr_2O_3	0.49	0.51	0.56	0.52	0.46
MnO	0.24	0.30	0.33	0.20	0.22
P_2O_5	0.30[c]	0.30	0.31	0.20	0.30
Co	0.08	0.08	0.06	0.05	—
TiO_2	0.12	0.14	0.14	0.15	0.13[c]
K_2O	0.16	0.20	0.13	0.23	0.05[d]
H_2O	0.34[d]	0.21[d]	0.27[d]	0.10	12.86
NiO	0.11[d]	—	—	—	1.53
CoO	—	—	—	—	0.07
C[e]	0.35	0.10	0.10	0.20	2.20

[a] After Mason (1962) and Schmitt *et al.* (1972).
[b] FeS is absent in C2 chondrites; this is the total content of sulfur reported as elemental S.
[c] See Mason (1971).
[d] Probably most reported H_2O is terrestrial contamination. The value for NiO in the E chondrites is surely an error.
[e] Moore and Lewis (1965, 1966, 1967).

2.3.3. Volatile Element Abundances, Carriers, and Isotopes in Chondrites

We shall here confine our attention to H, C, N, S, labile oxygen, and the rare gases. The halogens and P will be treated in more detail in Section 2.7, where the theory of condensation is explored.

The true intrinsic hydrogen content of ordinary (H, L, and LL) and enstatite (E) chondrites is not known, nor is any chemical carrier of hydrogen securely identified. The hydrogen carriers in the carbonaceous (C) chondrites are principally phyllosilicates such as chlorite or serpentine, in which water is bound as hydroxyl groups in the lattice. Interlayer or absorbed water is also possible, but it is so labile that it would exchange readily with water vapor in the atmosphere, so that identification is difficult. As much as half of the reported water in C1 and C2 chondrites may be due to terrestrial contamination or exchange. The other H carrier in the C chondrites is organic matter, which can provide almost as much hydrogen as the hydroxyl silicates, and hydrated water-soluble salts such as gypsum and epsomite, which also can exchange readily with the atmosphere.

In the absence of contamination, native bound hydrogen (calculated as equivalent weight percent water) probably comprises about 10% of the C1 chondrites and 5% of the C2 chondrites, although literature values for the present water contents of these classes often run twice this large. The true native water content of H, L, LL, and E chondrites has never been established (Kaplan, 1971). The importance of organic polymers as carriers of hydrogen in ordinary chondrites is not obvious. Polynuclear aromatics with atomic C : H ratios above 1.0 are present to some extent; however, even in the C3 chondrites it is claimed that at least 90% of the carbon is in the form of crystalline carbynes (Whittaker *et al.*, 1980). If we take 0.1 wt. % C as typical of ordinary chondrites and require that at least 90% of this material be carbynes, then the involatile aromatics would contribute about 8 ppm of hydrogen, equivalent to about 70 ppm of water if fully oxidized and outgassed. This is an order of magnitude smaller than the water content of the Earth.

Boato (1954) found that all water liberated by heating carbonaceous chondrites up to 180°C has terrestrial isotopic composition, whereas water released at higher temperatures is markedly different from terrestrial water. The deuterium enrichment in the high-temperature (native) fraction of water from the Ivuna C1 chondrite relative to average terrestrial water (Lake Michigan!) was observed to be +30%.

Isotopic fractionations are always calculated relative to a specific reference isotope and a specific standard isotopic sample of the element. Denoting the ratio of the atomic abundance of isotope i to that of the reference isotope j as $r = n_i/n_j$, and labeling the isotope ratio in the standard by subscript "st" and that in the unknown sample by subscript "x," the standard isotopic fractionation in permil (parts per thousand) is

$$\delta(i) = 1000 \left(\frac{r_x - r_{st}}{r_{st}} \right). \tag{2.28}$$

The isotopic standards usually used in investigations of the isotopes discussed in this section and the reference isotopes can be briefly summarized. For hydrogen, the reference isotope is ^1H and the standard is standard mean ocean water (SMOW). For carbon the reference isotope is ^{12}C and the standard is PeeDee Belemnites (PDB), which are fossil carbonate cephalopod shells from the PeeDee formation. For nitrogen the reference isotope is ^{14}N and the standard is the Earth's atmosphere. For sulfur the reference isotope is ^{34}S and the standard is troilite from the Canyon Diablo iron meteorite, the object whose impact formed Meteor Crater in Arizona. For oxygen the reference isotope is ^{16}O and the standard is again SMOW. For the rare gases it is not obvious what sample should be used as a standard, since so much variation is seen in nature. Ideally the solar isotopic composition would be used, since it presumably most faithfully reflects the gas composition of the solar nebula. The most common standards are the Earth's atmosphere and average carbonaceous chondrite (AVCC) gases. The reference isotopes are ^4He, ^{20}Ne, ^{36}Ar, ^{86}Kr, and ^{130}Xe.

Carbon contents in chondrites have been reported by Moore and Lewis (1965, 1966, 1967). Ordinary chondrites span the range from 0.02 to 0.60% carbon, with most individuals clustering rather closely about the mean of 0.1% given in Table 2.7. The E chondrites range from 0.056 to 0.56% with a mean near 0.35%. Gibson *et al.* (1971) find that the Vigarano-type (C3V) chondrites contain about 0.9% C, the Ornans-type (C3O) chondrites about 0.4% C, the C2 chondrites about 2.2% C, and the C1 chondrites about 3.3% C. Belsky and Kaplan (1970) have shown that the isotopic composition of carbon in the E, C1, and C2 chondrites is similar to terrestrial crustal carbon, whereas the ordinary chondrites are markedly isotopically lighter than terrestrial carbon. The isotopic differences in the C, H, and N in the chemical components of carbonaceous chondrites have been studied by Robert and Epstein (1982) and by Becker and Epstein (1982). The available data on the abundance and isotopic composition of total carbon in chondrites are summarized in Figure 2.8.

The most reasonable interpretation of the carbon content of ordinary and E chondrites is that crystalline forms of C, possibly carbynes (long-chain polyacetylenes) and graphite, are dominant, with a trace of Fe_3C (cohenite) possible.

The abundance and isotopic composition of nitrogen in chondrites has been studied carefully by Kung and Clayton (1978). The nitrogen in C1 and C2

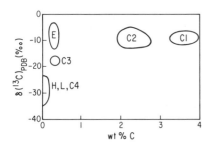

FIGURE 2.8. **Carbon abundance and isotopic composition in chondrites. Carbon 13 fractionation is in parts per thousand (permil) relative to Pee Dee Belemnite standard. (After Kung and Clayton, 1978.)**

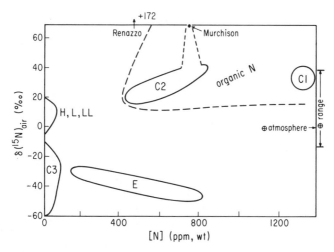

FIGURE 2.9. Nitrogen abundance and isotopic composition in chondrites. Nitrogen 15 fractionation in permil is relative to terrestrial atmospheric nitrogen, which dominates the Earth's known N inventory. Note the extraordinary ^{15}N enrichment in the Renazzo carbonaceous chondrite, a unique anomalous metal-bearing meteorite which otherwise is similar to C2 chondrites. (After Kung and Clayton, 1978.)

chondrites is dominantly that in the organic matter, as can be seen from Figure 2.9, although a two-component mixing line interpretation is possible for the C2 chondrites. The very different nitrogen isotopic composition of the E and C3 chondrites suggests a wholly different nitrogen carrier, and again the elongation of the E chondrite field (which does *not* seem to show sorting by metamorphic grade) may reveal mixing of two isotopically similar components of different origin.

The high total nitrogen content in the E chondrites is due to the presence of nitrides, notably sinoite Si_2N_2O (Anderson *et al.*, 1964). Kung and Clayton report that total extraction of N from the E chondrites requires complete dissolution of the minerals, strongly suggesting the presence of silicate–nitride solid solutions. Hayatsu *et al.* (1980) have shown that carbynes in the Allende meteorite contain substantial amounts of nitrogen in the form of nitrile (—C≡N) groups, presumably terminating polyacetylene chains. It would not be surprising to find such a nitrogen carrier contributing to the E chondrites as well. We should bear in mind, however, that the C isotopic compositions of the C3 and E chondrites do not overlap, and that different sources of elemental C may contribute to these two classes. Herndon and Rudee (1976) find that virtually all the C in the E chondrites is associated with the metal phase, apparently as Fe_3C. Conditions conducive to formation of polyacetylenes or Fe_3C can be found at both high (Lewis and Ney, 1979) and low (Hayatsu *et al.*, 1980) temperatures in astrophysically plausible settings.

Sulfur is, as we have seen, usually found exclusively as troilite in the E, H, L, LL, and C3 chondrites. Nickeliferous pyrrhotite and pentlandite are found in slightly more oxidizing circumstances, and elemental sulfur (apparently small crystals of rhombic S) is the dominant form of S in the C2 chondrites. The C1 chondrites are rich in calcium and magnesium sulfates, but also contain small amounts of highly oxidized and corroded sulfide and some elemental sulfur. In addition, some sulfur is present in the organic complex. The bulk sulfur isotopic composition in every class of meteorites seems to be about the same, with a $^{34}S : ^{32}S$ ratio indistinguishable from that in the troilite of iron meteorites. There are, however, very large isotopic fractionations seen between coexisting phases in the Cl and C2 chondrites. These isotope effects are definitely not compatible with equilibrium isotope fractionation at any temperature, and are almost certainly due to kinetic isotopic fractionation during unidirectional oxidation of troilite (Lewis, 1967). Laboratory simulations of this process show that oxidation of troilite by low concentrations of strong oxidizing agents at temperatures near 0°C can match the isotopic compositions seen in the sulfate, sulfur, and sulfide phases in carbonaceous chondrites (Lewis and Krouse, 1969). Herndon *et al.* (1975) have presented arguments that the morphology of magnetite grains in Cl chondrites is due to their formation by oxidation of sulfides in a low-temperature aqueous environment, in agreement with this scenario. The sulfur from this oxidation process was found by Lewis and Krouse to be isotopically lighter than in FeS, and the sulfate was in turn isotopically much lighter than the sulfur, by fully 6% [$\delta(^{34}S)$].

The main carrier of chemically labile oxygen in the H, L, LL, and C3 chondrites is FeO in the olivine and pyroxene phases. The E chondrites generally contain only negligible traces of FeO, usually far less than 1%. The C2 chondrites, as we have described above, also contain large amounts of magnetite and bound water and trace amounts of carbonates, as well as carboxylic acid groups or other less reactive structural units in both the water-soluble and polymeric organic matter. The Cl chondrites are noteworthy for their large abundances of bound water, magnetite, sulfates, and carbonates.

The isotopic composition of oxygen shows variations from class to class which are as fascinating as they are inexplicable. As we show in Figure 2.10, the data on the relative abundances of oxygen 16, 17, and 18 clearly separate the chondritic meteorites into classes that cannot be interconverted by normal chemical or physical fractionation processes (Clayton *et al.*, 1976, 1977).

High-temperature mineral grains from the C2 and C3 chondrites are found to be enriched in ^{16}O by very large amounts, due to differences in the nucleosynthetic histories between these grains and the volatile-rich matrix material. These isotopic differences make it virtually certain that the substantial majority of preplanetary solids in the solar system were *not* fully vaporized during the solar nebula stage, but that these solids were instead partially equilibrated with nebular gases at temperatures that were probably always below about 1000 K. The oxygen isotopic differences between the chondrite classes cannot presently be attributed to or correlated with any particular chemical or petrological differences. This may be due to nebular equilibration temperatures being high enough for solid-state diffusion to average out any distinctive chemical or mineralogical carriers that may have been present initially, but not high enough for full isotopic equilibration with the nebula. Alternatively, small and variable amounts of chemically reactive, isotopically

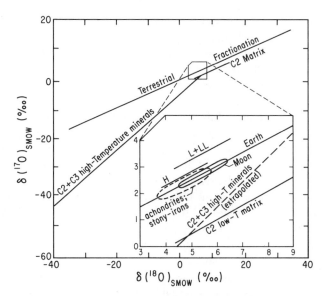

FIGURE 2.10. Oxygen isotope variations in the Solar System. The lines of slope ~$\frac{1}{2}$ are due to chemical and physical fractionation effects (equilibrium partitioning between minerals; distillation; diffusion, etc.). The high-temperature anhydrous mineral grains in the C2 and C3 chondrites lie on the line of slope ~1.0, reflecting mixing between some fairly tame endmember and another extremely ^{16}O-rich component, almost certainly a distinct nucleosynthetic component. The ordinary chondrites lie off the Earth fractionation line; the moon, achondrites, stony irons, and E chondrites lie on it (E chondrites not graphed for clarity). (Data from Clayton et al., 1976, 1977.)

anomalous material may have been added *after* the cooling of the nebula, but prior to mild thermal metamorphism within the meteorite parent bodies.

The retention of rare gases in the ordinary chondrites has long been a problem. Suess (1949) suggested that the observed strong depletion of rare gases and strong fractionation favoring nearly complete loss of the lighter rare gases on Earth was due to fractionation and preferential loss in an asteroidal gravity field. This same pattern of rare gas abundances is seen in most chondrites, and has in the last few years also been observed in the atmospheres of Mars and Venus.

Lancet and Anders (1973) studied the solubilities of the rare gases in magnetite, but were unable to explain the observed severity of elemental fractionation in meteorites, or the observed total rare gas contents (unless unbelievably large and variable rare gas pressures were assumed). In general, rare gas solubilities in melts depend on the volume of the rare gas atom: A hole must be formed in the liquid, in opposition to the surface work of the liquid. In such a system, the light rare gases are expected to be easier to insert into the liquid, and hence more soluble. Zaikowski and Schaeffer (1976) have studied rare gas solubility in the other major mineral in C1 chondrites, the phyllosilicates, and again do not find them a satisfactory rare gas host. Very low-temperature adsorption can provide the proper heavy rare gas abundances and fractionation in C chondrites (Fanale and Cannon, 1972), but a separate mechanism is still required to implant these adsorbed gases into mineral grains.

Clearly it is important to know what minerals actually contain the fractionated "planetary" rare gases in chondrites.

The discovery of carbynes in meteorites (Hayatsu *et al.*, 1980; Whittaker *et al.*, 1980) may, if confirmed, advance our understanding of the retention of rare gases in meteorites. It had previously been known that a HNO_3-soluble phase (or phases), part of the insoluble residue from dissolution of bulk Allende meteorite (C3) material, carries a very large part of the total abundance of primordial (nonradiogenic; nonsolar) rare gases. This HNO_3-soluble component, named "phase Q," makes up only about 0.04% of the mass of the meteorite, but bears the majority of its heavy rare gases (Gros and Anders, 1977). It was presumed that, because Fe and Cr were found in the "Q" solution, the rare gas carrier was an Fe–Cr–S–O compound or chromite; in fact, Šrinivasan *et al.* (1978) had leaned strongly in the direction of identifying Q as chromite or at least some compositionally similar inorganic phase rather than carbon. Hayatsu *et al.* (1980) have, however, shown that chromite catalyzes the formation of carbynes from CO and H_2. Thus, the close association of these phases has a sensible explanation.

A detailed description of the isotopic variations of the rare gases in meteorites is beyond the scope of this book, especially since many aspects of the problems presented by meteorites have no apparent counterpart in the study of planetary atmospheres. It is necessary to know, however, that the gases present in the atmospheres of the terrestrial planets are not only strongly depleted relative to solar abundances, but they are also strongly fractionated in favor of the heavy rare gases. Krypton and xenon are the least severely depleted rare gases on Earth, being depleted by a factor of about 10^{-7} relative to their solar abundances. The several problems raised by the elemental and isotopic compositions of the rare gases on the terrestrial planets will be treated in detail in Chapter 4. The reader interested in a broader perspective on the rare gases is encouraged to refer to the very useful review by Podosek (1978).

An extensive summary of rare gas data on meteorites has been presented by Heymann (1971). We have already given the cosmic abundances and isotopic compositions of the rare gases in Table 2.2, for which the isotopic abundances are heavily reliant on analyses of C chondrites. Neglecting the radiogenic rare gases (^4He from U and Th decay, ^{40}Ar from ^{40}K decay, ^{129}Xe from ^{129}I decay, and fissiogenic Kr and Xe), the "trapped" component of rare gases in C chondrites has a mean $^4He : ^{20}Ne$ ratio near 300, a mean $^{20}Ne : ^{36}Ar$ ratio near 0.27, and a mean $^{36}Ar : ^{84}Kr$ ratio near 80. The light isotopes of krypton are less abundant in C chondrites than in Earth's atmosphere due to a smooth and nearly linear fractionation effect of about 0.3%/amu (Figure 2.11). The trapped xenon abundance is given by a $^{84}Kr : ^{132}Xe$ ratio of about 0.8 ± 0.2, and there is a very large mass-dependent fractionation of the light Xe isotopes 124, 126, 128, 130, 131, and 132 relative to terrestrial atmospheric Xe with a slope of about 3.8%/amu. This fractionation has the opposite sign of the Kr fractionation: C chondrite Xe is isotopically lighter than the Earth's atmosphere. In addition, the C chondrites have an excess of the heavy isotopes 136 and 134 over the smooth, linear fractionation trend estab-

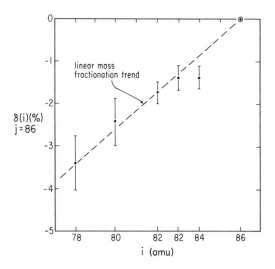

FIGURE 2.11. **Krypton isotopic composition in carbonaceous chondrites relative to Earth's atmosphere. (After Heymann, 1971.)**

lished by the lighter Xe isotopes, but still have lower 136:130 and 134:130 ratios than the atmosphere. This is apparently due to the presence of a fission Xe component that is present in the C chondrites but not discernable in the Earth's atmosphere. The C1 xenon isotopic composition relative to Earth's atmosphere is given in Figure 2.12.

The ordinary chondrites contain small and somewhat variable amounts of trapped, fractionated rare gases. The H, L, and LL chondrites average about 0.6×10^{-8} ml^3 gm^{-1} of trapped ^{36}Ar, with a scatter of about a factor of 3. The ^{36}Ar:^{84}Kr:^{132}Xe ratios are close to $80:1:0.8$. Trapped Ne is only occasionally

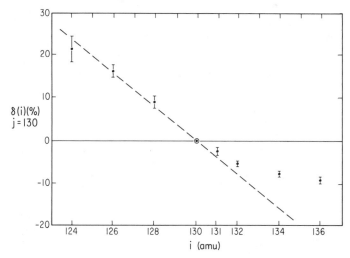

FIGURE 2.12. **Xenon isotopic composition in carbonaceous chondrites relative to Earth's atmosphere. (After Heymann, 1971.)**

observed. The isotopic composition of trapped Ar in the ordinary chondrites is indistinguishable from that in the C chondrites ($^{36}Ar : {}^{38}Ar = 5.35$), whereas Kr has a similar but somewhat steeper fractionation than in the C chondrites. The Xe isotopic composition is nearly the same in all cases.

With these data on the composition of the chondrites in hand, we may now turn to the specific compositional models for the planets which are based on analogies with the chondrites.

2.4. CARBONACEOUS CHONDRITE PLANETS

Ringwood (1966) postulated that the primitive material present throughout the inner solar system was of the same composition as the primitive Type I Carbonaceous Chondrites (CI). All of the terrestrial planets, their satellites, asteroids, and meteorite parent bodies are pictured as being ultimately derived from such material. The properties of the terrestrial planets are qualitatively attributed to varying degrees of heating and outgassing of the primitive CI material. Even with the degree of heating taken as a free parameter for each planet, a large additional external heat source is required in order to explain the anomalously high density of Mercury.

TABLE 2.8
Volatile Element Abundances in CI Chondrites

Element	Weight fraction	Abundance (Si = 10^6)	References
H	1%	2.3×10^6	Kaplan (1971)
He	0.02 ppm	1.2	Heymann (1971)
O (H_2O)	10%	1.3×10^6	Kaplan (1971)
(FeO)	30% (Fe_3O_4)	1.3×10^6	
C	3.19%	0.7×10^6	Vdovykin and Moore (1971)
N	0.26%	4.3×10^4	Moore (1971a)
Ne	0.3 ppb	4.0×10^{-3}	Heymann (1971)
S	6.04%	0.5×10^6	Moore (1971c)
Ar (36–38)	2 ppb	9.7×10^{-3}	Heymann (1971)
(40)	9 ppb		
P	0.12%	1.05×10^4	Moore (1971b)
Cl	0.06%	4.74×10^3	Dreibus et al. (1979)
F	0.01%	7.80×10^2	Dreibus et al. (1979)
Kr	0.004 ppb	1.9×10^{-4}	Heymann (1971)
Br	1 ppm	9.2	Dreibus et al. (1979)
Xe	0.003 ppb	1.0×10^{-4}	Heymann (1971)

TABLE 2.9
Isotopic Composition of Volatile Elements in CI Chondrites

Element	Isotope ratios		References
H	H/D	-3.2 to -12.7% SMOW	Boato (1954)
He	$^4He/^3He$	1800:1	Heymann (1971)
C	$^{13}C/^{12}C$	-8% PDB	See Fig. 2.8
N	$^{15}N/^{14}N$	$+35\%$ air	See Fig. 2.9
O	$^{17}O/^{16}O$	$\rightarrow$	See Fig. 2.10
	$^{18}O/^{16}O$		
Ne	$^{21}Ne/^{20}Ne$	0.026	Heymann (1971)
	$^{22}Ne/^{20}Ne$	0.138	
S	$^{33}S/^{32}S$	$+0.3\%$ CDFeS	Hulston and Thode (1965)
	$^{34}S/^{32}S$	0.0 CDFeS	
	$^{36}S/^{32}S$	$+0.2$ CDFeS	
Cl	$^{37}Cl/^{35}Cl$	?	No data
Ar	$^{38}Ar/^{36}Ar$	0.192	Heymann (1971)
	$^{40}Ar/^{36}Ar$	$\rightarrow$	See Table 2.8
Kr		$\rightarrow$	See Fig. 2.11
Xe		$\rightarrow$	See Fig. 2.12

Because of the great dissimilarity between the volatile-element abundances of a CI proto-Earth ($\sim$10 wt. % H_2O) and the present Earth ($\sim$0.03% H_2O), complete removal of the primitive atmosphere is again required. In this case the amount of gas which must be removed is $\sim$0.2 Earth mass, or about the combined total masses of Mercury and Mars. Since this gas is largely CO and H_2O, with a mean molecular weight about 10 times that of a solar-composition gas, the problem of removal of the gas is formidable.

The total primitive volatile content of each planet is fixed by this model to be that found in CI chondrites. The chemical and isotopic compositions defined for these volatiles are described in detail in Tables 2.8 and 2.9, respectively.

We shall have little more to say about this specific model in its pure form, since the original proponent of it has now modified his ideas extensively in the direction of making carbonaceous chondritic matter a minor constituent of the Earth (Ringwood, 1979). This revision alleviates but does not remove many of the problems with the original model; however, the present version is properly a multicomponent (but homogeneous) model and will accordingly be treated in Section 2.6.

2.5. ORDINARY CHONDRITE PARENTAGE

There has, for several decades, been an expectation among many geochemists that the composition of the Earth would prove similar to that of the most abundant meteorites observed to fall on Earth, the ordinary chondrites. Although no single known class of meteorites matches the known composition of the Earth in all respects, the high-iron (H-group) chondrites plus an admixture of metallic iron would provide a close fit. Goldschmidt (1922), Washington (1925), and Rankama and Sahama (1950) have given important discussions of chemical relationships between planetary and chondritic preplanetary material, and Taylor (1964b,c) has more recently shown the plausibility of a compositional similarity between Earth and H-group chondrites. A more detailed discussion of such a compositional model will be given in Section 4.3.1.

The volatile content of the primitive Earth in the chondritic model can be determined from analytical data on the abundances of many elements in ordinary chondrites. Table 2.10 summarizes the abundances of a number of volatiles in these meteorites. Wasson (1969) and Fanale (1971b) have stressed the remarkable agreement between the nonradiogenic rare gas abundances on Earth and in ordinary chondrites. The isotopic composition of C, N, and O has been discussed in Section 2.3. Sulfur in all ordinary chondrites studied is indistinguishable from Canyon Diablo FeS, and the isotopic abundances of the rare gases closely resemble those in Cl chondrites (see Sections 2.3 and 2.4). Figure 2.13 presents the rare gas elemental abundances in the major chondrite classes, from the discussion by Anders and Owen (1977).

TABLE 2.10
Volatile Element Abundances in Ordinary Chondrites[a]

Element	Mass fraction	Abundance (Si = 10^6)
H	~0.02%	
O (as H_2O)	~0.2%	
C	0.10%	1.3×10^4
N	20 ppm	2.4×10^2
He	0.2 ppb	8×10^{-3}
Ne	0.003 ppb	2×10^{-5}
$^{36}Ar + {}^{38}Ar$	0.02 ppb	4×10^{-5}
Kr	0.0004 ppb	7×10^{-7}
Xe	0.0003 ppb	4×10^{-7}
S	2.1%	1.0×10^5
P	0.11%	5.5×10^3
Cl	0.01%	4.0×10^2
F	0.013%	1.2×10^3
Br	0.5 ppm	1
I	0.05 ppm	7.0×10^{-2}
Ge	10 ppm	21
Se	8 ppm	16
As	1.8 ppm	3.8
Te	1.0 ppm	1.2
Hg	~0.1 ppm	~1×10^{-1}
Sb	0.10 ppm	1.3×10^{-1}
Tl	~7 ppb	~6.0×10^{-3}

[a] After Mason (1971).

FIGURE 2.13. **Chondritic rare gas abundances. The C2 and C1 (▲) chondrites have virtually identical elemental abundances for the rare gases, except that the C2 abundances are about 30% lower. The Ornans-type (C3O, ○) and Vigarano-type (C3V, △) chondrites are quite distinct. The ordinary chondrites, represented here by an average over a number of H (●) chondrites of different degrees of recrystallization, have lower overall abundances and distinctive depletion of Ne relative to the other rare gases. (After Anders and Owen, 1977.)**

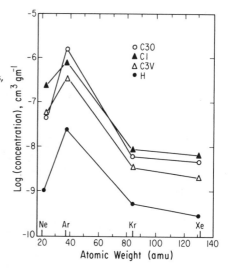

The existence of such analytic data should not create the presumption that these abundances are understood, or that the mechanisms for retention of the rare gases in chondrites are well known. In particular, it is clear that several apparently independent processes are responsible for rare gas retention. Leaving aside for the moment the distinctive radiogenic (^{4}He, ^{40}Ar, ^{129}Xe), fissiogenic (Kr and Xe), and cosmic ray spallogenic (characterized by high ^{3}He : ^{4}He, ^{21}Ne : ^{20}Ne, and ^{38}Ar : ^{36}Ar ratios) gases, there are two dominant types of rare gases in chondrites. The first of these, similar in elemental and isotopic composition to the present solar wind, is termed "solar." The other, generally similar to the rare gases found in Earth's atmosphere and characterized by severe depletion of the light rare gases, is called "planetary" (Suess, 1949), or "trapped" by some authors.

Etching studies have confirmed that the solar-type rare gases are located in weakly bound sites in the immediate surface regions of mineral grains (Eberhardt *et al.*, 1965). Suess *et al.* (1964) first proposed that the solar-type rare gases were embedded by the solar wind in mineral grains in the regoliths of meteorite parent bodies.

The situation with respect to the planetary-type gases is far less clear. These gases are found in highly retentive sites evenly distributed throughout each mineral grain (Stauffer, 1961). One explanation for the elemental abundance and isotopic composition of planetary-type gases has been sought in studies of the low-temperature equilibrium solubility of these gases in meteoritic mineral grains (Lancet and Anders, 1973), although the results are very unpromising as a means of explaining the observed elemental fractionations. Organic polymers are more promising as rare gas carriers.

Fanale and Cannon (1972) have shown that the rare gases *adsorbed* on meteorite powder from a source gas with solar abundance ratios fit the planetary gas composition well. However, this is very loosely bound gas that is released immediately upon reduction of the external pressure. Deep implantation of this adsorbed gas in the mineral grains, perhaps by mechanical shock, is then necessary.

Retention of carbon and nitrogen by ordinary chondrites has been attributed by Anders and co-workers to a highly polymeric or "amorphous" carbon phase derived from Fischer–Tropsch reactions of CO and H_2, a hypothesis discussed in more depth in Section 2.3. Labile oxygen is carried by FeO, and sulfur by FeS.

2.6. MULTICOMPONENT AND NONHOMOGENEOUS MODELS

One possible scheme for explaining the major- and volatile-element composition of the Earth was suggested by Anders (1968), who considered the possibility that the Earth was accreted largely from highly volatile-deficient achondritic material, with a small late addition of a trace ($\sim 1\%$) of very volatile-rich C1 or cometary material. Larimer (1971) has explored the compatibility of the major-element abundances in the Earth with Ca-poor achondrite parentage, and has concluded that, *neglecting the volatile elements*, it is not possible to decide conclusively between chondritic and achondritic Earth models. Of course, by this model, the relative abundances of the volatile elements are fixed by their proportions of C1 chondrites.

Lewis (1974b) has examined the effects of infall of cometary and C1 meteoritic debris on volatile-element inventories on Venus, and has pointed out several diagnostic *in situ* measurements which could be made on Venus to distinguish between *ab initio* gas retention and late injection of cometary volatiles. Since we have already given, in Table 2.8, the volatile-element abundances in C1 material, we need only present for comparison the expected composition of cometary materials. For this purpose, we use the major composition classes of ices produced by low-temperature equilibrium condensation from the solar nebula (Lewis, 1972a). These compositions are given in Table 2.11.

It should be noted explicitly that such a model source of volatiles gives abundances in solar proportions *for those materials that are condensed*, and *zero* abundances for those materials *not* condensed. The latter condition is especially relevant for NH_3, CH_4, H_2, and the rare gases. Direct condensation of He, Ne, Ar, Kr, and Xe from a solar nebula with a total pressure of $\sim 10^{-6}$ bar occurs at 0.7, 8, 25, 29, and 40 K, respectively, whereas H_2 condensation occurs at 7 K. Fegley (1982) has shown that the clathrate hydrates of Ar, Kr, and Xe will form at 38, 42, and 59 K, respectively.

Sill and Wilkening (1978) have suggested that the rare gases, carbon, and nitrogen in the terrestrial planets were brought in by cometary infall. Basing their argument on the presumed inability of these volatile elements to be retained at formation temperatures of several hundred degrees, they postulate that the most plausible single source of these gases would be cometary ices, in which rare gases would be carried as clathrate hydrates. They calculate the formation temperatures of the rare gas clathrate hydrates for a constant-density model of the solar nebula,

TABLE 2.11
Volatile Element Abundances Predicted in Cometary Material[a,b]

Element	Weight fraction (%)			
	Class 1	Class 2	Class 3	Class 4
H	7.06	8.25	9.04	12.33
C	—	—	3.6	18.23
O (as H_2O)	56.44	50.22	47.82	38.04
N	—	9.22	8.73	7.00
He	—	—	—	—
Ne	—	—	—	—
Ar	—	—	—	—
Kr	—	—	—	—
Xe	—	—	—	—

[a] After Lewis (1972a, 1974b).

[b] The elements S, P, Cl, F, Br, I, Ge, Se, As, Te, Hg, Sb, and Tl have solar abundances. The following cometary classes are considered: (1) rock + H_2O ice (condensation temperature $T \sim 120$ K), (2) rock + H_2O and $NH_3 \cdot H_2O$ ices ($T \sim 80$ K), (3) rock + $NH_3 \cdot H_2O$ and $CH_4 \cdot 7H_2O$ ices ($T \sim 35$K), (4) rock + $NH_3 \cdot H_2O$, $CH_4 \cdot 7H_2O$, and CH_4 ices ($T \sim 15$ K).

and show that no single formation temperature suffices to give the observed planetary rare gas abundances. It is worthy of mention that physically plausible nebular models (adiabatic, not isopycnic) allow rare gas clathrate formation only at extremely low temperatures. Thus, Sill and Wilkening are led to propose a multicomponent model with individual components originating at 400, 100, and 40 K. They do not make clear how the rare gas abundances in chondrites, which are strikingly similar to those in the atmospheres of the terrestrial planets, would be explained by this mechanism.

The general idea of nonhomogeneous accretion of the terrestrial planets during falling temperatures in a solar nebula has recently been revived by Turekian and Clark (1969) for Earth. They envision essentially sequential condensation and rapid accretion of metal and magnesium silicates, topped off by a veneer of C1 or cometary debris containing almost the entire complement of volatiles in the Earth. Each layer is accreted with arbitrary efficiency. This model has not been developed quantitatively for application to the terrestrial planets, although it has been applied to Venus (see Section 4.2). Because of the large number of degrees of freedom in the model, it may not be possible to make quantitative predictions.

A strictly quantitative nonhomogeneous accretion model can in fact be proposed for any final nebular temperature *if* the efficiencies of accretion are the same for all condensates, and *if* each condensate formed by cooling of the solar nebula gas accretes at once onto a single massive body. Solids, once condensed, do not react further with the nebular gas because they are promptly segregated from the gas and buried by accretion. The structure of such a body for a final nebular temperature of ~500 K, determined solely by the condensation sequence and elemental abun-

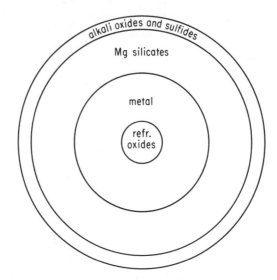

FIGURE 2.14. **Structure of a nonhomogeneously accreted planet. Rapid accretion of condensates as the form in a cooling gas of solar elemental composition will yield a refractory oxide mineral core, a massive Fe–Ni metallic outer core, a mantle of sequentially intergrading Mg$_2$SiO$_4$, MgSiO$_3$, and SiO$_2$, and a "crust" of alkali metal oxides, sulfides, and halides. Further cooling (below 200 K) would add H$_2$O ice, NH$_4$SH, solid NH$_3$, and solid CH$_4$ in that order. (After Lewis, 1974a.)**

dances in the nebula, is given in Figure 2.14. The relevant condensation steps for the major elements are (Lewis, 1974a; Fegley and Lewis, 1980):

(1) Condensation of refractory oxide minerals, rich in Al$_2$O$_3$, CaO, Ti$_2$O$_5$, rare earth oxides, etc. ($\sim$1700 K),
(2) metallic (Fe, Ni) alloy ($\sim$1500 K),
(3) Mg$_2$SiO$_4$, MgSiO$_3$, and SiO$_2$ ($\sim$1450 K), and
(4) alkali metal halides and sulfides ($\sim$700 K).

Then, only below 200 K, ammonium salts including NH$_4$Cl, NH$_4$Br, NH$_4$SH, and NH$_4$H$_2$PO$_4$ and ices such as H$_2$O, NH$_3$, and CH$_4$ form. Such common meteoritic constituents as Fe oxides, FeS, and water-bearing silicates are wholly absent.

Late addition of a cometary ice "veneer" provides a rich source of H, C, N, S, and rare gases: In this class of models, these elements are supplied solely by the last-stage material.

A less strict (and less self-consistent) interpretation of the nonhomogeneous accretion model assumes extremely rapid accretion of new condensates, but at the same time requires that a substantial proportion of "early" (high-temperature) condensates, such as Ca and Al silicates and iron metal, remain in intimate contact with the nebular gas down to very low temperatures. It is further required that, despite the high accretion efficiency and the (relatively) long times available for accumulation of these species into large bodies, the late-stage solids must be similar to C1 chondrites. It is hard to see, then, why C1s in fact have somewhat *higher* abundances of Ca, Al, Fe, and other refractory elements than the ordinary chondrites.

The nonhomogeneous accretion model has been severely criticized by Ringwood (1979), who has considered in some detail the application of this model to the planet about which we know the most, Earth. The six major criticisms leveled by Ringwood are briefly summarized as follows:

1. Earth's mantle is strikingly homogeneous, not layered.
2. The model puts no FeO in the lower mantle and a large but very uncertain amount in the upper mantle: The seismic data on the structure of the mantle contradict this distribution of FeO.
3. The upper mantle should be severely depleted in high-temperature condensates, contrary to observation.
4. FeO and FeS, and other likely light elements, would be excluded from the core, where they are in fact needed to explain the outer core density.
5. The model requires extremely rapid accretion of the Earth, within 10^4 yr at most, in order to accrete on the cooling time scale of the nebula and preserve the volatility sequence of accreting material.
6. Derivation of the NiO content of the upper mantle from metal-free low-temperature condensate would also provide many other moderately volatile elements in solar proportions, contrary to observation.

These arguments and their counterarguments will be critically assessed when we discuss the Earth (Section 4.3).

One recent suggestion goes farther than any of the above postulates in finding a source for a volatile-rich component. Butler *et al.* (1978) have shown that encounters of the solar system with dense interstellar clouds could lead to infall of roughly solar-composition gases and dust in masses perhaps large enough to provide the observed amount of neon on Earth, and to influence strongly the abundances of several volatile elements on Mars. Our ignorance of the chemistry of these interstellar clouds, including their rare gas abundances and isotopic composition, causes the predictions of this model to be poorly defined. This makes it difficult for experimental data on planetary atmospheres to distinguish this source from certain others, such as early solar wind or nebular gas capture.

Despite the difficulty of formulating a simple, unambiguous, quantifiable model for formation of a multicomponent planet, it is quite clear that the terminal stages of planetary accretion must sample the population of small bodies rather widely (Safronov, 1972; Weidenschilling, 1974; Wetherill, 1975; Hartmann, 1976; Cox and Lewis, 1980). Two-component models involving a volatile-poor high-temperature component mixed with a few percent of volatile-rich low-temperature condensate (Anders, 1968; Ringwood, 1979) are perhaps too simplistic to do full justice to any real accretion process, but models exhibiting a satisfactory degree of complexity tend to be underdetermined by observational constraints. Nonetheless, the search for a single meteorite class to serve as the carrier of volatiles in a planet is a worthwhile endeavor if only to see how far the idea can be stretched before breaking. Such a two-component model has been considered for Mars and Earth by Anders and Owen (1977), and will be discussed in Section 4.4.

Accretion models using realistically shaped sampling functions (Barshay and Lewis, 1981) yield planetary compositions that can only very approximately be fitted by a two-component model.

Anders and his co-workers, in an extended series of articles, have developed a complex phenomenological model to fit the available compositional data on chondritic meteorites, and have applied it to various planetary bodies (Ganapathy and Anders, 1974; Hertogen et al., 1977; Morgan et al., 1978). This model, although flexible in form, basically assumes that solid solar system materials comprise mixtures of seven fixed components. These are *early refractory condensate* (similar to the white high-temperature inclusions in the Allende meteorites), *remelted silicate, unremelted silicate* (basically the remelted silicate plus alkalis and trace moderately volatile lithophiles), a *volatile-rich* component (here usually taken to mean C3V-like composition), *metal, troilite,* and *remelted metal.* Subject to the assumption that these components are fixed in composition and that there are no other components, it becomes possible to use analyses of a suite of geochemically diverse elements to calculate the contribution of each of the six algebraically independent composition variables, and thus to predict the full planetary content of all other elements. Such a model is used to interpret data, not to make predictions about hitherto unvisited planets. The application of this model to Mercury, Earth, Mars, the Moon, and the eucrite parent body (Vesta?) will be discussed in Sections 4.1, 4.3, 4.4, 4.6, and 4.7 respectively.

2.7. EQUILIBRIUM CONDENSATION

Finally, there is a very simple (and therefore predictively powerful) model for the variation of composition of condensed preplanetary material with condensation temperature. In essence, the model calculates the major element composition, mineralogy, and volatile element content of primitive condensates by chemical equilibrium methods as general functions of both temperature and pressure, assuming that the parent gaseous nebular material had solar elemental composition (Lewis, 1972a,b). In conjunction with models for the dependence of pressure and temperature on heliocentric distance in the solar nebula [such as that of Cameron (1978)], this model provides explicit detailed predictions of the volatile element inventories of preplanetary condensates as a function of distance from the Sun (Lewis, 1971a,b, 1972a).

In order to describe the composition of planets resulting from accretion processes, it is necessary to specify the way in which each accreting planet samples the solids available over a range of heliocentric distances.

By far the simplest accretional model for the planets (and by far the least physical) is for each planet to sample only those solids present at the exact heliocentric distance at which the planet is accreting. This is the accretion scenario shown in the first panel of Figure 1.6. Each planet then faithfully reflects a single condensation temperature: The planet forms from material in exact internal chemical equilibrium, and no radial mixing of nebular condensates is allowed.

Curiously, this extremely simple model already does a remarkably good job of fitting the observed densities of the terrestrial planets. We will therefore summarize its assumptions and predictions in some detail.

First, this is a chemical equilibrium model: Its results are path independent. We are unable to distinguish models in which the presently observable chemical state was reached in a heating event or a cooling event. Although it is conventional to represent the results of such a model in terms of a "condensation sequence" beginning at very high temperatures, it is by no means sure or even likely that solid solar system materials were fully vaporized in the nebular phase. It is easier from a kinetic point of view to rationalize the "condensation temperatures" of the solids as being really "equilibration temperatures," probably reflecting the *highest* temperature at which gases and solids were intimately mixed. This interpretation is reinforced by the existence of distinct differences in oxygen isotopic composition between meteorite classes (Section 2.3). These differences are due to different amounts of an ^{16}O-rich component and are definitely not due to physical or chemical fractionation processes acting on average solar system oxygen: ^{16}O can vary greatly while $^{17}O : ^{18}O$ remains constant (Clayton *et al.*, 1976).

Second, as we have seen, a single nebular (P, T) point is associated with the orbit of each planet. This gross oversimplification is intended to enhance the predictive power of the model for the purpose of comparing its predictions to observation. There certainly is no reason to believe that planets really accreted entirely out of primitive material formed at well-defined (P, T) points. This model defines the composition of the preplanetary condensate versus heliocentric distance in the central plane at that point in time when the condensates were last in intimate chemical contact with the solar nebular gases. Accumulation of a real planet will involve not only sampling material formed at different heliocentric distances, but also at different times and different distances from the central plane, and probably occurs long after the end of the nebula's lifetime. Further, kinetic inhibition of certain reactions can inject a path-dependent character that is wholly absent in equilibrium thermodynamic calculations.

Third, no mechanism for physical fractionation between metal and silicates is included. Every previous model, no matter how liberally endowed with adjustable parameters, has been unable to explain the densities of the terrestrial planets without introduction of yet more assumptions. As will be seen, the equilibrium condensation model does admirably without such a special effect, but must strain to fit Mercury.

Historically, the equilibrium condensation concept was developed to explain the elemental and mineralogical composition of meteorites. Urey (1951, 1952b; many later articles) first sought to explain the compositions of meteorites and planets by such a model, but was frustrated by the erroneously low value for the cosmic abundance of iron then accepted: All calculated mineral assemblages were found to be less dense than any of the terrestrial planets. In recent years there has been intensive investigation of equilibrium condensation models for meteorites (Lord, 1965; Larimer, 1967; Larimer and Anders, 1967, 1970; Grossman, 1972, 1973; Grossman and Larimer, 1974), with principal emphasis on the high-temperature sequence of condensates and on the condensation behavior of moderately volatile trace elements. Generalized major-element condensation calculations for solar material over wide ranges of temperature (0–2200 K) and pressure (10^{-9}–10^3 bar) have also been done for correlation with available observational data on the compositions and structures of the terrestrial planets and the natural satellites, and on the chemical and isotopic compositions of planetary and satellite atmospheres (Lewis, 1972a,b, 1973a,b, 1974a,c; Barshay and Lewis, 1975, 1976, 1978; Fegley and Lewis, 1979, 1980). The result of these calculations contain numerous explicit and as yet untested predictions regarding oxidation state, major-element mineralogy, melting behavior, thermal evolution, and volatile content of preplanetary material as functions of one independent variable: temperature of condensation. We shall review those predictions of the equilibrium model which most strongly impact our understanding of the origin and evolution of the atmospheres of the terrestrial planets.

The sequence of condensation reactions for the major elements at temperatures above 300 K is briefly summarized as follows:

(1) Refractory oxide condensation: $CaTiO_3$, $Ca_2Al_2SiO_7$, $MgAl_2O_4$, rare earth element (REE) oxides, etc., above ~1600 K.
(2) Metallic Fe–Ni alloy, containing traces of Co, P, C, N, etc., condenses from atomic Fe and Ni vapors, near 1400 K.
(3) $MgSiO_3$ (enstatite) condensation by reaction of SiO, Mg, and H_2O gases, near 1350 K.
(4) Alkali aluminosilicate formation by reaction of atomic alkali metal vapors with high-temperature Ca aluminosilicates, near 1000 K.
(5) Corrosion of Fe° to form FeS (troilite) by reaction with nebular H_2S, at and below 680 K.
(6) Progressive oxidation of Fe° to form FeO-bearing silicates, as detailed in Figure 2.16.
(7) Formation of tremolite, the calcic endmember amphibole $Ca_2Mg_5Si_8O_{22}(OH)_2$, from enstatite, diopside, and water vapor (~400 K).
(8) Formation of serpentine, talc, etc., by reaction of water vapor with ferromagnesian minerals near 350 K.

This sequence should be compared to that discussed in Section 2.6, where minerals once formed were isolated from the gas.

Figure 2.15 displays the condensation and reaction sequences for the major components of solar material over the temperature range 0–2200 K and the pressure range 10^{-8}–10^{+2} bar, with emphasis on the $T > 300$ K region. Figure 2.16 shows the condensation of iron vapor.

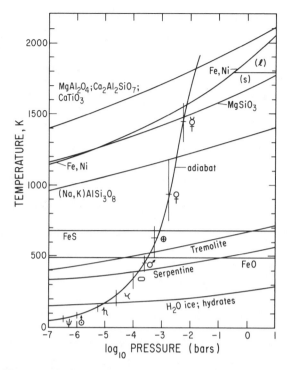

FIGURE 2.15. **Equilibrium condensation sequence for a system of solar elemental composition. The high-temperature refractory oxide condensate is rich in calcium, aluminum, titanium, rare earth elements, etc. Next, the very abundant and dense element iron condenses as a metal, bringing with it nickel, cobalt, and traces of dissolved carbon, phosphorus, nitrogen, chlorine, etc. Immediately below the condensation temperature of metal, magnesium silicates become stable. Initially, the Mg endmember of olivine, Mg_2SiO_4 (forsterite), condenses, to be quickly replaced by $MgSiO_3$, the pyroxene enstatite, at slightly lower temperatures. At temperatures near the condensation point, these minerals contain only miniscule traces of FeO. The alkali metals condense next, by reacting with aluminum-rich early condensate grains to make alkali feldspars, $NaAlSi_3O_8$ (albite) and $KAlSi_3O_8$ (orthoclase). This step causes retention of the very important radioactive heat source [40]K in the solid component of the nebula. Next, sulfur-bearing gases are exhausted from the nebula by the conversion of part of the metallic iron to FeS (troilite), and the remaining iron metal is oxidized to FeO, which enters into the ferromagnesian minerals olivine and pyroxene. These processes are summarized in Figure 2.16. Next comes formation of the amphibole tremolite and then serpentine by reaction of water vapor with calcium-rich and ferromagnesian minerals, respectively. At lower temperatures the rock-forming minerals are relatively inert, and ice-forming materials such as H_2O, the ammonia hydrate $NH_3 \cdot H_2O$, and the methane clathrate hydrate $CH_4 \cdot 7H_2O$ condense.**

The specific reactions by which the volatile elements are predicted to be retained by minerals are of importance to us, since we can check the predicted mineralogy against the suites of minerals observed in chondrites (see Table 2.12).

Volatile element retention, if defined as the condensation of minerals capable of providing atmospheric gases on planets, begins with the condensation of refractory oxides. Both U and Th are retained in this component, which places nearly half the long-lived radioactive heat source in a minor portion of the mass of the rock-forming condensates, giving a heat source density in this early condensate about 10

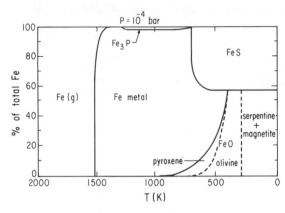

FIGURE 2.16. **Chemistry of iron during isobaric cooling of a solar-composition system. Condensation of iron vapor to form metallic iron is followed by FeS formation at ~680 K and progressive oxidation to FeO in the silicates. The mole fraction of FeO in the olivine and pyroxene phases is close to 10^{-4} at 1350 K, 10^{-3} at 800 K, 10^{-2} at 650 K, and 10^{-1} at 550 K.**

times higher than in an ordinary chondrite. Both the α,β decay chains of ^{232}Th, ^{235}U, and ^{238}U and the spontaneous fission of ^{235}U are available as sources of atmospheric gases. The former, of course, provide radiogenic ^{4}He, whereas the latter provides fission components with distinctive mass spectra which overlap the atomic weight ranges of both Kr and Xe. The heavy-Xe fission component alluded to above in our discussion of rare gases in carbonaceous chondrites is *not* due to U or Th.

Upon condensation of the metal phase, small concentrations of a number of elements may be retained dissolved in the metal. Some of these may exsolve as discrete minerals upon further cooling. The most important of the volatiles hosted by the metal phase is carbon. Lewis *et al.* (1979) have shown that the activity of carbon along a standard solar nebula adiabat reaches a maximum between the orbits of Earth and Venus: Closer to the Sun, CO is dominant, and farther out CH_4 is

TABLE 2.12
Reactions whereby Volatile Elements Are Condensed in the Thermochemical Equilibrium Nebular Model[a]

Elements	Retention mechanisms
O	H_2O (g) + Fe → FeO + H_2 (g)
N	Fe + $1/2N_2$ (g) → (Fe, N)
	$Ti_2O_3 + N_2$ (g) + $3H_2$ (g) → 2TiN + $3H_2O$ (g)
	NH_3 (g) → ammonium aluminosilicates
	NH_3 (g) + H_2O(s) → $NH_3 \cdot H_2O$(s)
C	CO (g) + H_2 (g) + Fe → (Fe, C) + H_2O (g)
	CH_4 (g) + $7H_2O$(s) → $CH_4 \cdot 7H_2O$(s)
S	H_2S (g) + Fe → FeS + H_2 (g)
H (and O)	H_2O (g) → tremolite
	H_2O (g) → serpentine
	H_2O (g) → H_2O(s)

[a]Details of the oxidation reaction of Fe are given in Figure 2.16. The other reactions proceed as indicated here for temperatures at and below those illustrated in Figure 2.15.

stable. Using real-solution activity coefficients for ternary Fe–Ni–C solid solutions, they find that the concentration of dissolved carbon reaches a maximum of about 30 ppm in the same region if strict equilibrium between gases and solids is assumed. If fully oxidized, this amount of carbon is 2 to 3 times too small to provide the known CO_2 inventories of Venus and Earth. This predicted amount, about 10 ppm in the planet overall, is also far smaller than the mean C content in the ordinary chondrites, which is about 1000 ppm.

Studier *et al.* (1968) had proposed some years ago that the bulk of the carbon in meteorites was produced by reactions between CO and H_2 gases in that region of the nebula where methane is thermodynamically more stable than CO. It can be seen from Figure 2.17 that the crossover between these gases occurs at about 700 K: It is below that temperature that, according to Studier *et al.*, the sluggish reaction between CO and H_2 would be catalyzed by mineral grains. Involatile species intermediate in oxidation state between CO and CH_4 would build up, coating the grains with amorphous carbon and involatile polynuclear aromatic molecules. Hayatsu *et al.* (1980) have added carbynes to this list of products from nebular Fischer-Tropsch-type reactions. The problem of attainment of equilibrium between CO and H_2 in the cooler outer parts of the nebula is severe, and a very similar problem

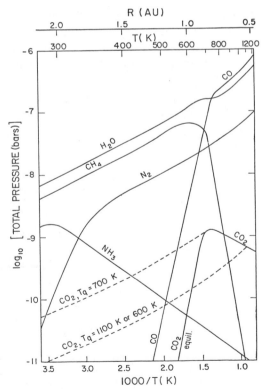

FIGURE 2.17. Partial pressures of major C, N, and O gases along a solar nebula adiabat. Here R is distance from the sun and T is temperature. These pressures are calculated for strict thermodynamic equilibrium (solid curves). The dotted lines illustrate the effects of quenching CO reduction reactions at 600, 700, and 1100 K on the CO_2 pressure profile. (From Lewis and Prinn, 1980.)

exists with N_2 and H_2. As Figure 2.17 shows, N_2 becomes thermodynamically unstable below about 330 K.

The kinetics of CO and N_2 reduction reactions catalyzed by metal grains were studied by Lewis and Prinn (1980), who found that only limited amounts of NH_3 and CH_4 could be made over the entire lifetime of the nebula. For reasonably rapid turbulent mixing of the nebula it is unlikely that more than a few percent of the N_2 and CO can be converted to NH_3 and CH_4. This leads to the outer portion of the nebula containing an amount of CO, CO_2, and N_2 which is grossly larger than the equilibrium amount. One result is that, at and beyond Earth's orbit, the thermodynamic activity of carbon is higher than previously calculated. The concentrations of dissolved carbon in the metal phase can then reach a maximum near the orbit of Mars, just above the disappearance temperature of metal. The concentration of dissolved carbon in the metal phase for this new case, taking kinetic inhibition of CH_4 formation into account, is shown in Figure 2.18.

Additional carbon retention can take place at lower temperatures, since CO, CO_2, and CH_4 are all present in the gas phase. One of the more interesting results of this kinetic model is that, at the temperature at which water ice condenses in the nebula (about 150 K), two other compounds may also condense: These are NH_4HCO_3, ammonium bicarbonate, and NH_4COONH_2, ammonium carbamate. The latter is of great interest, since it contains a covalent C—N bond, and is a possible precursor of HCN and other organic molecules. It was precisely this compound that was used by Wöhler in the demonstration of the synthesis of urea from inorganic materials, a landmark in the development of science. Depending upon the exact quench temperature (the temperature at which reduction reactions cease

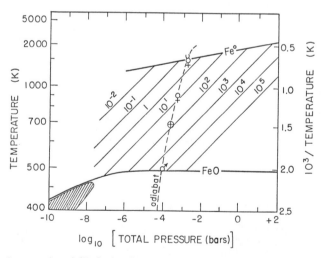

FIGURE 2.18. Concentration of dissolved carbon in the metal for the case of inhibition of CO reduction. Nucleation of graphite is prohibited. The concentration unit is ppm by mass. The cross-hatched area is the *equilibrium* stability field of graphite for solar material after Lewis et al. (1979). (From Lewis and Prinn, 1980.)

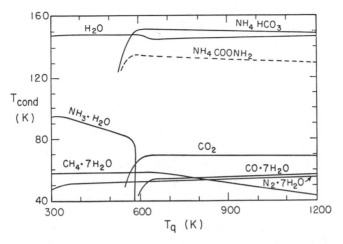

FIGURE 2.19. Condensation temperatures of carbon, nitrogen, and oxygen compounds as a function of the quench temperature for gas reactions. Gas hydrate data are from the compilations of Davidson (1973) and of Ripmeester and Davidson (1977), except for CO·6.75H₂O, which was assumed to have the same decomposition pressure as N₂·6.75H₂O. Leftover or unclathrated CO and N₂ would condense directly only at 25 and 22 K, respectively, for all reasonable quench temperatures. (From Lewis and Prinn, 1980.)

to have significant rates over the mixing time of the nebula), different carbon- and nitrogen-bearing low-temperature condensates are possible. These condensates are summarized in Figure 2.19. Note especially the formation of the methane clathrate hydrate $CH_4 \cdot 7H_2O$ and the condensation of CO_2 at higher temperatures than the formation temperature of the CO_2 clathrate hydrate.

Among the disequilibrium products from CO reduction on grain surfaces, carbynes are fairly promising as carriers of nitrogen, which is incorporated in the form of $-C \equiv N$ groups on chain ends. The solubility of N_2 in metal is so small as to be negligible for our purposes, while the difficulty of forming NH_3 suggests that NH_4 silicates should also be negligible. At low temperatures the condensation behavior follows any of the condensation sequences shown in Figure 2.19. Note that, in the extreme case of successful equilibration down to low temperatures, the condensates become the same as those found in the exact equilibrium case in Figure 2.15.

High-temperature retention of nitrogen has been studied by Fegley (1982). Dissolved N_2 and Fe_4N in the metal phase were found to be negligible, while nitride retention in solid solution in corundum and in taenite remains quantitatively plausible.

Sulfur is retained as FeS in all the E, H, L, and LL chondrites, although the S : Si ratios in these meteorites vary over a factor of two. It is conceivable that exhaustion of exposed metal-grain surfaces may lead to the persistence of H_2S in the gas phase at lower temperatures. If so, then NH_4SH may be a condensate near 140 K (Lewis, 1972b).

Phosphorus, which plays an important role in the chemistry of the halogens, also can be present in small concentrations in the first metallic condensate. However, significant amounts of phosphorus do not condense until the temperature drops to the condensation temperature of schreibersite, Fe_3P, at 1225 K (Fegley and Lewis, 1980). Below about 714 K phosphorus is fully oxidized to the phosphates fluorapatite and whitlockite, $Ca_3(PO_4)_2$.

Alkali metal retention as $KAlSi_3O_8$ and as $NaAlSi_3O_8$ in plagioclase solid solutions occurs near 1000 K. This step provides the isotope ^{40}K in the condensate and makes possible not only heating by the most important of the long-lived radionuclides, but also the production of the gaseous daughter isotope ^{40}Ar. As a result of these two processes of heating and ^{40}Ar production, this isotope may be released into the atmosphere of a planet. Note that primordial argon(^{36}Ar and ^{38}Ar) will be released at the time of first melting, along with very little ^{40}Ar, which accumulates in the mineral matrix with time.

The first significant chlorine-bearing condensate in the equilibrium condensation sequence is sodalite, $Na_4(AlSiO_4)_3Cl$, which forms at 895 K on the standard adiabat. This remains the dominant chlorine carrier down to low temperatures.

Fluorine has been shown by Fegley and Lewis to be retained at 766 K by formation of fluorapatite, $Ca_5(PO_4)_3F$.

It is possible that the first hydrogen-bearing mineral is the calcic amphibole tremolite, as shown in Figure 2.15. However, there is a second interesting possibility. Fegley and Lewis found that, depending on the assumed solid solution behavior between the apatite minerals, hydroxyapatite $Ca_5(PO_4)_3OH$ should form between 460 and 510 K.

Cooling to much lower temperature (350–300 K) would make hydration reactions of ferromagnesian minerals thermodynamically possible, so that serpentine, chlorite, and other water-rich phases could be formed at equilibrium. On the other hand, the kinetic feasibility of making these minerals in the absence of liquid water, in the solar nebula, by reaction of water vapor with a partial pressure of about 10^{-8} bar with mafic (ferromagnesian) crystalline minerals, and doing it at or only slightly above room temperature, is extremely hard to accept. This scepticism about the feasibility of serpentinization in the nebula is greatly reinforced by the mounting evidence that the C1 and C2 matrix material has undergone extensive equilibration with liquid water in a parent body environment, in which formation of these phyllosilicates is quite plausible (du Fresne and Anders, 1962; Nagy et al., 1963; Boström and Fredriksson, 1966; Herndon et al., 1975; Day, 1976; Richardson and McSween, 1976; Richardson, 1978; Kerridge et al., 1979; Bunch and Chang, 1980). If phyllosilicates, which can serve as catalysts in the formation of organic matter, are secondary alteration products in meteorites, then perhaps the complex organic species found in the C1 and C2 chondrites are also secondary in origin. If so, then nebular catalytic processes may have produced chiefly solid polymorphs of carbon.

It is of interest to calculate how much water may be retained in preplanetary solids by each of the three types of minerals we have considered. Hydroxyapatite and tremolite contain essential calcium, and hence are formed in limited amounts compared to the amount of serpentinized material derivable from the ferromagnesian silicates. Conversion of 25% of the phosphorus in an ordinary chondrite containing 0.3 wt. % P_2O_5 (Table 2.7), as predicted by Fegley and Lewis below 460–507 K, will provide 0.18 wt. % hydroxyapatite, or 49 ppm water by weight. Tremolite formation, if it used the entire CaO abundance of the meteorite, could provide up to 0.2 wt. % H_2O. Serpentinization of the olivine and pyroxene component could provide up to about 10 wt. % H_2O.

Finally, we must mention that the minerals predicted by the equilibrium condensation approach bear a very close similarity to the assemblages actually observed in the E, H, LL, L, and C3 chondrites. Whitlockite is a ubiquitous mineral in ordinary chondrites, and Cl-, F-, and OH-apatite are all known in ordinary chondrites. The major-element minerals can be seen to be identical to those predicted. Tremolite has not been found in meteorites. A detailed discussion of the behavior of the major elements in nonhomogeneous and equilibrium condensation models has been presented by Lewis (1974a), and the volatile elements have been similarly discussed by Fegley and Lewis (1980). They conclude that most of the minerals predicted for strict nonhomogeneous accretion are never found in meteorites.

Clearly once the volatile content and oxidation state of a protoplanet are specified, the most important remaining unknown is the thermal history of the body. A complete itemization of heat sources and transport mechanisms is required, beginning with knowledge of the accretion history of the planet, the concentration and initial distribution of the radionuclides ^{20}Al, ^{40}K, ^{235}U, ^{238}U, and ^{232}Th, the melting and viscous-creep behavior of bulk planetary material, and the partitioning of the radioactive nuclides between coexisting phases during primary differentiation of the planet. The equilibrium condensation model provides explicit predictions regarding the concentrations of K, U, and Th. Mercury should be essentially devoid of K, but have solar U : Fe and Th : Fe ratios; Venus and Mars should have the same K, U, and Th content as Earth and H-group chondrites *relative to Si*. Because of the depletion of Si relative to Fe in Mercury, the absolute U and Th weight fractions are about twice as high on Mercury as on the other terrestrial planets (Lewis, 1972b).

With regard to melting behavior, the presence of the Fe + FeS pair in Earth makes early melting easy through formation of an Fe–FeS eutectic melt (Murthy and Hall, 1970). This results in extraction of most of the chalcophile elements in the Earth into the core, and permits core formation instantly upon accretion of the Earth. The melting and differentiation of planetary bodies is discussed in general terms in Section 3.1.

The significance of the "condensation temperature" used in this particular model deserves a final emphasis. Although the "condensation sequence" is most easily visualized as the result of slow cooling from very high temperatures, starting with all materials fully vaporized, it must be realized that the condensate compositions described herein are *equilibrium* results, and hence path-independent. It would be just as legitimate to regard the composition of condensate at a given distance from the Sun as being designated by the *highest* temperature at which the solids and gas were in intimate contact. This is probably a good first approximation even if later cooling of the gas–dust mixture did in reality occur, since the time required to achieve a close approach to equilibrium is dramatically lengthened by modest cooling. If the kinetics of the relevant reactions were known, it would be possible to calculate the effects of such cooling on the mineralogy. It is, however, by no means clear that the thermal history of a given sample of the nebula was even this simple.

The accretion of equilibrium condensates into planets necessarily involves sampling over a wide range of formation temperatures and heliocentric distances. Models of the effects of convolving the equilibrium condensation model with various accretion sampling functions have recently been carried out (Barshay and Lewis, 1981), using sampling functions of the sort illustrated in Figure 1.5. These models predict a number of specific interrelationships between the bulk composition, major element mineralogy, oxidation state, and volatile-element content of massive bodies as a function of heliocentric distance in the terrestrial planet region. Because of the necessity of comparing these predictions to the observed properties of planets, we defer a more detailed presentation of the results of this model to Chapter 4.

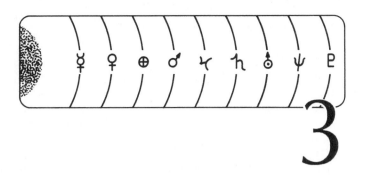

Evolutionary Processes

This chapter is devoted to general descriptions of the types of processes that are responsible for producing, depleting, cycling, and regulating planetary atmospheres.

We have seen in Chapter 2 that a wide variety of bulk compositional models have been proposed for the terrestrial planets. From this basis we shall explore the principal ideas regarding the accretion of the raw preplanetary material into planets; the ways in which planets may heat up, melt, differentiate, and release volatiles; and the ways in which these processes are modeled.

Once a planet begins to form an atmosphere, and possibly polar caps and a hydrosphere as well, it becomes essential to treat in detail the chemical behavior of the volatiles. This necessarily includes the chemistry of magmatic gases, thermochemical reactions in the atmosphere, dissolution of gases in solvents such as water, photolysis of reactive gases, and weathering reactions which tie up volatiles in minerals. Mechanisms for escape of light gases from the upper atmosphere must then be explored, and similarly, where relevant, so must the effects of life on atmospheric chemistry.

Having established some familiarity with chemical equilibrium and steady-state modeling of these processes, we will then be in a good position to combine them, using them as the functional elements of complex cyclic and evolutionary processes which regulate the present state and govern the future fates of planetary atmospheres.

Our discussion in this chapter will be concerned with the processes themselves; the specific applications of these principles to solar system bodies will be found in Chapter 4.

We shall begin our discussion of evolutionary processes by considering the fate of a temporary solar-composition protoatmosphere of the sort which may be retained (after dissipation of the nebula) by any massive bodies which may have been present in the nebula. We shall assume solar composition for the captured gas and simply remove hydrogen from it, simulating the effects of escape of light gases (H_2; He) and retention of species such as CH_4, H_2O, NH_3, etc. This will constitute nearly our entire discussion of primordial atmospheres, excepting only our brief survey of the Jovian planets in Section 4.5.

The other sections of this chapter will all be relevant to "solid" bodies such as terrestrial planets, large satellites, and asteroids.

3.1. HYDROGEN-LOSS EVOLUTION OF SOLAR PRIMORDIAL ATMOSPHERES

A primordial atmosphere captured from the solar nebula will have, to first approximation, the same elemental composition as the nebula. In the inner solar system, where rock forming elements are nearly fully condensed and accreted into macroscopic bodies, the remaining more volatile elements constitute the gas phase available for capture. Similarly, at large heliocentric distances, where the ice-forming elements are fully condensed, the remaining gas in the nebula will be mainly H_2, He, and Ne. There is thus a *complementarity* between the compositions of the solid body and the captured gas. When a solid body becomes massive enough to capture all the nebular gas available to it, then this complementarity may ultimately resolve itself by reconstituting a planet with layered structure, but with overall solar composition. The Jovian planets presumably result from extensive gas capture, approaching overall solar composition most closely in the case of Jupiter.

For moderately massive (lunar-sized or larger) bodies present in the inner regions of the nebula, the mass of captured gas is, as we have seen, at most a very minor proportion of the mass of the solid body. In such a case, the escape of hydrogen and helium from the body is possible because the escape velocity of the planet is comparable to the thermal speed of the light gases. In addition, the amount of hydrogen that must be lost in order to change the atmospheric composition drastically is not prohibitively large.

At temperatures of a few hundred degrees Kelvin a captured nebular atmosphere would initially consist of H_2, He, CH_4, H_2O, NH_3, N_2, H_2S, HCl, HF, PH_3, etc. Loss of hydrogen gas will permit equilibria such as

$$CH_4 + H_2O \rightleftarrows CO + 3H_2, \tag{3.1}$$

$$2NH_3 \rightleftarrows N_2 + 3H_2, \tag{3.2}$$

$$H_2S + 2H_2O \rightleftarrows SO_2 + 3H_2, \tag{3.3}$$

$$8H_2S \rightleftarrows S_8(s) + 8H_2, \tag{3.4}$$

$$CO + H_2O \rightleftarrows CO_2 + H_2, \tag{3.5}$$

$$CH_4 \rightleftarrows C(gr) + 2H_2, \tag{3.6}$$

$$4PH_3 + 6H_2O \rightleftarrows P_4O_6 + 12H_2 \tag{3.7}$$

to shift to the right. Although a few of the most stable hydrogen compounds, such as HCl and HF, will remain abundant, there will generally be a tendency for highly reduced gases to give way to oxides and elements.

The details of the evolutionary track followed by the gas as hydrogen is lost depend upon the temperature and pressure in the atmosphere. Since condensible phases such as water, sulfur, graphite, and phosphorus oxides are produced, clearly precipitation of these materials can occur under certain conditions of temperature and pressure.

Rather than attempting to explore a wide variety of such models, we shall present a single representative example.

Table 3.1 shows the relative elemental abundances of a few important elements selected from Table 2.1. The dominant chemical species of these elements are given for both the extreme cases of a hydrogen-rich (solar) gas and a severely hydrogen-depleted gas.

The detailed evolutionary history of a solar-composition atmosphere at 800 K and 100 bar initial pressure is given in Figure 3.1. The "time" axis is replaced by a simple index of the degree of hydrogen loss: Each step represents depletion of hydrogen gas by a factor of two. The time required to effect such a reduction is a

TABLE 3.1
Abundances and Chemical Compounds of Selected Elements before and after Hydrogen Loss

Element	Cosmic abundance $(Si = 10^6)$	Dominant compounds	
		When $X_{H_2} \simeq 1$	When $X_{H_2} \ll 1$
H	2.66×10^{10}	H_2	H_2O
O	1.84×10^{7a}	H_2O	CO_2
C	1.12×10^7	CH_4	CO_2, CO, C(gr)
N	2.32×10^6	NH_3	N_2
S	5.0×10^5	H_2S	COS, SO_2, S_8(rh)
P	6.50×10^3	PH_3	P_4O_6(g)
Cl	4.74×10^3	HCl	HCl
F	9.80×10^2	HF	HF

[a]The abundance of O after allowance for incorporation of oxygen into silicate and oxide minerals is approximately equal to 1.5×10^7. The abundances are from Cameron (1980).

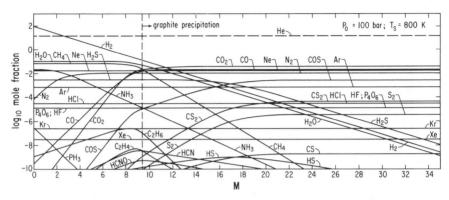

FIGURE 3.1. Compositional evolution of a solar primordial atmosphere during hydrogen loss. The partial pressures of the gases are given as a function of the hydrogen loss factor M, where $p_{H_2}(t) = (0.5)^M p_{H_2}(0)$. Note the well-defined conversion points for the oxidation reactions $CH_4 \rightarrow CO + CO_2$, $NH_3 \rightarrow N_2$, etc. Maximum chemical complexity and maximum graphite activity occur when $p_{CH_4} \approx p_{CO}$: It is at this point in evolution that the production of interesting organic molecules would be easiest. Precipitation of C (graphite) occurs in this example, calculated for $T_S = 800$ K and $P_S(t = 0) = 100$ bar. At lower surface temperatures, S_8 and NH_4Cl may be present as initial condensates; at somewhat lower temperatures H_2O may condense as water or ice; and near 180–220 K (depending on the pressure), NH_3 and H_2S gases may react to precipitate solid NH_4SH. Such an evolutionary history provides graphite as a crustal mineral and leaves behind large amounts of neon and nitrogen. Helium may escape or persist, depending on the details of the escape physics (see Section 3.7).

very sensitive function of upper atmospheric temperature and composition and of planetary mass, radius, and atmospheric structure. The relevant principles are discussed in general terms in Section 3.5, and in specific detail in the relevant portions of Chapter 4.

The equilibrium pressures given in Figure 3.1 neglect the chemical consequences of any reactions with the planetary surface beyond the necessary direct precipitation of substances such as graphite and sulfur onto the surface. Such surface reactions will be introduced in Section 3.7. It should be kept in mind that sulfur, chlorine, and fluorine may in reality be severely depleted by reaction with the surface, thus reducing their abundances far below those given in Figure 3.1, even at a few hundred degrees Kelvin. At lower temperatures, formation of carbonates, sulfides, sulfates, and hydroxyl silicates may also occur. All of these processes have the effect of raising the mole fractions of neon and the heavy rare gases.

Figure 3.1 clearly shows the stepwise transition from reducing to oxidizing gases described by reactions (3.1) to (3.7). The transition from an ammonia-rich to a nitrogen-rich gas causes a change from the basic and reducing case, where the NH_3 abundance exceeds that of all acidic gases (HCl, HF, etc.) combined, to one in which the atmosphere is nitrogen-rich and acidic. When CO_2 becomes the dominant gas, the atmosphere becomes almost entirely composed of acid anhydrides and inert gases. Thus, the pH of any aqueous solution present would change from very basic to very acidic as p_{H_2} drops. This, of course, causes a radical change in the nature of the weathering reactions by which the atmosphere attacks surface minerals.

The first universal trait of such a system is that, forced by the fact that the abundance of carbon is more than half the oxygen abundance, removal of hydrogen to the point of conversion of most carbon to CO_2 will invariably cause the precipitation of graphite upon the surface of the planet. Second, neon is a very important atmospheric constituent, more abundant than nitrogen. Third, water is wholly absent from the final atmosphere. The ultimate consequence of removal of H_2 is decomposition of water and oxidation of carbon to the extent of tying up all oxygen as CO_2. These results are summarized concisely in Table 3.2.

The sequence of oxidation of element hydrides may be seen from Figure 3.1 to begin with PH_3, then NH_3, then CH_4, and finally H_2S. Very extreme depletion of hydrogen is required in order for H_2O, HCl, and HF to give way to O_2, Cl_2, and F_2. It can be seen from Figure 3.1 that a 10^{10}-fold reduction in p_{H_2} ($M \simeq 33$) does not suffice to give any of these species a mole fraction as high as 10^{-9} of the CO_2 abundance.

The dominant sulfur compound in the resulting gas is, not surprisingly, COS.

In the 100-bar, 800-K example, about 15% of the total carbon originally present precipitates as graphite. At higher pressures CO is less important and more graphite is precipitated. At sufficiently low total pressures CO is abundant and can accommodate enough carbon in the gas phase so that no graphite forms.

The effects of lower temperatures are complex. First, CO_2 is stabilized relative to CO, which enhances the stability of graphite at every pressure at which CO and CO_2 are more abundant than methane. In the methane-rich regime (low T, high P, low degree of H_2 loss) lowering the temperature enhances the stability of methane relative to graphite. For temperatures of 300–500 K any total pressure greater than $\sim 10^{-5}$ bar will evolve so as to precipitate graphite. A captured atmosphere with a pressure near or above 1 bar will evolve to precipitate graphite at any temperature below ~ 800 K. Figure 3.2 displays in simplified form the way in which the stability field of graphite expands over P, T space as a result of hydrogen loss.

The six most abundant spectroscopically active gases in Figure 3.1—CO_2, CO,

TABLE 3.2
**Residual Atmospheric Composition after Hydrogen
and Helium Escape from a Solar-Composition Gas**

Species	Mole fraction
CO_2[a]	0.632
Ne	0.218
N_2	0.097
COS	0.042
$^{36}Ar + ^{38}Ar$	0.009

[a]Thirty-three percent of total carbon is found as graphite. The solar oxygen abundance has been corrected to allow for prior formation of SiO_2, MgO, CaO, and Al_2O_3. A temperature below ~ 400 K was assumed.

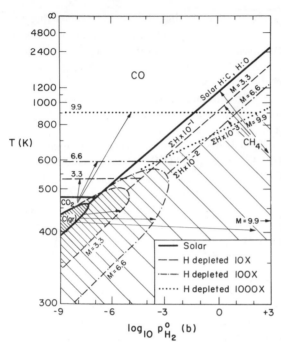

FIGURE 3.2. **Stability field of graphite versus degree of hydrogen loss. The thick solid lines show the boundaries of the regions in which the most abundant carbon-bearing gases in a solar-composition gas are CO, CO_2, and CH_4 respectively. CO is most abundant at high temperatures and low pressures, CO_2 at low temperatures and low pressures, and CH_4 at high pressures and low temperatures. Graphite is present within the cross-hatched region. The thin lines show the expansion of the stability field of graphite to higher temperatures and pressures as p_{H_2} falls: The curves bounding the graphite region are labeled with the depletion index M ($p_{H_2} = p_{H_2}^o/2^M$). Note that the axes of the graph give the surface temperature and *initial* surface pressure of the atmosphere.**

COS, CS_2, HCl, and HF—are evocative of the atmospheric composition of Venus, although, as we shall see in Section 4.2, the Ne and N_2 abundances on Venus are very much lower.

3.2. ACCRETION, THERMAL HISTORY, AND OUTGASSING

The first essential ingredient of a thermal history model is a compositional model for the planet. This model must specify the bulk composition, the major-element mineralogy, and the abundances of radioactive elements which could provide significant amounts of heat.

Second, it is necessary to specify certain features of the accretion process. For example, models with identical bulk composition may be either accreted completely homogeneously, or the minerals may be assumed to accrete in the sequence in which they condense from the solar nebula. The latter assumption, of course, begins with planets which are already layered, although not in exactly the same way that an initially homogeneous planet would become layered after melting and geochemical differentiations. The kind of layering assumed has a strong impact on thermal history models because it may place very high concentrations of radionuclides in very restricted portions of the planet, such as its core or crust. Another very important kind of assumption must be made regarding accretion: the time scale over which accretion occurs. Since the time scale for cooling of the solar nebula is of order 10^4–10^5 years all condensation steps in the cooling sequence must occur in that short an interval of time. If protoplanets are present and are extremely efficient at sweeping freshly condensed mineral gains from the nebula, then layered (non-homogeneously accreted) planets may result. If, however, the time required for growth of planets is much longer than 10^4–10^5 years, the dust available for accretion will be the *total* condensate. Accretion will then result in more or less homogeneous planets, within which early refractory condensates and late volatile-rich condensates will not be segregated into already separated layers. A direct logical consequence of this consideration is that any nonhomogeneously accreted planet must have accreted in $\lesssim 10^4$ years. By geological standards 10^4 years is a twinkling of the eye.

3.2.1. Accretion Heating

Consider an Earth-sized planet ($M_\oplus \simeq 6 \times 10^{27}$ gm) accreted in 10^4 years (3×10^{11} sec). The gravitational potential energy released has an order of magnitude of

$$\Delta GPE = GM_\oplus^2/r_\oplus, \tag{3.8}$$

where G is the universal gravitational constant, and $M_\oplus$ and $r_\oplus$ are the planetary mass and radius, respectively. Thus ΔGPE will be $\sim 7 \times 10^{-8}(6 \times 10^{27})^2/(6.4 \times 10^8) \simeq 4 \times 10^{39}$ erg. If this heat were all stored internally, each piece of rock would be heated to a peak temperature near 60,000 K! Clearly accretion of such a planet depends on its ability to radiate off the accretion energy. The rate of energy emission per unit area which exactly balances the heat production rate is

$$F = \frac{4 \times 10^{39}}{4\pi r_\oplus^2 \, \tau_{accr}} = \sigma T^4, \tag{3.9}$$

where σ is the Stefan–Boltzman constant and τ_{accr} is the accretion time scale: The flux is then 2.5×10^9 erg cm^{-2} sec^{-1} and $T \simeq 2600$ K, the mean surface temperature required to radiate off so much heat in a steady state. Of course, during the terminal stages of accretion the energy brought in per gram of material will be much larger than the average. A late arrival brings in as an absolute minimum a kinetic energy per gram of $\frac{1}{2}v_{esc}^2$, where v_{esc} is Earth's present escape velocity. This gives 7.6×10^{11} erg gm^{-1}. Since the heat capacity of most rocks is near 0.25 cal gm^{-1} K^{-1}, this is enough energy to heat the particle to 7.6×10^4 K!

An accretion time scale as long as 10^8 years would imply a flux of only 3×10^4 erg cm^{-2} sec^{-1} at steady state. Since Earth is maintained near 280 K by solar irradiation, this would increase the emitted flux of $\sigma T^4 = 4.67 \times 10^{-5}(280)^4 = 3.48 \times 10^5$ erg cm^{-2} sec^{-1} by only 8.6%. This would in turn require an increase of the Earth's surface temperature dT given by

$$\frac{dF}{F} = \frac{4\sigma T^3 \, dT}{\sigma T^4} = 0.086 = 4\frac{dT}{T}, \tag{3.10}$$

whence $dT = 0.0215T = 6$ K.

Clearly, slow accretion of a planet will have negligible influence on its internal temperatures, whereas rapid accretion in the presence of the solar nebula would have the most profound consequences.

Theoretical consideration of the sedimentation of particles toward the central plane of the nebula by Goldreich and Ward (1973) suggests that accretion of asteroid-sized bodies could have occurred during the lifetime of the nebula, but that terminal accretion of planets from these fragments may have taken far longer. Safronov (1972), Weidenschilling (1974), and Cox and Lewis (1980) all estimate a terminal accretion time scale of $\sim 10^8$ years.

Accretion of a lunar-sized body on a time scale of 10^4 years will, by the argument given above for Earth, give a steady-state emitted flux near 1.8×10^7 erg cm^{-2} sec^{-1}, or a temperature of ~ 750 K, close to the present surface temperature of Venus. For Ceres with $M = 1.175 \times 10^{24}$ gm and $r = 536$ km, the steady-state energy flux due to accretion in 10^4 years could maintain a temperature about 40 K above the present temperature.

Thus, the accretion energy of asteroidal to lunar-sized objects is marginally important for *very* short accretion times.

3.2.2. Early Solar Luminosity and the T-Tauri Phase Solar Wind

An interesting question is what surface temperature a body would have immediately after the dissipation of the solar nebula. Available observational and theoretical evidence [see especially Larson (1974) and Kuhi (1964)] suggests that the early Sun may have been superluminous by a factor of 2 or 3 relative to present conditions for about 10^7 years after ignition of thermonuclear reactions. During this era, a few tenths of a solar mass passed through the inner solar system as a very dense plasma wind carrying an energy flux comparable to the electromagnetic luminosity. A total solar energy output four times the present level would raise all radiative steady-state temperatures by 40% relative to present values.

The energy carried by the dense plasma wind during this early pre-MS (T-Tauri) era is not, however, typically deposited in the surface of a body. The plasma magnetic field lines carried by the T-Tauri phase mass flux can be dragged through a poor electrical conductor such as silicate rocks. The moving magnetic field induces electric currents in the weakly conducting medium, which in turn produce ohmic heating (Sonett *et al.*, 1968). Thus, the energy deposited by the T-Tauri phase mass flux may be localized deep inside a body, where the time required for thermal conductive cooling is even longer than the duration of the T-Tauri phase itself.

The model for melting of asteroid-sized and larger bodies proposed by Sonett *et al.* (1970) required sufficient gas and dust blanketing to provide "hohlraum" temperatures of nearly 800 K in the asteroid belt, in order to achieve sufficiently high electrical conductivity for efficient coupling with the solar wind. This requirement arose artificially from the assumption that the electrical conductivity of asteroid material was similar to that of terrestrial olivine. Since most asteroids are found to have surface reflection spectra similar to carbonaceous chondrites (which clearly have not experienced 800 K), and since virtually all other meteoritic material contains metallic iron, it is clear that realistic electrical conductivities must be employed in the model. Brecher *et al.* (1975) and Briggs (1976) have studied the electrical conductivities and dielectric constants of a suite of carbonaceous chondrite samples, including the effects of heating and recooling. They find conductivities so high that the ~800 K temperatures required in the model of Sonett *et al.* (1970) can be reduced to the actual (~160 K) radiative steady-state temperatures of belt asteroids. Thus, strong coupling with the T-Tauri solar wind by C2 chondritic asteroids seems to be easy.

Since the coupling between the solar wind and an asteroid depends on the heliocentric distance, on the electrical conductivity of the asteroid (and thus on its composition and temperature), and on the size of the asteroid, some interesting results are found in realistic heating models. Asteroids of identical size and heliocentric distance may behave very differently if their compositions are different, whereas bodies of identical size and composition may behave very differently at different heliocentric distances (Briggs, 1976).

3.2.3. Short-Lived Radionuclides

There is yet another source of energy for very early melting. The most abundant radioactive elements—K, U, and Th—have half-lives far too long to produce a quick early heat pulse, but there is very strong evidence for the presence of short-lived radionuclides in the early solar system. It has long been known that the isotope ^{129}Xe, the decay product of ^{129}I, is anomalously abundant in many meteorites and varies strikingly in abundance relative to all the other xenon isotopes (Reynolds, 1960), apparently reflecting different primordial I:Xe ratios and different times of formation for these bodies. The half-life for β^- decay of ^{129}I is 17×10^6 years. Another xenon isotope anomaly consists of the strong enhancement of the heaviest Xe isotopes in some meteorites. After subtraction of the usual meteoritic xenon mass spectrum, the component responsible for the heavy isotopes is found to have a mass spectrum that is very distinctive, somewhat resembling that of xenon from fission of uranium and thorium (Reynolds and Turner, 1964). A plausible fit to the isotopic abundances can be found in the gases produced by fission of the isotope ^{244}Pu (Rowe and Kuroda, 1965); however, the heavy isotopes ^{134}Xe and ^{136}Xe are often enriched *simultaneously* with the light isotopes ^{124}Xe, ^{126}Xe, and ^{128}Xe, relative to ^{130}Xe and ^{132}Xe (Manuel *et al.*, 1972). Manuel and co-workers (Manuel *et al.*, 1972; Sabu *et al.*, 1974; Sabu and Manuel, 1976) have interpreted this phenomenon as evidence for supernova processing of Xe ~10^8 years prior to the origin of the solar system; Clayton (1975) has suggested that this component was trapped by prompt grain condensation in supernova envelopes. Anders and co-workers (Anders, 1969; Anders and Larimer, 1972; Anders *et al.*, 1975; Takahashi *et al.*, 1976) prefer a derivation of the anomalous "fission" Xe from decay of superheavy nuclides of Z $\simeq$ 116. In every case, decay of parent nuclides with half-lives of ~10^6–10^8 years is required. The half-life for fission of ^{244}Pu is 83 $\times$ 10^6 years.

As compelling as these lines of evidence may be with regard to the presence of short-lived nuclides in the early solar system, none of these deduced parent species would have been abundant enough to constitute an important heat source.

In an early study of the possibility of heating by short-lived nuclides, Urey (1952c) proposed ^{26}Al as the most promising candidate. Despite several negative attempts to discover evidence of ^{26}Al decay product (^{26}Mg) variations in bulk meteoritic material (see Schramm et al., 1970), efforts have persisted. The discovery of highly refractory-rich inclusions in C3 and C4 chondrites provides us with very ancient material with a high Al:Mg ratio, wherein ^{26}Mg from ^{26}Al decay could more easily be discernable. Lee and Papanastassiou (1974) and Lee et al. (1976) have found that the most Al-rich grains in these inclusions do in fact display enhanced ^{26}Mg abundances corresponding to an original ^{26}Al:^{27}Al ratio of ~4 $\times$ 10^{-5}. The half-life for ^{26}Al decay is only 0.72 $\times$ 10^6 years, hence it is necessary to accrete objects in a few million years in order to take advantage of this heat source. On the other hand, the amount of energy available is so large that any Gold-reich–Ward planetesimals accreted to a size of ~10 km or larger would partly melt. This, of course, is dependent upon the very short accretion times for such planetesimals, of order 10^4 years.

A fundamental logical problem arises directly from the acceptance of such a heat source: the observed prevalence of massive bodies with carbonaceous chondritic surfaces in the asteroid belt. How could such delicate volatile-rich material have possibly survived such a powerful melting episode in such vast quantity? We would be compelled to believe that all chondritic meteorites come from parent bodies of radii less than ~10 km, or that the surfaces of large asteroids have remained wholly undisturbed since the first few million years of history of the solar system.

Radiogenic ^{26}Mg anomalies can only be detected in very Mg-poor and Al-rich materials, and the only minerals well suited to the search are high-temperature condensates such as corundum (Al_2O_3). Thus, the only evidence for short-lived ^{26}Al is in a small subpopulation of chemically and isotopically unusual grains, which belong to a population of refractory particles, which in turn are a minor component of a single subclass of carbonaceous chondrites. Since we do not know the exact isotopic composition of primordial Mg, it is possible to argue that ^{26}Al was actually omnipresent in the early solar system, that it was an important heat source in all classes of meteorites, and that nebular homogenization of isotopes by high-temperature reactions has rendered it impossible to separate the radiogenic ^{26}Mg from primordial ^{26}Mg.

Our opinion is that it is very difficult to envision any nebular process that can isotopically and chemically homogenize refractory grains which would not even more readily homogenize oxygen isotopes: However, the latter has not occurred (see Section 2.3). A second, less appealing prospect is that the mixing and homogenization processes were not nebular, but rather took place after accretion of parent bodies. In this case, refractory grains containing both Al and O isotope anomalies were accreted onto the parent bodies and then equilibrated, both chemically and isotopically, by metamorphism. Special explanations would then be needed if, as seems likely, the H3, L3, C2, and C1 chondrites are essentially devoid of such anomalous grains.

A review on this subject has recently appeared (Lee, 1979).

Thus, although the presently available evidence points in the direction of the importance of one or more short-lived early heat sources, notably the T-Tauri phase and ^{26}Al decay, the quantification of these heat sources for purposes of thermal modeling is extremely difficult.

3.2.4. Long-Lived Radionuclides

Over much longer periods of time, the most important heat source is decay of the long-lived radionuclides ^{40}K, ^{232}Th, ^{235}U, and ^{238}U. The abundances and half-lives of these nuclides in ordinary chondrites are given in Table 3.3. Note that ^{40}K and ^{235}U have half-lives several times smaller than the age of the Solar System. This, of course, means that their primordial abundances 4.5×10^9 years ago were an order of magnitude above the present values, and their rates of heat production were proportionately greater. The evolution of these heat sources with time is shown in Figure 3.3.

3.2.5. Differentiation Heat Release:
Stored Potential Energy

Since the release of gravitational potential energy during formation of a planet is enormous, it is not surprising that the sinking of dense core material during differentiation of a planet can also contribute an important amount of heat. This heat release is, however, clearly conditional upon the prior warming of the planet up to the melting point, and cannot in any way alter whether or not a planet differenti-

TABLE 3.3
Present Chondritic Abundances of Long-Lived Radionuclides

Element	Nuclide	Present abundance (ppm wt.)	$t_{1/2}$ (10^9 yr)
K		815	
	^{40}K	0.097	1.47 (to ^{40}Ca)
			11.8 (to ^{40}Ar)
Th	^{232}Th	0.04	13.9
U	^{235}U	0.00009	0.72
	^{238}U	0.012	4.51
Rb	^{87}Rb	2.5	50

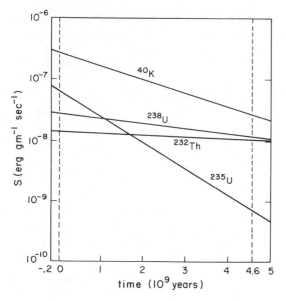

FIGURE 3.3. **Rates of heat production of long-lived radionuclides versus time. Chondritic abundances of K, U, and Th are assumed. Note the great early importance of ^{235}U and ^{40}K. The short-lived radionuclide ^{26}Al has a half-life of only 0.72×10^6 yr, and thus would decay by a factor of 6×10^5 over a time of only 0.1×10^9 yr.**

ates. The significance of this contribution is that it provides a way by which melting, once begun, can catalyze itself.

A final possible heat source can be provided by stored chemical potential energy. In a planet that accretes in layers which are not in chemical equilibrium with one another, mixing or melting can permit various exothermic chemical reactions to occur. An example can be seen in Figure 2.11 (page 49) where alkali sulfides are not in equilibrium with Al-rich silicates in the refractory oxide layer.

3.2.6. Heat of Phase Changes Caused by Changing Temperatures

Whenever any substance in the planet is melted (or solidified) by temperature changes, the heat of fusion is absorbed (or released). In the case of large-scale melting, substantial amounts of heat may be absorbed. Melting is therefore a fairly effective thermostating agent during heating. Melting of a pure phase A with melting point at 1000 K will cause the temperature to cease rising and remain at 1000 K until enough additional heat has been produced to exhaust solid A. The melting temperature, volume change of melting, and heat of melting are all to some degree functions of pressure, especially the melting temperature.

Production of a vapor phase, such as by dehydration of a hydroxyl silicate mineral M:

$$M(OH)_2 \rightarrow H_2O(g) + MO, \qquad (3.11)$$

or by oxidation of carbon by iron oxides:

$$C(gr) + FeO \rightarrow Fe^\circ + CO(g) \qquad (3.12)$$

and

$$C_{(gr)} + 2FeO \rightarrow 2Fe^\circ + CO_{2(g)}, \tag{3.13}$$

absorbs or releases heat to some degree, but the fluid phase is perhaps most significant to thermal history modeling because it is of very low density and viscosity: It is able to transport heat upward very rapidly by bulk motion.

The melting temperature T_m as a function of pressure of any pure phase is given by the Clausius–Clapeyron equation as

$$dT_m/dP = \Delta v_m/\Delta S_m = T\,\Delta v_m/\Delta H_m, \tag{3.14}$$

where Δv_m is the molar volume change upon melting, and ΔS_m and ΔH_m are the molar entropy and enthalpy of melting, respectively.

3.2.7. Melting in Multicomponent Systems

The melting temperatures of most pure substances rise with increasing pressure because the molar volumes of almost all liquids are larger than for the corresponding solids. There is, however, a phenomenon which works in the opposite direction, tending to lower the temperature of melting: the effect of solutes dissolved in the melt.

The temperature of melting T of a solution with a solvent mole fraction of X is related to the melting temperature of pure solvent, T, by

$$\frac{1}{T_x} = \frac{1}{T} - \frac{R \ln X}{\Delta H_m}, \tag{3.15}$$

where R is the gas constant.

Consider two pure solids, A and B, which are mutually soluble only in the liquid state. Addition of small amounts of B to A permits melting to begin at lower temperatures, as described above, whereas addition of some A to pure solid B will likewise lower its melting temperature. The resultant of these two effects is pictured in Figure 3.4, which shows the portion of the Fe–S system with compositions between pure Fe and pure FeS, as in ordinary chondrites.

Addition of other solid phases which are soluble in the Fe–FeS melt leads to ever-greater depression of the freezing point. In the ternary (three-component) system Fe–FeS–FeO the ternary eutectic temperature is ~910°C, some 70° below the binary Fe–FeS eutectic.

The most profound effects can be realized by addition of water and other volatiles to a silicate system, thus forming water-rich low-temperature magmas such as carbonatites.

3.2.8. Outgassing Reactions in Warming Planets

The simplest scenarios for outgassing of planet are to arrange the initial conditions in such a way that either (a) temperatures never approach the melting point any-

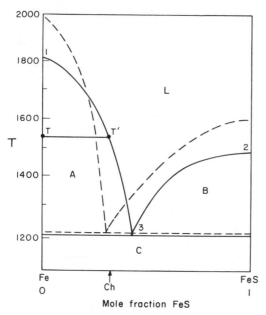

FIGURE 3.4. **Phase diagram of the Fe–FeS system. The melting points of pure Fe and FeS at $p = 1$ bar are marked 1 and 2, respectively. Addition of FeS to Fe, or of Fe to FeS, lowers the melting temperature along the solid curves to a minimum temperature (marked 3) called the eutectic temperature. In region L the only phase present is an FeS–Fe melt. In region A, solid metallic iron coexists with the melt. Coexisting phases are connected by tie lines (isotherms) such as T–T'. In region B, solid FeS coexists with the melt, whereas in region C, below the eutectic temperature, the pure solids Fe and FeS alone are present. At elevated pressures (dashed line) the melting temperatures of both Fe and FeS go up, the eutectic composition shifts toward pure Fe, and the eutectic temperature remains essentially unchanged. The typical Fe : FeS ratio in chondrites is marked Ch.**

where inside the body, or (*b*) early (T-Tauri phase, accretion ^{26}Al) heating is so intense that the entire planet is thoroughly outgassed. In the former case, no atmosphere ever forms. In the latter, the entire volatile budget of the planet can be delivered at once to the surface.

We have, however, no positive evidence that prompt total outgassing occurred for any planet, and we have strong evidence that it did *not* occur on some planets. This forces on us a closer look at mechanisms for outgassing from planets with intermediate thermal histories.

We have already commented on the importance of vertical heat transport by a volatile-rich fluid, and on the depression of melting points by volatiles. We now state the obvious: The *most* important consequence of production of a rising volatile-rich fluid is that it is the raw material for the planetary atmosphere.

Release of gases from mineral assemblages occurs readily when the equilibrium pressure of gases is at least equal to the static pressure due to the weight of overburden. This is a good approximation when gas release occurs at great depths, where the rock is well consolidated and cracks and pores do not communicate effectively with the atmosphere.

The partial pressures of the individual reactive gases are computed by the condition of local thermodynamic equilibrium. It is important to emphasize that, in general, near-surface regions of a young planet must have temperature gradients and large heat fluxes: From the thermal point of view, this region is probably not even in a steady state. The assumption of equilibrium is, therefore, applicable only over a narrow interval of depth and a short period of time.

The principal materials responsible for the production of a gas phase and representative reactions for them include the following:

$$(Mg, Fe)_3Si_2O_5 (OH)_4 \rightleftarrows (Mg, Fe)_2SiO_4 + (Mg, Fe)SiO_3 + 2H_2O(g) \qquad (3.16)$$
serpentine

$$Ca_2Mg_5Si_8O_{22} (OH)_2 \rightleftarrows 2CaMgSi_2O_6 + 3MgSiO_3 + SiO_2 + H_2O(g) \qquad (3.17)$$
tremolite

$$(Fe, Ni, C) + FeO \quad \rightleftarrows (Fe, Ni) + CO(g) \qquad (3.18)$$
metal

$$(Fe, Ni, C) + 2FeO \quad \rightleftarrows (Fe, Ni) + CO_2 (g) \qquad (3.19)$$
$$FeS + 2FeO \quad \rightleftarrows 3Fe + SO_2 (g) \qquad (3.20)$$
troilite

$$FeS + CO (g) \quad \rightleftarrows Fe + COS (g) \qquad (3.21)$$
$$FeS + H_2O (g) \quad \rightleftarrows FeO + H_2S (g) \qquad (3.22)$$
$$(Fe, Ni, N) \quad \rightleftarrows (Fe, Ni) + N_2 (g) \qquad (3.23)$$
$$Fe + H_2O (g) \quad \rightleftarrows FeO + H_2 (g) \qquad (3.24)$$
$$(Fe, Ni, C) + 2H_2 (g) \rightleftarrows (Fe, Ni) + CH_4 (g) \qquad (3.25)$$
$$N_2 (g) + 3H_2 (g) \quad \rightleftarrows 2NH_3 (g) \qquad (3.26)$$
$$(^{232}Th, ^{235}U, ^{238}U) \quad \rightarrow {}^4He (g) \qquad (3.27)$$
$$^{40}K \quad \rightarrow {}^{40}Ar (g) \qquad (3.28)$$

In addition, reactions for release of HCl, HF, and other trace reactive gases must be included, along with radioactive decay of ^{129}I to ^{129}Xe, fission production of krypton and xenon, and release of trapped primordial rare gases. These chemical reactions and release of rare gases from inside mineral grains require high temperatures that were not necessarily present on the exposed surface of a newly formed planet, but which can be reached at depth inside the body. Unfortunately, this means that the production of a fluid phase is likely to occur at levels where the pressure is over 1 kbar, and the gas is likely to be strongly nonideal. This greatly complicates the thermodynamic treatment of the system.

The volatile-rich fluid either may seep slowly toward the surface, reequilibrating as it moves upward into regions of lower temperature and pressure; or the gases, accompanied by a wet low-density magma, may erupt directly onto the surface. In the latter case, temperatures are sufficiently high for very rapid equilibration between the fluid phase and the magma, and the composition of the volcanic gas reflects the temperature and pressure conditions in the throat of the volcanic vent quite near the surface, where nonideality corrections are unimportant. In this latter case, the main problem is assessing the activity coefficients of the major components of the magma and the solubility relationships for the chemically active gases.

It is relatively easy to compute the equilibrium composition of an isolated gas phase if the elemental abundances in the gas are specified: In the absence of adequate thermodynamic data on the magma, this is the best that can be done. The procedures and results of such modeling will be presented in Section 4.3.

The first substantial melting of ordinary chondritic material would occur at the Fe–FeS–FeO ternary eutectic temperature, near 1200 K. At this temperature, carbon and nitrogen from the metal phase, carbon and nitrogen in carbynes, sulfur in FeS, and oxygen in FeO would all be able to react readily with each other in a

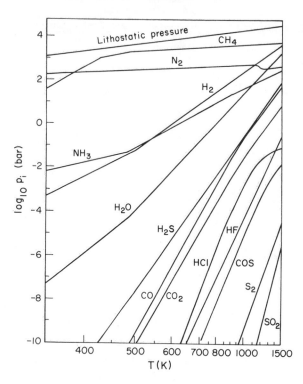

FIGURE 3.5. Gas composition versus depth for an H-chondrite model on a log P versus $-1/T$ plot. Note that the gas pressure in each layer is always lower than the lithostatic pressure. Exact chemical equilibrium is assumed at all depths. (After Bukvic and Lewis, 1981.)

common melt. At lower temperatures equilibration is much more difficult. Indeed, heating of a chondrite to 500–1000 K may cause sequential loss of volatiles, with that phase which has the highest decomposition pressure (hydroxyl silicates?) outgassing first.

In dense hot rock, at such depths that selective venting of early vapors does not occur, we may have the gas phase truly governed by chemical equilibrium. Figure 3.5 shows how the composition of the gas phase in equilibrium with H chondrite rock would vary with depth along an early Earth geotherm (Bukvic and Lewis, 1981).

The release of volatiles to form a secondary (as opposed to primary or primordial) atmosphere is clearly dependent on the detailed thermal history of the planet, and should be modeled only in the context of particular assumptions regarding the available heat sources.

3.2.9. Holistic Thermal History and Atmospheric Origin Modeling

In order to model the thermal history of a planet properly, the temperature and composition of the materials available for accretion must be specified. The time dependence of the accretion rate, the composition of the infalling material, and the radiative steady-state temperature at that planet's distance from the Sun must be specified. The composition must be given in some detail, including the major-element abundances and mineralogy, volatile abundances, and radionuclide abundances. In order to model the transport of heat, the thermal conductivity, heat capacity, and thermal opacity of the bulk material must be known. To compute the effects of T-Tauri phase solar wind heating, the electrical conductivity must be known both as a function of composition and of temperature.

The model is assumed to have spherical symmetry, so that in a sense the computations may be said to be one-dimensional. The planet is divided into a large number of (order 10^2) concentric spherical shells. Heat production rates are computed at each depth due to each of the possible heating mechanisms, including both long- and short-lived radionuclides, accretion energy, solar wind induction heating, and conversion of gravitational potential energy into heat by sinking of dense melts. In sophisticated models, it may be found necessary to allow for the storage of compression energy in the deep interior due to the accretion of added weight on the planetary surface (adiabatic compression energy).

The temperature of first melting must be calculated as a function of depth, taking into account the local composition and pressure. Whenever the temperature reaches the local melting temperature, heat must be put into melting, the composition of the melt must be estimated, the presence or absence of a gas phase must be determined, and the fluids and solids must be permitted to separate in accordance with their relative densities.

At temperatures within a few hundred degrees of the melting temperature, convective heat transport begins to become important, and simple conductive heat transport no longer suffices. At temperatures in excess of 1000 K for some minerals the infrared opacity is sufficiently low that radiative transport also becomes important.

Finally, as melting begins it becomes possible for high- or low-density liquids to percolate through the rock and begin the process of chemical differentiation. Of course, when solids and one or more melts coexist, the solubility of the minerals in each melt must be known, and the partitioning behavior of minor elements (especially the volatile and heat-producing ones) must be treated with care.

It is fair to say that this is an enormously complex undertaking, and that even mastery of the modeling of these processes on Earth might not readily yield equally successful models of another planet, such as Mars.

No thermal model of any planet has yet been published that meets all of these requirements; indeed, most models contain seriously unrealistic assumptions and oversimplifications. Since the computational tools for treating most of these effects are of relatively recent vintage, substantial progress in thse modeling techniques is to be expected in the near future.

3.3. ATMOSPHERIC CHEMISTRY

It should now be clear that volatile molecules that were roughly in thermochemical equilibrium with magmas or mineral assemblages at a particular level or levels in the crust can be outgassed to contribute to a secondary atmosphere. We must next address the question of the *stability* of these various volatile molecules in order to assess their actual contribution to the observed atmosphere of a planetary body. We must therefore be concerned with chemical reactions in the atmosphere, atmospheric escape, phase transformations, gas–surface reactions, and finally, where applicable, with biological transformations. These important phenomena that collectively determine the *atmospheric cycle* of a particular volatile element will be discussed in the following sections, beginning in this section with considerations of atmospheric chemistry. We should also add that massive primary (i.e., solar composition) atmospheres will likewise be influenced by certain of these phenomena, in particular by disequilibrating gas-phase chemical reactions. These will be discussed in Section 4.5.

3.3.1. Thermochemical Equilibrium in the Atmosphere

Once the volatile-element inventory for the atmosphere has been determined either by considerations of outgassing or by assuming solar composition, it is very instructive to first ask what the gas phase composition would be in thermochemical equilibrium at the appropriate atmospheric temperatures and pressures. At equilibrium there are no irreversible reactions occurring and the Gibbs free energy G and the entropy S of the air parcels are time invariant. For example, *homogeneous* reactions between *ideal* gases with partial pressure P_i (atm) we have

$$dG = -S\,dT + V\,dP + \sum_i \left\{ \left(\frac{\partial G}{\partial[i]} \right)_{1\ atm} + RT \ln P_i \right\} d[i] = 0. \quad (3.29)$$

Integrating over a reversible reaction path keeping temperature T and total pressure P constant, we therefore have

$$\Delta G = \sum_i \left\{ \left(\frac{\partial G}{\partial[i]} \right)_{1\ atm} \Delta[i] + RT \ln (P_i)^{\Delta[i]} \right\}$$
$$= \Delta G\,(T,\ 1\ atm) + RT \ln(P_i)^{\Delta[i]}$$
$$= 0. \quad (3.30)$$

That is

$$\Pi(P_i)^{\Delta[i]} = \exp\text{-}\Delta G(T,\ 1\ atm)/RT = K(T), \quad (3.31)$$

where $K(T)$ is the *equilibrium constant*, which determines the partial pressures of the various reactants in thermochemical equilibrium. This constant varies with temperature according to the Van't Hoff equation:

$$\frac{d[\ln K(T)]}{dT} = \frac{\Delta H\,(1\ atm,\ T)}{RT^2}. \quad (3.32)$$

Here $\Delta H(1\ atm,\ T)$ is the standard heat or enthalpy of the reaction, which is generally a very weak function of temperature. Similar equilibrium constants can be defined to relate gas partial pressures in *heterogeneous* reaction systems where gas–liquid and gas–solid reactions occur. For example, the Clausius–Clapeyron equation mentioned earlier [Eq. (2.4)], which gives the temperature dependence of the partial pressure of a gas over its condensed phase, is simply a form of Eq. (3.32) appropriate to a simple heterogeneous reaction.

In order to carry out calculations of atmospheric composition in thermochemical equilibrium, we clearly require knowledge of the variation of temperature and pressure with altitude, the relative abundances of volatile elements, and, for the terrestrial planets, the activities of chemically active surface minerals. It is not particularly difficult to make at least rough estimates for all these required quantities, and it is therefore not only very instructive to consider the equilibrium atmospheric model but relatively easy to construct it. Indeed, as we will see in Chapter 4, the thermochemical equilibrium assumption leads to models of the Venusian and major planetary atmospheres that look to a considerable degree like the real thing. In other cases, the assumption is poor.

3.3.2. Disequilibrating Processes in the Atmosphere

When the Gibbs free energy or the entropy of an air parcel changes significantly with time, the assumption of thermochemical equilibrium becomes invalid. In general, chemical disequilibration is due to one or both of two causes: irreversible chemical reactions in the air parcel (e.g., dissociation caused by photons, shock waves from thunder, lightning discharges) and net exchange of material in the air parcel with its surroundings (e.g., changes in composition resulting from atmospheric mixing, diffusion, and particulate sedimentation).

The extent and cause of chemical disequilibration for each species i can be readily ascertained from a scale analysis of its three-dimensional continuity equation:

$$\frac{\partial[i]}{\partial t} = \left(\frac{d[i]}{dt} \right)_{\text{chem}} - \nabla\cdot(\mathbf{V}[i]) - \frac{\partial(w_i[i])}{\partial z}. \tag{3.33}$$

This equation describes the time (t) changes in $[i]$ *at a particular position*. The first term on the right-hand side describes variations in $[i]$ due to chemical reactions, the second term describes concentration changes due to atmospheric motions with velocity $\mathbf{V}$, and the third term gives the changes in $[i]$ resulting from gravitational sedimentation with vertical velocity w_i (for gases $w_i \simeq 0$). From this equation we see that the variability of gas or aerosol concentrations is generally due to a combination of net transport and net chemical production or destruction. It is clear that a solution of this equation generally requires a very accurate knowledge of atmospheric motions obtained either by utilizing observations (e.g., Prinn *et al.*, 1975) or the results of numerical circulation models (e.g., Cunnold *et al.*, 1975). Unfortunately, but not surprisingly, we possess extremely little knowledge of transport processes in the Earth's ancient atmosphere and even less of motions in the present and ancient atmospheres of other planets and satellites. An exact solution of Eq. (3.33) is therefore precluded by our lack of knowledge.

This apparently difficult situation can fortunately be considerably relieved if we recognize, first, that exact solutions are not really required here, and second, that horizontally averaged solutions are adequate for most of our purposes. We can therefore simply consider an approximate *one-dimensional* continuity equation:

$$\frac{\partial [i]}{\partial t} = \left(\frac{d[M]}{dt} \right)_{\text{chem}} + \frac{\partial}{\partial z} K[M] \frac{\partial f_i}{\partial z} - \frac{\partial (w_i[i])}{\partial z}, \qquad (3.34)$$

which will give us the *altitude* (z) distribution of species i. Here the net vertical flux of i is assumed to be proportional to the vertical gradient in its mixing ratio f_i ($= [i]/[M]$), multiplied by the total number density $[M]$ (cm^{-3}). The constant of proportionality is called the vertical *eddy diffusion coefficient*, K $(\text{cm}^2 \text{ sec}^{-1})$. This diffusion approach cannot be generally justified on physical grounds but serves as a useful engineering device (see, for example, Hunten, 1975). It is quite adequate for our purposes providing we recognize that K is determined by atmospheric motions ranging from microscale to the planetary scale. In using Eq. (3.34), we can obtain $[M]$ by assuming hydrostatic equilibrium (no net vertical flux of the background atmosphere):

$$\frac{1}{[M]} \frac{\partial [M]}{\partial z} = - \frac{\bar{m}g}{kT} - \frac{1}{T} \frac{\partial T}{\partial z} = - \frac{1}{H_M}. \qquad (3.35)$$

Here $\bar{m}$ is the mean molecular mass, g is the acceleration due to gravity, k is Boltzmann's constant, H_M is the density *scale height*, and T is temperature. Since we are usually interested in long-term averages in planetary atmospheres, we therefore seek only the *steady-state* solution for Eq. (3.34) where $\delta[i]/\delta t = 0$.

In order to understand the chemical disequilibrium process, it is convenient to consider *time constants* for various competing processes in Eq. (3.34) and thus determine the dominant terms. We define this time constant as the period for which the process must operate to change $[i]$ by a factor of e. For example, the rates of first-, second-, and third-order chemical reactions removing i are given respectively by $J[i]$, $k[i][a]$, and $l[i][a][b]$, where a and b are other reactants, and J, k, and l are first-, second-, and third-order rate constants. The chemical time constants τ_{chem} for each reaction removing i are thus

$$\tau_{\text{chem}} = \left(\frac{1}{[i]} \frac{d[i]}{dt} \right)^{-1} = \begin{cases} J^{-1} \\ (k\,[a])^{-1} \\ (l\,[a]\,[b])^{-1}. \end{cases} \qquad (3.36)$$

The transport time τ_{trans} is generally a function of the vertical distribution of i (Prinn and Owen, 1976), but under many circumstances is given roughly by

$$\tau_{\text{trans}} \simeq H_M^2/K. \qquad (3.37)$$

Thermochemical equilibrium for i can only occur if the time constant for *thermochemical* reactions producing and removing i is much shorter than those for transport and other nonthermochemical reactions. Otherwise, we have a disequilibrium situation, and we are usually interested in the steady-state disequilibrium situation in particular. We clearly need estimates for K, J, k, and l in order to make such a distinction in each individual situation. It will therefore be helpful to review briefly our knowledge of these quantities here.

The relationship between K and actual atmospheric motions is unfortunately complex, if indeed any such relationship really exists. In some cases we can identify the principal type of motion causing mixing in a particular region and therefore utilize a nonempirical or at least a semiempirical estimate for K. For example, in low-pressure subadiabatic regions of atmospheres, transient, vertically propagating internal gravity waves and tides may occasionally play an important role in vertical mixing. In this case we can write roughly (Lindzen, 1970)

$$K(z) \simeq K(0)[[M]\ (0)/[M](z)]^{1/2}. \tag{3.38}$$

In adiabatic regions of atmospheres, free convection is often the dominant motion involved in vertical mixing, and we can then write (Stone, 1976)

$$K(z) \simeq [R\phi/C_p \bar{m}[M](z)]^{1/3} H_M, \tag{3.39}$$

where ϕ is the upward heat flux carried by the free convection, R is the molar gas constant, and C_p is the molar gas heat capacity at constant pressure. Formulations for K in rapidly rotating subadiabatic atmospheres may be obtained from Stone (1973). The most reliable method of obtaining K is, not unexpectedly, empirical. If we can observe the vertical distribution of a diffusing species, we can essentially invert Eq. (3.34) to obtain the appropriate K value. This procedure has been carried out for tracers in the atmospheres of Venus, Earth, Mars, and Jupiter.

For molecular reactants, thermochemical reaction rate constants are generally large only at temperatures well above room temperature. Measurements of these high-temperature rate constants are unfortunately rather sparse, although some information can be obtained from the literature on fossil fuel combustion research. Thermochemical reactions involving atomic, free radical, or ionic reactants can, in contrast, be very rapid even at temperatures well below room temperature. These reactions have been studied extensively in the laboratory, and values for a large number of reaction rate constants are now available. For an exhaustive compilation of high- and/or low-temperature reaction rate constants the reader is referred to Hochstim (1969) and CIAP, Vol. I (1975) (Grobecker *et al.*, 1974).

All atmospheres are irradiated by solar UV and X-ray photons and also by cosmic rays. The latter consist mainly of protons and alpha particles with extremely high speeds approaching that of light. These energetic photons and particles can cause dissociation of molecules, yielding very reactive atoms, free radicals, ions, electrons, and neutrons. They thus provide a mechanism whereby atmospheric gases, which would otherwise be in thermochemical equilibrium with each other and with the surface, can be rearranged into strictly nonequilibrium mixtures. The rates of these photodissociation or energetic-particle dissociation reactions depend on the incident intensities, I, of the dissociating photons or particles, and on the absorption or capture cross sections, σ, of atmospheric molecules for these photons or particles. For example, for photodissociation by UV radiation of frequency ν

$$J(z) = \int_0^\infty \sigma(\nu)I(\nu) \exp\left[-\frac{\tau(\nu)}{\cos\theta} \right] d\nu, \qquad (3.40)$$

where $\tau(\nu)$ is the vertical optical depth at frequency ν above altitude z and θ is the zenith angle of the sun. In addition to determining dissociation rate constants, the yields of the various possible dissociation products must also be determined.

Other processes for dissociating molecules do exist, but, unlike the phenomena discussed above, they are not ubiquitous in the Solar System. For planets with no intrinsic magnetic field, the energetic particles comprising the solar wind (see Section 3.6.2) may penetrate into the atmosphere and lead to dissociation. On planets such as the Earth and Jupiter, with strong intrinsic magnetic fields and consequently well-developed radiation belts, dissociation may be caused by dumping of energetic particles from these belts into the "auroral" zones near the magnetic poles. Molecular dissociation may also occur in thunderstorms. This dissociation occurs partly due to the high temperatures and electron impact rates occurring in lightning strokes and partly as a result of extreme heating induced by the subsequent thunder shock waves. Thunderstorms require strong convection to build up charges on cloud particles, and we presently have no a priori or observational reason to expect such storms in clouds other than water clouds.

Of the various methods for molecular dissociation discussed above, we should emphasize that UV radiation, which makes up about 1% of the solar energy flux, is by far the most important. On Earth, the only planet on which the thunderstorm energetic efficiency is well known, only about 0.001% of the incident solar energy is converted to energy in lightning as electric currents and in thunder as acoustic waves, and the efficiency on Jupiter appears to be even smaller (Lewis, 1980b). Even in the substantial radiation belts on Jupiter the peak particle energy flux (principally proton kinetic energy) is roughly 0.1% of the incident solar flux at Jupiter, and Lewis (1976) estimates that less than 1% of this particle energy flux may conceivably be dumped into the Jovian upper atmosphere. Particle dumping from the belts therefore corresponds to an average flux no larger than 0.001% of the incident solar flux. Finally, the solar wind energy flux is only about 10^{-6} of the solar radiation energy flux, and in addition, only a small fraction of the incident solar wind flux is usually absorbed by the atmosphere (see Section 3.6.2).

Studies of disequilibrium atmospheric chemistry in our own atmosphere have taught us that the only reliable approach is to identify and measure the rates of each *elementary* chemical reaction involved in the overall chemical conversions, and then to include these laboratory data in a computer model that solves the continuity equation [Eq. (3.33) or (3.34)] as a function of altitude (in a steady state) and/or time. A tabulation of the most important types of such elementary reactions occurring in the atmosphere is given in Table 3.4. In most of our subsequent discussions

TABLE 3.4
Principal Classes of Elementary Reactions in Planetary Atmospheres

Photodissociation	$h\nu + AB \rightarrow A + B$
Photoionization	$h\nu + AB \rightarrow AB^+ + e^-$
Cosmic ray dissociation	$C^+ + AB \rightarrow$ ions, protons, neutrons, etc.
	(often a chain reaction is initiated)
Dissociative recombination	$AB^+ + e^- \rightarrow A + B$
Radiative recombination	$AB^+ + e^- \rightarrow AB + h\nu$
Unimolecular decomposition	$AB \rightarrow A + B$
Bimolecular reaction	$A + B \rightarrow C + D$
Termolecular reaction	$A + B + C \rightarrow D + E + F$
Recombination	$A + B + M \rightarrow AB + M$ (M an inert third body)

of the chemistry of the primitive and present atmospheres on each planet, some version of this general approach is utilized; there are a few exceptions and these will be made explicit where they occur.

In order to illustrate in a more concrete manner the many concepts introduced in this section, it is appropriate at this point to study a simple but particularly important disequilibrating process in the atmosphere. Let us consider an atmospheric gas i which is photodissociated by solar UV radiation. Say this dissociating radiation penetrates only to a particular level in the atmosphere which for convenience we will define by $z = 0$. As a first approximation we can write for $z > 0$

$$J_i \simeq \int_0^\infty \sigma_i(\nu)I(\nu)\,d\nu, \tag{3.41}$$

which is independent of z. Thus, the continuity equation for i in the region $z > 0$ is from Eq. (3.34)

$$\frac{\partial[i]}{\partial t} = J_i[i] + \frac{\partial}{\partial z}\left(K[M]\frac{\partial f_i}{\partial z} \right) = 0. \tag{3.42}$$

This equation can be easily rearranged to produce a linear second-order differential equation for the mixing ratio of i, f_i (see Prinn, 1975):

$$\frac{\partial^2 f_i}{\partial z^2} - \frac{1}{H_M}\frac{J_i f_i}{K} = 0. \tag{3.43}$$

For constant H_M and K, Eq. (3.43) has an analytical solution with two arbitrary constants to be determined by two boundary conditions. Clearly one of these boundary conditions is that f_i must be finite at $z = \infty$; another could be a prescribed value for f_i at $z = 0$. The latter prescribed value will roughly be the mixing ratio of i in the region below $z = 0$ where i is chemically stable. Utilizing these two boundary conditions we have

$$f_i(z) = f_i(0)\exp(-z/h)$$

and

$$h = \left[-\frac{1}{2H_M} + \left(\frac{1}{4H_M^2} + \frac{J_i}{K} \right)^{1/2} \right]^{-1}, \tag{3.44}$$

where h is the scale height for the *mixing ratio* of i. We see the importance now of comparing time constants. If $1/J_i \ll 4H_M^2/K$, then $h \simeq (K/J_i)^{1/2} \ll H_M$ and the mixing ratio of i decreases very rapidly above $z = 0$; upward transport of i is too slow to prevent depletion by photodissociation. The case which we have considered here is, of course, somewhat idealized, but nevertheless provides considerable physical insight into the interplay occurring between transport rates and chemical rates in determining atmospheric composition.

Before leaving the general topic of disequilibrium atmospheric chemistry, a comment on attempts to simulate the atmosphere in laboratory containers is in order. We would first caution that is is an exceedingly difficult task and can provide entirely misleading results (Lewis and Prinn, 1971; Prinn and Owen, 1976). How-ever, it is almost inevitable that our selection of elementary reactions in the general method outlined in this section is incomplete. Laboratory experiments, if properly conceived and interpreted, may sometimes lead to identification of important omis-sions from these chemical schemes. We shall refer to simulation experiments in the appropriate sections of Chapter 4 and give a critical assessment of the utility of each.

Because the rates and primary products of photodissociation are strongly depen-dent on the incident UV spectrum, the spectrum used in the laboratory should closely approximate that encountered in the region of the atmosphere under con-sideration. There is also no obvious validity in equating the results obtained from high-temperature experiments, energetic particle bombardment experiments, or electrical discharge experiments, with the chemistry induced by UV radiation. In addition, it is important that the temperature, pressure, and gas composition used in the laboratory mimic the real atmosphere. Temperature is important both for controlling reaction rates and also for determining whether condensation of a par-ticular product (which essentially removes it from further reaction) will occur. The pressure is important for it controls three-body reaction rates and also the quenching rates for excited atoms and radicals. Great care should also be taken to ensure that reactions on the walls of the apparatus are not influencing the results.

The most difficult aspect of the atmosphere to simulate in the laboratory is the effect of air motions. Ignoring this aspect may result in successively more compli-cated species being produced during an experiment conducted without their re-moval, or without replacement of their precursors, by transport processes. Our earlier discussion of Eq. (3.44) makes this abundantly clear. Imaginative use of a flow system in which the chemical and transport times are comparable to those in the real atmosphere seems a feasible way to overcome this problem. Unfortunately, laboratory simulations of atmospheric chemistry have almost invariably been per-formed without due consideration to at least some of the above requirements. They have therefore been of very limited use to our studies of the chemistry of planetary atmospheres.

3.4. CLOUD PHYSICS AND CHEMISTRY

In the previous section we were concerned principally with *homogeneous* gas-phase chemistry. There are, in addition, some important *heterogeneous* reactions in the atmosphere which we will need to consider. In these reactions, atmospheric gases may form or interact with liquid and solid particles suspended in the air, and produce or react with liquid and solid phases on the planetary surface. We can conveniently discuss these diverse phenomena under four major headings relevant to the nature of the reactions involved: condensation, dissolution, photochemical smog formation, and gas–crust reactions.

3.4.1. Condensation

3.4.1.1. General Considerations

For a general condensation process by which a gaseous substance A forms droplets or particles of liquid or solid A, the condition of thermodynamic equilibrium yields

$$A(s, l) \rightleftarrows A(g) \tag{3.45}$$

$$K_T = P_A/a_A \tag{3.46}$$

where a_A is the thermodynamic activity of the pure condensate ($a_A = 1$ for pure solid or liquid A), and p_A is the equilibrium vapor pressure of A at temperature T. The equilibrium constant for vaporization, K_T, is related to the Gibbs free energy, enthalpy, and entropy changes of evaporation by the relation

$$\log K_T = \frac{-\Delta G^\circ_{vap}}{2.303RT} = \frac{T \, \Delta S^\circ_{vap} - \Delta H^\circ_{vap}}{4.576T}, \tag{3.47}$$

whence, for pure condensate A,

$$\log p_A = \frac{\Delta S^\circ_{vap}}{4.576} - \frac{\Delta H^\circ_{vap}}{4.576T}. \tag{3.48}$$

The slope of the log vapor pressure curve versus the natural thermodynamic temperature variable, $-1/T$, is

$$\frac{\partial(\log p_A)}{\partial(-1/T)} = \frac{\Delta H^\circ_{vap}}{4.576} \simeq \text{constant.} \tag{3.49}$$

This expression should be compared to Eq. (3.32).

If temperatures in the atmosphere or at the surface are sufficiently low, then condensation of one or more gases to form a pure solid or liquid condensate may occur, much as we have assumed above. However, the physics of condensate cloud formation (see, e.g., Mason, 1957) is a particularly complex subject whose details are unfortunately beyond the scope of this book. From a simple thermodynamic viewpoint, we would expect condensation to occur whenever the vapor pressure of a gas exceeds its equilibrium value [see Eq. (2.4)]. This criterion should always be valid for direct condensation onto a planetary surface. However, condensation in the atmosphere is complicated by the fact that the vapor pressure of a gas above a condensate surface increases significantly as the radius of curvature of the surface decreases, especially for particle sizes less than ~ 0.1 μm. For these small particles, the surface energy is an important fraction of the latent heat of vaporization. Consequently, unless condensate particles or foreign particles (called *condensation nuclei*) are available upon which the gas may condense, the vapor pressure must considerably exceed the equilibrium value before atmospheric condensation takes place. In the Earth's atmosphere, microscopic mineral, soil, sea-salt, and soot particles suspended in the air provide ideal nuclei for water vapor condensation in cloud-free areas. Supersaturation is therefore not required for terrestrial water-cloud formation to be initiated. In other planetary atmospheres, we often cannot deduce a priori the presence of sufficient condensation nuclei to preclude the possibility of some regions in the atmosphere being supersaturated with respect to particular gases. The point should be kept in mind, since we generally assume as a first approximation that condensation occurs once saturation is reached in the atmosphere. We should be aware, however, that condensation by adsorption (such as water onto hygroscopic condensation nuclei) can occur at pressures *below* saturation.

Another aspect of cloud formation to which we should also give some attention involves their differing formation mechanisms (see, e.g., Willett and Sanders, 1959). At one extreme are *convective clouds* which develop in convectively unstable (wet superadiabatic) regions of the atmosphere, where large vertical velocities develop. The rapidly ascending buoyant air parcels cool essentially adiabatically as they expand. Condensation then occurs in these cooling air parcels. The resultant cloud particles may then grow by collisions with other particles (i.e., coagulation and coalescence) or by further condensation. If these various particle growth processes are strong enough, very large particles are produced, and these fall rapidly, giving rise to the familiar rainstorm. Once at the top of the cloud, the now dry and no longer buoyant air parcels descend and heat up as they are compressed. Condensation therefore does not occur in these downward-moving regions (see Section 2.2.5).

Because the uplifting of air and the resulting condensation is usually highly localized, convective clouds are generally horizontally *very inhomogeneous*. The vertical motion in such clouds may be induced by the latent heat released during spontaneous condensation, by air being forced up mountain sides, by radiative heating of the air parcel, or by heating at the lower boundary (e.g., the ground) of the air column. At the other extreme are *stratiform clouds* formed in convectively stable (wet subadiabatic) atmospheric regions. Condensation usually results from rapid cooling of the air due to emission of thermal infrared radiation or to physical contact with cold atmospheric or surface layers. These clouds are usually much less dense than convective clouds and horizontally much more *homogeneous*. Since growth of large particles is often very slow in these clouds, they do not usually produce rain; particles may either slowly sediment out or simply evaporate.

At this point it is appropriate to consider a question of particular relevance to our discussion of atmospheric evolution: What is the ultimate fate of condensation clouds once formed in the atmosphere? One obvious case occurs when the surface temperature is above the sublimation temperature or boiling point appropriate to the surface pressure of the condensing and background gases. It is then clear that cloud particles formed in the air will evaporate again before they reach the surface. We will see in Chapter 4 that this simple recycling process plays an important role in the deep, warm atmospheres of Venus and the major planets.

The contrasting case occurs when the ground temperature of the planet is low enough for condensed material to collect on the surface. As such a condensible species is slowly outgassed from the planet, it thus accumulates as a liquid or solid surface layer (e.g., oceans, lakes, polar caps) rather than contributing to the atmosphere. Only that quantity of the gas permitted by its vapor pressure equation will remain in the gas phase. This type of phenomenon occurs on the Earth, Mars, and several icy satellites of the major planets.

3.4.1.2. Planetary Temperature Profiles

At this juncture, the reader can readily recognize the importance of a knowledge of atmospheric and surface temperatures in any description of atmospheric evolution. In the absence of direct observations, what indeed can we deduce about even the planetary-average vertical temperature profile on a planet or satellite at a particular distance from the sun? In this respect, it is particularly instructive to first consider the so-called *"radiative equilibrium"* approximation, which is in reality a *radiative steady state*. In this approximation the solar radiation is usually considered to be deposited at an appropriate lower boundary of the atmosphere such as a thick absorbing cloud layer or the ground itself. The net average inward solar energy flux, ϕ, is simply the total solar power received by the planet divided by the heated surface area. For a rotating body, the cross-sectional area intercepts sunlight, but the entire surface is heated:

$$\phi = \frac{\pi a^2 (1 - A)F\odot}{4\pi a^2 R^2} = \frac{(1 - A)F\odot}{4R^2}, \tag{3.50}$$

where a is the planetary radius, A is the planetary albedo (i.e., fraction of solar radiation scattered back out of the planet), $F_\odot$ is the solar energy flux at unit radial distance from the sun, and R is the radial distance from the sun. Equation (3.50) is appropriate for a planet where the solar heating is deposited evenly over the surface. Such a situation may occur as a result of rapid planetary rotation or of rapid atmospheric motions which effectively distribute the solar heating over illuminated and dark sides. For slowly rotating planets effectively heated only on one side the heated surface area, $4\pi a^2$, in Eq. (3.50) should be reduced to $2\pi a^2$.

The heated lower boundary will be emitting thermal IR radiation, and in most cases we can consider this emission to be blackbody emission. Some of this emitted thermal IR radiation escapes to space while some is absorbed by the atmospheric gases and clouds, leading to heating of the atmosphere. The heated atmosphere can in turn radiate back down to the ground, to other part of the atmosphere, or to space. *Radiative steady state* occurs when the IR radiative heating rate equals the IR radiative cooling rate at each point in the atmosphere. We can roughly model the IR absorbing properties of the atmosphere by the dimensionless "grey" optical depth $\tau(z)$ above each level z. If we then combine Eq. (3.45) and the radiative steady-state assumption, we obtain the simple formula

$$[T(z)]^4 = (\phi/\sigma)[\tfrac{3}{4}\tau(z) + \tfrac{1}{2}], \qquad (3.51)$$

to describe the temperature $T(K)$ at each altitude z (see, e.g., Goody, 1964). Here σ is the Stefan–Boltzmann constant. By similar reasoning the ground or lower boundary temperature (where $z = 0$) is given by

$$T_g^4 = (\phi/\sigma) [\tfrac{3}{4}\tau(0) + 1]. \qquad (3.52)$$

Note that since $T_g > T(0)$, there is a temperature discontinuity at the ground. This discontinuity is in practice generally removed by convection near the ground.

Equation (3.52) implies that in a very weakly absorbing atmosphere [i.e., $\tau(0) \ll 1$], the blackbody ground temperature $T_g = (\phi/\sigma)^{1/4}$. We refer to the quantity $(\phi/\sigma)^{1/4}$ as the *effective* temperature, T_e, of the planet. It is simply that temperature of a (hypothetical) blackbody planet which would radiate thermal IR energy at the same rate that the *actual* planet gives off energy. For the case of a thermal steady state in which *all* the emitted radiation is absorbed sunlight, Eq. (3.50) gives

$$T_e = \left[\frac{(1 - A)F\odot}{4R^2\sigma} \right]^{1/4} \qquad (3.53)$$

and we see that T_e varies inversely as the square root of the distance from the sun.

Equations (3.51) and (3.52) imply that we can obtain atmospheric and surface temperatures in considerable excess of T_e only when $\tau(0) \gg 1$, that is, when the atmosphere is very opaque to thermal IR radiation. Generally, as a first approximation we can consider $\tau(z)$ as being proportional to pressure. Thus, when $\tau(0) \gg 1$, Eq. (3.51) tells us that the atmospheric temperature will increase from the so-called *Gold–Humphrey's skin temperature*, $T_e/2^{1/4}$, at very high altitudes to values well in excess of T_e near the ground. This increase in temperature is often referred to as the *greenhouse effect*. It is particularly important for causing the opaque atmospheres of Venus, the major planets, and Titan to exhibit temperatures far in excess of T_e. In this connection, we should add that Jupiter and Saturn have *internal* heat sources, and that this flux of internal energy through the lower boundary must be added to ϕ in computing the radiative steady-state temperatures defined in Eqs. (3.51)–(3.53). In Table 3.5 we tabulate values of T_e for a number of solar system bodies, and these can be compared with the highest observed surface or atmospheric temperatures in each case.

The radiative steady-state temperature profile will sometimes provide a reasonably satisfactory fit to the observed planetary-average temperature profile. However, because it is derived by ignoring the vertical flux of energy resulting from atmospheric motions, it must be used with caution. Corrections can be made in certain circumstances to allow for the effects of vertical motions.

The temperature gradient in a planetary atmosphere has a critical value at which

TABLE 3.5
Effective Temperatures of Solar System Bodies

Body	Heliocentric distance R (AU)	Visual albedo A	Calculated grey-body T (K)	IR observed T_e (K)	T_s (K)
Mercury	0.387	0.15	433	430	433
Venus	0.723	0.72	240	240	745[a]
Earth	1.000	0.45	233	235	275[a]
Moon	1.000	0.15	270	270	270
Mars	1.523	0.25	211	210	220[a]
Ceres	2.767	0.06	166	165	166
Jupiter	5.203	0.49	103	128	128[a,b]
Io	5.203	0.7	91	111	111[b]
Europa	5.203	0.5	103	103	103
Ganymede	5.203	0.3	112	112	112
Callisto	5.203	0.2	116	116	116
Saturn	9.535	0.62	71	98	98[a,b]
Titan	9.535	0.2	86	86	90[a]
Uranus	19.170	0.38	56	56	56[a]
Neptune	30.055	0.3	47	57	57[a,b]
Pluto	39.423	0.5	38	~38?	38

[a]Known greenhouse effect.
[b]Known large internal heat source.

convection becomes important and beyond which convective transport of heat is so rapid as to damp out significant increase in the temperature gradient.

Consider a parcel of gas undergoing a general change in altitude (z) and in pressure and temperature so rapidly that radiative exchange of heat with its surroundings is negligibly slow. To very good approximation, the P–V work done by (or on) the gas as it changes volume is exactly equal to the change in internal energy of the gas. By the first law of thermodynamics,

$$C_V \, dT + P \, dv = 0, \qquad (3.54)$$

where C_V is the molar heat capacity at constant volume and v is the molar volume.

In hydrostatic pressure balance in a gravitational field,

$$dP = -\rho g \, dz, \qquad (3.55)$$

where ρ is the gas density and g is the local acceleration due to gravity.

Assuming ideal gas behavior, for a gas of molecular weight m,

$$Pv = RT, \qquad (3.56)$$

$$\rho = m/v = Pm/RT, \qquad (3.57)$$

and

$$P \, dv + v \, dP = R \, dT, \qquad (3.58)$$

whence

$$C_V \, dT + R \, dT - P \, dv = 0, \tag{3.59}$$

but since $C_V + R = C_P$ for ideal gases,

$$C_P \, dT - v \, dP = 0 = C_P \, dT - v\rho g \, dz \tag{3.60}$$

or

$$C_P \, dT = - \frac{RT}{P} \frac{Pm}{RT} \, g \, dz = - mg \, dz, \tag{3.61}$$

whence

$$\frac{dT}{dz} = \left(\frac{\partial T}{\partial z} \right)_S = - \frac{mg}{C_P} \tag{3.62}$$

This is termed the dry adiabatic lapse rate. The subscript S indicates that, because the heat flow through the walls of the parcel is zero, the process is adiabatic (the entropy S is constant). It is called "dry" because the thermodynamic effects of condensation have been omitted.

When the temperature lapse rate $dT(z)/dz$ exceeds the adiabatic lapse rate, then free convection will tend to reduce the actual temperature lapse rate back to its adiabatic value (wet or dry as the case may be). Such an adjustment can be made to the radiative equilibrium profile where appropriate, and the resulting adjusted temperatures are referred to as *radiative–convective* profiles (see, e.g., Goody, 1964). Such adjusted profiles appear to provide good approximations to the real temperature profiles on Venus (Pollack, 1969) and Jupiter (Trafton and Stone, 1974).

Unfortunately, stable (i.e., subadiabatic) lapse rates do not necessarily imply negligible vertical transport of heat by air motions. This is particularly true in rapidly rotating, differentially heated atmospheres like the Earth and Mars where large-scale motions carry heat upward to maintain a distinctly *subadiabatic* lapse rate. A theory to predict temperature profiles in such cases is given by Stone (1972). In the general case it is certainly true that a prediction of the climatic mean or *radiative–dynamical* "equilibrium" temperature profile requires a careful study of the atmospheric dynamics. Some recent progress has fortunately been made in the study of the meteorology of planetary atmospheres, so that this task of prediction is not entirely beyond us. The reader is referred to reviews by Goody (1967), Stone (1973, 1975, 1976), and Gierasch (1975) for details.

3.4.1.3. Condensing Cloud Models

Studies of atmospheric evolution occasionally require more than a simple knowledge of where a particular gas will condense. In this case suitable cloud models must be devised. In constructing such models we are usually interested only in the horizontally averaged case. The question of whether the clouds are convective or stratiform can thus be avoided. Because cloud particles are continuously being formed and then removed by transport processes, we clearly have a disequilibrium situation. In principle, we should therefore solve Eq. (3.34) to obtain cloud densities, but in practice we can occasionally bypass this procedure.

Consider first a case where the cloud particle sedimentation velocity w_i is much less than the typical vertical velocity K/H associated with vertical motions. This will occur if particles are very small or vertical mixing is vigorous. Under such conditions the steady-state mixing ratio of the condensible material, including both the vapor and condensed phases, is essentially constant with altitude (e.g., Prinn, 1974, 1975). If we then assume that the partial pressure of the condensing gas at each level above the cloud base equals its equilibrium vapor pressure, we can immediately obtain the cloud density at each level. In reality, the air above the cloud base is usually somewhat below the saturation level, but the method is clearly useful for providing a semiquantitative picture. Of course, the assumption that $w_i \ll K/H$ implies that the cloud mass densities we compute represent the *maximum* densities which could occur at each level. The model clouds extend to considerable altitudes and do not have sharply defined tops.

At the other end of the spectrum is the cloud modeling procedure utilized by Lewis (1969a,b) and by Weidenschilling and Lewis (1973) for the clouds of Venus and the major planets. As in the method described in the previous paragraph, the air is assumed to be saturated with respect to the condensing gas at all levels above the cloud base. However, the cloud mass in a particular altitude increment is assumed to be given by the change in vapor pressure across the increment. The cloud density scale height is then essentially the same as the condensing gas density scale height. Although this modeling procedure appears somewhat arbitrary, the analytical solution for a general cloud continuity equation given by Prinn (1974) suggests it is valid providing $w_i \gg K/H$; that is, particles are very large or vertical motions very weak. Thus, the procedure used by Lewis leads to smaller densities and more sharply defined cloud tops than the procedure given in the previous paragraph. Because we often have no idea of the values of w_i and K in a particular cloud layer, the choice of modeling procedure is clearly arbitrary and the results in such cases should obviously be considered only in a qualitative sense.

We now have considered those processes by which the internal chemistry and physics of an atmosphere may be influenced by external heating, bulk motions, thermochemical reactions, solar UV photolysis, and condensation. The remaining processes are those in which the atmosphere functions as an open system whose walls are selectively permeable to particular gaseous components. These processes include gas–crust interactions and atmospheric escape.

3.5. DISSOLUTION, PHOTOCONDENSATION, AND REACTIONS WITH PLANETARY SURFACES

An atmsophere on a terrestrial planet will in general be open to mass exchange across its lower boundary. The only apparent exceptions to this rule will be the cases in which (*a*) the planet is fully outgassed and the surface is too hot for minerals containing volatile elements to be stable, or (b) the atmosphere is fully inert both chemically and photochemically, and no constituent of the atmosphere is condensible at the surface. Still, radiogenic rare gas fluxes from the interior are never wholly absent, and in the strictest sense all terrestrial planets will participate to some degree in such processes.

The most fundamental such processes involve condensation of atmospheric gases on the surface, formation of aerosols or surface materials as results of photochemical processes, dissolution of other atmospheric gases in solvent phases such as liquid H_2O, NH_3, or CH_4, and reactions between surface minerals and atmospheric gases. The thermodynamic basis for discussion of condensation and regulation of near-surface temperatures has already been given in Section 3.4 and will not be repeated. Dissolution of gases, photocondensation, reactions with surface mineral assemblages, and biological interactions will be treated in this section and the next.

3.5.1. Dissolution

Certain of the atmospheric gases appearing in planetary atmospheres are soluble in the liquid clouds or liquid surface layers produced by the process of condensation which we have just discussed. Such soluble gases can therefore be removed from the atmosphere despite the fact that their vapor pressures may be well below the saturated values over their *pure* condensed phases.

Gases may enter the solvent phase by three different methods:

(i) *Simple solution in the solvent.* Here we can utilize Henry's law provided the resulting solution is sufficiently dilute. This law states

$$[i](\text{soln.}) = K_p P_i(g) = K_c[i](g), \tag{3.62}$$

where K_p and K_c are constant at constant temperature. A good example is N_2O dissolving in terrestrial rain clouds, lakes, and oceans.

(ii) *Solution followed by **reversible** dissociation into ions and solvation of the resultant ions.* In this case we can assume that the undissociated part of the dissolved gas obeys Henry's law provided again that the resultant solution is sufficiently dilute. The concentration of the ions can then be found assuming thermochemical equilibrium. For example, for the reactions

$$AB(g) \rightarrow AB(\text{soln.}),$$
$$AB(\text{soln.}) \rightarrow A^+(\text{soln.}) + B^-(\text{soln.})$$

we can utilize Eq. (3.49) to write

$$[A^+] = [B^-] = (K_e[AB] (\text{soln.}))^{1/2}$$
$$= (K_c K_e[AB](g))^{1/2}$$

where K_e is the equilibrium constant for the ionization reaction. An important terrestrial example is CO_2, which dissolves in rainclouds, lakes, and oceans and subsequently ionizes to produce hydrated CO_3^{2-} and HCO_3^- ions (see, e.g., Junge, 1963, p. 303).

(iii) *Solution followed by* **irreversible** *chemical reactions.* Here no simple rules apply and each situation must be analyzed on its own merits. An example of particular importance in the terrestrial atmosphere is the oxidation of SO_2 to SO_4^{2-} within aqueous aerosols. Apparently dissolved SO_2 and O_2 gases can combine rapidly when certain anions (e.g., Mn^{2+}, Cu^{2+}, Fe^{2+}, NH_4^+) are present to act as catalysts (Junge and Ryan, 1958).

We should add a cautionary statement about our ability (or indeed, inability) to quantify these three dissolution processes. Clearly a reasonable amount of time is required for gases to reach equilibrium with solutions according to Henry's law. This time is usually available for dissolution in surface solvents, but may not be available for dissolution in rapidly falling cloud droplets. Chemical reactions within droplets can often be strongly influenced by trace impurities and by the pH of the droplets. We can occasionally be guided by terrestrial experience and by laboratory simulations, but the reader should be aware that estimates of volatile removal *rates* by dissolution are often order of magnitude estimates at best. This is particularly true for process (*iii*) above.

3.5.2. Photochemical Condensation

Another mechanism for removing species from the gas phase involves conversion of volatiles to involatiles by UV-induced chemical reactions. This process, which we will refer to here as "photochemical condensation," appears to be ubiquitous throughout the solar system: It results in sulfuric acid and sulfur clouds being produced from SO_2 and COS on Venus and Earth; it enables atmospheric O_2 to oxidize Fe^{2+} to Fe^{3+} on the surface of Mars; it appears to give rise to red phosphorus clouds from PH_3 and sulfur and polysulfide clouds from H_2S on Jupiter (and undoubtedly on the other major planets as well); it also results in the production of carbon and high-molecular-weight hydrocarbon particulates from CH_4 on Titan.

We can conveniently discuss photochemical condensation under two headings.

3.5.2.1. Photochemical Condensation in the Atmosphere

In this case we can picture a *precursor gas* being mixed up by atmospheric motions into a "cloud production" region. Here it is dissociated either directly by UV light or indirectly by radicals and ions produced by solar UV rays. Following a series of gas-phase photochemical reactions, an involatile or highly soluble species results which enters the particulate phase. The particles then sediment out and either accumulate on the surface or re-evaporate in a "cloud destruction" region. In the latter case, the precursor gas may be partially or wholly regenerated by thermochemical reactions from the products of the cloud particle evaporation (see, e.g., Prinn, 1974, 1975).

These photochemical clouds differ considerably in morphology from the simple condensation clouds presented earlier. They are formed at the top rather than near the cloud base; and because they are continuously being produced in the sunlit hemisphere, breaks in the clouds only occur under conditions of very vigorous downward transport of particles. The simplest way to illustrate the method for modeling these clouds is to consider the precursor gas, i, as being distributed in the "cloud production" region according to Eq. (3.44). If we choose a convenient element in the precursor gas as a tracer, we can then demand that in a steady state the net upward flux of this tracer in the precursor gas must equal the net downward flux of the same tracer in the photochemically produced particles. If f_j is the ratio of the number of condensed tracer atoms per unit volume of air, $[j]$, to the total number of air molecules per unit volume, $[M]$, we can state this flux equality condition as

$$\frac{\partial f_j}{\partial z} + \frac{w_j}{K} f_j = - \frac{\partial f_i}{\partial z}, \qquad (3.63)$$

where w_j is the mass-weighted mean sedimentation velocity for cloud material. It can be readily shown from Eq. (3.63) that the sedimentation term is important when $w_j \geq K/H$, and, since w_j varies as the square of the particle radius, then sedimentation increases rapidly with particle size. This dependence of Eq. (3.63) on particle size introduces a problem since it is very difficult to assess particle sizes from first principles. Observations in the terrestrial and Venusian atmospheres suggest that photochemically produced particles are small (i.e., radii ~ 1 μm), but this has not been shown to be universally true. Note that with K very large or w_j very small, Eq. (3.63) implies that $f_i + f_j$ is constant with altitude. This general property was also mentioned when we discussed condensing cloud models under Section 3.4.1.

3.5.2.2. Photochemical Condensation at the Surface

This category is clearly reserved for terrestrial bodies with sufficiently thin atmospheres for dissociating UV radiation to reach their surfaces. This latter requirement restricts this process to Mars and to certain satellites of the major planets. This process can be considered to begin with adsorption of atmospheric molecules on surface minerals followed by chemical reactions instigated by absorption of UV photons by the mineral crystals. Partial or complete incorporation of the absorbed gases into the regolith may then occur.

3.5.3. Atmosphere–Lithosphere Reactions

The above considerations regarding reactions of photolysis products with planetary surfaces lead us naturally to consideration of the general circumstance in which the atmosphere of the planet near the surface is not in thermochemical equilibrium with the exposed surface rocks or sediments. There are numerous reasons why such a condition is extremely common. First of all, because of the geometry of irradiation by the Sun on a spheroidal surface, horizontal temperature gradients are a nearly universal feature of terrestrial-type planets. These temperature gradients drive winds that transport heat from dayside to nightside and from equator to poles, carrying into cooler regions an atmospheric composition which usually reflects more closely the results of equilibration at the warmest temperatures experienced. If the atmosphere is arbitrarily assumed to be in precise chemical equilibrium at some local temperature T_{eq}, characteristic perhaps of the daytime equatorial temperature, then the atmosphere *cannot* in general be in equilibrium with the rest of the planetary surface.

Injection of volcanic gases into the atmosphere is also an obvious source of gases whose composition reflects conditions radically different from those seen on the surface; and as we have seen, photochemical or even thermochemical products produced in very hot, high-altitude regions may also be available.

The reactions that transpire between the surface and reactive gases are governed by the principle that the Gibbs free energy change for the reactions must be negative; in common language, we would say that the system approaches (but, for kinetic reasons, need not attain) equilibrium.

As we discussed in Section 3.3.1, the equilibrium case is the easiest to study and provides a meaningful benchmark against which the actual chemical state of the system may be assessed. Equilibrium is most closely attained where the time scale for chemical reactions is short compared to the time scale for outgassing, escape, photolysis, and all other perturbing and disequilibrating processes. Among the terrestrial planets, the body to which the assumption of atmosphere surface equilibrium is most relevant is Venus. Indeed, the application of the equilibrium model to Venus will be discussed at some length in Chapter 4. Since we have already discussed the general nature of thermodynamic equilibrium, it is necessary only to describe the way in which equilibrium theory is applied to this particular type of heterogeneous system.

Chemically reactive gases participate in reactions with minerals exposed at the planetary surface via balanced reactions such as the Urey reaction:

$$CaCO_3 + SiO_2 = CaSiO_3 + CO_2. \tag{3.64}$$
$$\text{(calcite) (quartz) (wollastonite)}$$

Wherever pure solid calcite, quartz, and wollastonite are exposed to the atmosphere, reaction to liberate or consume carbon dioxide will occur unless the partial pressure of CO_2 is precisely that corresponding to thermochemical equilibrium. The complete equilibrium expression for the above four species in reaction (3.64) is

$$\log K_p = a_{woll}p_{CO_2}/a_{cal}a_{qz} = -G_{rx}^{\circ}/2.303RT, \tag{3.65}$$

where each a_i is the thermodynamic activity of phase i and p_{CO_2} is the equilibrium partial pressure of carbon dioxide. For the case in which all the mineral phases are present as pure solids, then all the a_i are exactly unity, and we obtain

$$\log p_{CO_2} = -G_{rx}^{\circ}/2.303RT, \tag{3.66}$$

whence it is clear that the partial pressure of CO_2 is a function of temperature alone. Reactions of this sort, in which a gas pressure depends on T alone (including the simplest case of vapor pressure equilibrium of pure condensed phases discussed in Section 3.4), are called *buffer reactions*.

In some circumstances, certain gases may be produced by the reaction of other gases with a mineral or mineral assemblage. For example,

$$CO_2 + FeS + MgSiO_3 = FeMgSiO_4 + COS \qquad (3.67)$$
$$\text{(troilite)} \quad \text{(enstatite)} \qquad \text{(olivine)}$$

by which the gas carbonyl sulfide is formed, or

$$H_2O + MgF_2 = MgO + 2HF \qquad (3.68)$$
$$\text{(sellaite)} \quad \text{(periclase)}$$

and analogous reactions by which water supplies the hydrogen to make the hydrogen halides.

Obviously there are two variables that determine the equilibrium pressure of a reactive gas: In addition to the simple quantitative dependence on temperature given above, there is the fundamental qualitative dependence on the identities of the mineral phases present. Thus, iron and its oxides may buffer the partial pressure of oxygen by any of the important reactions

$$Fe^° + \tfrac{1}{2}O_2 = FeO, \qquad (3.69)$$
$$\text{(metal)} \qquad \text{(wüstite)}$$
$$3FeO + \tfrac{1}{2}O_2 = Fe_3O_4, \qquad (3.70)$$
$$\text{(magnetite)}$$
$$2Fe_3O_4 + \tfrac{1}{2}O_2 = 3Fe_2O_3 \qquad (3.71)$$
$$\text{(hematite)}$$

for each of which there is a distinct $\log p_{O_2}$ versus temperature curve, with each buffer giving a higher p_{O_2} than its predecessor at any given temperature.

In addition to the obvious necessity that the minerals involved in a given buffer reaction actually be present, there is a second requirement that the reaction be kinetically feasible. Between these two requirements, then, we stand warned about the possibility that one participating mineral in a buffer (such as FeS above) may become exhausted before the calculated equilibrium partial pressure of COS is attained, that weathering reactions may deplete a reactant or coat it with an impermeable layer of alteration products, or that the rate of a particular reaction at low temperatures may be so low that it can no longer act as a buffer.

Exposure of fresh, unweathered rocks commonly occurs on Earth as a result of the effects of running water, wind sand-blasting, cracking of rocks by freezing of pore water, or volcanic eruption. On other planets of a lesser degree of internal tectonic activity, the dominant mechanism for exposure of fresh material may be bombardment by asteroidal or cometary debris.

The role of liquid water in weathering processes is complex and extremely important. First, there is direct extraction of soluble minerals, such as halite (NaCl), from exposed rocks. Second, there is chemical etching of many other minerals by dissolved acids, which we have seen to be mainly CO_2, HCl, and HF. Third, there is hydration and hydroxylation of minerals, such as the formation of serpentine from ferromagnesian silicates. Fourth, water serves as a medium in which complex ions form. These dissolved ions, such as bicarbonate (HCO_3^-), are often capable of effecting equilibration of mineral assemblages such as that in the Urey equilibrium at temperatures as low as 0°C, even though, in the absence of liquid water, equilibrium could not be attained in a reasonable time at temperatures less than 300°C. Fifth, water similarly provides an ideal medium in which oxidation of surface rocks by dissolved oxygen can take place. The oxidation of highly soluble reduced species such as Fe^{2+} by dissolved oxygen produces highly insoluble magnetite and hematite, and manganese similarly yields the very insoluble MnO_2. As a logical consequence of weathering, then, soluble oxidized species such as carbonates, sulfates, nitrates, and borates will accumulate in solution until saturation is reached, at which point these salts will begin to precipitate to form an *evaporite*. Very large quantities of mineral salts, especially carbonates, phosphates, and silicates, are precipitated in Earth's hydrosphere by biological activity, principally the formation of exoskeletons of simple marine organisms.

The insoluble residue from the liquid water weathering process, which is rich in species such as hydroxyl compounds of aluminum, is very characteristic and quite possibly diagnostic of prolonged exposure to liquid water. Certainly the discovery of extensive beds of such materials on Mars would be of great significance as a clue to the planet's ancient history.

Clearly the eventual fate of volatile-rich ocean floor sediments on Earth is also of great interest. Subduction or high-grade metamorphism of these sediments brought about by crustal motions will lead to extensive devolatilization and return vast quantities of volcanic gases to the atmosphere. The necessity of careful study of the geochemical cycles of the volatile elements is self-evident. In the case of Earth, an understanding of biological involvement in these cycles is rendered essential by the very short turnover times of biologically useful nutrients and minerals in the biosphere, so short that the abundances of many atmospheric and oceanic constituents are regulated largely by biological processes.

3.6. BIOCHEMICAL TRANSFORMATIONS

The critical role played by biochemical reactions in the cycles of almost *all* atmospheric gases on our own planet has been recognized for many years. The transformations effected are sometimes simple, sometimes very profound, while the organisms involved range from the most primitive to the most complex. The part played by man is largely indirect. Of particular importance are the burning of fossils fuels and the modulation of vegetation, soil conditions, and ocean ecologies by our farming, land development, and waste disposal practices.

This accepted role played by living organisms in our present atmosphere dictates that we also investigate the influence of biological activity on atmospheric evolution in general. On Earth we can fortunately turn to the geological and fossil records for many vital clues. These clues, coupled with studies of the structure, metabolism, and evolution of living terrestrial organisms themselves, enable us to draw at least rudimentary outlines of plausible evolutionary paths. Unfortunately, it is difficult to be very quantitative in assessing the roles of various biochemical transformations on atmospheric evolution. Indeed, as we will discuss in Chapter 4, certain quantitative aspects of the biochemical influences in our *present* atmosphere have yet to be elucidated. In contrast to many of the thermochemical and photochemical processes discussed earlier, we must therefore be prepared to accept a lack of definiteness in our discussions of biological processes; the margins for error are consequently much larger.

When we attempt to extrapolate from our Earth experience to other planets, we are, of course, faced with three of the most controversial questions in planetary science. Did life evolve on other planets? If so, were these living organisms present only in a transient epoch or do they still exist today? Finally, what role did they play, or do they still play, in modulating atmospheric composition and evolution? Without an answer to the first question we would truly be building houses on sand if we proceeded haphazardly to the second and third.

Our apparent dilemma might be at least partially avoided if we step back and ask a few broader questions. For example, what are the chemical elements which appear necessary for life? Studies in our own biosphere have shown that of the 90 naturally occurring elements, somewhat more than 20 appear necessary to sustain life in all the diverse forms in which we know it. In Table 3.6 we can compare the abundances of various elements in the human body with those found in abiotic systems (Frieden, 1972). Of particular interest is the enhancement of C and P and the lack of Si in living material when compared to its inanimate environment. Eight elements provide more than 98% by number of the Earth's crust but only 5 of these are among the 11 that comprise > 99.9% of the human body. In contrast, 9 of

TABLE 3.6
Percent Abundances of Various Elements in the Human Body Are Compared to Those in the Earth's Crust, Seawater, and the Solar System[a]

Human body		Earth crust		Seawater		Solar system	
H	63	O	47	H	66	H	91
O	25.5	Si	28	O	33	He	9.1
C	9.5	Al	7.9	Cl	0.33	O	0.057
N	1.4	Fe	4.5	Na	0.28	N	0.042
Ca	0.31	Ca	3.5	Mg	0.033	C	0.021
P	0.22	Na	2.5	S	0.017	Si	0.003
Cl	0.03	K	2.5	Ca	0.006	Ne	0.003
K	0.06	Mg	2.2	K	0.006	Mg	0.002
S	0.05	Ti	0.46	C	0.0014	Fe	0.002
Na	0.03	H	0.22	Br	0.0005	S	0.001
Mg	0.01	C	0.19	Others	<0.1	Others	<0.01
Others	<0.01	Others	<0.1				

[a]After Frieden (1972)

the latter 11 comprise the most abundant elements in seawater; a critical role for the oceans in prebiotic synthesis is certainly suggested.

Frieden (1972) has summarized a number of characteristics which he considers to be important in governing the selection of elements essential to terrestrial life: the ubiquity and solvating properties of water, the ability of carbon to form long chains and stable rings in comparison to silicon, the mobility of CO_2 in comparison to quartz, and finally the small size and ease of chemical bond formation by the lighter elements (excluding the inert gases). Of particular significance is the ability of six elements (C, N, H, O, P, S) to make up the basic "building blocks" of living material: amino acids, sugars, fatty acids, purines, pyrimidines, and nucleotides. Four metals (Na, K, Ca, Mg) and three nonmetals (Cl, S as SO_4^{2-}, P as PO_4^{3-}) form the principal cations and anions which maintain the acidity and osmotic pressures in body fluids and cells. Finally, a number of elements are required in trace quantities as key components of enzymes and proteins (Cr, Mn, Fe, Co, Cu, Zn, Mo, V) and of hormones (I).

Certainly, if we wish to evolve terrestrial-type organisms, the above essential elements should be present. In this respect Venus is deficient in one of the most important essential materials; namely, liquid water. Mars appears to suffer from the same deficiency although liquid water may be present below the surface. We have no reason to suspect inadequate amounts of C, N, P, S, Cl, and the various essential metals on Mars.

The next question to ask is whether the right conditions exist for *abiotic* synthesis of the important building blocks mentioned above. We will discuss this aspect further in Chapter 4, but a brief review of the principal requirements will be useful here. First, we must have some energy source (UV light, acoustic waves, high-energy particles) to break up simple compounds and enable C—C, C—O, C—N, and other bonds to form. Second, the reacting mixture must be chemically reducing in order for organic molecules to form with any efficiency. Third, there must exist some "haven" from solar shortwave UV light and from elevated temperatures, both of which would destroy the larger disequilibrium organic molecules as they form.

If we can obtain affirmative answers to all these questions, then we can at least decide whether life is possible; the question of whether life is inevitable, even under terrestrial conditions, is, of course, completely open.

3.7. ATMOSPHERIC ESCAPE

We have seen in Sections 3.3 and 3.4 how volatile elements are partitioned into various chemical compounds and physical phases depending on the conditions present on each planet. These partitioning processes are essentially rearrangement processes; they do not result in removal of these elements from the planetary environment. There are, however, several important mechanisms which can result in removal of volatile elements from a planet. In order to escape, the kinetic energy of a particle (with mass m_i, velocity v_i, and kinetic temperature T_i) must exceed its gravitational potential energy. That is,

$$\frac{1}{2}m_i v_i^2 = \frac{3}{2}kT_i \geq m_i g(R + z) = \frac{Gm_i M_p}{R + z}, \qquad (3.72)$$

where R *is the planetary radius*, z is the altitude, g is the local gravitational acceleration, G is the gravitational constant, k is Boltzmann's constant, and M_p is the planetary mass. The quantity $(2GM_p R)^{1/2}$ is called the *escape velocity* of the body. In *thermal escape* the required energy is obtained by collisions with other particles. Figure 3.6 compares the average thermal speeds of molecules of molecular weight m to the escape speeds of planets and other bodies. In *nonthermal escape*, this energy is obtained from exothermic chemical reactions, or by acceleration in planetary or solar wind electric fields. We will discuss these energizing processes under these two convenient headings.

3.7.1. Thermal Escape

We call the outermost region of the atmosphere where collisions are unimportant the *exosphere* and the lower boundary of this region the *exobase*. The boundary level is defined as where an escaping particle will suffer $1-1/e$ collisions above it. If the

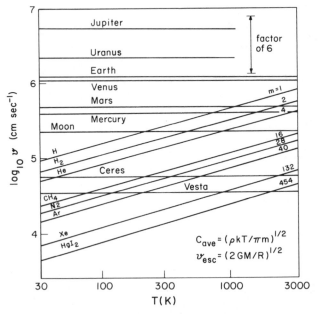

FIGURE 3.6. **Average thermal speeds of gases and escape velocities of solar system bodies. The populations of molecules with speeds 5 or 6 times the average speed are so low that, if the escape velocity of the planet exceeds the average speed of a particular exospheric gas by more than this factor, escape of that species will not be important over times of a few billion years.**

collision cross section of the escaping particle with background gases is σ_i, this level lies at an altitude z_c defined by

$$\sigma_i \int_{z_c}^{\infty} [M]\, dz = 1, \tag{3.73}$$

where $[M]$ is the total gas number density at altitude z. The background gas is essentially in diffusive steady state defined by

$$\frac{1}{[M]} \frac{d[M]}{dz} = -\frac{mg}{kT} - \frac{1}{T}\frac{dT}{dz} = -\frac{1}{H_M}, \tag{3.74}$$

where m is the molecular or atomic mass and H_M the density scale height of the background gas. Integrating (3.74), we can write approximately

$$[M] \simeq [M]_{z_c} \exp[-(z-z_c)/\bar{H}_M], \tag{3.75}$$

where $\bar{H}_M$ is an altitude-averaged density scale height. Substituting (3.75) into (3.73) we have

$$[M]_{z_c} \simeq (\sigma_i \bar{H}_M)^{-1}, \tag{3.76}$$

which effectively defines the exobase altitude.

Let us assume that collisions at and below the exobase can maintain a Maxwellian distribution of speeds for escaping particles. Thus the number of escaping particles with speeds between v_i and $v_i + dv_i$ in a direction defined by the vector $\mathbf{v}_i$ at the exobase is

$$f(v_i) = [i]_{z_c}\, v_i^2 \left(\frac{m_i}{2\pi k T_c} \right)^{3/2} \exp\left(-\frac{m_i v_i^2}{2 k T_c} \right), \tag{3.77}$$

where $[i]_{z_c}$ is the number density of the escaping particles and T_c the temperature at the exobase. The escape flux ϕ_i^c is obtained by multiplying the upward (i.e., escaping) component of v_i by $f(v_i)$ for all particles with v_i greater than the minimum escape velocity $[2g(R + z_c)]^{1/2}$, and integrating over an outward-facing hemisphere. This yields

$$\phi_i^c = \int_0^{2\pi} \int_0^{\pi/2} \int_{[2g(R+z_c)]^{1/2}}^{\infty} f(v_i) v_i \cos\theta \sin\theta\, d\theta\, d\phi\, dv_i$$

$$= [i]_{z_c} \left(\frac{kT_c}{2\pi m_i} \right)^{1/2} \left[1 + \frac{m_i g\,(R + z_c)}{kT_c} \right] \exp\left[-\frac{m_i g\,(R + z_c)}{kT_c} \right] \tag{3.78}$$

$$= [i]_{z_c} w_i,$$

where the angles θ and ϕ are illustrated in Figure 3.7, and w_i is the *net escape velocity*. This is the familiar Jeans formula (Jeans, 1928), and the particle loss is often referred to as Jeans escape. Small corrections ($\sim 25\%$) to this formula are required when the loss of high-velocity particles is sufficiently fast to cause a depar-

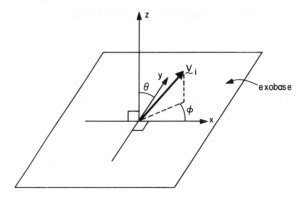

FIGURE 3.7. **Geometry of atmospheric escape. See text for explanations.**

ture from the Maxwell distribution of velocities in the exosphere (Brinkman, 1971a; Chamberlain and Smith, 1971). The Jeans formula implies that large net escape fluxes are expected for light particles around planets with small mass and high exospheric temperatures.

In order to utilize Jeans's formula, we require a value for $[i]_{z_e}$. We must therefore consider the source of escaping atoms; namely, outgassing or evaporation from the planetary surface followed by upward transport through the atmosphere to the exobase. This upward transport may be accompanied by various chemical transformations resulting in different chemical forms for the escaping particle at different altitudes. The upward flow can be described as due to a combination of molecular diffusion (Chapman and Cowling, 1952) and turbulent or eddy diffusion (Reed and German, 1965). The net vertical flux of the escaping particle through a stationary background gas is then given by Hunten (1973b),

$$\phi_i = -\frac{D_i[M]}{1 - f_i}\left\{\frac{df_i}{dz} + f_i(m_i - m)\frac{g}{kT}\right.$$
$$\left. + f_i\,(1 - f_i)\frac{\alpha_i}{T}\frac{dT}{dz}\right\} - K[M]\frac{df_i}{dz}. \qquad (3.79)$$

Here D_i and K are, respectively, the molecular and eddy diffusion coefficients, α_i is the dimensionless thermal diffusion parameter, and $f_i = [i]/[M]$ is the volume mixing ratio of the escaping particle. The level above which molecular diffusion dominates turbulent eddy diffusion (i.e., $D_i \geq K$) is called the turbopause.

Mange (1961) and Hunten (1973b) have shown the usefulness in dividing ϕ_i into two parts: a *countergradient diffusion* flux ϕ_i^d, dependent on df_i/dz, and a *limiting* flux ϕ_i^l, dependent on the buoyancy (i.e., $m - m_i$) and on the thermal diffusion of the escaping particle. That is,

$$\phi_i^d = -\left(K + \frac{D_i}{1 - f_i}\right)[M]\frac{df_i}{dz},$$
$$\phi_i^l = D_i[i]\left[\left(\frac{m - m_i}{1 - f_i}\right)\frac{g}{kT} - \frac{\alpha_i}{T}\frac{dT}{dz}\right], \qquad (3.80)$$
$$\phi_i = \phi_i^l + \phi_i^d.$$

At the turbopause Hunten (1973b) defines $\phi_i^l = [i]_{z^*} w_i^*$, where z^* is the turbopause altitude and w_i^* is the *limiting diffusion velocity*. When

$$(\alpha_i/T)\, dt/dz \simeq 0, \qquad m_i \ll m, \qquad \text{and } f_i \ll 1$$

(conditions which are often applicable for an escaping gas) we have

$$w_i^* \simeq D_i^*/H_M = K^*/H_M \tag{3.81}$$

where D_i^* and K^* are the values of D_i and K, respectively, at the turbopause. In the absence of chemical sources and sinks of escaping particles above the turbopause, continuity of flux clearly demands that

$$[i]_{z_c} w_i = [i]_{z^*} w_i^* + \phi_i^d(z^*).$$

In Spitzer's (1952) theory for atmospheric escape, he assumed that the escaping particle was in diffusive equilibrium above the turbopause. That is,

$$\frac{1}{[i]} \frac{d[i]}{dz} = -\frac{m_i g}{kT} - \frac{1}{T} \frac{dT}{dz} = -\frac{1}{H_i}. \tag{3.82}$$

Thus, for a light escaping gas ($m_i < m$) we have

$$\frac{df_i}{dz} = f_i \left(\frac{1}{H_M} - \frac{1}{H_i} \right) = \frac{f_i g}{kT} (m - m_i) > 0. \tag{3.83}$$

Thus, $\phi_i^d(z^*) < 0$ (i.e., the countergradient diffusion flux is downward in the heterosphere) and

$$[i]_{z_c} w_i < [i]_{z^*} w_i^*. \tag{3.84}$$

In most cases $\bar{H}_i > z_c - z^*$, where $\bar{H}_i$ is the altitude-averaged value of H_i. Thus,

$$w_i < w_i^* \exp \left(\frac{z_c - z^*}{\bar{H}_i} \right) \simeq w_i^*, \tag{3.85}$$

and we see the condition for Spitzer's case is simply $w_i < w_i^*$. The escape rate is limited by the net escape velocity of Jeans. In essence, the escape case described by Spitzer is true only when the vertical transport of the escaping species up to the exobase due to buoyancy and thermal diffusion is very rapid compared to the escape rate itself.

Hunten (1973b) has emphasized another case that occurs when vertical transport is not fast enough to maintain diffusive equilibrium (i.e., maintain $df_i/dz > 0$) for the escaping species. In particular, when $df_i/dz \simeq 0$ then $\phi_i^d \simeq 0$. Thus,

$$[i]_{z_c} w_i \simeq [i]_{z^*} w_i^*, \tag{3.86}$$

which implies

$$w_i \simeq w_i^* \exp \left[\frac{(z_c - z^*)}{\bar{H}_M} \right] > w_i^*. \tag{3.87}$$

The latter inequality holds because $z_e - z^* > \bar{H}_M$ for most cases. Thus, the condition for Hunten's case is simply $w_i > w_i^*$, and the escape rate is limited by the limiting diffusion velocity.

Finally, when $df_i/dz < 0$ (i.e., $w_i \gg w_i^*$), the value of $[i]_{z_e}$ is very small, and the escape flux $w_i[i]_{z_e}$ becomes negligible. We will give examples of the application of these informative concepts when we later discuss each atmosphere in turn.

3.7.2. Nonthermal Escape

3.7.2.1. Exothermic Reactions

Chemical reactions at and above the exobase whose exothermicity exceeds $Gm_iM/(R + z)$ may produce particles capable of escaping. For example, consider photoionization followed by an ion–electron dissociative recombination reaction

$$h\nu + i_2 \rightarrow i_2^+ + e^-$$
$$i_2^+ + e^- \rightarrow i + i \qquad (3.88)$$

in which the energy evolved exceeds $2Gm_iM/(R + z)$ and is equally partitioned as kinetic energy between the two products. Clearly the upward-heading particle may escape and the escape flux will be given by

$$\Phi_i^c \simeq \frac{1}{2} \int_{z_e}^{\infty} k_i[i_2^+][e^-]\, dz, \qquad (3.89)$$

where k_i is the recombination rate constant. Appropriate values for $[i_2^+]$ and $[e^-]$ required in (3.89) in general require solution of the continuity and flux equations for i_2^+ and i_2. Of course, as for thermal escape, the escape flux must not exceed the limiting flux of i_2 through the turbopause.

3.7.2.2. Polar Wind or Breeze

On planets like the Earth, Mercury, and Jupiter, which have well-developed magnetospheres, magnetic lines of force originating near or at the poles are not closed onto the planet, but open out into the interplanetary medium (see Figure 3.8). On Earth, field lines originating at geomagnetic latitudes greater than $75°$ are open. This corresponds to about $\frac{1}{40}$ of the Earth's surface. Ions and electrons are accelerated along these open field lines by escaping ion pressure gradient and by the ionospheric electric field which results from the gravitational separation of the heavy positive ions and the light electrons above the turbopause. The lightest ions and the electrons can reach supersonic speeds, and following the magnetic lines of force, either curve up and then back into the atmosphere at equatorial latitudes or escape the Earth's atmosphere at polar latitudes. On Earth this process causes the light ions H^+ and He^+, which are the dominant topside ionospheric ions at equatorial latitudes, to be severely depleted at polar latitudes. There is a correspond-

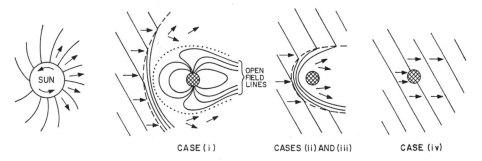

FIGURE 3.8. **Structure of planetary magnetospheres. See text for explanations.**

ing large increase in concentrations of the heavy ions O^+ and N^+ with increasing latitude.

Considerable controversy has arisen concerning an exact physical description of this escape process. The *polar wind* model proposed by Banks and Holzer (1969) implicitly emphasizes the importance of collisions between ions and electrons which gives a continuous nature to the escaping charged gas particles. They therefore use the hydrodynamic equations appropriate to a continuous fluid in close analogy to a theory developed previously for the solar wind. On the other hand, the *polar breeze* model proposed by Dessler and Cloutier (1969) argues that collisions are unimportant and utilizes equations similar to those used to describe the essentially collisionless Jeans thermal escape. Ion–ion collisions will be unimportant above the *ion exobase*, which by analogy to Eq. (3.76) lies at an altitude z_{ei} defined by

$$[M^+]_{z_{ei}} = (\sigma_i + H_{M^+})^{-1}. \tag{3.90}$$

Here σ_i^+ is the cross section for Coulomb interaction of the escaping ion i^+ with the background ion M^+, and H_{M^+} is the density scale height of M^+. A detailed discussion of these two viewpoints is given by LeMaire and Scherer (1973).

As for thermal escape, the ion escape rate cannot exceed the limiting flux of i^+ through M^+. This at least enables an upper limit to be imposed on the escape rate regardless of the detailed physics of the polar wind. By an analogous derivation to (3.80) the limiting ion flux (see Banks and Kockarts, 1973) of a minor escaping ion ignoring thermal diffusion is

$$\phi_i^l = D_{i^+}[i^+] (m_+ - m_{i^+})g/kT \tag{3.91}$$

where D_{i^+} is the diffusion coefficient for i^+ diffusing through M^+, and M_+ and m_{i^+} are the masses of M^+ and i^+, respectively.

3.7.2.3. Solar Wind Sweeping

The solar wind consists mainly of protons and electrons (about 4 particles cm^{-3} at 1 AU) flowing outward from the Sun at considerable velocities ($\sim$400 km sec^{-1} at 1

AU). These charged particles move in the interplanetary magnetic field which spirals out from the Sun. The flow is supersonic since the Mach number

$$Ma = \rho v^2 / \gamma p \simeq 7, \tag{3.92}$$

where ρ is the mass density in the solar wind, v is the solar wind velocity, $\gamma \simeq \frac{5}{3}$ is the ratio of the specific heats C_p / C_v, and p is the pressure in the solar wind. Four basic types of interaction between planetary bodies and this solar wind are possible.

(*i*) First, if the body possesses a sufficiently strong intrinsic magnetic field, it can effectively divert the flow of the solar wind around and away from itself. This intrinsic field must be strong enough so that the pressure it exerts on the solar wind electrons and ions exceeds the dynamic pressure of the incident solar wind particles. That is,

$$B^2 / 2\mu_0 \geq \rho v^2, \tag{3.93}$$

where B is the magnetic flux density of the intrinsic field and μ_0 is the magnetic permeability of free space. Earth-based observations at radio frequencies and also direct measurements by planetary flybys have provided us with estimates of B for a number of planets. In Table 3.7 we compare these observed values with the minimum magnetic flux density, B_{min}, which is required to divert the solar wind as defined by Eq. (3.93). It is clear that for present-day values of the various parameters Mercury, Jupiter, and the Earth, and probably Saturn and Uranus, have sufficiently strong intrinsic fields to cause diversion. For these planets the solar wind particles become thermalized; that is, Ma $\rightarrow$ O, at a *bow shock* (see Figure 3.7) which typically lies several planetary radii from the surface. These thermalized particles flow into the *magnetosheath* which lies between the bow shock and the magnetopause and are swept around the planet. They do not in general penetrate the magnetopause and thus have a negligible influence on the evolution of the atmosphere which lies well below the magnetopause.

(*ii*) Second, when the body possesses an atmosphere with a well-developed ionosphere but no intrinsic magnetic field, the electric field of the solar wind

TABLE 3.7
Ability of Planetary Magnetospheres to Stand Off the Solar Wind

Planet	$B_{surface}$ (G)	B_{min} (G)	Standoff?
Mercury	$\sim 8 \times 10^{-3}$	1×10^{-3}	Yes
Venus	$< 8 \times 10^{-4}$	8×10^{-4}	No
Earth	0.5	6×10^{-4}	Yes!
Moon	$< 10^{-4}$	6×10^{-4}	No
Mars	$\sim 4 \times 10^{-4}$	4×10^{-4}	Marginal
Jupiter	~ 10	1×10^{-4}	Yes!
Saturn	0.1	6×10^{-5}	Yes!
Uranus	0.1?	3×10^{-5}	Yes?
Neptune	0.1?	2×10^{-5}	Yes?

induces a current in the ionosphere resulting in an induced magnetic field. The induced current flows in a sense such as to increase the interplanetary field on the side facing the solar wind and decrease it on the other. Once again a bow shock is formed, but it lies very much closer to the planet compared to that formed in the strong-intrinsic-field case described above (see Figure 3.7). Known examples of this case are Venus and Mars. Thermalized solar wind particles flow into the magnetosheath and, due to the relative proximity of the bow shock to the surface, a direct interaction of the thermalized solar wind particles with the planetary atmosphere is possible. From an evolutionary viewpoint, this interaction consists of two competing processes:

(A) Solar wind ions exchange their positive charges with neutral atmospheric species resulting in *capture* of hydrogen and the other less cosmically abundant elements by the planet.

(B) The solar wind electric and magnetic fields accelerates atmospheric ions (produced by photon or solar wind bombardment) in the wind direction. At high altitudes near the limb of the planet these ions will be swept up into the solar wind resulting in *loss* of atmospheric gases by the planet.

A simple but sufficiently accurate method for estimating the capture or loss rates due to these two mechanisms has been proposed by Michel (1971). In general, there should exist a critical streamline for the flowing thermalized solar wind particles such that planetary ions produced on the solar wind side of this line are incorporated into the solar wind flow. Conversely, solar wind particles flowing on the planetary side of this line become neutralized and captured by the planet. As the initially supersonic solar wind flows into the planet it will accelerate atmospheric ions which then become part of the flow. However, Cloutier *et al.* (1969) show in a straightforward manner that simultaneous conservation of mass, momentum, and energy restricts the *maximum* flux of atmospheric ions that can be accommodated by the solar wind to the quantity $\alpha \rho v$, where

$$\alpha = \frac{(r+1)\mathrm{Ma}^2 \, [2 + (r-1)\mathrm{Ma}^2]}{(1 - \mathrm{Ma}^2)^2}. \tag{3.94}$$

The maximum loss rate of atmospheric ions will therefore occur if the solar wind flowing along the critical streamline is augmented by an atmospheric ion flux of $\alpha \rho v$. From our definition of the critical streamline, $\alpha \rho v$ also equals the integrated production rate of atmospheric ions along the streamline. This photoion production rate decreases upward with the scale height H of the parent molecule for the ion. Using the notation in Figure 3.8, the maximum total global loss rate of atmospheric mass due to solar wind sweeping is therefore roughly

$$(2\pi R \sin \theta) \int_0^{\pi/2} \int_{\theta_0}^{\infty} (\alpha \rho v) e^{-z/H} \, dz \, d\theta = 2\pi HR\rho v \int_{\theta_0}^{\pi/2} \alpha \sin \theta \, d\theta$$
$$\lesssim 2\pi HR\rho v. \tag{3.95}$$

The approximation $\int_{\theta_0}^{\pi/2} \alpha \sin \theta \, d\theta \lesssim 1$ in Eq. (3.95) is justified by the following arguments: Near the center of the planet ($\theta = \theta_0$) the solar wind has stagnated after passing through the shock layer and Ma ~ 0. Thus near $\theta = \theta_0$, $\alpha \gg 1$ from Eq. (3.94) and $\alpha \sin \theta \sim 1$. Conversely, near the limb of the planet the flowing solar wind still has an appreciable Mach number. That is, near $\theta = \pi/2$, $\alpha \sim 1$ using Eq. (3.94) and again $\alpha \sin \theta \sim 1$.

Finally, using the same notation as in Figure 3.8 the rate of mass addition to the planet from the solar wind will be roughly

$$\pi(R \sin \theta_0)^2 \rho v. \tag{3.96}$$

However, the vague definition of θ_0 makes this formula of limited use and we must generally resort to detailed computations to compute the solar wind *influx*. Such computations for Mars by Cloutier *et al.* (1969) imply $\sin \theta_0 \simeq 0.06$ or $\theta_0 \simeq 4°$. A similar small value for θ_0 should apply on Venus.

(*iii*) A third case occurs when the body has no ionosphere and no intrinsic magnetic field. A magnetic field can still be induced on the body by the solar wind providing the body itself has a high electrical conductivity. This could be due to a significant solid metallic, fluid metallic, or liquid electrolyte content. The interactions are essentially the same as in (*ii*) except that there will be no atmospheric ions to be swept away. Examples of such bodies might be some of the asteroids and the satellites of the major planets. The required minimum conductivity of the body or of the ionosphere in (*ii*) can be estimated as follows: If the time for the interplanetary field (which is flowing along at the velocity of the solar wind) to flow through the planet is greater than the time for the solar wind itself to flow past the planet, then the interplanetary field lines will "pile up" in front of the planet as illustrated in Fig. 3.8. Under these conditions, the bow shock required in (*ii*) and (*iii*) will form. Thus, for (*ii*) or (*iii*) we require

$$\mu_0 R^2 \kappa > R/v, \tag{3.97}$$

where κ is the conductivity and R the planetary radius. Thus, the conductivity must exceed $(\mu_0 R v)^{-1}$.

(*iv*) Finally, the fourth case occurs when the body has no ionosphere, no intrinsic magnetic field, and a conductivity less than $(\mu_0 R v)^{-1}$. The interplanetary field permeates the body itself and no bow shock forms. The supersonic solar wind collides directly with the surface of the body and accumulates at a rate of $\pi R^2 \rho v$. The moon is a good example.

3.7.3. Escape Time

It is particularly useful to define a suitable escape time for each constituent in an atmosphere to ascertain the stability of that constituent over the roughly 5×10^9 yr elapsed since planetary formation. It is important, however, to distinguish at least two different lifetime definitions. Consider an escape flux ϕ_i^e of a constituent i due to one or more of the mechanisms discussed earlier. Let us define the total number of particles per unit area on the planet (including those in the crust or condensed on the surface) as M_{tot}. When surface sources of the particle exist, and crustal outgassing (or evaporation) and vertical transport are constant, then ϕ_i^c will be independent of M. It can also be roughly independent of M_{tot} when no surface sources exist providing the escaping particles are the main atmospheric constituent and so determine the escape level conditions (McElroy and Hunten, 1969). In these cases the total lifetime of escaping particles on the planet is simply M_{tot}/ϕ_i^c. In contrast, when no surface sources exist and the particle is a minor constituent, ϕ_i^c usually decreases linearly with M_{tot}. Thus, M_{tot} decreases exponentially with time, and M_{tot}/ϕ_i^c is now the time for a $1/e$ decrease in M_{tot}. Care should be taken to define accurately which definition of the atmospheric lifetime is applicable in each case.

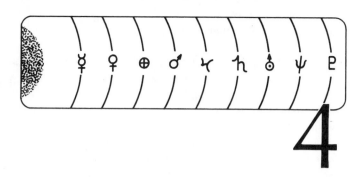

The Atmospheres of
the Planets

We shall now analyze the available data on the atmospheres of the terrestrial planets, and attempt to develop quantitative tests of the competing theories of atmospheric origin for each planet in turn. The sequence which we shall adopt is that of increasing heliocentric distance, since a number of crucial compositional determinants, such as formation temperature, present equilibrium temperature, incident UV intensity, and incident solar wind flux all vary smoothly with heliocentric distance. There is, however, *no* ranking of these bodies that preserves the same order for *all* important parameters: for example, mass, surface gravity, escape velocity, exospheric temperature, and atmospheric mass.

In order to provide the most extensive test possible of the various competing models, we shall consider not only Mercury, Venus, Earth, and Mars, but also the larger asteroids (notably Ceres) and the lunar-sized satellites (in particular, Earth's moon and Titan).

The authors of the present review are acutely aware of the frequency with which long-held qualitative assumptions are invalidated by new observational data or by even modest and simplistic quantitative theoretical considerations. Therefore, the emphasis in this portion of the review will be upon isolating those *facts* that we really *do* know, and identifying clearly the nature of the *assumptions* which have been brought to bear. We will attempt to raise critical questions regarding *all* competing theories (even our own), and will whenever possible make explicit statements regarding divergent predictions of different models in the hope of motivating crucial experiments for discriminating between these theories. To this end, we will in some cases be forced to quantify aspects of theories with which we do not agree. To date, none of the major alternatives to the equilibrium condensation hypothesis have yet been quantified. Therefore, as in a low-budget theatrical production, we must occasionally dart backstage and reemerge cloaked in the attire of one of our opponents. It is our ambition, through this artifice, to present the debate in such a form that the reader may be left unaware that we and our opponents never appear on the stage together. The zenith of this art would be to leave the reader in the end unsure regarding the true identity of the narrator; this effect is, however, unfortunately beyond our skill.

4.1. MERCURY

4.1.1. Atmosphere

The study of the atmosphere of Mercury has long been a thankless task. Earth-based observations have never succeeded in detecting even the faintest trace of an atmosphere on Mercury.

Near-infrared spectroscopic observations of Mercury were first reported by Adams and Dunham (1932), who failed to detect CO_2. Polarimetric studies by Lyot (1929) had already provided an upper limit of ~3 mbar for the total atmospheric pressure. These and other early studies are summarized in detail in the book *The Planet Mercury* by Sandner (1963).

Dollfus (1961) claimed the positive detection of a trace of atmosphere from polarimetric observations, by this estimate between 0.1 and 10 mbar. However, O'Leary and Rea (1967) have shown that the subtle polarimetric effects noted by Dollfus could have been produced by a solid surface only, without any atmosphere.

It is convenient to recognize a second era in the history of the study of Mercury beginning with the initial proposal by Field (1962, 1964) of an atmosphere rich in radiogenic ^{40}Ar, and the parallel spectroscopic studies by a number of astronomers. The latter efforts were largely devoted to the search for the heavy, abundant, spectroscopically active CO_2 molecule.

Moroz (1965) claimed tentative detection of the 1.6 μm CO_2 band in the reflection spectrum of Mercury; but Spinrad *et al.* (1965), in searching for evidence of the 8700 Å CO_2 band, were unable to detect any CO_2 with an absolute upper limit of ~4.2 mbar. Also, Binder and Cruikshank (1967) reexamined the 1.6-μm band and failed to detect the presence of any feature.

Belton *et al.* (1967) searched for the $\nu_1 + 2\nu_2 + 3\nu_3$ CO_2 bandhead near 1.049 μm without success, and placed an absolute upper limit on the CO_2 partial pressure of 0.35 mbar. This result requires that pressure broadening must be negligible in the 1.6 μm band and permits deduction of a comparable upper limit on CO_2 from Binder and Cruikshank's negative results. Bergstralh and Smith (1967) also failed to detect any sign of CO_2 absorption in the 1.05, 1.20, and 1.22 μm bands with a detection limit of roughly 0.04 mbar. These several results represent the best that could be expected from medium-resolution spectroscopy.

More recently, Poppen et al. (1973) obtained high-resolution interferometric IR spectra of Mercury and set a very low upper limit of $\sim 4 \times 10^{-4}$ mbar on the CO_2 abundance.

Kozyrev's (1964) claim to have detected an emission core due to Mercury superimposed over the Hβ solar absorption feature has been clearly shown by Spinrad and Hodge (1965) to be due to severe contamination of the (highly Doppler-shifted) reflection spectrum of Mercury with unshifted scattered sunlight from the sky.

Spinrad et al. (1965) also established upper limits of <1 m agt for O_2 and <30 μm of precipitable H_2O.

A third era in the study of Mercury was brought in by the Mariner 10 Venus–Mercury flyby. Mariner 10 flew by Mercury on 29 March 1974 and continued on a heliocentric orbit with a period of exactly 2 Mercury years, or 176 Earth days. The 3:2 spin–orbit coupling of Mercury causes it to present exactly the same face to the sun at intervals of 2 Mercury years. Thus, after one orbit of Mariner 10, it reencountered Mercury under conditions virtually identical to those of the first encounter.

The second encounter took place on 21 September 1974 (Strom et al., 1975a), and the third and final encounter was on 16 March 1975, after which the attitude control gas was exhausted. Mariner 10 then became unable to orient its high-gain antenna toward Earth.

By far the most stringent limits on the present abundances of atmospheric constituents on Mercury have been set by the ultraviolet spectrometer (UVS) experiment carried by the Mariner 10. Kumar (1976) has shown that no species indigenous to the planet has an abundance above 10^{-13} bar. Helium, hydrogen, and oxygen were detected by the experiment, with the latter perhaps subject to reinterpretation (Broadfoot et al., 1974b; Kumar, 1976). The helium surface pressure was estimated to be 3.6×10^{-16} bar and the hydrogen abundance is roughly 100 times smaller. The oxygen abundance may be comparable to that of helium (Broadfoot et al., 1976). Since helium may be provided either by the solar wind or by outgassing of radiogenic ^4He from uranium and thorium decay, its detection fails to provide unambiguous evidence of either source. The tiny trace of hydrogen seen could easily be attributed to inefficient input of solar wind gases through the Mercurian magnetosphere, whereas oxygen, if confirmed, may arise from sputtering, photolysis, and chemical processes involving interaction of hydrogen with surface silicaceous rocks.

It has been suggested several times that the most likely gas to persist in the atmosphere would be ^{40}Ar (e.g., Field, 1964), but studies by Rasool et al. (1966) showed that argon-rich atmospheric models have the serious flaw that they are fair absorbers of solar UV energy, but very poor emitters. The steady-state temperature reached by an argon upper atmosphere was found to be over 10^4 K, easily sufficient to permit thermal escape of argon. Addition of large amounts of a good radiator such as CO_2 lowers the exospheric temperature by an order of magnitude, but it is easy to show that the photochemical lifetime of CO_2 is so short at Mercury's distance from the Sun that this is not a feasible means of achieving a stable atmosphere.

Banks *et al.* (1970) examined a more realistic low-mass atmospheric model, in which the density was so low that the exosphere rested on the surface of the planet. For reasonable dayside surface temperatures of 600–700 K the planet was found capable of retaining a small steady-state atmosphere of argon if the internal source were large and if the upper atmosphere were shielded against solar wind sweeping by a planetary magnetosphere. Thomas (1974), writing prior to Mariner 10, presented arguments to the effect that a lunar-like model for Mercury (that is, no magnetic field) would permit the retention of a trace atmosphere of hydrogen and water with partial pressure near 10^{-13} bar. Williams (1974) proposed a similar idea. Hodges (1974) also presented a model for a hydrogen/helium/neon mixture with its exobase resting on the ground, with a reasonable distribution of surface temperatures over the planet. He assumed saturation of the surface with solar wind gases, and thus also tacitly assumed the absence of a magnetosphere. His models gave maximum pressures near 10^{-12} bar, with hydrogen and helium more abundant at steady state than neon.

Ness *et al.* (1974b, 1975a,b, 1976) have reported that the Mariner 10 fluxgate magnetometer experiment shows the presence of a well-developed magnetosphere about Mercury, with the magnetopause standing well off from the surface. Surface fields are estimated to range from about 300 to 800 γ (1 γ = 10^{-5} G). Hartle *et al.* (1975a), in modeling the Mercurian atmosphere in light of the Mariner 10 data, find that the solar wind input of helium needed to preserve a steady state in balance with the computed escape flux is less than 10^{-3} of the incident solar wind helium flux. The capture probability for an alpha particle striking the shock front is found to be 6×10^{-4} if all the helium is derived from the solar wind, but it is worth noting that the required input rate of ^{4}He is also closely similar to that expected from steady-state outgassing of a planet with an approximately chondritic U and Th content. Thus, the efficiency of capture of solar wind helium may be any number less than 6×10^{-4}, and the present outgassing rate of radiogenic He may be any number *up* to this amount of flux.

Although the solar wind is generally held off from direct interaction with the surface by Mercury's magnetosphere, such interaction is possible under extreme conditions (Siscoe and Christopher, 1975). Experience with the production of a number of molecules such as H_2, CH_4, H_2O, N_2, and CO by solar wind irradiation of the regolith of the Moon has led Gibson (1977) to propose that a similar process may take place on Mercury.

No trace of CO_2, CO, ^{40}Ar, H_2O, hydrogen halides, nitrogen, or sulfur compounds was found by Mariner 10. Upper limits of about 10^{-12} bar were found for CO_2, H_2O, O_2, N_2, H_2, and atomic O by the UV occultation experiment. Zeilik and Dalgarno (1973) had shown that, for a suite of atmospheric models, UV dayglow emission lines of ^{40}Ar might be detectable by the Mariner 10 UV spectrometer. However, Kumar (1976) found that the ^{40}Ar abundance expected if Mercury had Earthlike composition would be rather larger than the observational upper limit. He therefore speculated that Mercury may be deficient in potassium relative to the Earth.

4.1.2. Surface and Interior

Most of our meager evidence regarding the composition of the surface layers of Mercury is provided by study of the visible and near-IR reflection spectrum of the planet. McCord and Adams (1972) found the steep, almost featureless reflection spectrum of Mercury to resemble closely that of mature lunar soils, and they therefore attributed the darkness and color to the presence of titanium and iron in a glassy soil layer. Vilas and McCord (1976), however, find no evidence of a depression in the Mercurian reflection spectrum near 0.9 μm, where ferrous iron in olivine and pyroxene has a strong absorption feature. They are able to give an upper limit on the FeO content of the Mercurian surface of 6%. McCord and Clark (1979a,b) prefer an FeO content near 6%. It should be noted that these estimates are based on whole-disk spectra of Mercury.

Gaffey and McCord (1978), in their extensive review of reflection spectra of meteorites and asteroids, find asteroids with a visible and near-IR reflection spectra similar to Mercury. These asteroids include 16 Psyche and 140 Siwa, which are assigned to class RR. Gaffey and McCord interpret these bodies as metallic Fe–Ni plus an optically neutral Fe-free silicate such as enstatite or forsterite. Matson et al. (1976) emphasize that all meteorites that are as red as these very red asteroids (and Mercury) contain free metal. Opinion is divided whether the redness of the Moon is primarily due to metal coatings (Gold et al., 1971), very small metal particles (Hapke and Wells, 1976), or Fe and Ti ions in impact-generated glass (Adams and McCord, 1971). Hapke (1977) attributes the reflection spectrum of the Mercurian regolith to an FeO content of 3–6%.

The bulk composition of Mercury has been known for many years to be different from that of the other terrestrial planets. Arguments based on the high zero-pressure density of Mercury concluded over a decade ago that Mercury is roughly 60% metallic iron–nickel alloy, compared to about half that amount in Venus and Earth (see, e.g., MacDonald, 1963; Anderson and Kovach, 1967; Kozlovskaya, 1969; McCrea, 1969). Accordingly, dilution of the long-lived radionuclides ^{40}K, ^{232}Th, ^{235}U, and ^{238}U relative to chondritic material has often been assumed, leading to low heat production rates in some models.

It is desirable to seek constraints on the thermal evolution of Mercury from geological evidence whenever possible, and this has been an important activity since the Mariner 10 flybys. Very heavy cratering, similar to the lunar highlands, was found on the very first Mariner 10 flyby (Murray, 1975b; Murray et al., 1974b,c). Planetary differentiation before the end of accretion was suggested. No evidence for erosion by liquids or winds could be found (Murray *et al.*, 1975), and an overall history strikingly similar to that of the Moon was found (Trask and Guest, 1975). Cordell and Strom (1977) have studied the nature and distribution of scarps, and attribute them mostly to compressive tectonic stresses on the crust. They attribute these stresses to cooling and shrinkage of the planet early in its history.

Strom *et al.* (1975a,b) have attributed the smooth plains to early volcanism, and Trask and Strom (1976) have presented further stratigraphic and morphological arguments in defense of this hypothesis. Alternatively, Wilhelms (1976) has proposed that the plains material is basin ejecta or possibly impact melts. Schultz (1977) interprets some observed modification of impact craters to endogenic volcanism.

Theoretical studies of the thermal history of Mercury may also be constrained by the observation of a planetary magnetic field by Mariner 10 (Ness *et al.*, 1974b, 1975a,b, 1976). It was quickly concluded that the observed field could not be induced by the solar wind, but must instead be intrinsic to the planet (Hartle *et al.*, 1975b; Herbert *et al.*, 1976). Stephenson (1976) and Solomon (1976) found quantitative difficulties with present-day dynamo generation of the field in Mercury's core, unless a high core radioactivity is assumed. Cassen (1977) also require that a heat source density at least as large as that in the Earth's mantle must be present in order to keep the core molten and permit a present-day dynamo.

Stephenson (1976) and Sharpe and Strangway (1976) consider the possibility that the present field is due to remanent magnetism surviving from an ancient dynamo-produced field. Finally, Gubbins (1977) has shown that an Earth-like dynamo is plausible even at Mercury's very low spin rate.

Earth-analogy thermal history models, such as those by Majeva (1969), have given way to detailed consideration of the feasibility of core formation and of a present-day liquid state for the core (Siegfried and Solomon, 1974; Fricker *et al.*, 1976; Peale, 1976; Cassen, 1977). The lines of evidence presently available favor early core formation, and it is now reasonably clear that the core can be kept melted only if the heat source density is significantly above chondritic (or terrestrial) levels.

4.1.3. Origin

Now let us inquire into the possible modes of origin of Mercury and its atmosphere.

The first two theories considered in Chapter 2, which involve an early gas-rich protoplanetary phase, are both solar-composition models, in which the initial relative atomic abundances of H, He, O, C, N, Ne, Ar, etc. are fixed. Since H and He escape will, by any model, be very rapid on Mercury, we would expect formation of a residual high-molecular-weight atmosphere on a short time scale. The residual atmospheric composition expected after total helium and hydrogen loss is 47% CO_2, 34% Ne, 18% N_2, and 1% Ar. If these gases were present in solar proportions relative to iron, then the mass of neon alone would be ~90% of the mass of the solid planet, and a total atmospheric pressure of about 1 Mbar would result. Taking 10^{-12} bar as an extreme upper limit on the atmospheric pressure, we can see at once that no more than 1 part in 10^{18} of the cosmic proportion of heavy gases could be present. Gross (1973) has suggested that $Ne-N_2$ atmospheres are likely on lunar-sized bodies on the basis of the large cosmic abundances of these elements. However, we know that Ne and N_2 retention by the Earth was extremely inefficient, and that satellites (and Mercury) are smaller by far than the Earth. We suggest that any proposed mechanism for retention of particular gases by a planet should be backed by a clear statement of both the suggested mechanism and the abundances of other retained volatiles logically attendant upon that mechanism.

The capture of an adiabatic temporary atmosphere, as discussed in Section 2.2, results in retention of only 10^{-10} of the abundances of the volatile elements relative to rock-forming elements such as iron. This is enough to supply only 5×10^{-6} bar of CO, 4×10^{-7} bar of N_2, and 2×10^{-6} bar of Ne. Further, removal of the nebula would lead to rapid loss of gas.

By Ringwood's original 1966 model (discussed briefly in Section 2.4), the anomalously high density of Mercury is inexplicable without recourse to a special mechanism for vaporizing over 50% of the mass of silicates in the crust and mantle of the planet. One measure of the difficulty is that the density of CI material is 2.4 gm cm^{-3} compared to Mercury's density of 5.5. Ringwood proposes that this fractionation was caused by a hypothetical superluminous phase of the early Sun, as suggested by Hayashi (1961). Accretion and differentiation of Mercury (and by reasonable extension, the other, more massive, terrestrial planets) *prior* to the Hayashi phase is necessary for this model, since heating of unaccreted protoplanetary material *in vacuo* by a highly superluminous Sun would vaporize silicates and metal without discrimination, and hence effect no change in Mercury's density. Such rapid accretion would have a profound effect on the internal composition of each planet. Since CI chondrites contain a substantial excess of the oxidants H_2O and Fe_3O_4 over the reducing agent C, heating to temperatures over ~800 K leads to quantitative conversion of carbon to CO and CO_2 gases and oxidation of all iron to magnetite. A CI chondrite containing 10 wt. % H_2O and 40% magnetite, if reduced to the Fe/FeO buffer, will liberate 0.116 gm of O per gram. A carbon content of 5.5% would be fully oxidized to CO by only 63% of this amount of oxygen. Ringwood asserts the contrary: that equilibrium is somehow not attained and iron oxides are reduced to metal by carbon.

If an ordinary chondrite composition (Section 2.5) is postulated for Mercury, then the situation is similar but less extreme. Again, Hayashi phase heating or some other effect specific to Mercury is needed to vaporize silicates and alter the metal–silicate mass ratio.

In a two-component model for volatile retention (Section 2.5), we must estimate the mass fraction of volatile-rich material added to Mercury. In addition, we must specify whether this material is similar to CI chondrites or to one of the ice-bearing cometary composition classes. The best fit to the observed atmospheric properties of Mercury is to set the mass fraction of volatile-rich material equal to zero. However, the ease of atmospheric escape from Mercury gives us reason for caution.

The equilibrium model (Section 2.7) gives a qualitatively plausible explanation of the high density of Mercury without introduction of ad hoc postulates or unknown processes: In it, Mercury accretes at a temperature so high that silicates are incompletely condensed. Direct consequences of this model are the predictions that the planet Mercury accreted vertically devoid of water, sulfur, iron oxides, and alkali metals (and thus ^{40}Ar). Retention of nitrogen and carbon as nitrides and carbides in solid solution in the metal phase remains plausible, but only a small source of oxidizing agent should be available to make N_2, CO, and CO_2 out of nitrides and carbides.

Lewis et al. (1979) have investigated the availability of carbon as a function of condensation temperature for a solar-composition system, for the purpose of estimating the carbon content of preplanetary condensed solids as a function of heliocentric distance. The equilibrium activity of graphite calculated for a condensation temperature of 1400 K, which provides a very iron-rich condensate capable of explaining the high observed density of Mercury, was $a_{gr} = 10^{-5}$. Using metallurgical thermodynamic data on the activity coefficients of dilute carbon solutions in metallic iron–nickel alloys, an equilibrium carbon concentration of 1 ppm by weight in the metal phase was found. For the most extreme case of purely local accretion of Mercury's material, all with a condensation temperature of 1400 K, an extremely low FeO content in the silicates, $x_{FeO} = 10^{-4}$, is found. From even these tiny carbon and oxygen sources CO or CO_2 partial pressures near 1 bar could be generated by extensive outgassing. Other meteorite-based models for the origin of Mercury provide substantially larger quantities of carbon oxides, reaching roughly 16% of the mass of the planet for a carbonaceous chondritic model.

If Mercury experienced a violent and thorough outgassing event early in its history, a mass of carbon oxides ranging from 10^4 to 10^8 gm cm^{-2} would have been released onto the surface. Here the lower extreme is the amount produced by equilibrium condensation and the upper is that for a CI composition model. Photoionization and solar wind sweeping of CO_2, even in the absence of a magnetosphere, must have been several orders of magnitude slower than photodissociation of CO_2 to CO and atomic oxygen. Therefore, massive deposition of oxygen (and perhaps graphite) in the surface layers of the planet may have occurred. Daytime surface temperatures are, however, so high that carbonate formation and persistence to the present is not feasible. Oxidation of metallic iron to FeO in the surface of Mercury cannot be permitted to produce in excess of about 6% FeO because of the virtual indetectability of the 0.9 and 0.95 μm absorption feature of Fe^{2+} in the spectrum of the planet (Vilas and McCord, 1976; McCord and Clark, 1979a,b). Thus no more than ~1% of the soil could be ascribed to O in FeO, and absorption of 10^4 gm cm^{-2} of oxygen without visible evidence would require dispersal of Fe oxides through 10^6 gm cm^{-2} of regolith. For a density of 3 gm cm^{-2}, this requires a stirred regolith at least 3 km thick, even for that atmospheric origin model which provides the least gas.

Several possible materials are therefore feasible for providing a low surface albedo on Mercury, including graphite and magnetite if photodecomposition of an appreciable mass of carbon oxides occurred, or metallic iron (and even solid carbides and nitrides) if it did not. Greatly increased knowledge of the surface mineralogy and oxidation state of the Mercurian crust would therefore be of great interest with respect to the possible prior existence of a significant atmosphere on Mercury. Unfortunately, as we have seen, nature has conspired to remove virtually all of the most direct lines of evidence by the extreme efficiency of the mechanism for atmospheric escape. We must keep in mind the likelihood of large-scale production of radiogenic ^{4}He and of minor production of fissiogenic nuclides of krypton and xenon by U and Th decay. Further, any high-temperature condensation model explicitly predicts that lead on Mercury would be almost completely radiogenic, whereas strontium would contain an unusually large nonradiogenic component.

Unfortunately, these rather clear-cut features of high-temperature equilibrium condensation become seriously blurred by considering the process of planetary accretion. Preplanetary solids originating well outside Mercury's orbit will inevitably be swept up by Mercury during accretion, and the concentration of these lower-temperature condensates should be higher near the surface than in the deep interior. The detection of potassium, sodium, or sulfur and an accurate assay of FeO would permit estimates of the amount of such low-temperature material present.

Kaula (1976) has suggested that scattering of planetesimals by more distant planets may have led to the comminution and loss of silicate material from near Mercury, resulting in the selective enrichment of refractory and dense metallic materials in an undersized planet. Weidenschilling (1978) has pointed out that the strict equilibrium condensation approach gives a metal–silicate ratio as high as Mercury's over only a very narrow temperature (and heliocentric distance) range; so narrow, in fact, that it is very unlikely that it could collect to form Mercury without severe dilution by more silicate-rich material from slightly larger heliocentric distances. He suggests metal–silicate sorting by aerodynamic drag on small particles in the solar nebula early in the accretionary era: Low-density solids would more readily spiral into the Sun due to drag.

The condensation–accretion models by Barshay and Lewis (1981) bear out the validity of Weidenschilling's contention. All plausible accretion sampling functions result in a density for Mercury that is lower than that observed.

A further result from these condensation–accretion models is that the FeO content of Mercury is always found to be less than 1% for reasonable accretion sampling functions. The reason for this is simple: The high density of Mercury can only be approximated by very narrow sampling functions that select material condensed in the narrow interval between metallic iron and magnesium silicate saturation. But narrow accretion functions fail to sample the low-temperature condensate FeO. The best approximation, a Gaussian sampling function with extensive "wings," cannot provide 3% FeO while maintaining the high density of Mercury.

In summary, we conclude that no present hypothesis accounts fully for even our limited body of knowledge of Mercury. The appealingly (but unrealistically) simple equilibrium condensation hypothesis is unable to satisfy both the observed density and the deduced FeO content of 3–6%.

It is clear that some additional hypothesis is required, such as a mechanism for fractionation of metal from silicates near Mercury's orbit.

Perhaps the strongest conclusion we can reach is that the body of data on

Mercury is woefully inadequate. Geochemical data would be of particular value, but are essentially absent. The FeO, FeS, alkali metal, carbon, etc. abundances in the regolith would all help greatly in constraining the formation temperature of the preplanetary solids, while major-element abundances and mineral phase assemblage data would be of immense comparative value.

Readers interested in a more detailed historical account of the study of Mercury should consult the book by Sandner (1963) or the later short review by Vetukhnovskaya and Kuz'min (1967b). Morrison (1970) reviews the radiometric data and thermal models of the planet. Recent general reviews include the popular article by Murray (1975a) and the technical post-Mariner 10 overviews by Strom (1979) and by Murray (1975b).

4.2. VENUS

The first evidence for the presence of an atmosphere on Venus was found by M. V. Lomonosov during his observation of a transit of Venus across the disc of the Sun on 26 May 1761. Aside from the observed refraction of sunlight by the atmosphere, nothing further was learned of the intrinsic properties of Venus aside from its mass and cloud-top diameter until well into the present century.

All modern observations of the planets are superimposed upon a sort of scientific mythology, which, though without any significant supporting evidence, has served as the background for speculation about planetary evolution for the last century. This concept seems to be based on the notion that the planets were formed, at widely separated points in time, in close proximity to the Sun. The orbits of the planets are sometimes taken to move outward with time; however, in every case Mars is interpreted as a much older planet than Earth, while Venus is in a far earlier stage of development. Mercury, by this view, is still forming. Science fiction is replete with ruined Martian cities peopled (or no longer peopled) by the degenerate remains of once-great cultures. Venus, of course, must be a steaming jungle, a teeming swamp, or at least a global ocean with simple life forms. Mars has always been pictured as having "lost its oceans" by hydrogen escape into space, whereas Venus was usually assumed to have at least as much water as the Earth. These preconceptions have had so great a hold on the imaginations of authors and scientists that readily recognizable features of this fantasy can be seen in scientific literature as recent as 1970.

Telescopic observations of Venus in visible light show a dense, featureless cloud deck, unchanging from century to century. The clouds exhibit a pale lemon-yellow tint. At UV wavelengths faint features can be seen, apparently passing around the planet in the retrograde sense (opposite to the sense of the orbital revolution of the planets) in about 4 Earth days. The clouds were generally assumed, by terrestrial analogy, to be made of water droplets or water ice particles.

Since both water vapor and molecular oxygen were expected to be abundant on Venus, several spectroscopic searches were made for these species. The suspicion that something was wrong began to dawn when St. John and Nicholson (1922) established rather restrictive upper limits on the H_2O and O_2 abundances, showing that there must be less than 1 m agt of O_2 and less than 1 mm of precipitable water vapor in the line of sight (in and above the main cloud layer). Adams and Dunham (1932) were later able to identify several bands in the near-IR spectrum of Venus as due to carbon dioxide, which remained the only identified constituent of the atmosphere until as recently as 1965.

It may fairly be said that the basis of our modern understanding of Venus was laid by Rupert Wildt. First, Wildt (1934, 1940a) showed that the presence of a CO_2 atmosphere was a sufficient condition for the development of high surface temperatures. Visible sunlight can penetrate such an atmosphere to the surface of the planet, where it is absorbed and reemitted as Planckian radiation with much longer wavelengths, in the thermal IR. Such radiation is strongly absorbed by CO_2 and cannot readily leak out into space. As a result, according to Wildt's calculations, the surface temperature would be built up by this "greenhouse effect" to over 400 K. [Menzel (1923) and Adel (1937b, 1941) also predicted temperatures over 320 K.] This is well above the normal (1 bar) boiling point of water. It is noteworthy that this is due to a CO_2 greenhouse alone: Inclusion of water vapor adds considerable IR opacity between the CO_2 bands and makes the effect even stronger, providing, of course, that the water does not produce a cloud layer dense enough to block the downward passage of visible sunlight. Wildt (1942) further concluded that it was impossible to have condensed water on the surface, and that the dryness of the atmosphere strongly implied that Venus was very deficient in water relative to Earth. He suggested that the principal reservoir of water on the surface would have to be hydrated minerals. Finally, he noted that the best estimates of the total terrestrial CO_2 inventory available to him gave about 10 kg cm^{-2} of gaseous CO_2 equivalent, almost all from carbonate rocks. This figure was comparable to the CO_2 abundance Wildt expected to find in the atmosphere of Venus.

The next important insight into the role of geochemical processes in determining the atmospheric composition came in the classic book *The Planets: Their Origin and Development*, by H. C. Urey (1952c). Urey suggested that volatiles such as CO_2 and H_2O could be *buffered* by chemical reactions with mineral assemblages present on the planetary surface. As a general and somewhat schematic representation of these processes, he wrote what is now referred to as the Urey reaction:

$$CO_2 + CaSiO_3 \rightleftharpoons CaCO_3 + SiO_2. \tag{4.1}$$

A buffer reaction is defined as one in which a gas pressure is a function of temperature alone. Given the mineral assemblage, the CO_2 pressure can be readily calculated from thermochemical data (see Section 3.5.3). There are a number of such possible reactions, each with its own temperature dependence, and further progress in understanding this interaction requires more knowledge of both the bulk composition of crustal rock types and the surface temperature. Such buffer reactions will be discussed in detail in Section 4.2.3.

A very different interpretation of the dry, CO_2-rich atmosphere was suggested by Hoyle (1955). He postulated that, at an earlier stage in the history of Venus, massive oceans of liquid water and of liquid hydrocarbons (sometimes irreverently called Hoyle oil) coexisted. Photodissociation of water vapor gave rise to free hydrogen, which escaped from the planet, and free oxygen, which oxidized the oil to make CO_2. If Venus began with a stoichiometric excess of oil over water, then a dry CO_2 atmosphere (with residual oceans of crude oil) would be present today. Clearly a search for gaseous hydrocarbons would be worthwhile!

Menzel and Whipple (1955) provided yet another interpretation: that the surface of Venus was cool enough so that vast oceans of liquid water rich in dissolved carbonic acid could be present on the surface. The failure of the Urey reactions to turn all the CO_2 into carbonates would then have to be due to the absence of weathering, whereby reactive minerals would normally become exposed to the atmosphere and hydrosphere. This can be effected by hiding the crust beneath a deep global ocean.

A crucial new piece of evidence which was to constrain such theories very strongly was brought forward by the observations of Mayer *et al.* (1958), who found a brightness temperature of 600 K near a wavelength of 3 cm. The simplest interpretation was that this was thermal emission from the surface of Venus, transmitted largely unattenuated through the atmosphere. Numerous later observations have provided nearly complete wavelength coverage from about 2 mm to beyond 21 cm, as shown in Figure 4.1.

In response to these new observations, Öpik (1961) developed a model in which a global, parched desert at about 600 K is scoured by strong winds. Frictional dissipation of wind energy is assumed to provide the source of heat to keep the surface hot. The motions in the lower atmosphere could be forced from above by motions resulting from temperature gradients at higher altitudes. This was termed by Öpik the "aeolosphere model" after the crucial role of winds in the energetics.

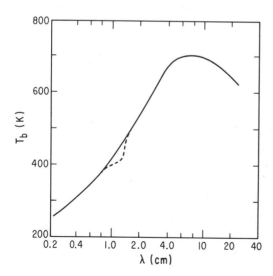

FIGURE 4.1. **Microwave brightness temperatures of Venus. A CO_2–H_2O–N_2 atmosphere is nearly transparent at wavelengths beyond about 4 cm, implying a surface temperature of 700 K or a little higher. Water vapor has a rotational transition at 1.35 cm. The dashed line indicates the approximate shape of the brightness curve that would be expected if the atmosphere contained 0.5% water vapor. Note that no H_2O absorption feature can in fact be seen in the data. (After Pollack and Morrison, 1970.)**

Simultaneously a number of suggestions were made of possible sources of microwave radiation that would not require a hot surface. These sources, including electrical discharges between cloud droplets, ionospheric emission, and cyclotron or synchrotron emission from dense radiation belts, were copiously proposed and energetically debated for several years (Jones, 1961; Tolbert and Straiton, 1962; Danilov and Yatsenko, 1963; Scarf, 1963; Danilov, 1964; Vakhnin and Lebedinskii, 1965; Applebaum et al., 1966; Drake, 1967). As the radio brightness temperature data accumulated, one after another of these hypotheses was found to be incapable of explaining the wavelength dependence of the observed emission (Pollack and Sagan, 1967b).

Coincident with all this activity, Sagan (1962) was reviving the greenhouse theory of Wildt (1940a). Opacities calculated for CO_2-rich gases containing minor amounts of water vapor were found to provide a very satisfactory fit to the observed emission spectrum, in agreement with a surface temperature at or above 700 K and a surface pressure on the order of 100 bar.

Since the solar carbon abundance is about half the oxygen abundance, exhaustive hydrogen loss from a captured solar-composition protoatmosphere would leave a residuum dominantly composed of CO_2. Cameron (1963b) and Suess (1964) proposed that the present atmosphere of Venus was the residue from just such a process, and predicted that large and roughly equal amounts of both nitrogen and neon would be present as well (see Section 3.1).

Holland (1964) favored the idea that Venus accreted from hot, dry planetesimals, and never had oceans of water.

The last attempts to dismiss the microwave emission as evidence of high surface temperatures were presented by Newell (1967) and by Libby (1968a,b). Libby engaged in a fascinating debate with Weertman (1968) regarding whether the massive (10 km thick) Venus ice caps reached only down to latitude 60° or all the way to 45°. In either case, the absence of liquid water, and hence the absence of a solvent and catalyst for the Urey reactions, and hence the dominance of atmospheric CO_2, was secured by the simple expedient of freezing all the water. Dauvillier (1968), on the other hand, postulated a supercritical fluid "ocean" of salt water rich in dissolved silicates.

A complete revolution in our knowledge of Venus has occurred since the time of these speculations, brought about by both the development of extremely high-resolution interferometric IR spectroscopy, Earth-based radar mapping of the inner planets, and the initiation of planetary probe missions. From the first attempts by the Soviet Union in 1961 to launch scientific spacecraft to Venus until the first successful entry mission in 1967, it was clear that a spacecraft would provide an enormously increased understanding of the atmosphere and surface, providing in many cases information which could not possibly be gathered from Earth-based studies. A brief chronological summary of the publicly announced spacecraft missions intended for Venus is given in Table 4.1. Numerous reviews of the principal spacecraft missions and experiments are available. For the first successful flyby mission, Mariner 2, the most important experiment was the microwave radiometer, which clearly confirmed that the high brightness temperature was due to emission

TABLE 4.1
Announced Spacecraft Missions to Venus

Launch window[a]	Spacecraft	Arrival at Venus
Jan. 1961	Venera 1	Early failure
Aug. 1962	Mariner 2	Flyby Jan. 1963
Apr. 1964	Zond 1	Early failure
Nov. 1965	Venera 2	Failed before entry
Nov. 1965	Venera 3	Failed before entry
Jun. 1967	Venera 4	Entry 18 Oct. 1967
Jun. 1967	Mariner 5	Flyby 19 Oct. 1967
Jan. 1969	Venera 5	Entry 16 May 1969
Jan. 1969	Venera 6	Entry 17 May 1969
Aug. 1970	Venera 7	Landed 15 Dec. 1970
Apr. 1972	Venera 8	Landed 22 Jul. 1972
Nov. 1973	Mariner 10	Flyby 5 Feb. 1974
Jun. 1975	Venera 9/bus	Landed 22 Oct. 1975
Jun. 1975	Venera 10/bus	Landed 25 Oct. 1975
Jan. 1977	No launches	— —
Aug. 1978	Venera 11/bus	Landed 25 Dec. 1978
Aug. 1978	Venera 12/bus	Landed 21 Dec. 1978
Aug. 1978	Pioneer Venus orbiter	Orbited 8 Dec. 1978
Aug. 1978	Pioneer Venus probes	Landed 8 Dec. 1978
Apr. 1980	No launches	— —
Nov. 1981	Venera 13	Landed 1 Mar. 1982
Nov. 1981	Venera 14	Landed 5 Mar. 1982
Jun. 1983		
Jan. 1985[b]		

[a]Launch windows are roughly 1 month long for most opportunities and open in the month given.

[b]Note that, due to the fact that 13 sidereal Venus years are almost exactly equal to 8 sidereal Earth years (to within 1 day), launch windows repeat every 8 yr. Note also that, since 1972, the Soviet Union has used *alternate* launch windows.

from the surface or deep atmosphere, not from the upper atmosphere or magnetosphere (Bareth *et al.*, 1963). The simultaneous Venera 4 entry probe and Mariner 5 flyby missions are summarized by Vakhnin (1968), by Hunten and Goody (1969), by Avduevskii *et al.* (1968), and by Snyder (1967), respectively. The Venera 5 and 6 missions, using spacecraft virtually identical to Venera 4, also entered the Venus atmosphere successfully, but were, like Venera 4, crushed by the atmospheric pressure while still high above the surface. The chemical composition and atmospheric structure data from Venera 5 and 6 are summarized by Avduevskii *et al.* (1970a,b,c). The improved spacecraft used for the 1970 Venera 7 and the 1972 Venera 8 missions had sufficient strength and thermal insulation to survive landing on the surface. Venera 7 succeeded only in measuring the temperature profile with very coarse digitization (Avduevskii *et al.*, 1971; Marov *et al.*, 1971), but confirmed the high surface temperature directly and the high pressure indirectly (by means of Doppler tracking of the decelerating probe as it fell through the atmosphere).

Venera 8 also returned photometric data on the penetration of sunlight into the atmosphere and carried out new chemical analyses of trace surface and atmospheric constituents (Avduevskii *et al.*, 1973; Vinogradov *et al.*, 1973; Surkov *et al.*, 1973a). The Mariner 10 Venus–Mercury flyby was designed principally to study Mercury, but was able to carry out photographic and both IR and UV spectroscopic measurements on Venus and to examine its particles and fields environment as well (Dunne, 1974a). The Venera 9 and 10 photographic lander missions of 1975 (Keldysh, 1977) also carried an impressively wide complement of other instruments, including mass spectrometers and gamma spectrometers to analyze the atmosphere and the radioactive constituents of the crust, respectively.

In a period of several days in December of 1978 Venus was beset by a horde of terrestrial spacecraft. Six atmospheric entry probes (four American and two Soviet) penetrated to the surface, while three satellites were placed in orbit around Venus (two Soviet and one American). These were the Venera 11 and 12 spacecraft, a Pioneer Venus orbiter, and the Pioneer Venus multiprobe package. The last-named satellite comprised one large and heavily instrumented deep entry probe, three small probes to sound the atmospheric structure at widely separated points on the planet, and the spacecraft bus, devoid of a heat shield, which was designed to carry out upper-atmosphere measurements prior to burning up during entry. The Venera 11 and 12 data are still under intensive study, but a number of very interesting reports have been published. For a general overview of Venera 11 and 12, see Barsukov *et al.* (1979) and Kurt and Zhegulev (1979). For the Pioneer Venus mission, there are reviews by Colin (1979a,b) and by Donahue (1979). A preliminary overview of the recent Venera 13 and 14 missions is also available (Sagdeev and Moroz, 1982).

Although there have been numerous scientific reviews of Venus, few are very general and even fewer are recent. Sagan (1961) and Holland (1964) each provided early selective reviews of certain aspects of the study of Venus, and Briggs (1963) presented a short general review of a semipopular nature. Roberts (1963), in his review of radio emission from the planets, pays considerable attention to the Venus microwave data and its interpretation. A general review of Venus on the eve of the Venera 4 and Mariner 5 encounters by Bronshtén (1967), and a brief conference review by Johnson (1968) may also be consulted for a perspective on the state of our knowledge at that point in history. A number of the most interesting aspects of the study of Venus are exposed in the brief review by Hunten and Goody (1969). Venus is discussed in a review on the origin of planetary atmospheres by F. S. Johnson (1969), while Kuz'min (1978) provides a brief review of the atmosphere, and Vetukhnovskaya and Kuz'min (1970) give a more general review of the state of our knowledge of Venus, but with emphasis on the radiophysics of the planet. A major review of the composition and structure of planetary atmospheres by Hunten (1971) touches only lightly on upper atmospheres, photochemistry, geochemistry, and evolutionary considerations. A semipopular review by Lewis (1971d) and a review of high-resolution spectroscopic studies of Venus by Young (1972) are available.

The best early review of Venus is by Marov (1972), who considered the state of our knowledge after the Venera 4–7 and Mariner 2 and 5 missions. This was updated in part by a brief conference review by Hunten (1975). The review on Venus by Marov (1978) provides an excellent summary up to the time of the Venera 11 and 12 and Pioneer Venus missions. Relevant reviews following these latter missions have been provided by Prinn (1980), by Moroz (1981), and in particular, by a series of papers from a recent international conference on Venus contained in a book edited by Hunten and Colin (1982).

With this broad historical perspective we may now begin an analysis of results of recent experimental and theoretical investigations of Venus. We shall begin with the mean static thermal structure and composition of the atmosphere, the effect of atmosphere–surface interactions in regulating the composition of the atmosphere, and the composition and morphology of the surface. We then will consider the structure and composition of the upper atmosphere with its major photochemical and escape processes, and the composition and structure of the clouds. We conclude with an assessment of competing theories for the origin and evolution of the atmosphere of Venus, including several recently proposed ideas that address the discoveries of the most recent generations of Venus entry probes.

4.2.1. Pressure and Temperature Structure

Our understanding of the thermal structure of the atmosphere of Venus is based on three principal types of evidence: radiometry, radio occultation measurements, and *in situ* probe studies. We begin with the Earth-based IR and microwave measurements of the dependence of the emitted flux on latitude, longitude (solar phase angle), wavelength, and time. At shorter IR wavelengths the radiation is very severely attenuated by the clouds and by molecular absorption, so that the flux observed is that appropriate to thermal emission from the cloud tops. At longer wavelengths, beyond the Planck peak for thermal emission from the surface (4 μm for a 750 K blackbody), absorption and cloud particle scattering remain very important out to about 1 mm. We have seen in Figure 4.1 that the atmospheric opacity is very small beyond about 4 cm, permitting almost unattenuated observation of the emission from the surface. The wavelength dependence of brightness temperature may be analyzed theoretically to determine the temperature and pressure profiles through the atmosphere. In addition, such models can constrain the abundances and vertical distribution of gases such as H_2O and NH_3, which are strong microwave absorbers.

Early attempts to measure the temperature in the cloud layer by IR photometry (Murray *et al.*, 1963) and by interpretation of the line strengths in the CO_2 near-IR spectrum (Spinrad, 1962a) developed in parallel with the earliest microwave observations (Mayer *et al.*, 1958) and their interpretation. Barrett's (1960) interpretation of the microwave emission suggested a surface temperature of at least 580 K at a pressure of about 1.5 bar, with an atmospheric composition of about 25% N_2 and 75% CO_2.

The measured IR fluxes generally suggest a temperature in the vicinity of 240 K

(Gillett et al., 1968; Hanel et al., 1968), in good agreement with the gray-body steady-state temperature which would be maintained by absorption of sunlight with a planetary albedo of 0.77 ± 0.07 (Irvine, 1968). The brightness temperature estimated from Mariner 10 IR radiometry is higher, about 255 K, but is based on observations at 45 μm, well beyond the Planck peak of about 12 μm (Chase et al., 1974). The region of the Planck maximum was well covered by the Venera 9 and 10 IR radiometer experiment (Ksanformaliti et al., 1976a,b), and, subject to calibration uncertainties, seems concordant with a lower temperature. Wright (1976) has studied the calibration of thermal IR measurements and suggests a best estimate of 240 ± 8 K for Venus. Diner et al. (1976) have documented spatial variations in the thermal emission in the 8–14 μm terrestrial atmospheric transmission window, including several regions showing temperature departure of 1–3° from the planetary average. The dayside cloud layer was found to average 2 K warmer than the night side. This requires that the cloud tops be quite flat and well defined over the planet. Cess (1972) concluded that the tropopause is essentially at constant altitude irrespective of latitude, in contradiction with the claim by Moroz (1971) that the tropopause in the polar regions was 5–7 km lower than at low latitudes.

There exists a very substantial literature on the observation and interpretation of the microwave emission, and we shall only cite a few of the more important estimates of the surface temperature derived from these studies. Pollack and Sagan (1967a), based on the Mariner 2 microwave radiometer observations, placed an upper limit of 150 bar on the surface pressure and an upper limit of 640 gm cm^{-2} for the total water content of the atmosphere. Wood et al. (1968) used radio emission and radar data on Venus to estimate a surface pressure of 90 bar and a temperature of 750 K. Closely similar results were found by Slade and Shapiro (1970), Sagan and Pollack (1969), Pollack and Morrison (1970), Hall and Branson (1971), and Vetukhnovskaya et al. (1971).

One of the more puzzling problems in the history of the study of Venus was raised by the in situ measurements of temperature and pressure by the Soviet Venera 4, 5, and 6 entry probes, in conjunction with the results from the radio occultation experiment carried out by Mariner 5. The radio occultation experiment (Kliore et al., 1968) measured the attenuation and phase shift of the radio waves transmitted through the atmosphere as Mariner passed behind Venus. The refractive index profile can be extracted, and thence the scale height. Knowledge of either the molecular weight or the local temperature permits calculation of the other. In this way, a reasonable compositional assumption (mostly CO_2) can be used to generate a detailed and reproducible vertical temperature profile (Fjeldbo et al., 1971). The spacecraft was observed during entry into occultation and again at exit from occultation, so the density of the atmosphere was obtained in two locations on opposite sides of the planet. Tracking of the spacecraft provided an accurate determination of the location of the center of mass of Venus, and thus the absolute distances of the surfaces of known density from the barycenter of the planet could be calculated.

In addition, Earth-based radar tracking of Venus had provided various estimates of the radius of the solid body of the planet. Representative estimates at the time included 6049.5 ± 3 km from Pollack and Morrison (1970), 6052.5 km from Anderson et al. (1968), and 6054.8 km from Gale et al. (1969), all of which compare rather well with the most recent consensus of 6051 km from a number of sources.

The problem arose because the Venera 4 spacecraft (and later Venera 5 and 6) were reported to have reached the surface and found a temperature near 540 K at a pressure of about 12 bar (Reese and Swan, 1967; Avduevskii et al., 1968, 1969; Mikhnevich and Sokolov, 1969). The problem with this claim was that, according to the best estimates from the microwave emission, and also according to the Mariner 5 occultation results combined with the radar radius of Venus, the point at which the Venera "found" the surface was at an altitude of approximately 26 km above the true surface. The Venera data were severely criticized by Ash et al. (1968), Kliore and Cain (1968), Eshleman et al. (1968), and Smith (1970), all of whom concluded that the radius was close to 6050 km, and that the surface temperature and pressure were 750 K and 100 bar.

Faced with the apparent failure of the radar altimeter on these three Venera craft, Soviet interpretation was changed. The observed pressure and temperature profiles were extrapolated down to the radar surface, and the cessation of transmission of Venera 4, 5, and 6 at 540–600 K was attributed to catastrophic failure of their pressure vessels, not collision with the surface. The revised Soviet models (Avduevskii et al., 1970a,b) thus came eventually into agreement with the other lines of evidence. The landing of the Venera 7 probe on 15 December 1970 verified the high surface temperature, even though the only instrument which sent back data was a crude temperature gauge digitized in 20° increments. The surface temperature measured directly by Venera 7 was 747 ± 20 K, and the surface pressure estimated from the Doppler record of the speed of fall of the capsule was 90 ± 15 bar (Marov et al., 1971; Avduevskii et al., 1971). The very successful Venera 8 spacecraft found 741 ± 7 K and 93 ± 1.5 bar (Marov et al., 1973a). A composite pressure–temperature profile from the Venera and Mariner data is given in Figure 4.2.

The major features of atmospheric structure delineated by these missions, the very high surface temperature and pressure, were thus in very good agreement with the values deduced from Earth-based studies of the radio-frequency thermal emission of the Venus surface. The atmospheric pressure–temperature profile was found to be quite close to a dry adiabat for pure CO_2 (see, e.g., Staley, 1970), but some small and possibly significant deviations are seen. Most notable is an apparently subadiabatic region between the ≈375 and ≈400 K levels (Fjeldbo et al., 1971). This region correlated with a layer of abnormally high density as determined by the β-ray transmission densitometers on the Venera probes (Avduevskii et al., 1968, 1970a,c). The simplest interpretation seemed to be that this feature is a condensate cloud layer (Lewis, 1969c; Rasool, 1970), or possibly a localized dust layer (Marov, 1972). In the latter case, however, the apparent absence of dust below that level

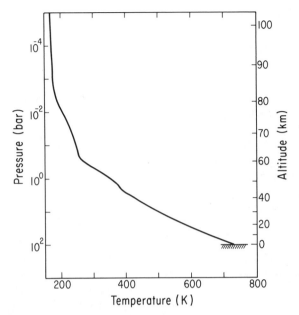

FIGURE 4.2. **Pressure–temperature profile for the atmosphere of Venus. For cloud models, see Fig. 4.4. Note the offset in temperature near 350 K and the extensive subadiabatic layer above the 240-K cloud tops. The temperature begins to rise again above 100 km on the day side, but falls on the night side (see text).**

(Avduevskii *et al.*, 1968) and the abrupt change in temperature gradient (Fjeldbo *et al.*, 1971) in that region are both unexplained.

The optical properties of the lower atmosphere were investigated by a photometer experiment aboard the Venera 8 lander (Avduevskii *et al.*, 1973). Interpretation of these data by Lacis and Hansen (1974) and by Lacis (1975) located the level of maximum cloud density near the 40 km level, with a sharp cloud base near the 30 km level (10 bar), a clear region between 10 and 30 km, and a moderately dense ground haze of about 10 1-μm particles per cubic centimeter below ~13 km (40 bar, 675 K). The identity of the lower layer of particulate material is unknown, but a number of geochemically plausible materials with appropriate vapor pressures and abundances are known (Lewis, 1969c). It is a serious problem to determine which of these moderately volatile elements, if any, might in fact be present on Venus. Any assertions about specific elements, such as arsenic, mercury, and antimony, would be conditioned by our preference of cosmogonic models for Venus: A refractory Venus might contain none of these materials, whereas an Earth-like Venus would have enough of them to put easily observable amounts of opacity in the lower atmosphere (Lewis and Fegley, 1982).

Mass spectrometric data on the composition of the lower atmosphere are of crucial importance, and there is little likelihood of understanding the origin and geochemical history of Venus without use of such data. We shall devote Section 4.2.7 to a review of the present state of our knowledge.

There are a number of interesting possibilities suggested by the high surface temperature. First, there is the likelihood that the atmosphere and lithosphere closely approach the state of chemical equilibrium at low altitudes, as was first suggested by Mueller (1963). This would suggest that knowledge of the composition of the lower troposphere would permit deduction of some features of the geochemistry and petrology of the surface rocks. Second, it is likely that an entire class of moderately volatile elements, which on Earth are released from volcanoes and fumaroles and condense as minerals, would on Venus be incapable of condensing, and might therefore reside largely in the atmosphere. Third, the lower atmosphere, since it is dominated by very different chemical processes than the atmosphere above the cloud tops, may differ substantially in molecular composition from the portion which can be observed by IR spectroscopy from Earth. Fourth, condensation of vapors present in the lower atmosphere, or photochemical production of condensates, could provide important sources of opacity in the atmosphere at both visible and IR wavelengths. This opacity is, of course, of great importance to the greenhouse effect. There is thus a curious bootstrapping effect: The high surface temperatures are due to a strong greenhouse effect in a dense, chemically complex atmosphere which is present because of the high surface temperatures. It is immediately obvious that the stability of this system and the evolutionary path by which it was reached are both questions of very fundamental importance.

It is clearly of great importance to our investigation to have detailed knowledge of the composition of the atmosphere as a function of altitude. Spectroscopic studies from Earth cannot penetrate the clouds, and sample to different depths depending on the strengths of the absorption bands being examined, and also depending somewhat on wavelength. The spectroscopic line-forming region in and above the main clouds is complex, and a very good optical and structural model of this region is needed in order to interpret the IR absorption spectrum and deduce the relative abundances of the spectroscopically detectable species which are present.

Attempts to deduce the pressure and temperature at the Venus cloud tops go back over 50 years. A review of the earlier literature by Chamberlain (1965) covers progress up to the time of the first spacecraft missions, just before the first successful application of very high-resolution interferometric spectrometric techniques. A large number of papers have interpreted the relative populations of the rotational levels in CO_2 bands in the near-IR in terms of the temperature of an equivalent isothermal, isobaric atmosphere that best simulates the observed line strengths. Chamberlain and Kuiper (1956) found 285 ± 9 K, and Kaplan (1961) reported 235 K with a cloud layer pressure close to 90 mbar. Kaplan found that the pressure broadening of the CO_2 lines seemed to require a total gas pressure several times larger than could be provided by the observed column abundance of CO_2, and estimated that the mole fraction of CO_2 was about 0.15. Spinrad (1962a) reported an analysis of some spectra originally taken by Adams and Dunham 30 years earlier, and found effective pressures ranging up to 18 bar and rotational temperatures to over 400 K. Attempts were made by Sagan (1962) and by Lewis (1968a) to produce atmospheric models which could be reconciled with Spinrad's P, T points. Lewis pointed out that no physically possible atmospheric model could be fitted to Spinrad's data, and that the closest approximation could be attained only for atmospheres improbably rich in rare gases. At the same time, more modern spectra were being analyzed by Gray and Schorn (1968) and by Belton (1968), who estimated a gas temperature of 200–250 and 240–270 K, respectively. A lengthy series of studies of individual weak CO_2 bands by L. D. Gray Young, R. A. Schorn, E. S. Barker, and co-workers (Gray et al., 1969; Schorn et al., 1969b, 1970, 1971, 1975; Young, 1969a, 1970; Young et al., 1970a,b, 1977) and a similar study of extremely high-resolution interferometric spectra by Dierenfeldt et al. (1977) have shown a temperature of 241 ± 3 K at the cloud tops. Dierenfeldt et al. estimate an effective pressure of about 120 ± 60 mbar at the same level. Measurements of the deceleration of the Venera 8 spacecraft while passing through the stratosphere (Cheremukhina et al., 1974) found a mean stratospheric temperature of 210–217 K above the cloud tops. Similar deceleration measurements for the Pioneer Venus probes (Seiff et al., 1979a,b) found a temperature of 240 K to lie between the 100 and 200 mbar levels for all four probes. The entire atmosphere above the 400 mbar level was found to be stable against convective overturn. The modest horizontal variability in the cloud level found by Seiff et al. (1979b) is consistent with the observed variations in the CO_2 abundance above the clouds, which fluctuates with a 4-day period (Hunt and Schorn, 1971; L. G. Young et al., 1973; A. T. Young, 1975).

That the region of the cloud tops partakes of rapid motion about the planet, with a mean speed of about 100 m sec^{-1} (the famous "four-day retrograde rotation"), has been well established from Earth-based UV observations of the clouds (see, e.g., Boyer and Guérin, 1969). An *irregular* ~4-day variation in the CO_2 abundance above the cloud tops has been reported by L. D. G. Young et al. (1975). Scott and Reese (1972) find a range of circulation speeds from 87 to 127 m sec^{-1}, which is confirmed by Boyer (1973) and by Caldwell (1972). Marov et al. (1973b) showed that the horizontal component of wind measured *in situ* during the descent of the Venera 8 entry probe decreased from about 100 m sec^{-1} in the upper troposphere where the pressure was under 1 bar to very low speeds at the surface. Independent analyses of the Venera 8 wind and thermal structure data by Andreev et al. (1974) and by Ainsworth and Herman (1975) confirm this general picture, and the zonal winds deduced from the *in situ* Pioneer Venus pressure profiles by Seiff et al. (1979b) indicate similar wind speeds at the same altitude. The most accurate and detailed *in situ* measurements of horizontal wind velocities on Venus were obtained through radio interferometric tracking of the Pioneer Venus probes (Counselman et al., 1979a,b, 1980). Zonal wind speeds are ≤1 m sec^{-1} near the surface rising to values ~100 m sec^{-1} above 60-km altitude. Meridional wind speeds are generally less than a few meters per second; maximum values are seen at ~50-km altitude where equatorward motions at ~5 m sec^{-1} are evident.

The Mariner 10 imaging experiment provided high-resolution coverage of the cloud motions, again confirming the same general picture (Murray *et al.*, 1974a). Doppler IR measurements of wind speeds were carried out by Traub and Carleton (1975). A. T. Young had claimed that the 4-day circulation is "illusory" (Young, 1975), a conclusion contradicted by the earlier direct observations by Venera 8, and refuted by Ainsworth and Herman (1977). All the available data on Doppler tracking of Soviet entry probes require winds of at least 40–50 msec^{-1} at and above 50 km (Kerzhanovich, 1972; Kerzhanovich *et al.*, 1972; Antsibor *et al.*, 1975; Avduevskii *et al.*, 1976a; Kerzhanovich and Marov, 1977). An interesting dynamical model of the 4-day circulation has been presented by Schubert and R. E. Young (1970), by Schubert *et al.* (1971), and by Young and Schubert (1973). A critical review of this and other proposed mechanisms has been written by Stone (1975).

Young and Pollack (1977) reported the results of a three-dimensional global circulation model in which they had simulated several observed features of the upper atmosphere of Venus, including the 4-day circulation and the Y-shaped pattern seen in UV photographs. Several assumptions in this model have, however, recently been strongly criticized (Rossow *et al.*, 1980; see also reply by Young and Pollack, 1980).

The static thermal structure of the atmosphere of Venus is thus quite well established. The very low temperatures in the upper troposphere and stratosphere serve to condense moderately volatile constituents out of the gas phase, and thus, to limit the species which may be spectroscopically detected from Earth. Thus, our attempts to understand the composition of the lower atmosphere must rest not only on remote-sensing data, but also on *in situ* entry probe data.

The dynamical structure of the atmosphere of Venus is less certain than the static structure but, nevertheless, a self-consistent picture of the general circulation of the atmosphere is emerging. A detailed review of the status of our knowledge after the Pioneer Venus mission is provided by Schubert *et al.* (1980). There is ample evidence for a thermally direct equator-to-pole Hadley cell in the 50–70 km altitude region. More tentative are frictionally driven thermally indirect cells immediately above and below this latter region and a thermally direct Hadley cell near the surface (0–30 km). Strong turbulence exists in thin layers at 45- and 60-km altitudes. At the cloud tops planetary-scale waves are evident.

4.2.2. Atmospheric Composition

The first species detected in the spectrum of Venus was CO_2 (Adams and Dunham, 1932). These spectra did not show any evidence of water vapor. R. Wildt (1934), in his earliest consideration of the chemistry of Venus, argued for an atmospheric composition which was surprisingly modern: "These three facts, lack of water, high carbon dioxide pressure, and temperature near the boiling point of water, make the Venus surface an uncongenial substrate for known life processes, and it is plausible under these circumstances that free oxygen be absent." The CO_2 abundance above the clouds was estimated by Adel (1937a) as 400 m agt. These early spectroscopic studies (see, e.g., Richardson, 1960) failed to detect species other than CO_2, although early claims of detection of water vapor were made (Dollfus, 1963). Spinrad and Richardson (1964) established an upper limit on oxygen, a suggested atmospheric constituent (Prokofyev and Petrova, 1963).

The development of interferometric techniques for planetary spectroscopy (Gebbie et al., 1962) led eventually, with the growth of the power of high-speed digital computers, to the very high-resolution spectra of Connes et al. (1967). The Connes spectra of Venus were found to contain a large number of weak isotopic and combination bands of CO_2. Connes et al. were able to show that the abundance of CO_2 above the cloud layer, as deduced from the strengths of individual absorption lines, was approximately capable by itself or providing the observed pressure broadening. Thus, they were able to suggest that CO_2 was the dominant atmospheric constituent. Further, they concluded that the carbon and oxygen isotopic composition on Venus were indistinguishable from the terrestrial values.

The CO_2 mole fraction has been even more precisely determined by in situ chemical analyses by the Venera 4 (Vinogradov et al., 1968a,b) and Venera 5 and 6 (Vinogradov et al., 1970a,b) entry probes. These determinations, summarized by Marov (1972), bracketed the CO_2 abundance between 93 and 100%, and the most recent data give 96–97% CO_2.

Water vapor, unsuccessfully sought by Spinrad (1962b), was first detected in high-altitude balloon-borne spectroscopy of Venus by Bottema et al. (1965a,b) and was confirmed by Belton and Hunten (1966) and by Spinrad and Shawl (1966). A criticism of this work by Owen (1967) was answered by Hunten et al. (1967). Schorn et al. (1969a) have reported a large number of water vapor observations, showing evidence for substantial variability in the abundance of water vapor above the clouds. The H_2O mole fraction is seen to rise at least as high as 5×10^{-5} and possibly 1×10^{-4}. Fink et al. (1972) and Barker (1975) have found X_{H_2O} to be close to 10^{-6} on a number of occasions, which is compatible with the upper limits on H_2O established by Cruikshank (1967), Owen (1968), and Gull et al. (1974).

The water vapor abundance below the level of the visible clouds is subject to even more uncertainty.

One of the techniques for determining the total water content of the atmosphere down to the surface is capable of execution from Earth, without atmospheric entry probes. This technique takes advantage of the fact that the wavelength dependence of the microwave opacities of CO_2 and H_2O are different, so that the wavelength dependence of the thermal emission or the opacity determined from the radar cross section (Evans and Ingalls, 1968) can be used to constrain the water vapor mole fraction. There is a strong rotational transition in the water vapor spectrum at 1.35-cm wavelength which has seemed to some observers (see Strelkov, 1968) to be discernable in the thermal emission spectrum. Theoretical spectra by Naumov and Strelkov (1970) show that water vapor at low pressures can be distinguished from water deeper in the troposphere because of the strong pressure broadening of the transition. See Figure 4.1 for the observed emission from Venus.

Radio interferometric brightness temperature measurements by Sinclair et al.

(1972), which first showed that the temperature of the poles of Venus was indistinguishable from the equatorial temperature, estimated a water vapor content of about 0.65% for the lower atmosphere. Measurements with higher spectral resolution but without spatial resolution by Jones et al. (1972) and by Janssen et al. (1973) both failed to detect any evidence of water, and both established an upper limit of 0.1–0.2% H_2O. Smirnova and Kuz'min (1974) reached the same conclusion on the basis of their analysis of Soviet microwave observations. De Bergh (1973) showed that no species other than CO_2 and H_2O were needed to fit the microwave brightness temperature data. A theoretical treatment of the loss of thermal IR radiation from the lower atmosphere (Shari, 1976) found that a mole fraction of 5×10^{-5} for H_2O would be capable of producing the observed greenhouse effect.

These inferences from remote-sensing data flatly contradicted the in situ measurements carried out by the extremely crude analytical devices on Venera 4, 5, and 6 (Moroz and Kurt, 1968; Vinogradov et al., 1968a,b; Mikhnevich et al., 1976). The Venera probe measurements referred to levels where the pressure was at least 600 mbar and the temperature at least 300 K. Amounts of water vapor ranging up to a mole fraction of 1–3% were claimed, with the reported vapor abundances decreasing downward into the atmosphere. Lewis (1969c) suggested that the method of detection of water by the Venera probes would be rendered very unreliable by the presence of HCl, and hence that the reported abundances of water may be illusory. Marov (1972), although accepting the likelihood of dissolved HCl, rejected the importance of the effect, apparently on the sole grounds that the abundance of HCl would have to be higher lower down in the clouds than at the cloud tops. This, however, is precisely what would be expected: The mole fraction of HCl would be larger prior to its condensation as a solute. We now know that the Venus clouds are dominantly composed of concentrated sulfuric acid rather than hydrochloric acid, but the detection of water by these early Veneras is made equally unreliable by the presence of either acid.

More recent spacecraft data on the water vapor content of the lower atmosphere have been contributed by Venera and 11–14, and by the Pioneer Venus large probe. A narrow-band photometer carried by Venera 9 measured the opacity of the atmosphere in a water absorption band (Ustinov et al., 1979) and found a mole fraction of water of 3×10^{-4}. Istomin et al. (1979) report only that the Venera 11 and 12 mass spectrometers (described by Grechnev et al., 1979) found a "trace" of water vapor, whereas the multichannel solar radiometer experiment on Venera 11 and 12 (Golovin et al., 1981) found a mole fraction of water vapor of only 2×10^{-5} at altitudes below 15 km (Moroz et al., 1979) with an abundance no higher than 10^{-4} up to an altitude of 42 km (Gel'man et al., 1979b). The IR radiometric data given by Boese et al. (1979) from Pioneer Venus measurements suggest a mixing ratio near 10^{-4} below 47 km. The Pioneer Venus gas chromatograph (Oyama et al., 1979a) reported a mixing ratio of $(1.35 \pm 0.015) \times 10^{-3}$. The preliminary results from the Venera 13 and 14 gas chromatograph experiments (Mukhin et al., 1982) give a water vapor mole fraction of $(7 \pm 3) \times 10^{-4}$ below 58 km. The visible–near-IR spectrophotometer experiments on Venera 13 and 14 find about 1 or 2×10^{-4} between 40 and 60 km, with the amount of water vapor less by a factor of 10 near the surface (Moroz *et al.*, 1982). Venera 13 and 14 also carried a dedicated device for measurement of the relative humidity of the atmosphere, using a variable-temperature LiCl absorber. This experiment indicates a water vapor mole fraction of about 2×10^{-3} in the 46–50 km altitude range (Surkov *et al.*, 1982). The ability of this device to function reliably in the presence of concentrated sulfuric acid, rather than pure water, remains to be established.

CO was identified by Connes *et al.* (1968) in their high-resolution Venus spectra. Belton (1968) has reanalyzed the Connes data in terms of a more detailed and realistic cloud scattering model and finds the CO mixing ratio ($X_{CO}/X_{CO_2} \simeq X_{CO}$) to be $(2 \pm 1) \times 10^{-4}$. The Venus airglow contains detectable emission from CO (Slysh, 1976), and CO has been detected in the microwave region due to emission in a line at 115 GHz (Kakar *et al.*, 1976). The results of the Pioneer Venus gas chromatograph experiment (Oyama *et al.*, 1979a,b) placed an extremely restrictive and altogether improbable upper limit of 0.6 ppm of CO at 24-km altitude, simultaneous with a reported observation of an incredibly large amount of oxygen. It was later announced that the peaks in the gas chromatogram were misidentified, and that oxygen is absent and CO present (Oyama *et al.*, 1980a,b). The Venera 12 gas chromatography experiment found a CO mole fraction of $(2.8 \pm 1.4) \times 10^{-5}$, several times lower than the spectroscopic data for the cloud-top atmosphere (Gel'man *et al.*, 1979b).

In the most recent analyses of the Pioneer Venus chromatographic data, Oyama *et al.* (1980b) tentatively attribute a partially unresolved peak to O_2. Other than this there is no convincing evidence for the presence of oxygen or its photochemical by-product ozone. O_3, on Venus. Spectroscopic searches beginning with those of Beckman (1967) and of Belton *et al.* (1968b) and culminating with the work of Barker (1975) and of Traub and Carleton (1974) agree that there must be less than 1 ppm of O_2 above the clouds. Ultraviolet observations, which are extremely sensitive to tiny traces of ozone, fail to find any at the level of a mole fraction of 3×10^{-9} (Jenkins *et al.*, 1969; Owen and Sagan, 1972). Venera 11 and 12 data (Gel'man *et al.*, 1979b; Moroz *et al.*, 1979) have not improved on these limits. As we shall see in the following pages, the oxygen partial pressure deduced from the thermochemistry of other atmospheric gases and from interaction of gases with surface rocks is many orders of magnitude below these limits. However, since oxygen is produced by UV photolysis of both CO_2 and H_2O, it appears that some reducing agent capable of rapid reaction with O_2 must be available within or below the clouds.

In addition to CO, the Connes spectra also provided detection of significant traces of two highly reactive compounds, HCl and HF (Connes *et al.*, 1967). Belton's (1968) estimates of the most probable mixing ratios are 10^{-6} for HCl and 2×10^{-8} for HF. The probable errors of these determinations are about a factor of two. No direct *in situ* detection of either species has been reported to date.

No organic molecule has ever been detected, despite the ability of high-resolution IR spectroscopy to detect even 1-ppm traces of many hydrocarbons (Connes *et al.*, 1968; Plummer, 1969a).

In situ analyses by Venera 6 placed an approximate upper limit of 2% on the abundance of molecular nitrogen (Vinogradov *et al.*, 1970a,b). The Venera 9 and 10 mass spectrometer data gave a positive detection of N_2 with an estimated abundance of $1.80 \pm 0.35\%$ in the 32 to 47 km altitude range (Surkov, 1977). The preliminary mass spectrometer data from Venera 11 and 12 (Istomin *et al.*, 1979) suggested a N_2 concentration of $4.50 \pm 0.5\%$, whereas the Venera gas chromatograph reported $2.50 \pm 0.3\%$ (Gel'man *et al.*, 1979b). The most recent report from the Pioneer Venus large probe mass spectrometer team quotes a N_2 abundance of $4 \pm 2\%$ (Hoffman *et al.*, 1980). Finally, from the Pioneer Venus gas chromatographic data Oyama *et al.* (1979a) deduce $4.60 \pm 0.09\%$ at an altitude of 54 km, $3.54 \pm 0.03\%$ at 44 km, and $3.41 \pm 0.20\%$ at 24 km, for a 40% variation with quoted error limits of about 1%. These latter puzzling differences with altitude, if real, are as yet unexplained.

Only two other nitrogen compounds have even been claimed to be present in the Venus atmosphere. The first was the claimed detection of ammonia with a mole fraction of 10^{-4}–10^{-3} by a crude and nondiagnostic technique, the change in color of Bromophenol Blue indicator (Surkov *et al.*, 1973b). Analysis of the microwave spectrum of Venus by de Bergh (1973) and by Smirnova and Kuz'min (1974) failed to find any evidence of the strong inversion line of NH_3 near 1 cm, permitting establishment of an upper limit of only 16 ppm on the ammonia abundance. Goettel and Lewis (1974) strongly criticized the NH_3 detection on the grounds of the incompatibility of the claimed amount of ammonia with the observed abundances of other species, especially HCl, which would have been wholly removed from the cloud-top atmosphere by NH_4Cl precipitation if the NH_3 abundance were even 1 ppm.

The second suggested constituent of the atmosphere of Venus which has not been observed is NO. Based on the presumption that lightning is common on Venus (Tolbert and Straiton, 1962; Meinel and Hoxie, 1962; Bottema *et al.*, 1965b; Applebaum *et al.*, 1966; Newell, 1967), it seemed that lightning-induced chemistry may be important on Venus. Indeed, this was first suggested by Dauvillier (1956). Scarf *et al.* (1980) have interpreted the detection of whistler mode signals by the Pioneer Venus orbiter as evidence of lightning. A specific lightning-detection experiment was flown during the Venera 11 and 12 missions (Ksanfomaliti, 1979), and a surprisingly large amount of electrical activity was detected (Ksanfomaliti, 1980; Krasnopol'skii, 1980). Assuming that the detected signals were in fact due to lightning and not to very local spacecraft-related activity, Bar-Nun (1980) calculated the chemical consequences of shock heating on a CO_2–N_2 atmosphere. He found that the most abundant distinctive product of high shock pressures would be NO. Also, Watson *et al.* (1979) have suggested $NOHSO_4$ (nitrosyl sulfuric acid) produced from lightning-synthesized NO as a component of the clouds. The evidence regarding lightning on Venus has recently been reviewed by Borucki (1982).

The abundances of sulfur compounds in the atmosphere were identified many years ago as important constraints on theories for the geochemical interaction between atmosphere and surface (Mueller, 1965). Sulfur forms stable gaseous compounds with a wide variety of oxidation states, ranging from the reduced gases H_2S and COS through S_2, S_8, and the other polyatomic sulfur species to SO_2, to the most oxidized, SO_3. Infrared observations had been unable to detect any of these sulfur-bearing molecules at about the 1 ppm level, and UV spectra taken on rocket flights (Jenkins et al., 1969) and by the Orbiting Astronomical Observatory (Owen and Sagan, 1972) were able to provide very substantial improvements in sensitivity. Upper limits were set on COS (less than 0.1 ppm), SO_2 (0.01 ppm), and H_2S (0.1 ppm). Comparable or even more restrictive upper limits could be put on CS_2 and SO_3 from the same spectra.

The first claimed direct detection of a sulfur gas on Venus was by the mass spectrometer experiment on the Venera 9 and 10 missions. This experiment found evidence for the presence of atomic S in the mass spectrum, in amounts which could be produced by fragmentation of either COS or H_2S with a concentration of 0.1% in the ion source region of the instrument (Surkov, 1977).

There is now ample evidence that sulfur compounds are present and play vital chemical and physical roles in the atmosphere of Venus. Sulfur dioxide gas, concentrated sulfuric acid droplets, and, more tentatively, hydrogen sulfide and elemental sulfur gases have now been identified on Venus with the initial discoveries being made, respectively, by Barker (1979), Young and Young (1973), Sill (1972), Hoffman et al. (1980), and Moroz et al. (1979). Chemical reactions involving sulfur species are intimately involved in the atmospheric oxygen balance and in the formation of sulfuric acid clouds with their concomitant effects on the deposition of solar and IR radiation in the atmosphere.

Sulfur dioxide was first detected above the clouds by Barker (1979) using ground-based UV observations. It was later detected in both the upper and lower atmosphere by instruments carried by the Pioneer Venus orbiter and large probe (Oyama et al., 1979a,b; Stewart et al., 1979) and by the Venera 11 and 12 probes (Gel'man et al., 1979b). Prior to the latter space missions, sulfur dioxide was first suggested as merely an intermediate in the photochemical formation of sulfuric acid from the presumed precursors COS and H_2S, and as a product of decomposition of SO_3 vapor (Prinn, 1971, 1973, 1975). Later work suggested that if vertical mixing between the surface and the clouds were sufficiently slow, then photochemically produced SO_2 and SO_3 could dominate thermochemically produced COS and H_2S throughout a significant portion of the atmosphere (Prinn, 1978). Under these circumstances SO_2 and SO_3 would be the principal precursors of the sulfuric acid clouds rather than COS or H_2S (Prinn, 1978).

Observations now show that SO_2 dominates over COS and H_2S at least above 22 km, and that the SO_2 mixing ratio has a value $\sim 1.5 \times 10^{-4}$ between 22 km and the cloud base (50 km) but decreases within the clouds to a value $\sim 10^{-7}$ at 70 km. This is consistent with SO_2 being converted to H_2SO_4 above 50 km. As discussed in Section 4.2.4, this in turn demands that SO_2 be produced from H_2SO_4 below 50 km at the same global rate.

Hydrogen sulfide has been detected by the Pioneer Venus neutral mass spectrometer (Hoffman et al., 1979b). This detection is not yet regarded as definite, but the data suggest a mixing ratio of $\sim 3 \times 10^{-6}$ in the lower atmosphere, decreasing to a value $\sim 10^{-6}$ in the visible cloud region (Hoffman *et al.*, 1980). This is consistent with H_2S being produced in the lower atmosphere and destroyed in the visible clouds. Carbonyl sulfide was not detected in the atmosphere, and the Pioneer Venus experiments suggest an upper limit of $\sim (2-3) \times 10^{-6}$ above an altitude of 22 km (Hoffman *et al.*, 1980; Oyama *et al.*, 1980a,b). No data on COS or H_2S are available for the region below 22 km.

The gas chromatograph experiment on Venera 13 and 14 (Mukhin *et al.*, 1982) has found both hydrogen sulfide [with a mole fraction of $(8 \pm 4) \times 10^{-5}$] and COS [mole fraction of $(4 \pm 2) \times 10^{-5}$], plus a possible marginal detection of SF_6, with the H_2S and COS abundances measured only in the 29 to 37 km altitude region.

Gaseous elemental sulfur has also been detected in the lower atmosphere. In particular, the Venera 11 and 12 spectrophotometers showed strong absorption in the 4000–6000 Å region between the altitudes of 10 and 30 km which is interpreted in terms of absorption by gaseous elemental sulfur (Moroz *et al.*, 1979). More specifically, the data can be fitted by allowing for the chemical transformation from the yellow absorber S_4, to the blue absorber S_3, and finally to the UV absorber S_2 with increasing temperature (i.e., increasing depth) in the atmosphere (Prinn, 1979; Sanko, 1980). Mixing ratios $\sim 2 \times 10^{-10}$ for S_3 are indicated (Sanko, 1980).

Finally, there is a host of convincing (yet still indirect) evidence that the principal constituent of the Venus clouds is concentrated sulfuric acid solution. This compositional evidence, together with a discussion of the vertical structure of the clouds and of other possible cloud constituents, is provided in Section 4.2.5.

The rare gases serve as extremely valuable tracers of processes and of sources of atmospheric constituents. The radiogenic isotopes ^{40}Ar (from ^{40}K decay) and ^4He (from the decay chains of U and Th) are produced throughout the life of a planet, whereas the primordial nonradiogenic rare gases (^3He, ^{20}Ne, ^{21}Ne, ^{22}Ne, ^{36}Ar, ^{38}Ar, and the many isotopes of Kr and Xe discussed in Sections 2.3–2.5) are simply released irreversibly by warm crustal rocks. Knudsen and Anderson (1969) predicted that, with an Earthlike radiogenic source and a negligible atmospheric escape rate, ^4He concentrations might be 200 ppm and ^{40}Ar concentrations might be 100 ppm on Venus. The UV spectrometer observations of the upper-atmosphere airglow emission of Venus from the Mariner 10 flyby (Kumar and Broadfoot, 1975) found a mixing ratio of $(1.0 \pm 0.5) \times 10^{-5}$ for ^4He just above the turbopause. Mass spectrometers aboard both the Pioneer Venus entry probe bus and the Pioneer Venus orbiter measured He and CO_2 abundance profiles above the turbopause (Niemann *et al.*, 1979a,b; von Zahn *et al.*, 1979a). Extrapolation of the high-altitude observations through the turbopause provides an estimated ^4He mole fraction of 2×10^{-5} in the lower atmosphere (von Zahn *et al.*, 1979b).

Neon, which has no radiogenic sources, was detected in the lower atmosphere of Venus by both the Pioneer Venus (Hoffman *et al.*, 1979a,b) and Venera 11 and 12 (Istomin *et al.*, 1979) mass spectrometer experiments. Very large concentrations, 10 ppm to nearly 30 ppm, were found by both experiments. Neon was also claimed by the gas chromatograph team on Pioneer Venus (Oyama *et al.*, 1979a) at the level of 1–10 ppm.

Nonradiogenic argon and radiogenic ^{40}Ar were both detected in large and comparable amounts by both Venera 11 and 12 and Pioneer Venus experiments. The first detection of argon on Venus, by the Venera 9 and 10 mass spectrometer experiment, showed large amounts of ^{40}Ar but no clear evidence for ^{36}Ar (Surkov, 1977). Istomin *et al.* (1979) found roughly equal amounts of ^{36}Ar and ^{40}Ar, with a $^{36}Ar : ^{38}Ar$ ratio of 5 : 1, and a total argon abundance of about 115 ppm. Gel'man *et al.* (1979b) found about 50 ppm total argon by gas chromatography. On Pioneer Venus, gas chromatography indicated total argon concentrations around 64 ppm after correction of earlier erroneous peak identifications (Oyama *et al.*, 1980a,b). The large probe neutral mass spectrometer (Hoffman *et al.*, 1980) indicates a total argon abundance of ~70 ppm with $^{40}Ar : ^{36}Ar$ and $^{38}Ar : ^{36}Ar$ ratios of 1.03 and 0.18, respectively. Considering the accuracy of these various experimental techniques, there is satisfactory agreement on both the total argon abundance and particularly on the argon isotopic ratios. The very large value of the $^{36}Ar : ^{40}Ar$ ratio on Venus relative to that on Earth and Mars was one of the more unexpected and exciting results of the 1978 Venera and Pioneer Venus missions.

Istomin *et al.* (1979) reported a positive detection of krypton, with a ratio of $^{36}Ar : ^{84}Kr$ of about 120 and a Kr abundance of 0.4 ppm. The initial Pioneer results failed to detect krypton at the level of about 1 ppm (Hoffman *et al.*, 1979a,b), and a more recent reassessment of this question has led to the establishment of an upper limit on krypton of about 5% of the amount claimed by the Venera mass spectrometer experimenters (Hoffman *et al.*, 1980; Donahue *et al.*, 1981). Xenon was not detected by any experiment, but the upper limit from the Pioneer Venus mass spectrometer measurements appears to be very restrictive. The Venera Kr : Ar ratio fits the fractionated ("planetary") rare gas component of meteorites fairly well, and would imply not only 20 times as much Kr as can be reconciled with the Pioneer Venus data, but also 40 times more Xe than the Pioneer Venus results would allow. Note that the Venera 11 and 12 mass spectrometers were incapable of detecting Xe because of design limitations, which permitted measurements only from 11 to 105 amu (Istomin *et al.*, 1979). The Venera 13 and 14 probes carried a modified mass spectrometer with a mass range of 11–138 amu, thus permitting coverage of both the krypton and xenon mass regions. In addition to providing another independent measurement of total argon (100 ppm), these experiments were able to detect both Kr and Xe. Both elements were found to have mole fractions within the limits $10^{-8}–10^{-7}$ (Istomin *et al.*, 1982).

The available data on the chemical composition of the troposphere and stratosphere of Venus are summarized in Table 4.2. The rare gas abundance debate is portrayed graphically in Figure 4.3.

The most recent (and perhaps the most provocative) isotopic data on the atmosphere of Venus have been provided by continuing analysis of the Pioneer Venus mass spectrometer data. During descent through the Venus atmosphere a droplet of

TABLE 4.2
Composition of the Venus Atmosphere

Gas	Volume mixing ratio	
	Troposphere (below clouds)	Stratosphere (above clouds)
CO_2	9.6×10^{-1}	9.6×10^{-1}
N_2	4×10^{-2}	4×10^{-2}
H_2O	$10^{-4}-10^{-3}$	$10^{-6}-10^{-5}$
CO	$(2-3) \times 10^{-5}$	$5 \times 10^{-5}-10^{-3}$
HCl	$<10^{-5}$	10^{-6}
HF	?	10^{-8}
SO_2	1.5×10^{-4}	$5 \times 10^{-8}-8 \times 10^{-7}$
S_3	$\sim 10^{-10a}$	?
H_2S	$(1-3) \times 10^{-6a}$	?
COS	$<2 \times 10^{-6}$	?
O_2	$(2-4) \times 10^{-5a}$	$<10^{-6a}$
H_2	?	2×10^{-5a}
^{4}He	10^{-5}	10^{-5}
$^{20,22}Ne$	$(5-13) \times 10^{-6}$	$(5-13) \times 10^{-6}$
$^{36,38,40}Ar$	$(5-12) \times 10^{-5}$	$(5-12) \times 10^{-5}$
^{84}Kr	$<2 \times 10^{-8}-4 \times 10^{-7}$	$<2 \times 10^{-8}-4 \times 10^{-7}$

[a]Single experiment; corroboration required.

sulfuric acid was ingested and blocked the instrument inlet. For several minutes, only droplet vapor (with a high concentration of water vapor) was observed. This fortuitous event permitted determination of the D:H ratio. The deduced value is ~0.016, compared to the terrestrial D:H ratio of 1.5×10^{-4} (Donahue et al., 1982). This 100-fold enhancement could be due to a D-rich source of Venus volatiles, such as polymeric organic material, or more likely to loss of about 99% of the Venus water inventory by Jeans excape (H escapes far more rapidly than D). This would require an initial inventory of water on Venus of at least 1 bar, which is 0.003 times the terrestrial inventory. The implications of this discovery are discussed in Section 4.2.6.

4.2.3. Atmosphere–Lithosphere Interaction

Since all the spectroscopically detected constituents of the Venus atmosphere (CO_2, CO, H_2O, HCl, and HF) known from 1967 to 1978 were chemically reactive, and since the surface of Venus was known to be at a dull red heat (750 K), it was reasonable to search for explanations of the atmospheric composition in terms of chemical equilibrium between the atmosphere and minerals in the surface rocks.

Urey (1952c), writing long before the observation of high surface temperatures of

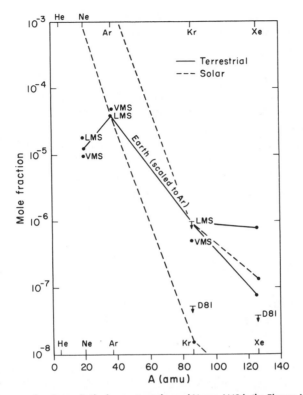

FIGURE 4.3. **Rare gas abundances in the lower atmosphere of Venus. LMS is the Pioneer Venus large probe mass spectrometer (Hoffman et al., 1979a,b); VMS is the Venera 11 and 12 mass spectrometer (Istomin et al., 1979); and D81 is the final interpretation of the LMS rare gas data (Donahue et al., 1981). The solid curve gives the terrestrial rare gas composition scaled to the abundance of ^{36}Ar on Venus from LMS. The lower point for Xe on the curve is atmospheric Xe; the higher is the presumed total terrestrial inventory of Xe. The dashed curves illustrate the solar rare gas relative abundances, as scaled to both the ^{36}Ar abundance and the ^{86}Kr upper limit. The Venera 13 and 14 mass spectrometer data on Ar agree with the earlier determinations given in the figure, whereas the mole fractions of both Kr and Xe are found to lie between 10^{-7} and 10^{-8} (Istomin et al., 1982).**

Venus, commented that, at room temperature, the following reactions of CO_2 with pyroxenes:

$$MgSiO_3 + CO_2 \text{ (g)} \rightleftarrows MgCO_3 + SiO_2 \qquad (4.2)$$

enstatite   magnesite   quartz

and

$$CaSiO_3 + CO_2 \text{ (g)} \rightleftarrows CaCO_3 + SiO_2 \qquad (4.3)$$

wollastonite   calcite   quartz

reach equilibrium at CO_2 partial pressures below 10^{-5} bar, but that catalysis by water is essential for the attainment of equilibrium on a reasonable time scale. He attributed the extreme overabundance of CO_2 (relative to the low-temperature equilibrium pressure) on Venus primarily to the absence of a suitable catalyst for enhancing the rate of reaction of CO_2 with minerals at low temperatures, and to the lack of erosive exposure of fresh silicate minerals. Both of these conditions were attributed to the absence of water on Venus.

After the discovery of the high brightness temperature of Venus at radio wavelengths (Mayer et al., 1958), many workers attempted to preserve the notion of a low-temperature surface because, apparently, of the preconception that the present Venus must be an identical twin of Earth, endowed with vast quantities of liquid water. We have mentioned that several proposals (Jones, 1961; Danilov, 1964; Vakhnin and Lebedinskii, 1965) placed the source of the microwave and decimetric radiation outside the neutral atmosphere, either in an extremely dense ionosphere ($\sim 10^9$ electrons cm^{-3}) or in hypothetical trapped radiation belts. Others (Tolbert and Straiton, 1962; Bottema et al., 1965a,b) argued that intense electrical discharge activity within a dense cloud layer produced the electromagnetic emission nonthermally. Of the presumed preconditions for these various mechanisms (dense ionosphere, planetary dipole magnetic field, van Allen belts, electrically conducting cloud droplets), only the latter is compatible with current knowledge of Venus. None of them can contribute a significant proportion of the radio emission, in contradiction to the cautionary note sounded by Lewis (1970a), who stressed the likelihood of substantial electrically conducting clouds.

Several early attempts to describe limited aspects of the atmospheric composition either involved erroneous assumptions regarding the true surface conditions and atmospheric composition of Venus, or overlooked some of the most stable compounds in the atmosphere, thus rendering their conclusions of very little use (Adamcik and Draper, 1963; Mueller, 1963, 1964a,b, 1965; Walter, 1964). Mueller's basic rationale for the assumption of strong atmosphere–lithosphere interaction has, however, remained unchanged and has served as the basis for all subsequent work.

A simple thermodynamic treatment of the observed CO_2, CO, and H_2O content of the atmosphere (Dayhoff et al., 1967; Lippincott et al., 1967) demonstrated that organic matter could not be stable in it at any level. Because interactions with the surface were ignored, no explanation of the observed composition could logically be offered.

Immediately subsequent to the discovery of CO, HCl, and HF on Venus by Connes et al. (1967, 1968), it became clear that the presence of these highly reactive gases *required* that the surface temperature of Venus be high. Using the relative abundances of CO_2, CO, H_2O, HCl, and HF then available, but without certain knowledge of the CO_2 mole fraction, the surface mineralogy, or the surface temperature and pressure, Lewis (1968a) was able to derive a *model–independent* absolute lower limit on the surface temperature 515 K. For CO_2-rich model atmospheres, the lower limit of the surface temperature became 560 K. In order to check the validity of these results by every available means, Lewis also calculated the abundances of several sulfur-bearing gases, including COS, H_2S, SO_2, S_2, and SO_3, for each model atmosphere. Of these, COS and H_2S were predicted to be the most abundant. A very interesting result was the conclusion that the temperature in the equilibration region could not be much higher than 560 K without producing an amount of COS that was larger than the spectroscopic detection limit. This suggested one of the following: that there was no sulfur in the crust–atmosphere system on Venus, that some mechanism existed for precipitating sulfur compounds as clouds below the level reachable by Earth-based IR spectroscopy, or that part of the surface was at a temperature very close to 560 K. Lewis (1969c) proposed that sulfides of volatile elements such as Hg, As, and Sb had vapor pressures in the right range to precipitate all sulfur from the visible portion of the atmosphere, and that sufficient amounts of these elements could reside in the atmosphere if, for example, Earth's crust were heated to the surface temperature of Venus. More recently, Prinn (1971, 1973, 1975, 1978) has shown that photochemical oxidation of COS is rapid enough to convert COS into SO_2 and clouds of sulfuric acid droplets and elemental sulfur and to keep the COS abundance unobservably small (see Section 4.2.4).

Lewis (1970c) showed that, given such a mechanism for removal of gaseous sulfur compounds, the observed atmospheric composition was compatible with a surface temperature of 750 K and a surface pressure of $\sim$100 bar, and that every atmospheric constituent could be buffered by plausible mineral assemblages on the surface. The reaction of plausible minerals with, for example, CO_2 gas at 750 K has been shown to occur on a geologically very short time scale ($\sim$100 years) (Mueller and Kridelbaugh, 1973), and thus such an equilibrium is kinetically reasonable.

Further, a detailed study of the abundances of about 100 possible gaseous compounds of H, C, N, O, F, Cl, S, Fe, and Si showed that the only chemically active gases then unobserved which were expected to make up more than 1 part in 10^{10} of the atmosphere were COS, H_2, H_2S, SO_2, S_2, and CS_2. Of these, the five sulfur compounds are subject to photochemical oxidation to elemental sulfur and H_2SO_4 (Prinn, 1973), and H_2 is not spectroscopically observable unless it is a dominant constituent of the atmosphere. Table 4.3 lists the *predicted* most abundant chemically active gases on Venus for a surface temperature of 750 K (after Lewis, 1970c). The value given for nitrogen was an upper limit and was assumed for the sole purpose of calculating upper limits on NH_3, HCN, and other nitrogen compounds, whose abundances were found to be negligible. These predictions can be compared with the Pioneer Venus and Venera results of a decade later (Table 4.2).

There are three important ways in which the tropospheric composition may vary with altitude. There may be condensation of certain constituents, or temperature-dependent shifts in equilibria involving species such as NH_3 or sulfur compounds, or photochemical destruction of some species and production of others not on the list. These processes are capable of affecting the global inventory of only one element, hydrogen, which may escape from the planet after photochemical dissociation of H_2O vapor or HCl. Lewis (1969c) and Prinn (1971, 1973, 1975, 1978) have examined the chemistry and photochemistry of the sulfur compounds, and Goettel and Lewis (1974) have studied the ammonia production and precipitation reactions as a function of altitude, including the stability of the possible condensates such as NH_4HCO_3 and NH_4Cl.

In addition to the species discussed above, compounds of several other volatile

TABLE 4.3
Predicted Abundances of Chemically Active Gases in the Lower Atmosphere of Venus[a]

Species	Mole fraction
CO_2[b]	9.6×10^{-1}
N_2	4×10^{-2} (assumed)
H_2O[b]	5×10^{-4}
CO[b]	2×10^{-4}
COS	5×10^{-5}
H_2S	5×10^{-6}
HCl[b]	1×10^{-6}
H_2	7×10^{-7}
SO_2	3×10^{-7}
HF[b]	2×10^{-8}
S_2	2×10^{-8}
CS_2	7×10^{-10}
NH_3	7×10^{-11}
$FeCl_2$	7×10^{-11}
CH_4	6×10^{-13}
HCN	2×10^{-14}

[a]After Lewis (1970c).
[b]Observed by IR spectroscopy above the clouds.

elements may be appreciable trace constituents of the atmosphere. Compounds of Br, I, As, Hg, Sb, etc. may form significant quantities of clouds, and vapor transport of elements such as Pb and Zn may profoundly affect their distribution on the surface of Venus. In particular, these elements may be greatly enriched in permanently cooler regions of the surface, such as mountain tops, even if these regions are only $\simeq 2$ km higher (and hence about 16 K cooler) than the mean surface. Indeed, Lewis (1968a) suggested that different buffer reactions may occur at different altitudes for major gases as well.

One result of geochemical modeling of Venus has been the identification of numerous minerals and mineral assemblages that might be stable on the surface, in equilibrium with the atmosphere (Lewis, 1970c). These minerals include pyroxene, quartz, magnetite, calcite, halite, fluorite, tremolite, diopside, jadeite, ankermanite, and andalusite. Other possibilities include olivine, dolomite, and troilite. Of course, not all of these minerals can coexist in a single assemblage. It is nonetheless impossible to rule out any of the above minerals on these grounds, since that would be equivalent to the assumption that the chemical and mineralogical composition of the crust is everywhere the same. This is almost certainly false, as Venera surface analyses have since shown.

The above minerals, which can participate in buffer reactions capable of regulating the fugacities of CO, O_2, HCl, HF, and H_2O in the atmosphere, suggest a geochemically differentiated crust, rich in silica and alkalis (Lewis, 1968a, 1970c).

This conclusion was strikingly confirmed by the gamma-ray spectrometer landed on Venus aboard the Venera 8 probe (Surkov *et al.*, 1973a; Vinogradov *et al.*, 1973). The measured abundances of the major radioisotopes were 4.0% K, 6.5 ppm Th, and 2.2 ppm U. These concentrations are reminiscent of terrestrial granitic rocks. The same experiment was also carried out at the Venera 9 and 10 landing sites (Surkov *et al.*, 1976; Surkov, 1977). These measurements show 0.47 ± 0.08 and $0.30 \pm 0.16\%$ K, 0.60 ± 0.16 and 0.46 ± 0.26 ppm U, and 3.65 ± 0.42 and 0.70 ± 0.34 ppm Th, respectively, typical of basalts.

Similar geochemical considerations also provided a list of minerals which are almost certainly *absent* from the surface. Such minerals as graphite, talc, serpentine, siderite, diaspore, brucite, sellaite, cryolite, fluorphlogopite, pyrite, anhydrite, kamacite and taenite, hematite, and probably magnesite are absent.

Among these early predictions, several are worthy of note:

(1) It seemed most probable that the pressure of CO_2 is controlled (buffered) by reaction (4.3) or some other similar reaction in the most calcium-rich portions of the surface of Venus.

(2) The partial pressure of water could be buffered by amphibole reaction equilibria such as

$$Ca_2Mg_5Si_8O_{22}(OH)_2 \rightleftarrows 3MgSiO_3 + 2CaMgSi_2O_6 + SiO_2 + H_2O. \qquad (4.4)$$

tremolite   enstatite   diopside   quartz

(3) The dominant sulfur-bearing gas is most probably COS, regulated by a reaction such as

$$3FeS + 4CO_2 \rightleftarrows Fe_3O_4 + 3COS + CO. \qquad (4.5)$$

troilite   magnetite

(4) Similar equilibria can be written for buffering of HCl and HF by reaction of water vapor with mineral assemblages containing fluorite (CaF_2), halite (NaCl), and other halide minerals. Sill (1975), pointing out the relative thermodynamic instability of solid bromides, proposed that large amounts of HBr are present in the atmosphere.

(5) The fugacity of oxygen, and hence the $CO:CO_2$ ratio, is most likely buffered by

$$3\text{"FeO"} + CO_2 \rightleftarrows Fe_3O_4 + CO, \qquad (4.6)$$

wüstite   magnetite

where "FeO" is the $FeSiO_3$ component of pyroxenes or the Fe_2SiO_4 component of olivine. Because of the likelihood that free quartz is abundant on the surface, coexistence of ferromagnesian pyroxenes with magnetite is likely.

Note that C, O, H, S, Cl, and F can all be stored as surface minerals at 750 K and ~100 bar in a CO_2-rich dry atmosphere. Therefore, the atmospheric inventories of these elements are unlikely to be their entire planetary inventories. No mineral sink for nitrogen is known, since the surface is far too oxidizing for nitrides and far too NH_3-poor for ammonium entry in substitution for potassium.

Speculations about surface materials which ignore thermodynamic and geo-chemical constraints (Sagan, 1975) or which are based on egregious misuse of basic thermodynamics (Sagan, 1962) understandably yield very different results.

The Venera 11 and 12 and Pioneer Venus analyses described in the previous section provided the basis for new interpretive models. An interesting computational approach by Khodakovskii *et al.* (1979a) attempts to reconcile the early discordant (and erroneous) results from both missions by postulating that Pioneer Venus entered in a descending, highly oxidized region containing substantial free O_2 (Oyama *et al.*, 1979a). Leaving aside this issue, Khodakovskii *et al.* treat the gas–mineral interaction as a closed multicomponent system, where the rocks are assumed to have the same composition as familiar terrestrial rocks. This assures that the various gas buffers are completely compatible with each other.

Such an approach is surely appropriate at every point on the surface of Venus. There are, however, several reasons for regarding the application of this model as premature. First, we do not know if any of the regolith of Venus really does have a major element composition and mineralogy identical to any terrestrial rock, since the only compositional data we have are the gamma-ray analyses at three points. K, U, and Th do not make a rock! Second, several of the reactions that seem to be the most promising buffers for CO_2, H_2O, HCl, and HF involve Ca, and the most Ca-rich region of the surface of Venus would tend to act as the buffer for these gases. This certainly need not be an *average* region. Third, as first pointed out by Lewis (1968a), permanently cooler regions such as high plateaus could act as "cold traps" for buffering gas pressures. Chemical weathering in the highlands may be a very important source of dust, which may form preferentially from Ca-rich minerals. The highland rocks would disaggregate, and the fine dust produced by chemical weathering would blow into the lowlands. Thus, wind-deposited sediments may have very different compositions (and vastly larger surface areas per gram) than igneous rocks (Nozette and Lewis, 1982).

Of course, the results of Khodakovskii *et al.* (1979a) are still of very great interest, since they inform us of the likely fate of known rock types when exposed to the Venus atmosphere. They considered several major rock types, from alkali basalts to andesites and rhyolites, and covered a range of surface temperatures from about 670 to 745 K, corresponding to altitudes of 0 to 10 km above the mean lowland surface. A CO mole fraction of 5×10^{-5} and a water vapor mole fraction of 3×10^{-4} were used, and compounds of S, Cl, and F were omitted. Alkali lavas, upon equilibration with the atmosphere, were shown to form pyroxene, magnetite, sphene, and variable but possibly large amounts of forsterite, nepheline, phlogopite, glauco-phane, and calcite. A trace of apatite and possible small amounts of epidote, dolomite, andradite, and hercynite were also predicted. Basalts were found to equilibrate with the Venus atmosphere to produce pyroxenes, anorthite, albite, microcline, quartz, glaucophane, annite, chloritoid, epidote, almandine, magne-tite, rutile, and apatite. Rhyolite would generate plagioclase, microcline, quartz, chloritoid, epidote, almandine, glaucophane, and possibly a trace of pyroxene. Andesite behaved similarly, except that a trace of sphene was also possible. The main hydroxyl silicates were found to be tremolite, phlogopite, glaucophane, and epidote, whereas the principal carbonate was calcite (a major rock-forming mineral in sodium alkali rocks). The only other carbonate phase found to be stable was dolomite. The deduced oxidation state of the surface rocks (from the assumed $CO:CO_2$ ratio in the lower atmosphere) was given by a $Fe_2O_3:(Fe_2O_3 + FeO)$ ratio of 0.5–0.7, slightly higher than for Earth (0.46 in basalt; 0.62 in rhyolite).

A treatment of the crust–atmosphere system including sulfur compounds has been presented by Khodakovskii et al. (1979b). Models were constructed for both the erroneously oxygen-rich case posed by the preliminary Pioneer Venus gas chromatograph data and for a lower troposphere in which the oxygen fugacity was fixed at extremely low levels by the $CO:CO_2$ mixing ratio of $(1.5–3) \times 10^{-5}$. COS was found to be an important sulfur gas, and both pyrite (FeS_2) and anhydrite ($CaSO_4$) were found to be plausible surface minerals. The (nonexistent) oxidized case was interpreted as a descending parcel of photochemically oxidized gas. Prinn (1973, 1975) and, in more detail, Florenskii et al. (1976) proposed a three-layer model of atmospheric composition, with a photochemically dominated oxidized layer at the top, a reduced geochemically equilibrated layer at the bottom, and a mixed region in between.

We have already commented on the great value of the sulfur-bearing gases as indicators of the oxidation state and mineralogy of the atmosphere–crust system. Based on the Venera and Pioneer Venus analytical results on sulfur compounds, several papers on the sulfur chemistry of the lower atmosphere and surface have been written. A. T. Young (1979a) suggests that the lower atmosphere should contain 5% by mass of sulfur, a state of affairs rendered extremely improbable by geochemical considerations. Prinn (1979) suggests an important role for sulfanes and sulfur chain species in the photochemistry of the lower atmosphere: Even fractional parts per million of S_3 etc. can release sulfur atoms upon exposure to visible light. A. T. Young (1979b) argues for CS_2 as a constituent of the upper troposphere with a concentration of 4×10^{-7}. The spectral feature he intends to explain is attributed correctly by Barker (1979) to SO_2, not CS_2.

A study of the relationship between sulfur gas abundances, surface oxidation state, and the iron oxide content of surface rocks has been carried out by Lewis and Kreimendahl (1980). They found that the iron content and oxidation state of the crust are separable variables, in that the *total* abundance of sulfur gases is inversely proportional to the FeO content, whereas the *relative* abundances of the sulfur gases are determined by the oxygen fugacity.

In order to avoid overconstraining the system by starting assumptions, it is useful to sketch out in a general way how the mineralogy of the surface and the atmospheric composition are related to the oxygen partial pressure and the total oxidized iron content of the surface rocks. Figure 4.4 identifies the stability fields of the major iron and sulfur minerals as a function of the O_2 fugacity and the activity of FeO. Magnetite is clearly seen to be stable relative to the other oxides of iron if the total concentration of oxidized iron is rather high. The oxidation sequence for sulfur minerals is from troilite to pyrite to anhydrite ($CaSO_4$).

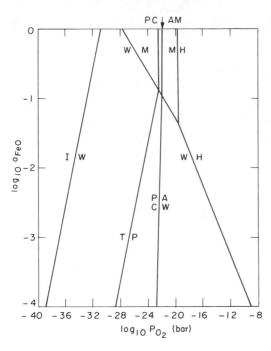

FIGURE 4.4. Sulfur and iron mineral stability fields at the surface of Venus. Buffer reactions involving iron compounds at 750 K are shown as a function of O_2 partial pressure and wüstite (FeO) activity. Pure magnetite disappears for a_{FeO} less than about $10^{-1.3}$. A is anhydrite ($CaSO_4$), C is calcite ($CaCO_3$), H is hematite (Fe_2O_3), I is iron (Fe°), M is magnetite (Fe_3O_4), P is pyrite (FeS_2), T is troilite (FeS), and W is wüstite (FeO). (After Lewis and Kreimendahl, 1980.)

Lewis and Kreimendahl (1980) discussed the equilibrium composition of the atmosphere, especially the sulfur-bearing gases, over this entire range of conditions. Unfortunately, even after the Pioneer Venus and Venera 11 and 12 missions, we have no direct information on the oxidation state of sulfur-bearing gases below 22 km so that we cannot objectively use the results of this study to provide clues on surface composition. We do know that the total mixing ratio of sulfur gases (SO_2, COS, H_2S) at the surface is $\sim 1.5 \times 10^{-4}$, which does provide at least one useful constraint. However, since 80% of the atmospheric mass of Venus lies below 22 km, our present ignorance of the detailed composition of this lower region prevents any truly firm conclusions concerning the validity of equilibrium calculations and their application to the deduction of surface mineralogy.

The predicted abundances of a number of important atmospheric constituents are shown in Figure 4.5. Limits on the oxidation state of the atmosphere can be deduced from the CO abundance, the stability of the atmosphere against graphite precipitation, and the failure to observe CH_4 and NH_3. The calculated abundances of sulfur-bearing gases for the case of an FeO activity of 10^{-3} are shown as a function of the oxygen fugacity in Figure 4.6. Note the narrow region in which comparable amounts of H_2S, COS, and SO_2 coexist. The pyrite–calcite–anhydrite–wüstite buffer (Barsukov et al., 1980b; Lewis and Kreimendahl, 1980):

$$FeS_2 + 2CaCO_3 + \tfrac{7}{2}O_2 = 2CaSO_4 + FeO + 2CO_2 \qquad (4.7)$$

is the oxygen buffer (assuming that the CO_2 pressure is fixed).

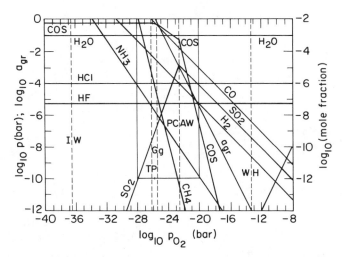

FIGURE 4.5. Predicted abundances of gases in the Venus atmosphere as a function of the O_2 partial pressure. The temperature is 750 K. The mole fraction of H_2O is fixed at 10^{-3} and the activity of FeO in the surface rocks is 10^{-3}. IW is the iron–wüstite buffer (Fe coexisting with silicates containing an FeO activity of 10^{-3}); WH is wüstite–hematite (Fe_2O_3) buffer. TP is the line where troilite (FeS) converts to pyrite (FeS_2), Gg is the line where graphite is in equilibrium with the gas, and PCAW is the pyrite–calcite–anhydrite–wüstite oxygen buffer, which regulates p_{O_2} by changes in the oxidation state of sulfur. The region inside the rectangle from 10^{-28} to 10^{-20} bar O_2 pressure and above a mole fraction of 10^{-12} is detailed in Figure 4.6, with emphasis on the sulfur gases. (After Lewis and Kreimendahl, 1980.)

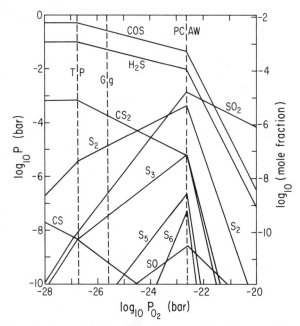

FIGURE 4.6. Predicted abundances of sulfur-bearing gases as a function of the O_2 partial pressure: $T = 750$ K, $X_{H2O} = 10^{-3}$, $a_{FeO} = 10^{-3}$. The mole fraction of SO_3 reaches its maximum value of $10^{-14.2}$ in the sulfate stability region, and atomic sulfur peaks in abundance at $X_S = 10^{-15.6}$ at the PCAW buffer. The mole fractions of S_4, S_7, and S_8 (not shown) also peak at the PCAW buffer at $10^{-10.5}$, $10^{-10.4}$, and $10^{-11.3}$, respectively. (After Lewis and Kreimendahl, 1980.)

The early treatment of sulfur chemistry on Venus by Mueller (1965) unfortunately omitted the thermodynamically most stable sulfur species, COS, but correctly concluded that the upper limit on the abundance of sulfur gases would be near 10^{-3} bar if the crust had Earth-like composition. In other words, an Earth-like crustal composition would imply a total sulfur mixing ratio $\leq 10^{-5}$, which is clearly at odds with the observations.

Note from Figure 4.6 that the suggestion by A. T. Young (1979b) of several percent sulfur in the lower atmosphere is geochemically impossible; indeed, even if the sulfur were present, it would produce vast amounts of COS by reaction with CO_2. We also note from Figures 4.5 and 4.6 that the total mixing ratio of sulfur gases (SO_2 + COS + H_2S) generally decreases rapidly with increasing O_2, that is, with increasing abundances of FeO and $CaSO_4$ in the crust. In particular, a total mixing ratio for sulfur gases $\geq 1.5 \times 10^{-4}$ requires FeO activities $\leq \frac{1}{10}$ those on earth (Lewis and Kreimendahl, 1980). Thus, even a midlly oxidized Venus crust is incompatible with significant amounts of any sulfur gases in the atmosphere. Finally, the observed total sulfur gas mixing ratio $\sim 1.5 \times 10^{-4}$ requires COS and H_2S to be more abundant than SO_2. Indeed, the maximum SO_2 mixing ratio predicted under any conditions at the surface is $\frac{1}{10}$ that observed (Lewis and Kreimendahl, 1980).

The above results depend on a chemical equilibrium assumption between the atmosphere and surface minerals, and they are clearly at odds with SO_2 remaining the principal sulfur species near the surface. Thermochemical reactions at the surface would tend to either incorporate SO_2 into anhydrite until the SO_2 mixing ratio was decreased to $\leq \frac{1}{50}$ of that observed or transform SO_2 to COS and H_2S until the SO_2 mixing ratio was $\leq \frac{1}{10}$ that observed. Therefore, two conclusions are possible. The first is that COS and H_2S are indeed the dominant sulfur gases at the surface (i.e., equilibrium prevails). The second is that sulfur is first outgassed as COS and H_2S and then these gases are oxidized to SO_2 and sulfuric acid by photochemical reactions. In this second case disequilibrium prevails, and the reaction times for incorporation of SO_2 etc. into $CaSO_4$ and for conversion of SO_2 into COS and H_2S are so long that SO_2 is enhanced, and COS and H_2S are depleted, relative to equilibrium. Which of these two cases prevails depends on the rates of the reactions involved.

Although laboratory studies on the rates of relevant gas–rock reactions are not available, there are some data on the rates of the gas reactions. These suggest the possibility of gas-phase equilibrium, but only very close to the surface (Prinn, 1978; Krasnopolsky and Parshev, 1980a). Calculations can obviously be carried out assuming gas-phase equilibrium which use elemental abundances derived from observations above 22 km rather than those determined by atmosphere–surface equilibrium. Calculations using atmospheric composition determinations from Venera 11 and 12 (Krasnopolsky and Parshev, 1980b) and Pioneer Venus (Oyama et al., 1980) predict SO_2, COS, S_2, and H_2S mixing ratios $\sim 10^{-4}$, 10^{-5}, 10^{-7}, and 10^{-6}–10^{-7}, respectively. The mixing ratios of SO_2 (and O_2) in these calculations are too large to be compatible with equilibrium with surface minerals, but the SO_2,

COS, S_2, and H_2S abundances are all roughly compatible with observations. This obviously implies that the slow or "bottleneck" reactions preventing complete equilibrium are the incorporation of SO_2 into anhydrite and O_2 into anhydrite plus FeO at the surface. These reactions may simply proceed too slowly when compared to photochemical production of SO_2 and O_2 from COS and CO_2.

What is the overall cycle for sulfur compounds on Venus? Prinn (see von Zahn *et al.*, 1982) has recently suggested the existence of three important cycles which operate to govern the global sulfur budget and these are summarized in Figure 4.7. The first cycle, which he refers to as the *"fast atmospheric cycle,"* involves the following net photochemical reactions in the upper atmosphere:

$$CO_2 \rightarrow CO + \tfrac{1}{2}O_2,$$
$$\tfrac{1}{2}O_2 + SO_2 \rightarrow SO_3,$$
$$SO_3 + H_2O \rightarrow H_2SO_4,$$

being reversed in the lower atmosphere by the following net thermochemical reactions:

$$H_2SO_4 \rightarrow H_2O + SO_3,$$
$$SO_3 + CO \rightarrow SO_2 + CO_2$$

This fast cycle is thus intimately involved with the photochemistry of the atmosphere and particularly with the formation of the sulfuric acid clouds and recycling of photochemically produced CO and O_2 (see Section 4.2.4).

The second cycle, which Prinn refers to as the *"slow atmospheric cycle,"* involves the following net photochemical reactions in the upper atmosphere:

$$6CO_2 \rightarrow 6CO + 3O_2,$$
$$\tfrac{3}{2}O_2 + H_2S \rightarrow SO_3 + H_2,$$
$$\tfrac{3}{2}O_2 + COS \rightarrow SO_3 + CO,$$
$$2SO_3 + 2H_2O \rightarrow 2H_2SO_4,$$
$$H_2S \rightarrow H_2 + \tfrac{1}{n}S_n,$$
$$COS \rightarrow CO + \tfrac{1}{n}S_n,$$

being reversed in the lower atmosphere close to and at the surface by the following net thermochemical reactions:

$$2H_2SO_4 \rightarrow 2H_2O + 2SO_3,$$
$$SO_3 + 4CO \rightarrow COS + 3CO_2,$$
$$H_2 + SO_3 + 3CO \rightarrow H_2S + 3CO_2,$$
$$H_2 + \tfrac{1}{n}S_n \rightarrow H_2S,$$
$$CO + \tfrac{1}{n}S_n \rightarrow COS.$$

This slow atmospheric cycle involves H_2SO_4 production, but it is not the dominant source for the sulfuric acid clouds. However, it does involve production of elemental sulfur which is an attractive candidate for the UV-absorbing component of the clouds (Prinn, 1971, 1975; Young, 1977).

An important aspect of both the fast and slow atmospheric cycles is that the thermochemical reactions in the lower atmosphere in each cycle apparently proceed fast enough so that the $SO_2 : H_2S$ and $SO_2 : S_2$ ratios, and by implication the

VENUSIAN SULFUR CYCLE

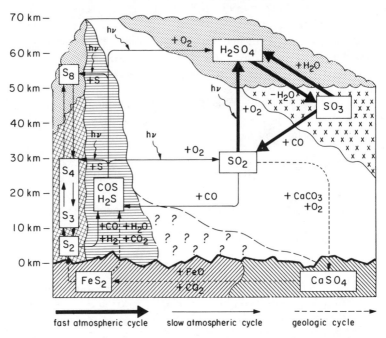

FIGURE 4.7. **The sulfur cycle on Venus. The regions of stability and relative abundance of the major sulfur compounds are qualitatively indicated by the position and size of the shaded area assigned to each one. Net transformations in the fast atmospheric, slow atmospheric, and geologic cycles are shown with thick solid arrows, thin solid arrows, and dashed arrows, respectively. (After Prinn, in von Zahn et al., 1982.)**

SO_2 : COS and SO_2 : SO_3 ratios, are maintained approximately at the gas-phase equilibrium values expected when chemical interaction with the surface is negligible.

The third cycle which Prinn refers to as the *"geologic cycle"* begins with outgassing (volcanic or otherwise) of the reduced gases COS and H_2S from the crust as a result of mineral reactions such as those involving pyrite:

$$FeS_2 + 2H_2O \rightarrow FeO + 2H_2S + \tfrac{1}{2}O_2,$$
$$FeS_2 + 2CO_2 \rightarrow FeO + 2COS + \tfrac{1}{2}O_2.$$

These gases are then cycled many times through the fast and slow atmospheric cycles. In the process, however, they do spend the majority of their time in the form of SO_2. Thus, continual outgassing will build up SO_2 in the atmosphere through the following net photochemical reactions (see Section 4.2.4):

$$2H_2S + 3O_2 \rightarrow 2H_2O + 2SO_2,$$
$$2COS + 3O_2 \rightarrow 2CO_2 + 2SO_2.$$

Once SO_2 builds up to levels in excess of its equilibrium concentration relative to surface rocks, a number of thermochemical reactions attempt to restore the system to equilibrium. These restoring reactions include formation of anhydrite:

$$4SO_2 + 2O_2 + 4CaCO_3 \rightarrow 4CaSO_4 + 4CO_2,$$

followed by conversion of anhydrite to pyrite to complete the cycle:

$$4CaSO_4 + 2FeO + 4CO_2 \rightarrow 2FeS_2 + 4CaCO_3 + 7O_2.$$

Unlike the thermochemical reactions restoring equilibrium in the two atmospheric cycles, there is no observational evidence to suggest that the restoring reactions in the geologic cycle are occurring fast enough to maintain atmosphere–surface equilibrium. Neither is there evidence that the geologic cycle is even in a steady state. Outgassing of sulfur gases may be dominated by sporadic volcanic activity or cometary impacts, whereas their reincorporation into minerals is dominated by surface weathering and crustal recycling.

The importance of the processes of volcanism, surface weathering, and plate tectonics in the cycles proposed by Prinn is very apparent. In addressing the weathering process, the tendency of volatile elements (e.g., S) to form compounds with calcium (e.g., $CaSO_4$) and the preferential weathering of high areas by reactive gas attack may cause fine-grained calcium-rich dust to blow into lower areas and dominate the buffering of atmospheric gases (Nozette and Lewis, 1982). This implies that the topography and surface wind speeds are both of geochemical significance. There are even recent suggestions that the compositions of the near-surface atmosphere and surface minerals are functions of the large-scale meterological (Barsukov et al., 1980a) and topographical (Khodakovsky et al., 1979a,b) patterns.

What knowledge do we have of the surface topography of this totally cloud-covered planet? Earth-based radar mapping of Venus has been actively pursued since the 1960s, and several regions of enhanced roughness were established by early observations. By simultaneous gating of the returned radar echoes by time delay (range) and Doppler shift (range rate), it is possible to map out the distribution of areas of enhanced reflectivity in two dimensions. There is, of course, a North–South ambiguity in any short run of data, but long-term observation or radar interferometry can resolve this ambiguity (Rogers and Ingalls, 1970). One of the bright features, Beta, is described by Zohar and Goldstein (1968): It is extremely difficult to attempt a geomorphological interpretation of such a map, but it is fair to say that Beta looks like a group of large mountains. Rogers and Ingalls (1969), by means of radar interferometry, have identified circular features which may possibly be the remains of large basins. In general, Venus is rather flat, with only a few percent of the surface area higher than 2 or 3 km above the planetary mean. Rogers et al. (1972) reported that the topographic variation around the equator spans a range of only 6 km. Campbell et al. (1972) describe a similar study of the equatorial region in which they find a mean equatorial radius of 6050.0 ± 5 km. A region with an elevation of 3 km above the mean and a size of several hundred kilometers was seen.

A number of publications have presented improved radar maps: The reader may be interested in consulting Goldstein and Rumsey (1972), Rogers *et al.* (1974), and Goldstein *et al.* (1976). Rumsey *et al.* (1974) describe an extensive region about 1500 km in diameter containing a number of very large 35- to 160-km-diameter craters with very subdued vertical relief. More recent geological interpretations of the radar data have been done by Malin and Saunders (1977) and by McGill (1979).

Altimetry and reflectivity measurements by the radar mapper experiment on the Pioneer Venus orbiter (Pettengill *et al.*, 1979a) have greatly improved our available coverage, including providing a sufficiently complete global map to permit the full appreciation of the uniqueness of the feature called Maxwell. Maxwell reaches to a maximum height of about 16 km and exposes a considerable surface area to temperatures as low as 640 K (Pettengill *et al.*, 1979b).

Mueller (1969) has opened the issue of the presumed decrease in strength of the lithosphere of Venus due to high surface temperatures. In addition to inhibiting the formation of large tectonic uplifts, this low strength would also have the effect of erasing the topographic expression of large impact features. Theory also suggests that there should be a *lower* limit to the size of impact craters due to the difficulty of penetrating the atmosphere (Tauber and Kirk, 1976).

Direct *in situ* observations of the structure of the surface have been made by panoramic cameras aboard the Venera 9 and 10 landers (Moroz, 1976d; Florenskii *et al.*, 1977). Both large rocks and dust are common features of both panoramas. Florenskii *et al.* (1977) attribute the fine-grained material to chemical weathering. Transport of dust produced by chemical weathering has been considered by Sagan (1976), and surface wind speeds were measured by the Venera 9 and 10 landers (Avduevskii *et al.*, 1976a).

It is, however, difficult to translate the little knowledge which we now have concerning the physical structure of the Venus surface into truly definitive statements on the rates of crustal outgassing and surface weathering. The most tantalizing recent information comes from the Pioneer Venus radar data. The high radar reflectivity of Theia Mons, a mountain in the elevated equatorial region called Beta Regio, which has long been considered to be volcanic, is apparently due to the presence of either iron-rich basalts or of FeS_2 of volcanic origin (Ford and Pettengill, 1983). The exposure of fresh FeS_2 and its subsequent slow conversion to FeO is an integral part of the proposed "slow" geologic cycle discussed earlier (Figure 4.7).

4.2.4. Photochemistry

As reviewed earlier, the Venusian cloud-top atmosphere is composed predominantly of CO_2 with a few percent N_2 and minor observable amounts of CO, H_2O, SO_2, HCl, and HF. In predominantly CO_2 atmospheres such as those present on both Venus and Mars a perplexing photochemical stability problem arises (cf. Donahue, 1968, 1971). Carbon dioxide is continuously photodissociated by solar UV radiation to produce CO and O:

$$CO_2 + h\nu \rightarrow CO(^1\Sigma_g^+) + O(^3P) \quad (\lambda < 2275 \text{ Å})$$
$$\rightarrow CO(^1\Sigma_g^+) + O(^1D) \quad (\lambda < 1670 \text{ Å}). \quad (4.8)$$

The excited $O(^1D)$ atoms are very rapidly quenched to $O(^3P)$ by collisions with CO_2. However, the direct path for CO_2 reformation

$$CO(^1\Sigma_g{}^+) + O(^3P) + CO_2 \rightarrow CO_2 + CO_2 \qquad (4.9)$$

is spin-forbidden and consequently about 50,000 times slower than the spin-allowed reaction

$$O(^3P) + O(^3P) + CO_2 \rightarrow O_2 + CO_2. \qquad (4.10)$$

As a result we might expect a pure CO_2 atmosphere to evolve slowly into a predominantly $CO-O_2$ atmosphere. The time required is roughly given by dividing the total amount of CO_2 on Venus ($\sim 1.5 \times 10^{27}$ molecules cm^{-2}) by the CO_2 column photodissociation rate ($\sim 10^{13}$ molecules cm^{-2} sec^{-1}) and is about 5×10^6 years. The times required to produce the presently observable quantities of CO and O_2 are even more telling; roughly 1000 yr for CO and less than about 5 yr for O_2 (since $X_{O_2} < 10^{-6}$). This striking problem has long been known: In fact, Wildt (1937) suggested CO oxidation by O_2 due to thermal reactions in the hot lower atmosphere.

In addition to this problem of overall photochemical stability, there is an associated problem emerging from the fact that the main removal mechanism for O atoms, namely reaction (4.10), is a three-body reaction requiring reasonably high total number densities before we expect efficient conversion of O to O_2. We therefore expect large O atom concentrations at high altitudes. There are a number of estimates of O atom concentrations at high altitudes which were predicted by ionospheric models: At 145 km these studies imply X_O values ranging from ~ 0.01 (R. W. Stewart, 1968; McElroy, 1969a; Kumar and Hunten, 1974; Liu and Donahue, 1975) to ~ 0.1 (Rottman and Moos, 1973; Bauer and Hartle, 1974). Even the largest estimated O atom concentrations can be built up from CO_2 photodissociation in less than 1 week at this atmospheric level. The Pioneer Venus observations actually show CO becoming the principal neutral species above 150 km near the terminator (Niemann et al., 1979a).

There are two obvious scenarios which we might suggest to relieve these problems. First, we could seek a very efficient catalyst for the CO–O reaction (4.9) so that reformation of CO_2 (rather than O_2 production) becomes the major sink for O atoms. Alternatively, we could allow O_2 formation to occur but seek a catalyst for oxidation of CO by O_2. In this case we must invoke rapid vertical transport (an eddy diffusion coefficient of, say, 10^7 cm^2 sec^{-1}) of O atoms from the 145 km level down to the lower altitudes where O_2 can be formed, if we are to keep O atom concentrations at 145 km low. Near the terminator less vigorous vertical mixing can allow the buildup of O. The O atom production time of less than 1 week mentioned above *requires* $K > 10^6$ cm^2 sec^{-1} in order to accomplish the latter objective. We are speaking here of altitudes near the Venusian turbopause, and such large values of K are certainly not improbable at these high levels.

The first proposed kinetic solution to the CO_2 stability problem on Venus was that by Reeves et al. (1966) which involved catalysis of the CO–O reaction by the following odd hydrogen reactions:

$$CO + HO_2 \rightarrow CO_2 + OH,$$
$$OH + O_3 \rightarrow HO_2 + O_2, \qquad (4.11)$$
$$O + O_2 + CO_2 \rightarrow O_3 + CO_2.$$

Hunten and McElroy (1970) later proposed a related catalytic scheme for the $CO-O_2$ reaction:

$$H + O_2 + CO_2 \rightarrow HO_2 + CO_2,$$
$$HO_2 + CO \rightarrow OH + CO_2, \qquad (4.12)$$
$$OH + CO \rightarrow H + CO_2.$$

Unfortunately, subsequent laboratory studies of the rate of reaction of HO_2 with CO (Baldwin et al., 1970) showed convincingly that it was much too slow for either of the above catalytic schemes to be effective. Recognizing this possibility, McElroy and Hunten (1970) also proposed an alternative mechanism for catalysis of the $CO-O(^1D)$ reaction by CO_2 itself:

$$CO_2 + h\nu \rightarrow CO + O(^1D),$$
$$O(^1D) + CO_2 \rightarrow CO_3{}^* \rightarrow CO_3 + h\nu, \qquad (4.13)$$
$$CO_3 + CO \rightarrow CO_2 + CO_2.$$

However, this scheme fails because $CO_3{}^*$ has much too short a lifetime; it simply decomposes to CO_2 and O again (Clark, 1971).

In conjunction with a study of HCl photochemistry on Venus, Prinn (1971, 1972, 1973) proposed a number of catalytic cycles for recombination of CO and O_2 which did not require the slow, HO_2-CO reaction. The first cycle involved catalysis by Cl atoms:

$$Cl + O_2 + CO_2 \rightarrow ClOO + CO_2,$$
$$ClOO + CO \rightarrow ClO + CO_2, \qquad (4.14)$$
$$ClO + CO \rightarrow Cl + CO_2.$$

A second cycle involved catalysis by odd hydrogen:

$$H + O_2 + CO_2 \rightarrow HO_2 + CO_2,$$
$$HO_2 + H \rightarrow OH + OH,$$
$$HO_2 + h\nu \rightarrow OH + O, \qquad (4.15)$$
$$OH + CO \rightarrow H + CO_2.$$

A third cycle involved catalysis by both Cl and H atoms:

$$H + O_2 + CO_2 \rightarrow HO_2 + CO_2,$$
$$Cl + HO_2 \rightarrow ClO + OH,$$
$$ClO + CO \rightarrow Cl + CO_2, \qquad (4.16)$$
$$OH + CO \rightarrow H + CO_2.$$

As an alternative, McElroy et al. (1973) proposed the following two cycles for Venus based on work by Parkinson and Hunten (1972) and by McElroy and Donahue (1972) for the analogous Martian situation:

$$H + O_2 + CO_2 \rightarrow HO_2 + CO_2,$$
$$O + HO_2 \rightarrow OH + O_2, \qquad (4.17)$$
$$OH + CO \rightarrow H + CO_2,$$

$$H + O_2 + CO_2 \rightarrow HO_2 + CO_2,$$
$$HO_2 + HO_2 \rightarrow H_2O_2 + O_2,$$
$$H_2O_2 + h\nu \rightarrow OH + OH, \tag{4.18}$$
$$OH + CO \rightarrow H + CO_2.$$

Finally, Sze and McElroy (1975) suggested another cycle involving chlorine and ozone:

$$O + O_2 + CO_2 \rightarrow O_3 + CO_2,$$
$$Cl + O_3 \rightarrow ClO + O_2, \tag{4.19}$$
$$ClO + CO \rightarrow Cl + \dot{C}O_2.$$

Which, if any, of the cycles (4.14)–(4.19) dominates remains a controversial matter. Unfortunately, a number of crucial rate constants, for example, for cycle (4.14), have not yet been measured. In their particular study Sze and McElroy (1975) considered the cycles (4.16)–(4.19) to dominate. Another solution to the problem of recombination of CO and O_2 was proposed by Prinn (1973), who concluded that the principal sink for O_2 at the cloud level involved oxidation of COS, H_2S, and SO_2 to H_2SO_4. There is now no doubt that formation of H_2SO_4 plays a dominant role in the removal of O_2 from the atmosphere; the photochemistries of CO_2 and sulfur compounds are therefore intimately connected.

Before moving on to discuss sulfur photochemistry, we should add that, even if no photochemical scheme for catalyzing the CO–O_2 reaction existed, computations assuming thermochemical equilibrium imply $X_{CO} = 2 \times 10^{-4}$ and $X_{O_2} \simeq 10^{-25}$ at the hot surface (Lewis, 1970c). Provided O_2 can be mixed down to the surface in less than 5 yr (which requires $K > 3 \times 10^5 \text{ cm}^2 \text{ sec}^{-1}$), and provided the CO–$O_2$ thermochemical reaction time near the surface is less than 5 yr (which is certainly expected at 750 K), then we can obviously solve the CO_2 stability problem without invoking photochemical catalysis. This is, in fact, essentially the solution suggested by Wildt (1937)! This solution is, of course, not available for the analogous Martian problem, where purely thermochemical reactions are exceedingly slow at the cool surface.

An understanding of the role of sulfur compounds in the Venus atmosphere has been a relatively recent development. Indeed, for many years a surprising aspect of spectroscopic studies of Venus had been the apparent failure to detect any sulfur-bearing gases in the visible atmosphere. Prior to the discovery of SO_2 in 1978 the upper limits on mixing ratios of COS, SO_2, and H_2S at the cloud tops implied from observations were 10^{-6} (Cruikshank, 1967), $3 \cdot 5 \times 10^{-8}$ (Cruikshank and Kuiper, 1967), and 3×10^{-7} (Anderson *et al.*, 1969). However, the thermodynamic equilibrium studies of the very hot Venusian surface which we discussed in detail in the previous section implied expected mixing ratios for COS, SO_2, and H_2S of 5×10^{-5}, 3×10^{-7}, and 5×10^{-6}, respectively, for even an Fe-rich Earth-like crustal composition. These are well above the observational limits *above* the cloud tops.

One is immediately tempted to surmise that Venus is highly deficient in sulfur and, indeed, we discussed earlier why we expected Venus to receive little FeS during accretion from the primitive solar nebula (Lewis, 1972b). However, only very small amounts of sulfur are required to saturate the atmosphere. Indeed, the element is of sufficiently high cosmic abundance that cometary impact alone could probably provide all that is necessary (Prinn, 1973; Lewis, 1974b).

The first attempt at explaining the apparent absence of sulfur gases was by condensation as HgS, as discussed by Lewis (1969c). However, quantitative removal of the expected amounts of atmospheric sulfur gases would require amounts of Hg in the atmosphere equivalent to extensive degassing of a chondritic Venus model. Considerably less Hg is expected. Another possible removal mechanism is photochemical destruction. Both COS and H_2S are readily photodissociated by solar UV radiation with wavelengths ranging from 2200 to 2700 Å and this radiation penetrates well into the visible clouds on Venus. Looking at the most abundant sulfur gas, COS, Prinn (1971) concluded that it would be irreversibly converted to elemental sulfur, SO_2, and SO_3 above the visible clouds. There was a problem, however, in keeping SO_2 below its claimed upper limit of $X_{SO_2} < 3.5 \times 10^{-8}$ (Cruikshank and Kuiper, 1967). Barker (1979) has since made a positive detection of three times that concentration.

The sulfur cycle on Venus then gained new significance with an important discovery concerning the composition of the Venus clouds. Polarimetric studies of sunlight reflected from Venus at visible and near-IR wavelengths by many authors (Sobolev, 1968; Coffeen, 1968, 1969; Dollfus and Coffeen, 1970; Loscutov, 1971; Forbes, 1971; Kattawar et al., 1971; Biryukov and Titarchuk, 1973) have convincingly demonstrated that the clouds have the polarizing properties of Mie particles (spheres) with radii near 1 μm and refractive indices (at 0.55 μm) of about 1.44 (Hansen and Arking, 1971). Then, based on the reflection spectrum of the Venus clouds and laboratory studies, Sill (1972) proposed that the main cloud layer may be sulfuric cloud droplets. Young (1973), Martonchik (1974), Pollack et al. (1974), and Young and Young (1973) have presented substantial observational evidence in support of this conclusion. Hansen and Hovenier (1974) have shown the close agreement between the optical properties of H_2SO_4 and those of the Venus clouds.

The implications of this discovery for the chemistry of the Venus atmosphere were considerable. The photochemical nature of the source of the sulfuric acid was actually implicit in the earlier work by Prinn (1971) on COS and H_2S photochemistry. Prinn (1973, 1975) later proposed a detailed model for the Venusian clouds in which COS is mixed up into the visible atmosphere where the following reactions occur among others:

$$COS + h\nu \rightarrow CO + S(^1D)$$
$$S(^1D) + CO_2 \rightarrow S + CO_2$$
$$S + COS \rightarrow S_2 + CO$$
$$S + O_2 \rightarrow SO + O$$
$$SO + OH \rightarrow SO_2 + H \qquad (4.20)$$
$$SO_2 + HO_2 \rightarrow SO_3 + OH$$
$$SO_3 + H_2O + CO_2 \rightarrow H_2SO_4 + CO_2$$
$$nS_2 \rightarrow S_{2n}$$

For each COS molecule oxidized to H_2SO_4 we require $\frac{3}{2}$ O_2 molecules. In the region of the visible clouds these sulfur reactions in fact form the principal removal mechanism for the O_2 which has been produced by CO_2 photodissociation. For example, predicted lifetimes for O_2 are about 1 min with the above sulfur chemistry in contrast to lifetimes of a few days with the odd hydrogen chemistry discussed earlier (McElroy *et al.*, 1973).

Whether H_2SO_4 or elemental sulfur is formed by the reactions (4.20) is simply a question of the local O_2 and COS concentrations. If $X_{O_2} \gg 7 \times 10^{-4} X_{COS}$, then H_2SO_4 dominates, whereas if $X_{O_2} \ll 7 \times 10^{-4} X_{COS}$, then sulfur becomes the main product (Prinn, 1975). It was therefore tempting to ascribe the observed UV absorption by the Venus clouds in the 3000–5000 Å region to elemental sulfur formed in oxygen-deficient cloud regions (Prinn, 1971, 1975). Indeed, Hapke and Nelson (1975) have obtained a very good fit to the near-UV spectrum with incompletely polymerized sulfur particles mixed in with the sulfuric acid droplets. Pollack *et al.* (1979) have shown from Pioneer Venus data that SO_2 is also a candidate for the UV absorber but only at wavelengths <0.35 μm.

Prinn (1975) emphasized that if re-equilibration of the photochemically formed SO_3 to SO_2 and finally to COS in the hot lower atmosphere was not complete, a significant fraction of the upward flux of gaseous sulfur may be in the form of SO_2 or SO_3 rather than COS. Indeed, as mentioned earlier, if vertical mixing in the atmosphere is sufficiently slow, then SO_2 and SO_3 will in fact dominate throughout the photochemically active region of the Venus atmosphere. We now know that SO_2 is indeed the dominant gas, at least above 22 km.

Cloud models based on the photochemical hypothesis are relatively insensitive to assumptions as to the exact nature of the precursor sulfur gases. Thus, the early models computed by Prinn (1973, 1975), which included vertical mixing and particle sedimentation and were constructed prior to the availability of observations, were remarkably accurate when compared to the later observations (see Figures 4.8 and 4.9).

In summary, we see that the chemistry of sulfur gases on Venus is driven by two essentially competing influences (Prinn, 1973; Florenskii *et al.*, 1978). A photochemical region exists within and above the clouds in which the chemistry is driven by solar UV radiation and which produces oxidized, thermochemically unstable sulfur compounds. This photochemical region may also extend down conceivably to altitudes as low as 10 km with reactions driven by solar near-UV and visible radiation (Prinn, 1979). In contrast, a thermochemical region exists at and near the surface where high temperatures and pressures enable purely thermochemical reactions to proceed which produce reduced, thermochemically stable compounds. The observed composition of the atmosphere depends upon the relative rates of reactions in these two regions and upon the rate of vertical transport between them. Our interpretation is critically dependent upon the analytical data for sulfur gases: If the Venera 13 and 14 gas chromatograph results stand the test of time, then we would be obliged to conclude that equilibrium can dominate up to an altitude of 30–35 km.

The role of COS in producing H_2SO_4 on Venus through the reaction sequence (4.20) has now lost its importance due to the dominance of SO_2. However, it is

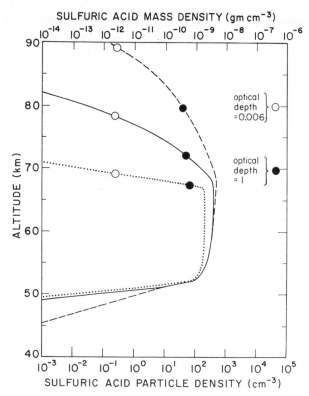

FIGURE 4.8. Vertical distribution models of the Venus clouds. The three curves reflect different assumptions regarding the dependence of the eddy mixing rate on altitude: the values of K (cm²sec⁻¹) are as follows:

	Below 57 km	57–67 km	Above 67 km
Dotted curve	$K = 10^5$	10^4	10^3
Solid curve	10^5	5×10^4	10^4
Dashed curve	10^6	2×10^5	7×10^4

The levels where the vertical optical depths are *observed* to be 0.006 (○) and 1 (●) are ~80 and 70 km, respectively. The points on the graph mark where the given cloud mass, assumed to be present as 1-μm particles, produces these optical depths for each of the three mixing models. The second model fits the observations best. (After Prinn, 1975.)

interesting to note that the very same reaction sequence was subsequently shown to be a major source of H_2SO_4 in the Earth's stratosphere (Crutzen, 1976) some 3 years after its ill-fated debut on Venus (Prinn, 1973).

The photochemistry of sulfur and chlorine compounds was readdressed by Winick and Stewart (1980) and by Krasnopol'skii and Parshev (1980a,b) after the 1978 American and Russian missions. These authors considered reactions involving chlorine and odd hydrogen in addition to sulfur compounds and included a number of chemical pathways not considered in the first studies of these three classes of compounds on Venus (Prinn, 1971, 1973; McElroy et al., 1973). Winick and Stewart (1980) successfully modeled the vertical profiles of SO_2 and cloud densities above 60 km, but their chemical model produced O_2 and CO abundances, respectively, 6 and >100 times the observed above-cloud values. Krasnopol'skii and Parshev (1980a,b) also successfully modeled the vertical SO_2 distribution and, in addition, predicted CO and O_2 concentrations which agreed much better with observations. However, their scheme for recombining CO and O_2 utilized the following reaction first proposed by Prinn (1971):

$$ClO + CO \rightarrow Cl + CO_2 \tag{4.21}$$

and assumed a rate constant for this reaction several orders of magnitude greater than that adopted by Winick and Stewart (1980). They also included formation of COCl, also proposed by Prinn (1971), and the subsequent reaction

$$COCl + O_2 \rightarrow CO_2 + ClO \tag{4.22}$$

in fact dominates recombination of CO and O_2 in their calculations. The kinetics of COCl are at present poorly defined.

Therefore, this latter reaction is speculative, but certainly worthy of further attention. Indeed, since the Winick and Stewart (1980) model represented a serious attempt to study the problem using only those reactions whose rates are reasonably well defined, it is clear that additional reactions must be postulated and the necessary laboratory work carried out to substantiate them. If one thing can be concluded from all of these investigations, it is that the question of how CO and particularly O_2 are kept at such low concentrations in the upper atmosphere of Venus is not yet convincingly answered despite over a decade of work on the problem. Four-center gas-phase oxidation reactions (e.g., between O_2 and COCl, SOCl, or SO_2Cl) and oxidation reactions in acidic cloud droplets (e.g., between dissolved H_2O_2, O_3, or chlorine oxides and SO_2 or H_2S) seem promising areas for laboratory investigations which might remove the present impasse. It is also apparent that the telescopic observation implying a cloud-top O_2 mixing ratio $<10^{-6}$ (Traub and Carleton, 1974) needs further corroboration in view of its great importance to our understanding of the upper-atmospheric chemistry.

We should add that although the importance of COS and H_2S is surely much less than proposed prior to the 1978 Pioneer Venus and Venera missions, these gases may nevertheless be playing important roles as O_2 sinks and elemental sulfur sources in the cloud region.

The photochemistry of H_2S has been studied by Prinn (1978). Depending on the abundances of O_2 and other oxidants in the cloud-top region, the SH radicals produced from H_2S photodissociation can lead either to H_2SO_4 or to elemental sulfur production. Our current perspective would suggest that H_2S is much more important as an O_2 sink and sulfur source than as a source of H_2SO_4:

$$\begin{aligned}
H_2S + h\nu &\rightarrow SH + H, \\
SH + SH &\rightarrow S + H_2S, \\
SH + O_2 &\rightarrow SO + OH, \\
S + O_2 &\rightarrow SO + O.
\end{aligned} \tag{4.23}$$

Formation of elemental sulfur from H_2S is also possible if H_2S dissolved in acidic cloud particles is oxidized by dissolved H_2O_2 (Satterfield et al., 1954).

Like H_2S, photodissociation of COS can lead to either sulfuric acid or sulfur formation. Current knowledge implies that COS present in the cloud region with a mixing ratio $\sim(2-3) \times 10^{-6}$ would be important as an O_2 sink and an elemental sulfur source through the following reactions:

$$
\begin{aligned}
COS + h\nu &\rightarrow CO + S, \\
SO + h\nu &\rightarrow S + O, \\
S + O_2 &\rightarrow SO + O, \\
S + COS &\rightarrow S_2 + CO,
\end{aligned}
\tag{4.24}
$$

but that, like H_2S, it is not an important contributor to H_2SO_4 formation.

Vertical mixing rates above the Venusian clouds are important in the photochemistry of both chlorine and sulfur species. Fortunately, they seem better understood than the chemistry. In the 60–80 km region, vertical eddy diffusion coefficients of $\sim10^4-10^5$ cm^2 sec^{-1} are compatible with the cloud particle scale height (Prinn, 1974), the SO_2 scale height (Winick and Stewart, 1980), and analyses of radio signal scintillations (Woo and Ishimaru, 1981). Above 80 km, diffusion coefficients increasing upward inversely as the square root of the density are reasonable on dynamical grounds (Prinn, 1975) and provide a good fit to minor constituent distributions in the thermosphere (von Zahn et al., 1979b; Stewart et al., 1980).

Finally, let us address the problem of the recycling of photochemically destroyed species. As discussed in Section 4.2.2, the observations definitely show that SO_2 is being destroyed in the visible atmosphere. According to Winick and Stewart (1980), the column SO_2 destruction rate (and column H_2SO_4 production rate) is $\gtrsim1.8 \times 10^{11}$ cm^{-2} sec^{-1}, and we must therefore reform SO_2 from H_2SO_4 in the lower atmosphere at this same rate. The products and rates of the thermochemical destruction of H_2SO_4 were discussed by Prinn (1978). The observed presence of carbon monoxide below the clouds makes the reaction

$$
SO_3 + CO \rightarrow SO_2 + CO_2
\tag{4.25}
$$

probable. In principle, this single thermochemical reaction may be all that is necessary to reverse the net tendencies dominant above the clouds, namely,

$$
\begin{aligned}
CO_2 &\rightarrow CO + \tfrac{1}{2}O_2, \\
SO_2 + \tfrac{1}{2}O_2 &\rightarrow SO_3.
\end{aligned}
\tag{4.26}
$$

Unfortunately, the rates of SO_3 reactions are essentially unknown, although SO_3 is undoubtedly more reactive than SO_2. In any case, reduction of SO_3 to SO_2 must proceed in some manner on Venus in order for SO_2 to survive.

The existence of H_2S, S_3, S_4, and perhaps COS below the clouds, together with recognized photochemical paths for converting these species into H_2SO_4 in the visible clouds, also demands that sources exist for these reduced species. As discussed above (Prinn, 1978), reduction of SO_2 by CO near the hot surface is suggested by laboratory experiments with S_2, COS, and H_2S being the probable products. The spectroscopy and probable sources of S_2, S_3 (thiozone), and S_4, and the consequences of their presence in the lower atmosphere have been discussed by Prinn (1979). Of particular interest is the photodissociation of S_4 and perhaps S_3 by near-UV, blue, and yellow light, and thus the existence of a photochemical regime below the visible clouds driven by solar radiation at much longer wavelengths than usually considered by atmospheric chemists. The presence of photochemically produced S and S_2 in the lower atmosphere would provide an important sink for carbonyl sulfide through the following chain reactions:

$$S + COS \rightarrow CO + S_2,$$
$$S_2 + COS \rightarrow CO + S_3, \qquad (4.27)$$
$$S_3 + COS \rightarrow CO + S_4.$$

Whether these photochemical sulfur atoms can also attack H_2S, H_2O, CO_2, or H_2 near the hot Venusian surface is not known. Laboratory studies are required before the importance of this photochemical regime below the clouds can be further assessed.

To this point we have assumed that the photochemical processes on Venus are in a steady state, and that all material cycles are closed. However, since H_2 gas can be produced by photolysis of HCl, H_2O, and possibly H_2S, it is important to see whether this gas can escape from Venus. If so, then the atmosphere is to some degree an open system: More precisely, the *atmosphere–lithosphere system* would be open to H loss. The bulk chemistry of the atmosphere and crust would undergo secular changes. In addition, ^{4}He from U and Th decay may either escape or accumulate, depending sensitively on the upper-atmosphere temperature. A quantitative discussion of atmospheric escape is therefore an essential element in our treatment of atmospheric evolution and is provided in Section 4.2.6.

4.2.5. Cloud Composition and Structure

A large and vigorous literature has grown up regarding the composition of the Venus clouds. Perhaps the principal reason for the attention lavished on this subject is that the clouds have often been regarded as evidence for large amounts of water in the Venus atmosphere, and hence of a particular (and interesting) class of models for the evolution and present state of the planet.

As we have indicated earlier, early spectroscopic studies of Venus uniformly failed to detect water vapor. As early as 1940, Wildt (1940b) concluded that the cloud-top region was too dry for H_2O condensation as ice or supercooled water and tentatively proposed that the clouds were a photochemical smog. He initially proposed that the cloud materials were polymers of formaldehyde, but upon review of the vapor pressure relations of those species he concluded that there could not be enough gaseous H_2CO present to permit polymers to exist without seriously violating the spectroscopic upper limit on formaldehyde (Wildt, 1942).

Dauvillier (1956), in the same vein, proposed that lightning discharges in a moist, nitrogen-rich atmosphere could generate a smog of ammonium nitrite

(NH_4NO_2) particles. Lewis (1969c) pointed out that an atmosphere moist enough for this mechanism to work would be able to precipitate much larger amounts of H_2O by direct condensation, and hence that there was no obvious utility to this process.

Yet another disequilibrium source of cloud particles was proposed by Robbins (1964), who suggested that reaction products from the partial reduction of CO_2 by solar wind hydrogen might form an optically thick haze.

The case for large amounts of water was argued by Menzel and Whipple (1955). Dense water clouds were employed in the interpretation of the microwave emission proposed by Deirmendjian (1964), in conjunction with disequilibrium (nonthermal) emission sources, which we discussed earlier. Bottema et al. (1964, 1965a,b) argued strongly for the presence of water clouds and a cool, wet surface. A microwave transmission model by Vetukhnovskaya and Kuz'min (1967a) required several grams of polar liquids, presumably water, per square centimeter of surface in the form of small cloud droplets. An enthusiastic appraisal of the prospects for Earthlike life in the clouds of Venus was offered by Morowitz and Sagan (1967) in the context of assumed liquid clouds near 280 K. Welch and Rea (1967), however, proposed a very restrictive upper limit on the water content of the clouds based on the microwave spectrum, and Rea and O'Leary (1967) interpreted the reflection spectrum of Venus in terms of the spectrum of gaseous CO_2 overlying a cloud free of ice and liquid water. Hansen and Cheyney (1968a) mildly criticized Rea and O'Leary's conclusions, and concluded that irregularly shaped micron-size ice particles could not be ruled out by current data (Hansen and Cheyney, 1968b).

Pollack and Sagan (1968) identified the cloud-top material as water ice crystals, in accord with an interpretation of the microwave emission spectrum by Pollack and Wood (1968) which claimed about 1% water vapor in the atmosphere. Potter (1969), interpreting the spectroscopic data on the water vapor abundance in light of new knowledge of the structure and optical properties of the cloud-top region, strongly concluded that the spectroscopically accessible clouds contain 20 times too little water vapor to be compatible with the presence of ice.

Plummer (1969b) argued for the presence of small (~ 2-μm diameter) ice crystals, and O'Leary (1970) changed his previous opinion as a result of an apparent photometric detection of a halo due to ice. Plummer (1970) reaffirmed his belief that the IR spectrum required the presence of ice. Fukuta et al. (1969) also argued strongly in favor of water ice clouds, which were an essential adjunct to Libby's (1968a) massive icecaps. Although criticized by Lewis (1970b) for omitting consideration of much telling evidence, they remained unshaken in their belief (Fukuta et al., 1970).

Veverka (1971) tested the conclusions of O'Leary (1970) by searching for polarimetric evidence of hexagonal ice platelets, but was unable to verify the presence of ice. Ward and O'Leary (1972), based on newer and more sensitive observations, were likewise unable to verify the presence of an ice halo by photometry. Schorn and Young (1971) offered a reinterpretation of Plummer's (1970) spectrum in which the data could be well fitted without the use of any ice clouds. Here we also should recall the polarimetric evidence for a very high refractive index (see, e.g., Hansen and Arking, 1971). Finally, Chamberlain and Smith (1972) showed that, for a range of light-scattering models of the cloud layer, water vapor would be very far from saturation at the observed cloud temperature. By 1972, therefore, the weight of evidence had shifted very strongly against the presence of H_2O clouds.

Lewis (1968a) pointed out that volatile solutes such as the observed species HCl might assist in condensing water at very low temperatures. Lewis (1968b) demonstrated that, in the presence of the observed proportions of water vapor and HCl, the first condensate to be formed upon cooling would be an extremely concentrated aqueous HCl solution, and that this solution could not condense above 216 K. He thus concluded that such a solution would have to be too cold to be identifiable with the main cloud layer. He suggested condensation of NH_4Cl below the 300 K level, in very small amounts.

The idea of aqueous solution acid clouds seemed promising as a working hypothesis. Arking and Nagaraja Rao (1971) pointed out that the refractive index of ordinary aqueous HCl at room temperature was noticeably lower than that deduced from polarization studies of the Venus clouds, although, of course, much closer to the observed refractive index than water or ice. Lewis (1971c) calculated the refractive index at realistic temperatures and concentrations for the high-altitude haze layer and found a much closer agreement with the observed value. Also, Lewis (1972c) suggested an interpretation of the spectra of Beer et al. (1971a) as due to aqueous hydrochloric acid solution containing Cl_3^- ion. Young and Young (1973) later proposed yet a third interpretation of Beer's data, in which CO_2 gas overlying aqueous sulfuric acid solution was shown to be capable of explaining the spectrum and the refractive index satisfactorily.

A useful review of the state of our understanding of the Venus clouds on the eve of the identification of sulfuric acid was published by Rea (1972).

Plummer (1969a) showed that the reflection spectrum of the clouds was devoid of any feature near 2.4 μm, where hydrocarbons have strong absorption features. He thus ruled out the proposals by Hoyle (1955), Mintz (1961), and Kaplan (1963) that hydrocarbon clouds might be present. He directed his arguments, however, against the speculative ideas of Velikovsky (1950), although the latter nowhere mentions hydrocarbon clouds, but often mentions gaseous hydrocarbons.

Kuiper and Sill (1969) and Kuiper (1969) "identified" the Venus clouds as made of solid hydrates of $FeCl_2$ near and above 240 K, with a thin haze of NH_4Cl at high altitudes. Lewis (1968b) had already shown that NH_4Cl is much too involatile to remain in the gas down to such low temperatures (near 200 K), and Lewis (1970c) showed that any geochemically plausible reaction for making $FeCl_2$ vapor at the surface would provide a mole fraction below 10^{-10} in the lower troposphere. Condensation would then occur at temperatures above 500 K, not near 240 K. A reanalysis of the reflection spectrum by Cruikshank and Thomson (1971) showed no positive evidence for the presence of $FeCl_2$ or its hydrates.

As we mentioned earlier, the geochemical models by Lewis (1968a, 1970c) had led to the conclusion that the sulfur content of the lower atmosphere should be substantially higher than the abundance of gaseous sulfur compounds above the clouds. Therefore, it followed that some mechanism for removal of sulfur from the atmosphere by formation of cloud particles must occur. Lewis (1969c) showed that volatile trace metals such as Hg, As, and Sb could possibly fill this role, and proposed a model in which the cloud tops are made of Hg_2Cl_2 overlying a possible layer of liquid mercury droplets and HgS clouds. Rasool (1970) adopted this model in his attempt to explain the observed attenuation history of the radio signals in the Mariner 5 occultation experiment. Potter (1972) showed that the Hg droplet clouds, if on top, would have too low an albedo to fit Venus. He did not treat the effects of a high-altitude haze or of a bright Hg_2Cl_2 frosting.

Rea and O'Leary (1968) suggested carbonate or silicate dust as the material of the main cloud deck, in keeping with the aeolosphere theory of Öpik (1961). Lewis (1969c) pointed out that the observed presence of HCl, which is extremely reactive with carbonates, and HF, which readily attacks silicates, rules out the presence of such dust. Beer et al. (1971a) revived the dust theory by proposing alkali bicarbonates as the main cloud material, a proposal with the same weakness as the foregoing. Hapke (1972) explained the visible and UV reflection spectrum by doping the aqueous HCl solution clouds suggested by Lewis with high concentrations of ferric chloride in order to give a yellow tint to the planet. However, $FeCl_3$ is neither volatile enough to reach such high altitudes as a gas, nor stable enough as a solid to exist as a dust near the surface. Further, the HCl solution clouds, if present, would have to be at much higher altitudes and lower temperatures than the main cloud deck.

Plummer and Carson (1970) showed that photochemically produced carbon suboxide (C_3O_2) was a most implausible cloud material because of the very restrictive upper limit on the vapor abundance.

Among suggestions of other cloud materials, Sill (1975) has proposed bromine in hydrobromic acid as a coloring agent for the clouds, exactly replacing Lewis' (1972c) suggested HCl_3 (from observed HCl) with HBr_3 (from unobserved HBr).

As we discussed in detail in Sections 4.2.2–4.2.4, there is now a general consensus that a major component of the Venus clouds is concentrated sulfuric acid formed photochemically from SO_2. This consensus should not, however, be automatically construed to imply that H_2SO_4 is the only or even dominant cloud constituent. Of some interest still are possible cloud constituents containing chlorine. Whereas the observed mixing ratio of HCl above the clouds is $\sim 10^{-6}$, we have only an *upper limit* for HCl $\sim 10^{-5}$ below the clouds (Hoffman et al., 1980). In addition, chlorine in an unknown chemical form was detected in the cloud particles themselves by the X-ray fluorescence experiment on Venera 12 (Surkov et al., 1981). The reported condensed chlorine density of 2×10^{-9} gm cm^{-3} refers to atmospheric pressures ~ 1 bar and thus implies a Cl : CO_2 ratio $\sim 10^{-6}$, which seems reasonable. The reported particulate Cl : S ratio $\gtrsim 10$ is surprising, but the analysis was apparently carried out at temperatures >350 K so that any H_2SO_4 would have evaporated from the particles. Thus, the large Cl : S ratio may simply imply that Cl is in a less volatile form than H_2SO_4 in the cloud particles. The X-ray data combined with the Pioneer Venus observations implying that crystals and not droplets may be dominant in the lower portion of the clouds (Knollenberg and Hunten, 1980) suggest that further consideration of chlorine species as cloud components is worthwhile.

Condensation of HCl itself in the main cloud layer is not viable. The phase relations in the H_2SO_4–HCl–H_2O system have been explored by Volkov et al. (1979), who find that condensation of the two acids is uncoupled. In order for HCl solution to condense in the main cloud layer, H_2SO_4 must be locally absent and H_2O abundant (a mole fraction of at least 10^{-3}). This is in agreement with the conclusions of Lewis (1968b) regarding the HCl–H_2O system, and relegates HCl–H_2O to the role of a high-altitude haze condensing below 215 K. As we have seen, the temperature minimum in the upper stratosphere is no higher than $\sim$180 K, so such a condensate may form if HCl has not been photochemically exhausted at this level.

As reviewed by Knollenberg and Hunten (1980), none of the other chlorine-containing candidates for the clouds suggested up until 1980 seem satisfactory. In this respect, Prinn (see von Zahn et al., 1982) has recently emphasized the importance of photochemically produced halogen compounds as possible cloud candidates. In particular, the photochemical production of reactive chlorine oxides from HCl suggests that more stable oxides (e.g., Cl_2O, OClO, Cl_2O_7) and oxyacids (e.g., HOCl, $HClO_3$, $HClO_4$) of chlorine may be produced as by-products. Of these possibilities, perchloric acid ($HClO_4$) appears most attractive. Like H_2SO_4, $HClO_4$ is a very strong acid with a low vapor pressure and a refractive index $\sim$1.38–1.42 depending on temperature and concentration. It also forms stable mixtures with concentrated sulfuric acid. In addition, as water is removed from the concentrated acid and/or as the temperature decreases, a number of crystalline $HClO_4$ hydrates are formed with successively higher and higher melting points. Of greatest interest for Venus is the monohydrate $HClO_4 \cdot H_2O$, which forms colorless monoclinic or orthorhombic crystals with a melting point $\sim$323°C. However, a major unresolved question concerning $HClO_4$ is its precise mechanism of formation in the clouds.

Although the question of the detailed composition of the Venus clouds is not at this point fully resolved, we do possess a considerable amount of information on the vertical structure of these clouds. Beginning at the top of the atmosphere, Aumann and Orton (1979) have deduced from their thermal IR spectra (12–20 µm) the existence of a haze layer near 1 mbar and near or slightly below 200 K. The Pioneer Venus orbiter photopolarimeter experiment (Travis et al., 1979) has also found a high-altitude haze layer. Preliminary analyses of their data place the haze very roughly near the 10 mbar level, and hence at or below 200 K.

The cloud structure in the lower regions of the atmosphere, inaccessible from above in visible light, must be deduced from spacecraft data. We have already seen the inflection in the temperature profile found by radio occultation and entry probe experiments near the 350 K level (see Figure 4.2), which may in principle be due to either an abrupt vertical change in the rate of deposition of sunlight, or to condensation of a minor constituent of the atmosphere, or to the large-scale circulation.

Direct probing of the transmission of both solar visible and planetary IR light through the atmosphere was carried out by the Venera 8, 9, and 10 spacecraft. Venera 8 measured the visible flux (Lukashevich *et al.*, 1974; Biryukov and Panfilov, 1974) and the near-IR component of the solar flux (Gutshabash and Safrai, 1975). Venera 9 and 10 not only measured the visible flux to the surface (Avduevskii *et al.*, 1976b), but also carried nephelometers to determine the cloud density as a function of altitude (Marov *et al.*, 1976). Some constraints on the cloud properties can also be deduced from the Venera 9 television images taken on the surface (Kerzhanovich *et al.*, 1979). The Pioneer Venus large probe carried a particle size spectrometer (Knollenberg and Hunten, 1979a) and all four probes carried nephelometers (Ragent and Blamont, 1979). The large probe also carried a solar flux radiometer (Tomasko *et al.*, 1979a) and an IR radiometer (Boese *et al.*, 1979).

Blamont and Ragent (1979) report four distinct regions of the atmosphere based solely on light-scattering behavior. Below an altitude of about 46 km there was no discernible cloud opacity. A sharply defined, thin dense cloud layer was observed between about 47 and 49 km, a less dense but deeper layer was found between about 49 and 58 km, and a diffuse haze extending from about 58 to 63 km, constituting the "visible cloud layer" seen from above, was also measured. The particle size results (Knollenberg and Hunten, 1979b) also found this same layered structure

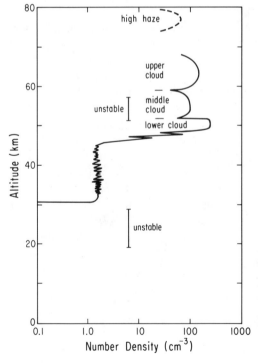

FIGURE 4.9. **Cloud layers and thermal structure observed by Pioneer Venus. Cloud particle density profiles are from Knollenberg and Hunten (1979a,b) and from Blamont and Ragent (1979). Thermal stability data are from Seiff et al. (1979b). The high haze of Ksanfomaliti (1979) is also schematically indicated. Compare to the theoretical predictions given by Prinn (1975) in Figure 4.8.**

with the same three identified cloud layers plus a haze layer from 31 to 46 km. The atmosphere was clear below 31 km. They report also the existence of three distinct size populations of particles. The smallest, with mean sizes well below 1 μm in diameter, are found at all altitudes and may be mainly elemental sulfur or some heretofore unidentified aerosol (Tomasko et al., 1979b). The second size class, with mean diameters of 2–3 μm, almost certainly is dominated by droplets of sulfuric acid. The largest size particles, mostly near 7–8 μm in diameter, are found only in the middle and lower cloud layers, and are absent in the lower atmosphere. These may be large H_2SO_4 droplets.

Dust was found to be extremely rare in the bottom 31 km, arguing against extensive volcanism or condensation of moderately volatile outgassed substances at temperatures in excess of 500 K.

Figure 4.9 summarizes the cloud and static stability data from Pioneer Venus. The regions marked "unstable" are those in which the vertical temperature gradient is steeper than the adiabatic gradient.

4.2.6. Atmospheric Escape

For many years the similarities in size and mass of Venus and the Earth, combined with their proximity in the Solar System, have been used as an argument for expecting considerable "commonality" in the primordial outgassing from these two planets (Ingersoll, 1969; Walker et al., 1970; Rasool and de Bergh, 1970; Smith and Gross, 1972; Walker, 1975). The conventional problem then became to explain why the Cytherean atmosphere possesses 10^4–10^5 times less water than is present in the terrestrial oceans. It is certainly clear from our earlier discussion of atmospheric–lithospheric interactions that this water could not be buried as hydrated crustal minerals under present surface conditions. Proponents of the "commonality" argument are therefore forced to seek schemes for removing an amount of hydrogen equal to that in the terrestrial oceans ($\sim 10^{28}$ atoms cm^{-2}) by escape from the exosphere. Averaged over geologic time, the required escape rate is $\sim 10^{11}$ H atoms cm^{-2} sec^{-1} compared to the presently observed escape rate of $\sim 10^4$–10^7 H atoms cm^{-2} sec^{-1}. At the same time $\sim 5 \times 10^{27}$ O atoms cm^{-2} had to be buried in the Venusian crust, presumably by oxidation of Fe° or Fe^{2+}, in order to keep O_2 below observable levels.

The above scenario may be contrasted with the hypothesis that Venus formed from material considerably depleted in water relative to the Earth (Holland, 1964; Anders, 1968; Lewis, 1972b). The latter hypothesis is, of course, the one which we preferred from the cosmochemical arguments presented earlier. It clearly does not make excessive demands on either H escape rates or surface oxidation rates. Nevertheless, the question of how much hydrogen could have escaped from Venus over geologic time is an important one for predicting both its primordial water content and the present degree of oxidation of its surface. The latter property could, for example, conceivably be measured by a surface lander.

Before discussing the proposed mechanisms for removal of very large amounts of water from Venus, a discussion of the present situation is in order. The following three important phenomena will have to be understood: (1) the chemical source for the escaping H atoms, (2) the rate of diffusion of hydrogen to the escape level, and (3) the rates of thermal and nonthermal excitation of hydrogen enabling escape.

Early studies of hydrogen source chemistry on Venus assumed that photodissociation of H_2O was the main source of H_2 for escape (e.g., Rasool, 1968; Donahue, 1968; McElroy and Hunten, 1969; Walker *et al.*, 1970). As partial validation of this assumption McElroy and Hunten (1969) utilized a theoretical temperature profile to argue against the possibility of a "cold trap" on Venus which would restrict H_2O mixing ratios in the Cytherean upper atmosphere. For example, the cool equatorial tropopause on Earth forms such a "cold trap" and restricts upper atmospheric H_2O mixing ratios to be $\sim 10^{-6}$. However, the theoretical temperature calculations used by McElroy and Hunten (1969) implied temperatures that were too high below 90 km in comparison to the later Mariner 5 measurements (Fjeldbo *et al.*, 1971). In addition, the influence of HCl and the (at that time unidentified) H_2SO_4 clouds on H_2O vapor pressures must now be taken into account.

The influence of the sulfuric acid clouds on upper atmospheric H_2O is particularly important. Although a value for $X_{H_2O} \simeq 10^{-4}$ is feasible below these clouds, the spectroscopic observations imply $X_{H_2O} \simeq 10^{-5}-10^{-6}$ within and above the clouds. These latter low mixing ratios are undoubtedly attributable to sulfuric acid formation (Young, 1973; Wofsy and Sze, 1975). Above the sulfuric acid clouds a further desiccating influence may be played by hydrochloric acid condensation above the 10 mbar level and ice and HCl-hydrate condensation above the 1 mbar level (Lewis, 1972c; Prinn, 1973).

A problem for any photodissociating minor species on Venus is that it must compete with the enormous excess of CO_2 for available UV photons. Absorption by CO_2 extends at least out to 2100 Å (Inn and Heimerl, 1971) and is particularly strong below 1900 Å. Nevertheless, if all the H fom such H_2O photodissociation were available for escape, then escape fluxes of $(0.3-3) \times 10^{10}$ atoms cm^{-2} could still be sustained. Such escape fluxes are well in excess of what is observed, but obviously fall short of the H escape flux needed to deplete a terrestrial-sized ocean over geologic time.

There is, however, one known atmospheric constituent, namely HCl, which absorbs at least out to 2700 Å and which can therefore capture a large fraction of the photons between the CO_2 absorption cutoff and ~ 2700 Å. In a study of the photochemistry of HCl, Prinn (1971, 1973) concluded that the rate of the photodissociation reaction

$$HCl + h\nu \rightarrow H + Cl \tag{4.28}$$

is 10^2-10^3 times that of H_2O for $X_{H_2O} = 10^{-5}-10^{-6}$. It thus now appears that HCl is much more important than H_2O as a source of H atoms in the Venus atmosphere. Some of these H atoms produce H_2 through the reaction

$$H + HO_2 \rightarrow H_2 + O_2, \tag{4.29}$$

and the H_2 in turn becomes another source of H atoms through the reaction (Prinn, 1971; McElroy *et al.*, 1973)

$$CL + H_2 \rightarrow HCl + H. \tag{4.30}$$

Some of the H_2 is transported upward and ultimately provides a source for the escaping H atoms.

We should add here that, despite the above chemistry, the ultimate source of the H for escape is still H_2O. If HCl is the principal H_2 source, the upward H_2 flux to the escape level must be balanced by an equal downward flux of Cl_2 into the lower atmosphere. There, HCl can be reformed from H_2O and Cl_2 by thermochemical reactions in the hot lower troposphere and at the surface, for example,

$$CO_2 + 3FeMgSiO_4 \rightarrow CO + 3MgSiO_3 + Fe_3O_4, \tag{4.31}$$

$$CO + H_2O \rightarrow CO_2 + H_2, \tag{4.32}$$

$$H_2 + Cl_2 \rightarrow 2HCl. \tag{4.33}$$

Reaction (4.33) is simply reversed in the upper atmosphere and, for every 2 H atoms which ultimately escape, reaction (4.31) implies one Fe^{2+} ion is oxidized to Fe^{3+}. Also note that reaction (4.32) implies that a thermochemical H_2 source exists in the lower atmosphere to augment the upper-atmospheric photochemical source (4.9). Indeed, the predicted H_2 mixing ratios from the two sources are quite similar (about 10^{-5}–10^{-6}; Donahue, 1968; Lewis, 1970c; Prinn, 1971). Loss of H_2O, from Venus may, therefore, in part involve only reactions (4.32) and (4.31) without the additional reactions (4.28)–(4.30).

Having established, as it were, that suitable methods for converting H_2O to H_2 and surface oxygen indeed exist, we can now discuss the specific problems of upward hydrogen diffusion and escape.

Several lines of evidence are available for estimating the upper-atmospheric temperature near the escape level. The electron scale height measured in the ionosphere by the Mariner 5 radio occultation experiments (Mariner Stanford Group, 1967; Kliore *et al.*, 1967) yielded dayside upper-atmosphere temperatures of 500–700 K (Stewart, 1968; McElroy, 1968b; Hogan and Stewart, 1969). A model by Whitten (1969) predicted that the ion temperature in the ionosphere ought to be the same as the neutral gas temperature. A theoretical energetic–dynamical model of the thermosphere developed by Dickinson (1971, 1972, 1973; Dickinson and Ridley, 1972, 1977) predicted large temperature differences between the day side and night side, with UV heating leading to the production of a large, hot, dayside thermospheric bulge. High-altitude winds would then transport photochemical products, including reactive atoms and radicals, to the night side, where general subsidence of the thermosphere would result due to radiative cooling. Temperatures in the nightside "thermosphere" were predicted to range as low as 170 K.

Izakov and Morozov (1975) presented a similar model with temperatures varying from 800 K at the subsolar point to 300 K at the antisolar point. Dickinson (1976) investigated the effect of a large eddy conductivity, whereby rapid forced turbulence leads to a nearly adiabatic temperature decrease in the nightside thermosphere, under conditions in which radiative cooling is very inefficient. In the wake of the Venera 9 and 10 radio occultation measurements on the nightside ionosphere (Aleksandrov et al., 1976), Izakov (1975) presented a detailed treatment of the eddy conduction cooling process. His results are especially important in light of the Pioneer Venus data on the upper-atmosphere structure (Niemann et al., 1979; Keating et al., 1979a,b; Knudsen et al., 1979), which reveal a nightside exospheric temperature in the range 85–110 K, far too low to be achieved by radiative cooling.

It has, thus, become very clear that thermal escape from the atmosphere can occur only in the immediate vicinity of the subsolar point (a solar zenith angle of $\lesssim 30°$).

The Mariner 5 Lyman-α (Barth et al., 1968; Anderson, 1976) and electron density (Eshleman, 1970) observations provide rough estimates of exobase H atom densities and exospheric temperatures (see McElroy, 1968a). We may therefore compute Jeans thermal escape rates for H atoms (McElroy and Hunten, 1969; Donahue, 1969). For exospheric temperatures of 700 and 1000 K (solar minimum and maximum, respectively) the escape fluxes deduced by McElroy and Hunten (1969) are 6.4×10^5 and 8.6×10^6 atoms cm^{-2} sec^{-1}, respectively. The mean flux should be about 25% of the maximum value. These numbers unfortunately cannot be considered very firm because of a number of difficulties which arise in interpreting the Lyman-α data (for a review, see Prinn, 1973). The leading interpretive problem is the presence of two components in the H Ly-α emission, one with a scale height about twice as large as the other (Barth, 1968; Anderson, 1976). These two components could be D and H or H_2 and H. In the former case, vast amounts of D would have to be present on Venus (Donahue, 1969), which in turn would suggest the loss of vast amounts of H in a selective (Jeans) escape process. High-resolution rocket UV spectroscopy was employed by Wallace et al. (1971) in an attempt to detect the H and D Ly-α components. Their count rates (relative to the background) in the H and D channels were 32 ± 8 and -8 ± 8, respectively. By Donahue's model D : H ratios of 1–10 would have been expected. Walker et al. (1970) used the detailed Lyman-α analysis of Wallace (1969) to deduce a Jeans escape rate of 2×10^5 atoms cm^{-2} sec^{-1}. This is somewhat lower than those deduced by McElroy and Hunten (1969), but the assumed exospheric temperature (640 K) was also lower. The escape flux for the heavier H_2 molecules is, as expected, much less than that of H: Walker et al. (1970) give an extreme upper limit of 5×10^3 molecules cm^{-2} sec^{-1} and Nedyalkova and Turchinovich (1973) prefer a much smaller flux.

Walker et al. (1970) also considered a number of possible nonthermal escape mechanisms (see Section 3.6.2). Ions may escape as a polar wind or by solar wind sweeping. An *upper limit* on the escape rate by either process is given by the ion production rate above the ionospheric peak which is $\sim 5 \times 10^4$ ions cm^{-2} sec^{-1} for H$^+$ and $\sim 6 \times 10^6$ ions cm^{-2} sec^{-1} for $H_2{}^+$. The actual escape rates are undoubtedly considerably smaller than these upper limits (see Nedyalkova and Turchinovich, 1975). Exothermic reactions may also play a role. Some of the hydrogen atoms produced by H_2 photodissociation have kinetic energies >0.54 eV, the energy required to escape from Venus (Barth, 1968). An upper limit to the escape rate is provided by taking one-half the "hot" H atom production rate from H_2 photodissociation above the exobase. The resultant H escape flux is therefore $< 5 \times 10^6$ atoms cm^{-2} sec^{-1}.

Radiogenic ^{4}He, which is marginally capable of escaping from Venus at a geologically significant rate, has been estimated to have an abundance of about 10–20 ppm below the turbopause from observations of the UV line emission of helium at 584 Å (Kumar and Broadfoot, 1975) and from mass spectrometry aboard the Pioneer Venus orbiter (von Zahn et al., 1979b). Knudsen and Anderson's (1969) Earth-like analogy Venus with no escape was predicted to contain about 200 ppm of ^{4}He. McElroy and Strobel (1969) likewise estimated $\sim$100 ppm of ^{4}He, and helium figures in the models of Whitten (1970) and Herman et al. (1971). Herman (1973) deduced a ^{4}He mixing ratio of 60 ppm in the troposphere, assuming a daytime exospheric temperature of 650 K. For higher exospheric temperatures, ^{4}He escape could become important.

The later Mariner 10 measurements of H Lyman-α and He 584 Å airglow intensities (Broadfoot et al., 1974a) and ionospheric electron densities (Howard et al., 1974; Bauer and Hartle, 1974) have also been used to shed light on the escape problem. The interesting point about this newer datum is that it is more compatible with exospheric temperatures of $\sim$350 K, or about half of that usually attributed to Mariner 5. Sze and McElroy (1975) have conducted a detailed analysis of both the Mariner 5 and Mariner 10 data. They conclude that the exospheric temperature was in actuality approximately 400 K during the flights of both Mariners and attribute the higher apparent Mariner 5 temperature to "hot" H atoms presumably produced by exospheric reactions (see also Hunten, 1973b). The H atom densities at the exobase were calculated to be 4×10^4 and 10^5 atoms cm^{-3}, respectively, for Mariners 10 and 5. For an exospheric temperature of 400 K the appropriate Jeans escape fluxes are 8×10^3 and 2×10^4 atoms cm^{-2} sec^{-1}, respectively. If these low thermal escape values are correct, it is entirely possible that nonthermal mechanisms dominate H atom escape from Venus.

In assessing nonthermal escape rates, Sze and McElroy (1975) emphasized the possible importance of a number of exothermic reactions producing hot H atoms for escape, in particular:

$$
\begin{aligned}
O^+ + H_2 &\rightarrow OH^+ + H, \\
OH^+ + e &\rightarrow O + H, \\
CO_2^+ + H_2 &\rightarrow CO_2H^+ + H, \\
CO_2H^+ + e &\rightarrow CO_2 + H, \\
O + H^+ &\rightarrow O^+ + H, \\
He^+ + H_2 &\rightarrow HeH^+ + H, \\
HeH^+ + e &\rightarrow He + H.
\end{aligned}
\tag{4.34}
$$

The resultant hydrogen escape flux could be as large as 4.4×10^6 atoms cm^{-2} sec^{-1}. They also estimate an additional H escape flux $\sim$10^6 atoms cm^{-2} sec^{-1}

generated by solar wind sweeping of H^+ ions. The latter value is significantly larger than that estimated by Walker *et al.* (1970).

However, the ability of the solar wind to sweep the ionosphere has been debated. The nature of the interaction between the planet and the solar wind depends on the presence or absence of a significant planetary magnetic field.

The nature of the magnetic field of Venus and the planet's interaction with the solar wind have been much debated since the initial measurements of magnetic field strengths and plasma behavior during the Venera 4 and Mariner 5 missions. The high altitude of Mariner 5 during the flyby resulted, not surprisingly, in no detection of a magnetic field or trapped particles. Van Allen *et al.* (1967) placed a firm upper limit on the dipole moment of Venus ($M_{\venus}$) of about 1% of Earth's moment and possibly less than 0.1%. Plasma and magnetic field measurements (Bridge *et al.*, 1967) disclosed a bow shock but no sign of a planetary field. The Venera 4 magnetometers (Dolginov *et al.*, 1968, 1969), operating much closer to the surface of the planet, placed an upper limit on the magnetic moment of $M_{\venus} < 11.3 \times 10^{-4} M_{\oplus}$. Plasma measurements on the same mission (Gringauz *et al.*, 1968) and on Venera 6 (Gringauz *et al.*, 1970) confirmed the existence of the shock front. Bauer *et al.* (1970) presented a model of the ionosphere–solar wind interaction. An excellent early review on the ionosphere is given by Gringauz and Breus (1968).

The Mariner 10 flyby magnetic field measurements (Ness *et al.*, 1974a) generally confirmed the Mariner 5 results, except that the shallower penetration of Mariner 10 into the plasmasphere and the detailed geometry of its encounter led to a complex record of planetary interaction with the solar wind. The plasma experiment on Mariner 10 (Brdige *et al.*, 1974) revealed a comet-like interaction between solar wind and ionosphere. Bauer and Hartle (1974) have presented a model of the interaction incorporating the Mariner 10 data.

A wholly different interpretation of the data taken by the closest measurement attempt, Venera 4, suggested a small planetary dipole moment of about $10^{-3} M_{\oplus}$ (Russell, 1976a), and a similar analysis of the structure of the magnetic wake of Venus observed by Venera 9 (Russell, 1976b,c) reinforced this suggestion. However, the thorough coverage of the plasmasphere by the Pioneer Venus orbiter magnetometer (Russell *et al.*, 1979a,b) suggests an upper limit on the dipole moment well below $10^{-4} M_{\oplus}$. The Pioneer Venus orbiter plasma analyzer has so far discerned no evidence for a planetary magnetic dipole with trapped energetic particles (Wolfe *et al.*, 1979; Intriligator *et al.*, 1979).

The safest conclusion at the moment is that any small planetary dipole which may exist is insufficient to affect the ionosphere–solar wind interaction, and hence is very unlikely to have any influence on escape rates.

Further studies of hydrogen escape rates on Venus were carried out following the acquisition of data by the Pioneer Venus mission. Kumar *et al.* (1981; see also Kumar *et al.*, 1978) suggested that the mass-2 ion detected by the Pioneer Venus ion mass spectrometer was H_2^+, which implies an H_2 mixing ratio $\sim 10^{-5}$ below the turbopause. The nonthermal H atoms (denoted as H*) produced by the reactions.

$$O^+ + H_2 \rightarrow OH^+ + H^*$$
$$OH^+ + e \rightarrow O + H^* \tag{4.35}$$

provided an H escape rate $\sim 10^8$ atoms cm^{-2} sec^{-1}. McElroy *et al.* (1982) alternatively suggested that the mass-2 ion was D$^+$, implying a D : H ratio $\sim 10^{-2}$ in the present atmosphere. The computed nonthermal H escape flux due to the reactions

$$O_2^+ + e \rightarrow O^* + O^*$$
$$O^* + H \rightarrow O + H^* \tag{4.36}$$

was $\sim 10^7$ atoms cm^{-2} sec^{-1}. The excited O atoms O* possess insufficient energy to eject D atoms from the exosphere. Extrapolating their H escape model back over geologic time, they estimated an initial H$_2$O abundance in the atmosphere about $\frac{1}{300}$ of the H$_2$O in the terrestrial oceans. Since escape of D is negligible, they compute a steady enrichment in the D : H ratio beginning with a value of 5×10^{-5} (an Earthlike value) 4.6 billion years ago and ending with their presently proposed value of 10^{-2}.

In summary, it would appear that the H escape flux from Venus is presently dominated by nonthermal processes although the precise nonthermal process which controls escape is still being debated. Current escape rates are calculated to be $\sim 10^7 - 10^8$ H atoms cm^{-2} sec^{-1} compared to the Jeans (thermal) escape rate $\sim 10^4$ atoms cm^{-2} sec^{-1}. For total H mixing ratios (H$_2$ + HCl + H$_2$O) at the turbopause of $(2-5) \times 10^{-6}$, Sze and McElroy (1975) estimated limiting fluxes ϕ_H^l as defined by Eq. (3.80) of $(4-8) \times 10^7$ cm^{-2} sec^{-1}. For a total H mixing ratio as large as 2×10^{-5} as suggested by Kumar *et al.* (1981), the value of ϕ_H^l should be $\sim 4 \times 10^8$ cm^{-2} sec^{-1}. Thus, the above nonthermal escape rates are not controlled by upward diffusion of H$_2$, but they are approaching this case.

One fact is particularly clear from our knowledge of the present H escape rates: They fall short of those required to deplete a terrestrial-sized ocean over geologic time by at least two orders of magnitude. The hypothesis developed to accomplish the latter purpose is colloquially referred to as the "runaway" greenhouse." As H$_2$O and CO$_2$ were outgassed into the atmosphere, the greenhouse effect caused surface temperatures to always exceed those required for oceans to accumulate as on the Earth (Rasool and de Bergh, 1970). This primitive Venus atmosphere was therefore composed principally of H$_2$O vapor with a somewhat smaller fraction of CO$_2$. Under such circumstances, Ingersoll (1969) argued that H$_2$O remains an important constituent even in the cool stratospheric regions. Thus, H$_2$O may absorb a very large fraction of the 10^{13} photons cm^{-2} sec^{-1} available in the 1600–2000 Å region. This is presumably an adequate source to sustain an escape flux of $\sim 10^{11}$ H atoms cm^{-2} sec^{-1} (Ingersoll, 1969).

In a brief study of the latter problem McElroy and Hunten (1969) argued that the thermosphere in a H$_2$O-dominated atmosphere would consist largely of atoms such as H and O which are poor radiators. Exospheric temperatures $\geqslant 2300$ K are thus quite possible and such temperatures would imply Jeans escape rates $\gg 10^{11}$ H atoms $cm^{-2} sec^{-1}$. The problem was pursued in more detail by Smith and Gross (1972). They constructed a photochemical model in order to predict the expected atmospheric composition at thermospheric levels. A radiative–diffusive thermospheric model was then used to imply exospheric temperatures in the 20,000–100,000 K range. These imply enormous Jeans escape rates, and the resultant fluid flow would in fact be inconsistent with the simple (nonhydrodynamic) assumptions of their model (see also Hunten, 1973b). In all probability one would expect the actual escape rate to at least be limited by hydrogen production and flow up to the exobase and by the rapidity of H_2O outgassing from the surface. If we allow a reasonable H escape flux of 10^{13} atoms $cm^{-2} sec^{-1}$, then the transition from the H_2O-dominated to an H_2O-poor atmosphere could have occurred in as geologically short a time as 5×10^7 yr.

Recently, Watson et al. (1981) have looked in more detail at the loss of H from water-rich atmospheres. Numerical solutions to the hydrodynamic equations for a solar wind-like outflow of gas indicate relatively low thermospheric temperatures <500 K. Such hydrodynamic outflow is expected when the internal energy of the gas exceeds its gravitational potential energy so that the Jeans escape theory is inapplicable. Watson et al. (1981) emphasize that the escape rate of H_2 from a water-rich Venus would be hydrodynamic and limited to a value $\sim 10^{12}$ molecules $cm^{-2} sec^{-1}$ since the energy for escape must be provided by the deposition of solar extreme UV radiation into the thermosphere. Also D is carried out along with the H until the H_2 escape flux decreases to $\sim 4 \times 10^{11}$ molecules $cm^{-2} sec^{-1}$. This latter flux is the diffusion-limited flux for a total hydrogen mixing ratio $\sim 2 \times 10^{-2}$. A plausible scenario for the early Venus atmosphere would therefore involve us beginning with the equivalent of an ocean of water followed by escape of hydrogen by hydrodynamic outflow at the energy-limited rate. In this manner the predominantly H_2O atmosphere could be reduced to one with only $\sim 1\%$ H_2O in $\sim 2.8 \times 10^8$ yr. During this phase the D:H ratio would remain unchanged since D would escape along with H. After this initial phase, escape of H would become diffusion-limited and escape of D would essentially cease. At this point enrichment of the D:H ratio could begin following the hypothesis of McElroy et al. (1982). It may, therefore, be impossible to differentiate between a water-rich and water-poor primordial Venus using the present-day D:H value. The oxidation states of surface minerals may provide the only definitive resolution of the primordial water controversy.

4.2.7. Models for Atmospheric Origin

We have seen how the evolution of solar-type primitive atmospheres progresses from H_2–He–H_2O–CH_4–Ne composition to CO_2–N_2–Ne as a direct consequence of thermal escape of hydrogen and helium. Cameron (1963b) and Suess (1964) have argued that the massive CO_2-rich atmosphere of Venus might be a remnant of such a solar-composition protoatmosphere, and others have echoed this suggestion. In the light of present compositional knowledge, we may place quantitative limits on the amount of such a residual atmosphere on Venus.

The most stringent limitation on the mass of such a solar-composition residue can be calculated from the observed abundances of the light rare gases. In the absence of knowledge about whether He may have once escaped readily from Venus, we will use the Ne and primordial argon abundances to do this calculation.

Taking the Ne partial pressure at the surface to be $\leq 2 \times 10^{-3}$ bar and the ^{36}Ar pressure to be $\leq 4 \times 10^{-3}$ bar, and using the Ne and primordial argon mole fractions from Table 3.2 for the residual gas after H and He loss, we calculate an upper limit on the total gas pressure of the residue of $\leq 1.0 \times 10^{-2}$ bar and $\leq 5.6 \times 10^{-1}$ bar, respectively. Clearly the more restrictive limit is the one derived from Ne [the ^{20}Ne : ^{36}Ar ratio on Venus (~ 0.5) is far smaller than the ^{20}Ne : ^{36}Ar ratio in solar material (~ 30)]. Thus, the CO_2 contribution from the solar component is $\leq 0.6 \times 10^{-3}$ bar, or 6×10^{-6} of the actual CO_2 inventory. It is therefore quite obvious that a captured solar component, whether a residue from a "primary" atmosphere of solar nebula origin or from the solar wind, cannot possibly be a significant source of chemically reactive gases. Indeed, the fact that the Ne : Ar ratio is close to that for fractionated ("planetary") rare gases and far lower than in solar material leads us to suspect that even the observed primordial rare gases are not solar. The observed Ne : N_2 ratio (Table 4.2; Fig. 4.3) is $\sim 6 \times 10^{-4}$, compared to the solar ratio of 2. Thus, like CO_2, N_2 must be overwhelmingly derived from some nonsolar source.

It is also entirely relevant to note that the solar C : O ratio is greater than 0.5, and hence that evolution of a solar gas to $CO_2 + N_2 + Ne$ composition would result in massive graphite precipitation. The presence of graphite in the crust of Venus at 750 K and a CO_2 pressure of 100 bar would require an equilibrium CO mole fraction of 6×10^{-3}, some 30–60 times higher than the observed value.

Since only 0.01% of the atmosphere could be attributed to the residue of a solar primitive atmosphere under the most extreme assumptions, we can estimate what fraction of the solar proportion of these volatiles was retained by Venus. The Ne abundance is $\leq 2 \times 10^{-3}$ bar, whereas a solar Ne : Si ratio would provide a surface Ne pressure near 0.5 Mbar. Thus, *at most* Venus retained 4×10^{-9} of its solar proportion of primitive volatiles. Therefore, we cannot rule out the possibility that a solar-composition primitive atmosphere with a surface pressure up to ~ 6 bar was once present on Venus. This is more than the amount of gas Venus could capture from a Cameron-type solar nebula in either the adiabatic or radiative–convective case.

Ringwood's (1966) carbonaceous chondrite model for the terrestrial planets provides an enormous quantity of chemically reactive volatiles. Wiik's (1956) figure of 20% H_2O in CI chondrites is probably too high by up to a factor of two due to adsorption of terrestrial water; Kaplan (1971) summarizes data suggesting that 10% H_2O is most reasonable for CI chondrites. A Venus model containing 10% H_2O, 3.5% C, 0.26% N, 6% S, and other volatiles in the same proportions as found in CI chondrites would give the atmospheric composition described in Table 4.4. Also

TABLE 4.4
CI Chondrite Model: Volatile Abundances

Species	Surface pressure (bar)	Loss factor
H_2O	1.16×10^5	$\sim 10^7$
CO_2[a]	1.4×10^5	1.4×10^3
(S)[b]	$(7.0 \times 10^5)^b$	—
N_2	3.0×10^3	7.5×10^2
(HCl)[b]	$(3.0 \times 10^2)^b$	—
(HF)[b]	$(3.0 \times 10^2)^b$	—
Hg	2.0×10^1	?
(HBr)[b]	$(2)^b$	—
(HI)[b]	$(6 \times 10^{-1})^b$	—
He	2×10^{-2}	~ 10
^{40}Ar	9×10^{-3}	~ 3
$^{36}Ar + ^{38}Ar$	2×10^{-3}	~ 1
Ne	3×10^{-4}	~ 3
Kr[c]	4×10^{-6}	$\sim 1-25$
Xe[c]	3×10^{-6}	$\sim 1-30$

[a]CI chondrites contain 25 wt.% of oxidized iron, including substantial amounts of magnetite. There is more than enough FeO alone to oxidize all C to CO, but not CO_2, without using H_2O.

[b]Sulfur is assumed to be found in sulfide minerals and the halogens are assumed to form halides. Only the small buffered gas pressures of COS, HCl, HF, etc. (Section 4.2.1) are expected in the atmosphere.

[c]Loss factors of 25–30 follow from the Venera measurement of Kr, while loss factors of ~ 1 would follow from the Pioneer Venus upper limits on Kr and Xe (see Fig. 4.3).

given in this table is the factor by which certain constituents would have to be depleted in order to be in agreement with the observed composition of the Venus atmosphere. The necessary amount of water is more than the mass of the planet Mars. Nonselective loss of atmosphere by the high-luminosity sun mechanism advocated by Ringwood would reduce the CO_2 abundance to 120 mbar before the H_2O abundance is reduced to present levels. Selective loss of H_2O would leave a residual atmosphere containing at least 2% N_2 (more if carbonate formation is permitted) and with a surface pressure of 140 kbar. Every atmospheric constituent must be independently depleted by a different correction factor in order to achieve agreement with the composition of the present atmosphere.

Ringwood (1966) claimed that excess graphite is formed by such an atmosphere formation process. If this is the case, then the evidence presented earlier for the absence of graphite in the crust of Venus would be relevant and very damaging.

It should also be noted in passing that disposing of unwanted carbon by burial as carbonates or graphite is no easy task. A surface layer of $CaCO_3$ 1800 km thick would be required! Of course, such burial of massive quantities of N_2 would be impossible.

In the ordinary chondrite model for Venus the abundances of all volatiles are less, often substantially so. Table 4.5 gives the atmospheric mass attainable by complete outgassing of an ordinary chondritic Venus. As before, we also give

TABLE 4.5

Ordinary Chondrite Model: Volatile Abundances

Species	Surface pressure (bar)	Loss factor
CO_2[a]	3.9×10^3	4×10^1
H_2O	2.0×10^3	10^5 (?)
N_2	6.0×10^1	$>10^1$
(S)[b]	(2.0×10^4)[b]	—
(HCl)[b]	(10^2)[b]	—
(HF)[b]	(10^2)[b]	—
Hg	10^{-1}	?
(HBr)[b]	(5×10^{-1})[b]	—
(HI)[b]	(5×10^{-2})[b]	—
He[c]	2×10^{-4}	10^{-1}
$^{36}Ar + {}^{38}Ar$[c]	2×10^{-5}	10^{-2}
Ne[c]	3×10^{-6}	2×10^{-3}
Kr[c]	4×10^{-7}	10^{-1}–3.33×10^{-3}
Xe[c]	3×10^{-7}	10^{-1}–3.33×10^{-3}

[a]Ordinary chondrites contain 6–17 wt.% FeO. Only 1.2% FeO is required to oxidize all C to CO, or 2.4% to oxidize all C to CO_2.

[b]Sulfur is assumed to form sulfide minerals and the halogens are assumed to form halides.

[c]All these rare gases have abundances on Venus which are *at least* 10 times the amount available in an ordinary chondritic Venus.

wherever possible the factor by which each volatile would have to be depleted to match the observed atmospheric composition.

Most H_2O determinations on ordinary chondrites are suspect; Kaplan (1971) stresses that even mild contamination could explain all the available observations. The true H_2O content could easily be a factor of 10 or more below the upper limit value used in Table 4.5. The CO_2 excess, while large, is only 3% of that faced in the CI model. It is by no means impossible that the core of Venus could presently contain 0.3% C. Similarly, 55 bar equivalent of N_2 could be accommodated in the core in a solution containing only 150 ppm of nitrogen. Some iron meteorites examined by Gibson and Moore (1971) contained this much nitrogen.

The abundances of primordial rare gases are also much smaller for an ordinary chondrite model than for CI material, but the great variability and complex nature of the rare gas abundance patterns in meteorites do not allow very precise comparisons. As shown in Table 4.5, the ordinary chondrite rare gas abundances run close to 100 times lower than their observed abundances on Venus.

In comparison with the predictions of the other models for atmospheric origin, the equilibrium condensation model is extremely simple. The expectations of this model are for accretion of Venus at a temperature substantially higher than that at which Earth accreted. Venus is predicted to be formed severely depleted in S and H_2O, with about 10 times less FeO than Earth, but with the same inventory of radioactive elements. The carbon content is attributed to the equilibrium formation of small traces of metal nitrides in solution in the metal. Table 4.6 gives the expected abundances of volatiles on Venus for the equilibrium model with no radial mixing.

Here we use the theoretically calculated value of 20 ppm (wt.) C in the metal, rather than Moore *et al.*'s (1969) values ~100 ppm, for the C concentration in octahedrites (Lewis *et al.*, 1979). The nitrogen abundance of 20 ppm in the metal phase due to Gibson's (1969) work on octahedrite irons is used because the theoretical treatment is not yet complete. For comparison, Gibson's value for N in octahedrites was ~10 ppm, and values for individual irons ranged from 1 to 215 ppm. The uncertainty in carbon content due to real variability among the meteorites measured is about a factor of 3. Gibson and Moore (1971) have more recently reported a median value of 18 ppm nitrogen in irons, with an observed range from 1 to 130 ppm. Thus, P_{N_2} on Venus could be 1–220 bar if outgassing were complete.

It is curious to note that the right magnitudes for CO_2 (and nitrogen?) are predicted by the equilibrium theory, which predicts *zero* hydrogen content for Venus unless an appreciable degree of radial mixing of solids occurs in the nebula, but up to 1% H_2O in the atmosphere for plausible accretion models (Barshay and Lewis, 1981).

In the case of oxidation of carbon, carbides, and nitrides to make CO_2 and N_2, the amount of FeO required is 0.03% of the mass of Venus. The equilibrium theory provides 30 times this amount in the initial condensate at the orbit of Venus.

TABLE 4.6

Equilibrium Condensation Model: Volatile Abundances

Species	Surface pressure (b)	Loss factor
CO_2	30.	1/3
H_2O^a	0.	0
N_2	0.1	<1/20
(S)[b]	(20000.)[b]	—
(HCl)[b]	(100.)[b]	—
(HF)[b]	(100.)[b]	—
Hg	0.1	?
(HBr)[b]	(0.5)[b]	—
(HI)[b]	(0.05)[b]	—
He[c]	?	?
$^{36}Ar + ^{38}Ar$[c]	?	?
Ne[c]	?	?
Kr[c]	?	?
Xe[c]	?	?

[a]Without sampling of material condensed at >1 AU from the Sun, no water would be present at the time of formation.

[b]Sulfur is assumed to form sulfides, and halogens to form halide minerals.

[c]The equilibrium condensation model is unable to predict the primordial rare gas content due to lack of laboratory data.

It is indeed interesting that the conventional question regarding Venus, namely, why it is so dry, is nicely inverted in the equilibrium model. Here we are led to ask why Venus, with $10^4–10^5$ times *less* water than Earth, is so *wet!* It is the presence of that tiny trace of water, less than 1 part in 10^8 of the mass of Venus, which requires explanation, and which requires that a small source of volatile-rich material be added.

With this perspective, let us now look in more detail at the question of how much cometary or C1 material might possibly be present in Venus.

Ringwood (1966), beginning as he does with volatile-rich C1 material, must produce the vast amount of iron needed to make the cores of Venus and Earth by an internal (disequilibrium) autoreduction process in which magnetite reacts with polymeric organic matter via the reaction

$$2\text{``CH}_2\text{''} + Fe_3O_4 \rightarrow 2CO + 2H_2O + 3Fe. \tag{4.37}$$

Turekian and Clark (1969) point out that vast quantities of carbon oxides and water must be somehow lost from the planet, and since Earth and Venus contain about 30 wt. % iron each, the mass of lost gas must be about that of the entire planet Mars. They further point out that Ringwood's model requires strongly reduced mantles (on the Fe/FeO buffer throughout) and thorough extraction of siderophiles, especially Ni, from their mantles and crusts. For these reasons, they reject the idea that the primordial composition of Venus and Earth was C1.

Instead, Turekian and Clark propose a nonhomogeneous accretion model for Venus in which only a small proportion of volatile-rich C1 material is deposited on the surfaces of Venus and Earth late in the accretionary era. They therefore adopt a model in which the interiors of both planets, being composed of early, high-temperature condensates, are quite free of volatiles (except, of course, radiogenic ^{4}He from U and Th decay). This vastly reduces the requirement for massive loss of volatiles and allows their upper mantles and crusts to be sufficiently oxidizing to retain Ni and the other siderophiles in the amounts actually observed on Earth. Turekian and Clark propose that the generation of volatiles from the carbonaceous veneer takes place via reactions such as

$$\text{``CH}_2\text{''} + 3Fe_3O_4 \rightleftarrows CO_2 + H_2O + 9FeO. \tag{4.38}$$

The observed large excess of water over CO_2 on Earth is taken to indicate that ~90% of the water is derived from outgassing of OH-bearing silicates.

However, the compositional model actually proposed by Turekian and Clark is rather unsettling in that the relative abundances of volatiles predicted by a C1 model are very different than those observed on Earth. Thus, these authors chose to make predictions for the ^{40}Ar and N_2 contents of Venus by ignoring the CI analyses and instead guessing that Venus has the same volatile-element inventory as in the Earth's atmosphere and crust! Thus, no predictions are made on the basis of the CI veneer model, and the eventual fate of these predictions does not bear on the validation of the model. (Their prediction of 1.86% N_2 is not bad, but their ^{36}Ar estimate of 0.75 ppm is about 40 times smaller than the observed amount.)

We shall now point out some of the consequences of the use of a CI veneer as a source of volatiles on Venus. We do so not because we like the model, but because it is impossible to talk sensibly about the idea without quantifying it and checking it for consistency with observations. For the moment we will examine its relevance to Venus only, leaving the assessment of its applicability to Earth and Mars to the following two sections.

First of all, any C1 model for water retention is likely to require deuterium abundances noticeably higher than in terrestrial water, as we saw in Boato's (1954) work. If massive escape of hydrogen from Venus is also required, then the $D:H$ ratio could be very large indeed, perhaps even approaching 1. The Connes IR spectra of Venus, which show numerous sharp, strong lines due to $H^{35}Cl$, $H^{37}Cl$, and HF, show no trace of $D^{35}Cl$, $D^{37}Cl$, or DF lines. An upper limit to $D:H$ on Venus of ~ 0.2 is suggested. Also, Wallace et al. (1971) have reported high-resolution UV spectra of the H Ly-α line in the Venus airglow, in which no deuterium component of the line could be detected. The Pioneer Venus mass spectrometry experiment has provided indirect evidence for the presence of deuterium in cloud droplets with a $D:H$ ratio of about 0.02: Confirmation of this measurement would obviously be of great interest (Donahue et al., 1982).

Next, from Table 4.2 we can see that the observed amount of CO_2 would be supplied by a mass of CI material equal to $M_{\venus}/1.4 \times 10^3$, or a layer 4 km thick. This amount of CI material will provide 80 bar of water from dehydration of layer-lattice silicates and 40 bar of water from reaction (4.38). Photolysis and loss of the hydrogen in this amount of water is of dubious plausibility (see Section 4.2.6); however, if it did occur, then 110 bar of O_2 must have been lost to the crust without saturating the oxygen sinks. Taking the FeO content of the crust to be of order 10% and using a crustal density of 2.5 gm cm^{-3}, we find that 130 kg O_2 cm^{-2} must be absorbed in making magnetite via

$$3FeO + \tfrac{1}{2}O_2 \rightarrow Fe_3O_4, \tag{4.39}$$

which requires 80 km of rock of Earth-like FeO content.

If FeO is in fact rarer on Venus (Lewis and Kreimendahl, 1980), then O_2 absorption simply becomes impossible. At this point we wonder how a planet which preserves a very heavily cratered ancient surface could possibly effect the incredible task of exposing $\sim 2\%$ of its mass to intimate contact with the atmosphere. Such exposure requires the extrusion or eruption of 60 km^3 of pristine, unoxidized crustal material each year, an amount ~ 15 times larger than provided on Earth by vigorous plate tectonic activity.

It is very important to realize that nitrogen and carbon can be accommodated to a much greater extent than this in the planet's core if these elements are permitted to interact with the core material. It is a matter of definition that Turekian and Clark's inhomogeneous accretion model never permits such contact, and thus C and N cannot be hidden. All C and N must reside in the crust–mantle–atmosphere system, where no nitrogen-bearing minerals are stable. For any homogeneously accreted planet, such as an ordinary chondrite model, C and N may be easily extracted into core material. Thus, an ordinary chondrite model or the like should not be rejected solely on the grounds that it provides more than the observed amounts of C and N.

One final point regarding Cl "veneering" should be made: The preservation of ferric iron (and absence of metal) in the near-surface regions is an essential part of the model. Thus, accretional mixing of the Cl material with ordinary chondritic or iron meteorite material must be closely restricted. Since Cl chondrites contain ~20 wt. % magnetite, an admixture of only 15 wt. % of ordinary chondrite material would introduce sufficient metal to reduce all magnetite to FeO and put the system on the Fe–FeO buffer. For a C2 veneer, a mere 1.8% of ordinary chondrite material would suffice. Such extraordinary selectivity of accretion is very difficult to accept.

Walker (1975) proposes a model which is rather similar to that of Turekian and Clark. He rationalizes the source of CO_2 to be "CH_2" (this is, saturated alkanes) by attributing to Lewis (1972a) and Grossman (1972) the conclusion that chemical equilibrium in the solar nebula produces condensed hydrocarbons, and by claiming that elementary carbon is rare in meteorites. Unfortunately, the actual conclusions of Lewis, Grossman, and others who have studied the chemistry of solar material is that condensed hydrocarbons *cannot* be present in a system of solar composition at equilibrium at any temperature above ~100 K. There the hydrocarbon is CH_4, not "CH_2." The presence of hydrocarbons in a few meteorites is attributed to *departures* from equilibrium due to sluggish kinetics and selection catalysis at low temperatures [see Studier *et al.* (1968) for one view]. Furthermore, organic matter is found only in meteorites of very low petrologic grades [1, 2, or 3 in the classification system of Van Schmus and Wood (1967)] and in ureilites. Wasson (1974) recognizes 36 carbonaceous chondrites and 6 ureilites. In addition, he lists 21 H, L, and LL meteorites of petrologic type 3. The remainder of the ~2000 known chondrites are volatile-poor and contain amorphous carbon or graphite as their dominant carbon forms, not "CH_2." About 80 achondrites in addition to the ureilites are known, and they contain no hydrocarbons. Some 600 iron meteorites are characterized, and these almost universally contain inclusions of graphite. Thus, 97% of the meteorites known are characterized by elemental carbon rather than hydrocarbons. Those which contain hydrocarbons contain organic matter that is very poor in H, far below Walker's "CH_2" (alkane) composition, with most hydrogen in the form of —OH silicates and water of hydration.

Walker's model encounters the same difficulties with water loss, oxygen absorption by the crust, and magnetite accounting that we mentioned with respect to Turekian and Clark's paper. Walker makes no predictions of atmospheric composition by which his model may be assessed.

Finally, the rare gas abundances in Cl chondrites are insufficient for Venus.

We are led to conclude that there are great difficulties associated with the derivation of the Venus atmosphere from a Cl source material. Clearly, if the volatile-rich component, including CO_2, were cometary ice-rich material, then the problem of the water content would be exacerbated: The probable assemblage in cometary solids, containing water ice and rock material in solar proportions, is over 60% water (Lewis, 1972a).

The present observed flux of volatile-rich material through the inner solar system has been shown by Lewis (1974b) to provide 10^{-8} of the mass of Venus over the age of the Solar System. This is adequate to provide all the hydrogen, halogens, and sulfur currently believed to be present in the atmosphere, but falls orders of magnitude short of supplying the observed CO_2 abundance. Thus, there exists a *known* source of injected volatiles strong enough so that *only* CO_2 need be considered as endogenous to Venus, and the equilibrium condensation model alone provides a reasonable $CO_2 : H_2O$ ratio.

It is unfortunately still true that we do not even know whether Venus is currently gaining or losing hydrogen. The current input flux of meteorite and cometary hydrogen from mass flux estimates through the inner solar system (Barker and Anders, 1968; Whipple, 1967; Öpik, 1969; Dohnanyi, 1972) is very similar to and quite possibly larger than the H escape flux discussed in Section 4.2.6. Thus, it is entirely possible that Venus has always been as dry or drier than it is now (Lewis, 1974b). Indeed, models requiring massive escape fluxes of H were rejected by Sagan (1961), Gold (1964), and Donahue (1968), and it was not until Ingersoll's (1969) formulation of the photochemical effects of a runaway greenhouse effect (also discussed in Section 4.2.6) that any mechanism capable of losing large amounts of hydrogen was suggested. Ingersoll's model is based on the idea that a troposphere with X_{H_2O} near unity will be so well thermostated by latent heat release that no true cold trap will form, and H_2O photolysis would indeed be very rapid until X_{H_2O} drops below ~0.1 (see Section 4.2.6). Wet-origin models for Venus with a runaway greenhouse effect have been presented by Shimazu and Urabe (1968), Rasool (1968), Palm (1969), Rasool and de Bergh (1970), Pollack (1971), Smith and Gross (1972), and Walker (1975). Walker (1975) argued for the need for early outgassing of the terrestrial planets by assuming a steady-state balance between the present H_2O outgassing rate of Venus and the observed H escape rate. They then further assumed that the same outgassing rate applied to Earth and showed that Earth could not have produced oceans from so little water. However, we see no reasons why any of these rates should be equal, and we certainly see no secure grounds for inferring the early presence of vast quantities of water on Venus, comparable to the terrestrial inventory.

Such massive loss of hydrogen will, as we have previously emphasized, lead to profound isotopic fractionations, and deuterium enrichment by large factors would be expected. The residual-oxygen problem is already clear.

Finally, consideration of the rare gases on Venus is in order.

Theoretical models for volatile retention have not yet advanced to the point where the rare gas abundances in meteorites can be explained. If the planetary rare gas component is due to the solubility of nebular gases in solid mineral grains, then the equilibrium condensation model would be expected to predict the same rare gas abundances seen in ordinary chondrites.

Radiogenic rare gases due to K, U, and Th decay, and to extinct ^{129}I decay would be produced at very nearly the same rates in Venus and Earth by either an ordinary chondrite or an equilibrium condensation model, with the ^{4}He, of course, susceptible to rapid escape. The total ^{40}Ar inventory would be the same for Venus as for Earth *if* the distribution of potassium is the same in both planets and *if* the outgassing efficiencies for both are the same. A chondritic Venus ($\sim$800 ppm K) of chondritic isotopic composition (^{40}K : K = 0.000119) with a partial decay half-life to ^{40}Ar of 11.8 × 10^9 yr would presently contain enough ^{40}Ar to yield a pressure of 56 mbar ($X_{40_{Ar}} \leq 6 \times 10^{-4}$). Thus, this is an *upper limit* for $X_{40_{Ar}}$ on Venus: An Earth-like K content would provide $\sim$9 mbar ^{40}Ar, whereas the observed amount is 2–3 mbar.

The capture of interstellar cloud material by Venus could have some effect on both the rare gas and hydrogen content of Venus (Butler *et al.*, 1978). Capture of interstellar material, comet and meteorite infall, and solar wind capture *each* seem capable of providing a mean flux of hydrogen into Venus larger than the Jeans escape rate (although not larger than the nonthermal escape rate). The most characteristic rare gas from the interstellar cloud source, neon, would have accumulated to a pressure of only $\sim$2 × 10^{-8} bar. Thus, even the neon provided in an ordinary chondritic Venus would swamp the interstellar material by nearly a factor of 100, and the observed Ne pressure of $\sim$2 × 10^{-3} bar clearly relegates the mechanism to insignificance.

Ringwood and Anderson (1977) have very strongly advocated a derivation of Venus, Earth, and Mars from extremely volatile-rich CI carbonaceous chondrites. Paradoxically, they accept the idea of a very high formation temperature for Mercury, but admit no elevation of the temperature at Venus relative to Earth or Mars. They then follow Ringwood's (1966) claim that the terrestrial planets underwent autoreduction and volatile-loss processes with a severity proportional to the mass of the planet. As a result, Venus is supposed to have ended up with a more volatile-rich composition than Earth, with a higher FeO content. The problem of attaining $\sim$1500 K at Mercury's orbit while preserving CI material (<300 K) at Venus is not addressed. Ringwood and Anderson also severely criticize a caricature of the model of Lewis (1972b, 1974b). Lewis (1974b) suggested that the bulk density of each planet was determined by the composition of local condensates while the volatile content could easily be dominated by a small mass fraction of low-temperature condensate. Lewis (1974b) also showed by a very conservative argument that the presently observed flux of cometary and meteoritic matter at Venus could easily supply the atmospheric inventories of several elements, including S, on Venus. Recent condensation–accretion models (Barshay and Lewis, 1981) estimate a sulfur abundance in Venus near 10% of the terrestrial abundance. Ringwood and Anderson claim that sulfur is absent on Venus in Lewis' models [a claim readily refuted by reference to Lewis (1968a, 1969c, 1970c, 1971d, 1972b, 1973b, 1974a)], and describe the known cometary source as an *ad hoc* rationalization. They apparently saw nothing ad hoc about the need to lose from Venus a mass of volatiles equal to the mass of Mars!

They then suggest that vast amounts of H_2O in a CI Venus have reacted with the $\sim 0.1\%$ S supposed to be present in the crust and with FeO to make H_2SO_4 and Fe_2O_3, with escape of H_2. They refrain from assessing this suggestion quantitatively, however. Let us assume that Ringwood and Anderson's estimated 13.4% FeO for the mantle of Venus is present throughout the lithosphere, along with 0.1% FeS. Let us calculate how large a mass of Venus rocks would have to be oxidized to Fe_2O_3 to use up the 10% H_2O in the CI parent material (a mass of 5×10^{26} gm of O_2). The answer is 60 gm of rock per gram of H_2O, or 3×10^{28} gm of rock. This is three times larger than the combined masses of the terrestrial planets. Using Ringwood's (1979) estimate of the H_2O content, the problem is three times as severe as this.

The actual amount of H_2O and S needed to make the sulfuric acid clouds is on the order of 10^{-9}–10^{-8} of the mass of Venus. The need for 10^7 times as much S and H_2O is less than obvious.

Another criticism of Lewis (1974b) was made by Ringwood and Anderson: that the actual density of Venus is 1.7% less than that of an Earth-like model, whereas an Earth-like model from which all S and FeO are removed is only 1.1% less dense than the Earth-like model. The discrepancy, only 0.6% of the planetary bulk density, is taken to favor FeO-rich models of Venus: Although they assume that metal/silicate fractionation at Mercury has been severe enough to effect a 100% enrichment over Earth of Fe, they are confident enough of the Fe:Si ratio in Venus that they do not allow even a 0.6% difference!

Goettel *et al.* (1982) have recently examined carefully a suite of compositional models for Venus to test the ability of the several competing models to explain the observed bulk density of Venus and the Venus–Earth density difference. They detail a number of deficiencies in the treatment by Ringwood and Anderson (1977), some of which (e.g., their assumption of $T \simeq 750$ K at the surface of Venus but with Earth-like temperatures at depth) are by themselves able to remove or reverse the sign of the density difference. Thus, careful consideration of the equilibrium condensation model for Venus and earth does not reveal any discrepancy between theory and the observed densities of the planets. For this and other reasons, we feel that the alternative model of Ringwood and Anderson should be considered with appropriate reservations.

Ringwood (1979) has extensively revised his ideas about the composition of preplanetary material, so that now only about 10% of the mass of the Earth or Venus is attributed to CI material. He first suggested that the light element in planetary cores is Si (Ringwood, 1966), then concluded that it is S (Ringwood, 1977), and has most recently settled upon O (Ringwood, 1979). These enormous interpretive changes were not forced by changes in our knowledge of Venus, but arose out of the great inherent difficulty in quantifying a model that is so little constrained and so reliant upon unknown processes. This makes it difficult to use observations to reach a confident conclusion about the correctness of Ringwood's models. The chief concern of many critics of CI-rich models of Earth and Venus is the purely *ad hoc* depletion and adjustment of the volatile-element abundances.

Three post-Pioneer Venus interpretations of the rare gas composition have attempted to find self-consistent models for atmospheric origin.

Pollack and Black (1979) have advocated a model in which rare gases are retained and carried by mineral grains, but in which the usual temperature gradients predicted by either radiatively controlled ($T \propto R^{-\frac{1}{2}}$) or adiabatic ($T \propto R^{-1}$) models are absent. They then postulate a very large pressure gradient in the inner solar system, with gas pressures at Venus 400–800 times as high as at Mars. Adsorption and subsequent implantation by other processes are assumed to account for the presumed fractionation of solar gas to produce a "planetary" abundance pattern. Since they were writing after the claimed Kr detection by Venera 11 and 12 but before the refinement of the Pioneer Venus large probe mass spectrometer data, they assumed that the rare gas composition on Venus was similar to that on Earth, but ~100 times more abundant.

Wetherill (1981) has pointed out a severe dynamical problem inherent in this model: The very large pressure gradient would lead to such large velocity differences between large and small solid bodies that accretion would not be possible. Rather, disruption of these bodies should occur. Later passage of the nebula through a high-temperature phase, as envisioned by Pollack and Black, would lead to loss of these rare gases.

Wetherill (1980) has suggested that the rare gases are retained by a wholly different mechanism than the much more abundant chemically active gases. He emphasizes the possibility that the latest Pioneer Venus upper limits on Kr and Xe may be valid and the Venera 12 claimed detection of Kr at the 1 ppm level may be incorrect. If this premise is granted, then the *heavy* rare gas abundance pattern closely reflects a *solar* source. The problem then becomes one of explaining the Ne abundance which, from a solar wind or nebula source, should be roughly 100 times as abundant as observed.

Wetherill proposes irradiation of an optically thick protoplanetary planetesimal swarm, after dissipation of the nebula, by an enhanced early solar wind. Implantation of gas ions occurs at elevated temperatures, and rapid diffusive loss of H, He, and Ne from the solids therefore occurs. At reasonable temperatures at 0.7 AU from the early Sun, Ne can be depleted without affecting the Ar abundance significantly. As Wetherill (1981) points out, it is very hard to accrete planetesimals without perturbing a large number of them into Earth-crossing orbits. Thus, the *abundance* of rare gases on Venus is more easily explained than the *selectivity* shown in forcing them to end up on Venus, but not on Earth and Mars.

McElroy and Prather (1981) have also proposed a solar-wind irradiation model for the origin of the rare gases on Venus. They explain the high rare gas abundance by delaying condensation and accretion of material close to the sun until the ^{26}Al content is low. Thus, planetesimals accreted at the orbit of, say, Mars would form with so large a radiogenic heat source that they would outgas and lose rare gases quite efficiently. Venus, condensing and accreting at least two or three ^{26}Al half-lives later, would not develop sufficiently high temperatures for melting and outgassing until the planet was too large for escape of rare gases. There are two major problems with this suggestion. First, the stable isotope anomalies discussed in Chapter 2, especially for oxygen, make it most unlikely that extensive vaporization of preplanetary solids occurred *after* injection of isotopically unusual material, since evaporation and recondensation would lead to thorough isotopic homogenization, contrary to observations on meteorites. Second, the idea that solids accrete more rapidly farther from the Sun is contrary to the results of all theoretical studies of accretion, as discussed in Section 3.2. Also, closely related to this second objection, we have the likelihood that most of the solids will accrete to many-kilometer bodies in a time less than a half-life of ^{26}Al.

It is interesting that the present-day flux of volatile-rich material onto Venus (Lewis, 1974b) also suggests a significant source of rare gases. Condensation of the ices in a comet almost certainly occurs within the stability field of solid clathrate hydrates of CH_4, CO, N_2, CO_2, and the rare gases (Lewis, 1972a, 1974b). Helium and neon, however, are essentially absent in such hydrates because of their ability to diffuse rapidly out of the ice lattice. We would expect that clathrate hydrates formed near and below 40 K would contain primordial Ar, Kr, and Xe in solar proportions, whereas Ne would be severely depleted and He would be absent (Fegley, 1982). Since this is merely a theoretical prediction and nothing whatsoever is known of the rare gas content of comets, we cannot do more than sketch out the consequences if this mechanism did indeed supply the rare gas abundances observed. Lewis (1974b) presented a suite of compositional models with volatile carriers ranging from CI chondrites (his model a) to comets (his model e). He gave a ^{36}Ar abundance of 1.5 ppm with an addition of 50 ppm of water as a lower limit, and favored several times more of both materials. Table 4.7 presents his lower-limit model and a scaled version of the same model in which all gases have been increased in abundance by a factor of 20 to fit the heavy rare gas abundances actually observed. Neon was not included in the original table, but Lewis (1974b) predicted Ne:Ar < 1 for a

TABLE 4.7
Abundances of Cometary Volatiles on Venus

Gas	Lower limit[a] (bar)	20× lower limit (bar)
H_2O[b]	5×10^{-3}	1×10^{-1}
N_2	1×10^{-3}	2×10^{-2}
COS[b]	6×10^{-4}	1.2×10^{-2}
^{36}Ar	1.5×10^{-4}	3×10^{-3}
HCl[b]	7×10^{-6}	1.4×10^{-4}
HF[b]	3×10^{-6}	6×10^{-5}
Kr	6×10^{-8}	1.2×10^{-6}
HBr[b?]	2×10^{-8}	4×10^{-7}
Xe	7×10^{-9}	1.4×10^{-7}
HI[b?]	1.3×10^{-9}	2.5×10^{-8}

[a]Cometary composition from Table 2e of Lewis (1974b).
[b]These gas pressures can be regulated by buffer interaction with surface minerals.

cometary source, in agreement with the Venus data (Ne:Ar $\simeq$ 0.5). Lewis and Prinn (1980) have proposed that the nitrogen and carbon content in cometary ices should be only a few percent of the amount assumed by Lewis (1974b) based on the assumption of NH_3 and CH_4 hydrate formation: Lewis and Prinn show that kinetic inhibition of NH_3 formation in the nebula at temperatures at which it is stable relative to N_2 ($\leq$330 K) should leave N_2 abundant and reduce the amount of NH_3 hydrate by a factor of at least 10. A second consequence of the kinetic inhibition of CH_4 and NH_3 formation in the nebula is that H_2O will not be fully altered to clathrate hydrates by reaction with CH_4. This leaves large amounts of ice free to take up the heavy rare gases as clathrate hydrates. In any event, whether we use the 2×10^{-2} bar N_2 in Table 4.7 or the more conservative value of about 10^{-3} bar which would follow from these kinetic considerations, it is clear that a cometary source capable of supplying the rare gases would not contribute importantly to the CO_2 or N_2 inventory ($\sim$90 and $\sim$3 bar, respectively).

The comet mass required to bring in 4×10^{20} gm (1×10^{-1} bar) of H_2O is about 10^{21} gm, or 2×10^{-7} of the mass of Venus.

The most implausible aspect of this scenario is in the selectivity of deposition of volatiles on Venus. Comets supply very nearly the same mass flux onto each of the terrestrial planets. Thus, in order to place such a large preponderance of the rare gases on Venus, we might be forced to postulate that the rare gases were brought in almost entirely by a single event. This requirement, though improbable, is not impossible. That objects of sufficient mass have hit Venus in the last few billion years is very likely: The large craters attest to that.

There is, however, a second consideration which mitigates this statistical difficulty: A major impact on Venus would result in efficient deposition of gases in the massive atmosphere, whereas a major impact on Mars would quite possibly result in a net *removal* of atmosphere. Earth, with its large gravity and intermediate atmospheric mass, would present a case somewhere between the extremes of Mars and Venus. In this scenario, then, numerous impacts of comets in the 10^{15}–10^{20} gm range might have occurred on each planet. Such impacts would bring in roughly 10^{28}–10^{33} erg of kinetic energy. According to the calculations by Lin (1966), an energy of about 5×10^{30} erg suffices to blow off the Earth's atmosphere above an impact site. A far lower energy, and hence a far smaller comet, would suffice for blowoff on Mars.

Clearly this mechanism requires further study. The issue is complex because removal of preexisting planetary-type rare gases can occur as well as capture.

We regret that, even after the last influx of new data from Pioneer Venus and the last two rounds of Venera probes, we cannot conclude this chapter with a single credible scenario for the origin and evolution of Venus which is responsive to all available data and theoretically impeccable. This situation arises in part because of important gaps and inconsistencies in our information concerning the rare gas abundances, lower atmosphere composition, surface oxidation state, and hydrogen balance. Also contributing to our uncertainty is the intrinsic complexity of Venus. We might hope that the range of possible genetic and evolutionary histories will be narrowed by comparison of Venus with Earth and Mars, and from a broad perspective on our planetary system. It is in this hope that we move on to the study of the atmosphere of our own planet.

4.3. EARTH

The problems of the origin, evolution, composition, and structure of Earth and its atmosphere occupy a literature of over 10,000 publications, and are spread over many disciplines. Since it is wholly impossible for us to summarize this vast literature in the space available, we must emphasize the data and interpretations that bear most directly on the global properties of Earth, its long-term evolution, and its relationship to other planets. Many interesting and relevant matters, such as a thorough review of the evidence from radiometric chronology, the detailed geochemical cycles of the volatile elements, short-time-scale chemical and climatic perturbations, etc. must be greatly compressed or omitted in favor of references to reviews.

We will begin our overview by summarizing the abundances of a number of gaseous constituents of the atmosphere, with abundances ranging down to a mixing ratio $f = 6 \times 10^{-20}$ for radon (Table 4.8). The known particulate constituents, both solid and liquid, have mass fractions ranging from about 10^{-3} for liquid water in raining clouds to about 10^{-6} for large hydrated ions in otherwise clear air. These constituents are involved in cyclic processes of varying complexity, which, in addition to the atmosphere, may involve the hydrosphere, biosphere, lithosphere, and even the deep interior of the Earth. As general references to the subjects covered in this section we recommend the books by Junge (1963), McEwan and Phillips (1975), Holland (1978), and Walker (1977).

TABLE 4.8
Composition of Tropospheric Air

Gas	Volume mixing ratio
Nitrogen (N_2)	7.81×10^{-1} (dry air)
Oxygen (O_2)	2.09×10^{-1} (dry air)
Argon (^{40}Ar)	9.34×10^{-3} (dry air)
Water vapor (H_2O)	$\leq 4 \times 10^{-2}$
Carbon dioxide (CO_2)	$2-4 \times 10^{-4}$
Neon (Ne)	1.82×10^{-5}
Helium (^4He)	5.24×10^{-6}
Methane (CH_4)	$1-2 \times 10^{-6}$
Krypton (Kr)	1.14×10^{-6}
Hydrogen (H_2)	$4-10 \times 10^{-7}$
Nitrous oxide (N_2O)	3.0×10^{-7}

(*continued*)

TABLE 4.8—*Continued*

Gas	Volume mixing ratio
Carbon monoxide (CO)	$0.1\text{--}2 \times 10^{-7}$
Xenon (Xe)	8.7×10^{-8}
Ozone (O_3)	$\leq 5 \times 10^{-8}$
Ammonia (NH_3)	$\leq 2 \times 10^{-8}$
Sulfur dioxide (SO_2)	$\leq 2 \times 10^{-8}$
Hydrogen sulfide (H_2S)	$0.2\text{--}2 \times 10^{-8}$
Formaldehyde (CH_2O)	$\leq 1 \times 10^{-8}$
Nitrogen dioxide (NO_2)	$\leq 3 \times 10^{-9}$
Nitric oxide (NO)	$\leq 3 \times 10^{-9}$
Hydrochloric acid (HCl)	$\leq 1.5 \times 10^{-9}$
Nitric acid (HNO_3)	$\leq 1 \times 10^{-9}$
Hydrogen peroxide (HOOH)	$\sim 1 \times 10^{-9}$
Methyl chloride (CH_3Cl)	$\sim 6 \times 10^{-10}$
Carbonyl sulfide (COS)	$\sim 5 \times 10^{-10}$
Freon-12 (CF_2Cl_2)	2.8×10^{-10}
Sulfuric acid (H_2SO_4)	$\sim 1 \times 10^{-10}$
Freon-11 ($CFCl_3$)	1.7×10^{-10}
Carbon tetrachloride (CCl_4)	1.2×10^{-10}
Methyl chloroform (CH_3CCl_3)	1.2×10^{-10}
Freon-12 ($CHCl_2F$)	$\sim 1.4 \times 10^{-11}$
Methyl iodide (CH_3I)	$\sim 1 \times 10^{-11}$
Chloroform ($CHCl_3$)	$\sim 9 \times 10^{-12}$
Methyl bromide (CH_3Br)	$\sim 5 \times 10^{-12}$
Radon	$\sim 6 \times 10^{-20}$

We have already mentioned the genetic and evolutionary processes that lead to an atmosphere rich in N_2, containing O_2 and the inert gases ^{40}Ar, Ne, ^4He, Kr, and Xe. Six of these seven gases have long atmospheric lifetimes, the shortest being about 10^6 yr for ^4He which, as we will discuss shortly, can escape from the top of the atmosphere. By contrast, all the other gases listed in Table 4.8 participate in relatively rapid chemical cycles and consequently have atmospheric residence times of a few decades or less.

Above the tropopause the composition of the atmosphere begins to change, primarily due to the dissociation of molecules by UV light and the resultant chemical reactions. The altitude distribution of the important atmospheric species above the tropopause is illustrated in Figure 4.10. We will discuss later the importance of the ozone layer at 25 km altitude for absorbing the biologically lethal radiation in the 2000–3000 Å wavelength region. In a pure $N_2 + O_2$ atmosphere most of this radiation would be transmitted to the surface.

One important aspect of the composition of the atmosphere is the large body of information that has been gathered concerning the isotopic composition of atmospheric constituents. These data are summarized in Table 4.9. In contradistinction

TABLE 4.9
Isotopic Composition of Earth's Atmosphere

Element	Isotopes	Abundances
Hydrogen	^{1}H	99.985
	D (^{2}H)	0.015
	T (^{3}H)	Var. trace ($t_{1/2} = 12$ yr)
Helium	^{3}He	0.00013
	^{4}He	~100
Carbon	^{12}C	98.63
	^{13}C	1.11
	^{14}C	Var. trace ($t_{1/2} = 5.7 \times 10^3$ yr)
Nitrogen	^{14}N	99.63
	^{15}N	0.37
Oxygen	^{16}O	99.759
	^{17}O	0.037
	^{18}O	0.204
Fluorine	^{19}F	100.00
Neon	^{20}Ne	90.51
	^{21}Ne	0.266
	^{22}Ne	9.22
Phosphorus	^{31}P	100.00
Sulfur	^{32}S	95.0
	^{33}S	0.76
	^{34}S	4.22
	^{36}S	0.014
Chlorine	^{35}Cl	75.77
	^{37}Cl	24.23
Argon	^{36}Ar	0.337
	^{38}Ar	0.063
	^{40}Ar	99.60
Bromine	^{79}Br	50.69
	^{81}Br	49.31
Krypton	^{78}Kr	0.354
	^{80}Kr	2.27
	^{82}Kr	11.56
	^{83}Kr	11.55
	^{84}Kr	56.90
	^{86}Kr	17.37
Iodine	^{127}I	100.00
Xenon	^{124}Xe	0.096
	^{126}Xe	0.090
	^{128}Xe	1.92
	^{129}Xe	26.44
	^{130}Xe	4.08
	^{131}Xe	21.18
	^{132}Xe	26.89
	^{134}Xe	10.44
	^{136}Xe	8.87

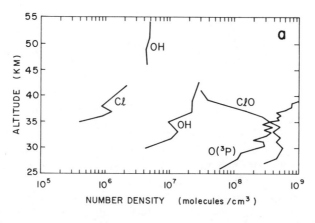

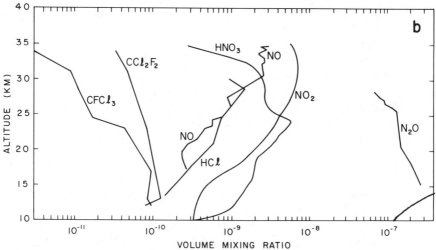

FIGURE 4.10. **Altitude distribution of important species above Earth's tropopause: (a) the abundances of several important reactive radicals; (b) the mixing ratios of a variety of species whose reactions are important in the regulation of the ozone layer. Ozone reaches a peak abundance of about 10^{-5} above 20 km, but rapidly falls to $< 10^{-7}$ below 10 km, and hence lies almost entirely off the diagram (except the line segment in the lower right corner). (After Prinn et al., 1978.)**

to the gaseous radioactive isotopes mentioned above, several of the nuclides found on this list are clearly products from the following reactions:

$$^{40}K \begin{cases} \xrightarrow[t_{1/2} = 1.47 \times 10^9 \text{ years}]{\beta^-} {}^{40}Ca, \\ \\ \xrightarrow[t_{1/2} = 11.8 \times 10^9 \text{ years}]{\beta^+; \, \epsilon} {}^{40}Ar, \end{cases} \qquad (4.40)$$

$$^{232}Th \xrightarrow[t_{1/2} = 1.39 \times 10^9 \text{ years}]{} {}^{208}Pb + 6{}^4He,$$

$$^{235}U \xrightarrow[t_{1/2} = 0.71 = 10^9 \text{ years}]{} {}^{207}Pb + 7{}^4He, \tag{4.41}$$

$$^{238}U \xrightarrow[t_{1/2} = 4.5 \times 10^9 \text{ years}]{} {}^{206}Pb + 8{}^4He,$$

$$^{235}U \text{ (S.F.)} \rightarrow \mathbf{Kr} + \mathbf{Xe} \text{ isotopes,} \tag{4.42}$$

$$^{129}I \xrightarrow[t_{1/2} = 1.6 \times 10^7 \text{ years}]{\beta^-} {}^{129}\mathbf{Xe}. \tag{4.43}$$

Of the important volatile elements, only P, F, and I are monoisotopic.

Temperature decreases with altitude in the troposphere, and precipitation of water vapor in this atmospheric region causes its mixing ratio, f_{H_2O}, to decrease substantially with altitude. However, above the tropopause the temperature begins to increase and f_{H_2O} remains roughly constant and equal to its tropopause value of $\sim 10^{-5}$–10^{-6}. Methane dissociation, which produces H_2O, may lead to an increase in f_{H_2O} in the upper stratosphere and mesosphere. The thermal structure of Earth's atmosphere is summarized in Figure 4.11. The *cold trap* for H_2O at the tropopause is particularly important for it severely restricts both the H_2O concentrations in the upper atmosphere and, as we will discuss later, the escape rate of H from the top of the atmosphere. About 100 km above the surface the photodissociation lifetime of O_2 becomes less than the advection time and the predominant atmospheric constituents become O (instead of O_2) and N_2.

A number of radioactive nuclides are formed naturally in the atmosphere by decay of the inert gas radon and by cosmic radiation. Radon, which is itself pro-

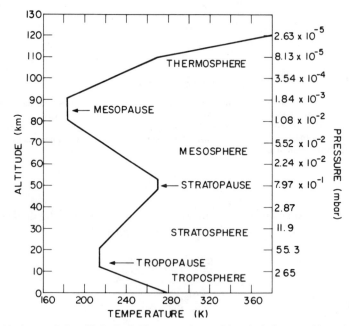

FIGURE 4.11. Average *T–P* profile for Earth. The nomenclature of the principal structural layers is also given.

duced by decay of U and Th in the crust, enters the atmosphere, where it in turn decays to produce a number of radioactive heavy metals. These metal atoms become attached to aerosol particles and ultimately sediment out of the atmosphere. Cosmic rays striking N_2, O_2, and ^{40}Ar, principally in the stratosphere, give rise to a number of radioactive isotopes, including ^{14}C, ^{7}Be, ^{10}Be, and ^{3}H. The incorporation of ^{14}C into organic matter, where it decays with a half-life of about 5600 years, forms the basis of the radiocarbon dating method. In addition to naturally occurring radioactivity, large quantities of radioactive material have been injected into the atmosphere as a result of nuclear bomb tests. The most dangerous isotope from these tests is ^{90}Sr, which can be incorporated into human bones where it decays with a half-life of about 28 years. Both natural and anthropogenic radioisotopes have been used as tracers for tropospheric, stratospheric, and oceanic motions.

Particulate matter suspended in the air for a reasonable lifetime forms another important component of our present atmosphere. Particles in the atmosphere range in size from about 10^{-3} to over 10^2 μm in radius. The term *aerosol* is usually reserved for particulate material other than water or ice. A summary of tropospheric aerosol size ranges and compositions is given in Figure 4.12. Concentrations of the very smallest aerosols in the atmosphere (*Aitken nuclei* with radii of 10^{-3}–10^{-1} μm) are limited by coagulation to form larger particles, whereas the concentrations of the *large particles* (radii of 10^{-1}–1 μm) and the *giant particles* (radii greater than 1 μm) are restricted by sedimentation. The size distributions of water droplets and ice crystals are, in addition, affected by evaporation, condensation, and coalescence processes. Some of the dry aerosols are water soluble and these can also grow by condensation.

Over the continents the principal aerosol materials are usually insoluble minerals injected into the air by wind erosion and volcanoes. These include $CaSiO_3$, SiO_2, Al_2O_3, FeO, MnO, MgO, and $CaCo_3$ and together they form the major components of airborne dust. In highly industrialized areas and in regions where forest fires are common, carbonaceous substances (soot) become important. A number of aerosols are produced in the air by gas reactions. For example, NH_3, NO, NO_2, NO_3, N_2O_5, SO_2, SO_3, H_2O, and CO_2 may under various conditions (e.g., in the presence of UV radiation) lead to ammonium sulfates, nitrates, chlorides, and carbonates and to sulfuric acid. Spores, pollen, and unsaturated organic compounds such as terpenes are common aerosols over forested regions. The bulk of aerosols in the atmosphere are not dust but marine aerosols ejected from the oceanic surface *microlayer* by the action of bubbles or wind. The common components of these water-soluble aerosols are the chlorides, carbonates, bicarbonates, sulfates, and bromides of Na, K, Ca, and Mg.

In addition to the tropospheric aerosols discussed above, there are also layers in the lower stratosphere at about 18–25 km altitude ("Junge" layer) and in the lower thermosphere at about 85 km altitude ("noctilucent" layer). The Junge layer is composed mainly of sulfuric acid, with smaller contributions from extraterrestrial dust and Aitken nuclei from the troposphere. The precise composition of the noctilucent layer is not known; H_2O condensing on dust, ions, or radicals is pres-

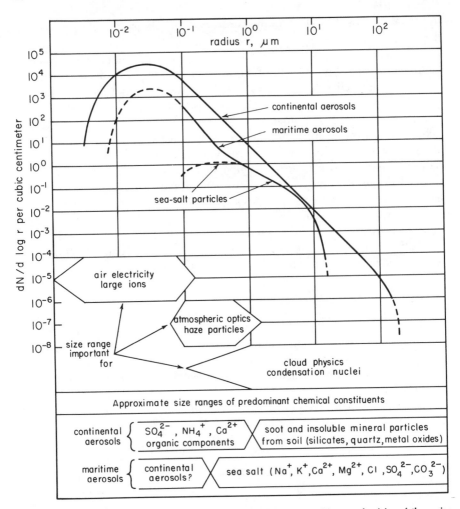

FIGURE 4.12. Tropospheric aerosol size ranges. The abundances, compositions, and origins of the major classes of atmospheric aerosols are also given. For examples of work on the formation and nature of sulfate aerosols, see Junge (1960); for continental sources and long-range transport of organic aerosol constituent, see Lunde and Bjorseth (1977).

ently popular (Turco *et al.*, 1982). The sulfuric acid in the Junge layer is produced by oxidation and hydration of SO_2 (Friend *et al.*, 1973) and COS (Crutzen, 1976). This same process, but strongly amplified, is the probable source of the sulfuric acid clouds on Venus which we discussed earlier. Indeed, the important role of COS in producing H_2SO_4 was recognized on Venus (Prinn, 1973) several years before it was also shown to be significant on Earth.

All aerosols play a significant role as absorbers and scatters of visible solar radiation and IR planetary radiation. They are also important as nuclei for the condensation of water and ice, the common particulates that are the principal ones governing the radiative transfer properties of the atmosphere.

The chemically active gases (and aerosols) are all also involved in complex geochemical cycles, the nature of which must be understood in order to attempt descriptions of evolutionary trends and predict future changes of composition. The rare gases, although chemically infinitely simpler, are nonetheless of enormous interest because of the involvement of several rare gas isotopes in radioactive decay processes which are useful for dating purposes. Also, the nonradiogenic rare gases are the least ambiguous indicators of the sources of terrestrial volatiles, and are crucial elements of any attempt to reconstruct the origin and early evolution of the atmosphere.

It is, however, a truism that no element of a complex system can ever be wholly isolated for study. In the present case, it is now very clear that the origin and composition of the atmosphere is intimately intertwined with the overall elemental composition, mineralogy, heat source strength, and melting and differentiation behavior of the entire planet. For this reason, the global properties and history of this planet must be surveyed with atmospheric genesis and evolution in mind.

4.3.1. Composition, Thermal History, Differentiation, and Outgassing

The Earth is a differentiated planet. The general structure of core, mantle, and crust originated before the oldest known rocks were formed. We have no samples of undifferentiated Earth material, and we are unable to analyze the core directly. A great deal of our knowledge of the composition of the mantle is inferential as well. The overwhelming majority of our detailed chemical analyses of Earth material refer to the 0.4% of the Earth's mass that makes up the crust, or to a very diverse set of samples of deep crustal and upper mantle origin, many of which are clearly not representative of the whole mantle. The composition of the Earth is usually interpreted in terms of bulk composition models (ordinary chondrite, carbonaceous chondrite, etc.) and presumptions about the processes of melting, geochemical differentiation, and minor element fractionation as they took place before the dawn of the geological record.

The melting history depends sensitively on the nature and availability of heat sources and the distribution of the released heat within the Earth. The history of the major radionuclides in the Earth and their present distribution are still a matter of active debate. A very diverse set of arguments, some astrophysical, some geochemical, some geophysical, and some meteoritical must be reviewed and assessed, and still we will be unable to settle upon a single definite history.

Is it possible that the volatile element inventory of the early Earth was much greater than it is today? Was there massive late addition of volatiles to Earth? How does the major-element composition relate to the volatile-element abundances? What was the geochemical setting of the origin of life? We begin to attack these problems by addressing the major ideas about the bulk composition of Earth and its volatiles, the abundances of heat sources and their distribution, and the melting and outgassing process that gave rise to the present structure of our planet.

4.3.1.1. Major-Element Composition

Variants of every one of the classes of genetic models discussed in Chapter 2 have been applied to the Earth. The number of different models published to date noticeably exceeds the number of their authors, and an exhaustive recounting of the history of this endeavor is neither necessary nor rewarding. For this reason, we shall concentrate principally on the literature of the past 20 years.

It is, however, worth mentioning that systematic comparison of the Earth's composition with chondritic meteorites has been under way for half a century. The classical geochemistry treatise by Goldschmidt (1954) has made the idea of comparative study of terrestrial and extraterrestrial materials familiar to generations of students. It has become widely accepted that the terrestrial planets were formed by low-temperature accretion of mineral grains from a nebular gas cloud, and that their atmospheres were formed from gases released by heating of these minerals inside the planets (Latimer, 1950; Urey, 1951, 1962).

As we shall see, two very different types of Earth histories arise as corollaries to adherence to the two most extreme types of "chondritic" models. The carbonaceous chondritic model envisions strong heating and gravitational escape of a vast volume of volatiles present in C1 chondrites but badly depleted on the present Earth. The ordinary chondrite model begins much closer in composition to the present Earth, but still requires explanation of several significant points of departure which we shall discuss.

Two more theoretically based approaches, the multicomponent and equilibrium condensation models, have in several ways complementary faults and strengths. Thus, the equilibrium condensation approach attempts to derive from a short list of simple chemical and dynamical postulates a first-order description of the bulk composition and density of the solid bodies in the Solar System, intentionally avoiding complexity in order to see whether the general conceptual framework of condensation and accretion modeling is useful. The evolutionary tendency of such a model is to refine, add, or reject working hypotheses of ever-greater detail and specificity in response to the growth in factual knowledge of the Solar System. On the other hand, the multicomponent model is concerned primarily with the categorization of the chemical elements into cosmochemically coherent families, based upon known condensation, reheating, melting, and crystallization processes. In this way, the ~90 compositional variables inherent in a solar-composition system can be collapsed into a small and manageable number of components, each ideally of fixed composition, which can then be combined by simple mixing to produce planets, asteroids, or meteorites *if enough* is already known of their composition to reconstruct the abundances of the various components in each body. In principle, analysis for a carefully chosen set of elements equal in number to the components would suffice to characterize the composition of a body fully. In practice, however, this is scarcely achievable even for undifferentiated bodies without more data; for differentiated and partly differentiated bodies, or for bodies that were not initially homogeneous, even more data are required to avoid ambiguity of interpretation. Since the application of this technique to bodies is partly *post hoc* (i.e., it cannot make any predictions until a quite significant body of analyses is already in hand), the identification of cosmogonic and evolutionary trends between solar system bodies is also rather post hoc: It requires that a sufficiently large number of planets has been studied so that the systematic variations in the abundances of the chemical components can be discerned. It appears that these two "competing" approaches are coming at the same goal from opposite directions. Indeed, it would be hard to regard either approach as a success until it has merged thoroughly into the other, so that the cosmochemical and dynamical provenance of meteoritic and planetary composition classes can be traced through condensation, accretion, melting, differentiation, and outgassing to yield the present diverse Solar System. As presently defined, the equilibrium condensation model is merely an intentionally over-simplified approximation, but it is quantifiable and makes explicit predictions. The multicomponent model *may* be complex enough, but it is incapable of making predictions about bodies not yet analyzed.

The question of the nature of the accretion mechanisms involved in assembly of a planet stands somewhat apart from the issue of bulk composition, since homogeneous and wholly nonhomogeneous accretion scenarios may yield planets with the same bulk composition. It is important to emphasize that "homogeneous," like "chondritic," means different things to different people. It can mean that the composition is the same throughout, on, let us say, the scale of a meter or a kilometer. On the other hand, even an ordinary chondritic planet may be said to be *non* homogeneous if one regards the minerals composing it as being fractionated components with different condensation temperatures or metamorphic histories. We shall generally take "nonhomogeneous" to mean "accreted in layers in the sequence of condensation," the sense intended by the proponents of such models.

Specific assumptions regarding both the bulk composition and the distribution of heat sources are needed in order to carry out thermal history calculations.

Chemical and thermal models of the Earth often assume for simplicity that the Earth has essentially the same elemental abundances as the ordinary chondrites, or more specifically, the H (high-iron) chondrites, and was accreted homogeneously. Hurley (1957) commented on the near equivalence of the average heat flow rate through the Earth's crust and that expected for steady-state heat loss from a chondritic Earth. MacDonald's (1959) thermal history models showed that, if convection were absent in the mantle, then the Earth could have less than the chondritic abundance of the major heat-producing radionuclides, ^{40}K, ^{235}U, ^{238}U, and ^{232}Th, but that a chondritic Earth model would require mantle convection. Gast (1960) showed that the present abundances of K, Rb, and Cs in the crust and presumed representative upper mantle rocks accounted for only about one-eighth to one-fifth their expected abundance in an ordinary chondritic Earth, and suggested that either Earth was deficient in these elements relative to chondrites or they had been lost "to the lower mantle."

Wasserburg *et al.* (1964) pointed out that the ratio $K:U$ is essentially invariant in the crust even where very large variations in absolute abundance occur. They take this ratio ($\sim 1 \times 10^4$) to be characteristic of the entire Earth, showing a marked difference between Earth and chondritic material, in which $K:U \simeq 6 \times 10^4$. They then assumed that the terrestrial heat flow could be accounted for by neglecting the K content of the deep interior of the Earth and adjusting the U and Th abundances upward to give the observed heat production rate. This adjustment attributes a much larger proportion of the Earth's heat production to radionuclides with longer half-lives than ^{40}K and, thus, decreases the estimated rate of heat production in the early Earth.

The issue of chondritic composition of the Earth was reexamined by Taylor (1964a,b,c), who again found that K, Rb, and Cs were several times less strongly enriched in the crust than were many other large-ion lithophile elements such as Ba, Sr, La, U, and Th. Gast (1965) found a $K:Rb$ ratio >1500 for the upper mantle from Rb–Sr isotope abundances (compared to a solar $K:Rb$ ratio of ~ 700). He concluded that the failure of the heavy alkalis to be strongly enriched in the crust was a major argument against derivation of the Earth from ordinary chondrite parent material. He proposed that Earth accreted from a mixture of metallic iron and Ca-rich achondrites, depleted in K, Rb, Cs, and other volatiles. Presumably, then, the observed volatiles on Earth must be attributed to yet another component. More recently, Hurley (1968) and Russell and Ozima (1971) have advocated a $K:Rb$ ratio of 710 for the bulk Earth.

A second major argument against an ordinary chondrite Earth model was based on the observed oxidation state of the crust and upper mantle. This argument achieved its greatest cogency when applied to the Ni^{2+} and Fe^{3+} content of the upper mantle. Analyses of upper-mantle xenoliths which have been delivered to Earth's surface by deep-seated volcanic activity show oxidized nickel and ferric iron contents far in excess of the H chondrite values. Differentiation of a homogeneous ordinary chondrite body at low pressures leads to all phases being equilibrated with Fe/FeO buffer, resulting in thorough extraction of nickel into the metal phase and complete reduction of iron to the metallic and ferrous states.

Ringwood (1966) provided Fe^{3+} and Ni^{2+} in the Earth by postulating formation of all of the terrestrial planets from the most oxidized and volatile-rich meteorites, Type 1 carbonaceous chondrites (C1). He then derived the present planets by extreme outgassing and reduction of iron oxides by carbon. He compared the Earth to the enstatite (E) chondrites, in which FeO-free silicates coexist with ~ 0.4 wt. % of elemental carbon. Ringwood argued that carbon reduction rather than hydrogen reduction is required because of the presence of abundant FeS in all E chondrites: A partial pressure of hydrogen sufficient to reduce FeO completely ($X_{FeO} \leq 10^{-3}$ in pyroxene) will destroy all FeS by reduction to Fe and H_2S. Ringwood then used the presence of $\sim 1\%$ elemental silicon in the metal phase of E chondrites to argue for the presence of ~20% Si in the core of the Earth, in apparent agreement with the low density of the outer core relative to that of pure iron (Birch, 1952, 1964). This process of reduction and outgassing of a Cl primitive Earth would have caused the release of ~0.2 $M_\oplus$ of volatiles, mostly CO, CO_2, and H_2O, which must be wholly lost during the hypothetical Hayashi phase of the Sun. He attributed the persistence of Ni^{2+} and Fe^{3+} in the crust and upper mantle to lack of attainment of equilibrium.

For a number of years, Si was generally assumed to be the light element in the core. Following the original suggestion by Birch (1952), MacDonald and Knopoff (1958) constructed Si-rich models of the core, but the latter authors also commented that S could not be ruled out as a major component of the core, in place of Si. Tuman (1964) treated the seismic structure of the core–mantle boundary in terms of freezing of a Si-rich core. Balcham and Cowan (1966) carried out experiments on Fe–Si alloys shocked to core pressure (2.7 Mbar) for their supposed relevance to the core. Ringwood (1966) proposed a highly disequilibrium "blast furnace" autoreduction process in Cl parent material as the source of both the iron and silicon in the core. Brett (1971) attempted to rescue this very ad hoc kinetic mechanism for elemental Si production by searching for conditions under which there might be some equilibrium partitioning of Si into the core. This approach was severely criticized by Ringwood (1971), who insisted that the Earth actually accreted nonhomogeneously and that core formation did not involve attainment of equilibrium. More recent work assuming Si in the core includes that by Stewart (1973).

Mason (1966), on the other hand, argued for the presence of sulfur rather than Si in the outer core on the basis of ubiquity of FeS in meteorites, the high cosmic abundance of sulfur, and the ease of melting of Fe–FeS mixtures. This idea was strongly supported by Murthy and Hall (1970) and by Lewis (1971a). In response to this suggestion, a number of papers examined the geophysical consequences of sulfur in the core (Tolland, 1973, 1974; King and Ahrens, 1973; Verhoogen, 1973), and Brett (1975) concurred with this conclusion. [See also the review by Brett (1976).] It is now rather generally accepted that any plausible meteoritic analogue for the composition of the Earth is likely to have an appreciable sulfur content. Indeed, it is noteworthy that the meteorites used by Ringwood as models of the reduction process contain ~5.7% sulfur, second only to Cl chondrites (~5.9%). The E chondrites would, if melted, produce a Ni–Fe–S–Si melt having a S : Si ratio of ~17 : 1, yet Ringwood rejected sulfur as a constituent of the core. The high sulfur abundance in the E chondrites cannot readily be explained away by postulating derivation of the sulfides, metal and major silicate phases from separate sources, since all phases show strong evidence of formation at very low oxygen fugacities. Further, the mineralogy of E chondrites is extremely unusual, being characterized by the presence of sinoite ($Si_3N_4 \cdot SiO_2$, an oxynitride of silicon), the magnesium–iron sulfide niningerite, and the potassium-bearing sulfide djerfisherite, $K_3CuFe_{12}S_{14}$ (Fuchs, 1966; Keil and Snetsinger, 1967; El Goresy et al., 1971). At oxygen fugacities as low as those needed to make E chondrites, many elements usually thought of as purely lithophile, including Ca, Mg, Mn, K, and even Na and Ti, display strong chalcophile tendencies.

The survival of these highly reduced sulfide species for billions of years in the enstatite meteorites is evidence for their remarkable stability. The original advocates of chalcophile behavior for potassium (Lewis, 1971a; Murthy and Hall, 1972) found that the available thermodynamic data on K_2S and the metallurgical and meteorite evidence on its occurrence and stability made it likely that high-pressure melting of a eutectic Fe–FeS mixture on the Fe°/FeO buffer would lead to substantial extraction of K into the sulfidemetal melt. The available experimental results which bear on this model (Goettel, 1972; Ganguly and Kennedy, 1977) extend only from 1 bar to moderate (lower crustal) pressures and leave the issue unresolved: Nevertheless, the matter has been vigorously debated (Oversby and Ringwood, 1972; Goettel and Lewis, 1973). None of these authors found evidence that *present* crustal conditions would be conducive to potassium sulfide formation; however, this reckons without the stabilization of K_2S by reaction with FeS, CuS, etc. to form complex minerals such as djerfisherite. Curiously, several reports of the occurrence of potassium sulfide minerals in heavy-metal deposits (Sokolova et al., 1970; Dmitrieva and Ilyukhin, 1975) and in kimberlites (Clarke et al., 1977) have now appeared. Simultaneously, theoretical treatment of the effects of core pressures on the chemical behavior of potassium indicates the plausibility that K will partition into a metallic core (Bukowinski, 1976; Knopoff and Bukowinski, 1977). It seems that the chalcophilic tendencies of potassium may have been underestimated. This may have a profound effect on the distribution of long-lived radioactive heat sources in the modern Earth (Goettel, 1976).

Recently, Ringwood (1979) has extensively modified the C1 composition model for the terrestrial planets in favor of an Earth model which is a homogeneous mixture of about 90 mass % of material similar to ordinary chondrites with about 10% of C1 material, which acts as the carrier of both volatile elements and ferric iron.

The interpretation of the present Fe^{3+} content of the crust and mantle is not, however, quite this simple. In the last few years, laboratory experiments on the phase stability of common geological materials at extremely high pressures typical of planetary interiors have opened a new perspective on the oxidation state of iron in the Earth (Bassett, 1978). Pressures in excess of several hundred kilobars have been produced by several groups, with the present record near 1.7 Mbar (Mao and Bell, 1978), corresponding to pressures within the Earth's outer core. Mao (1974) has suggested that, at pressures near 1 Mbar, FeO disproportionates to Fe metal and ferric oxides. Thus, a planet which is initially devoid of ferric iron may produce it internally if interior pressures are high enough. Ferric iron oxides may in turn enter into rising low-density melts or into the core-forming Fe–FeS melt, but will be excluded from mantle minerals because of the Fe^{3+} charge and ionic size. Disproportionation via

$$4FeO \rightleftharpoons Fe_3O_4 + Fe° \qquad (4.44)$$

yields a mass of metal equal to only 19% of the original mass of FeO. This process is therefore incapable of providing the entire metallic cores of Venus and Earth. Thus, metallic iron must have been a major primordial component of both planets, whereas Fe^{3+} need not have been present initially in order to be present today. Both the FeO disproportionation mechanism and the solution of iron oxides in a metallic Fe–Ni melt are invoked by Ringwood in order to enrich the outer core in FeO. Although silicon is no longer considered as a major component of the core, FeS is present as a component of the outer-core melt. Thus, Ringwood has adopted the suggestion by Murthy and Hall (1970) that the outer core is a melt in the Fe–Ni–S–O system.

The FeO disproportionation mechanism, however, seems to obviate the need to accrete large quantities of magnetite, while the vast differences between the observed volatile content of Earth and that of the model seem to require explanation. For example, this model provides at least 10 times the known terrestrial inventory of almost every volatile element. Finally, the very high sulfur abundance in the C1 and H chondrite components provides so much FeS for the outer core that there is no need of postulating an even larger ($\sim 50\%$) FeO component for the core. Although the masses of volatiles which must escape from the Earth to get down to present abundance levels is an order of magnitude smaller than in Ringwood's original model, there is still a requirement for escape of, for example, more than 95% of the primordial endowment of water. The total mass that must escape is roughly 2% of the mass of the Earth. Thus, the requirement for the existence of the Earth in the presence of a Hayashi-phase Sun is retained. Typical estimates of the duration of the Hayashi phase are $\sim 10^3$ years at most, whereas the time required to accrete the Earth is probably close to 10^8 yr, as we discussed earlier. Further, the very existence of a Hayashi phase of very high luminosity is dubious (Larson, 1974).

The available energy from gravitational collapse of the protosun ($\sim 10^{49}$ erg), which will provide a peak luminosity of $\sim 10^5 L_\odot$ if liberated in only $\sim 10^3$ years, will give luminosity of only $3 L_\odot$ if averaged over a T Tauri phase of duration 3×10^7 years. This energy can only be produced and liberated once by a given star. Since the T Tauri phase for a $1 M_\odot$ star has an observed luminosity of $\sim 3 L_\odot$ and a duration of $\sim 3 \times 10^7$ years, and since the highly superluminous Hayashi phase has never been observed, we are inclined to regard the Hayashi phase as at best a statistical rarity and at worst nonexistent. In either case, planets could not be subjected to the Hayashi phase.

It should be noted that the addition of a minor mass fraction of volatile-rich material, although occurring in both Ringwood's most recent model and in the nonhomogeneous accretion scenarios of Turekian and Clark (1969), plays a radically different role in the two models: In Ringwood's scenario, the accretion of the volatile-rich and volatile-poor material is *simultaneous* and the planet is initially *homogeneous*. The nonhomogeneous accretion model, loosely based on the condensation model of Eucken (1944), pictures sequential condensation of minerals during cooling of the primitive solar nebula, simultaneous with very rapid accretion of planetesimals. As a consequence of the postulated speed of accretion, each protoplanet accreted successive layers of new condensates as they formed in the nebula. Thus, each protoplanet has a nonhomogeneous structure. Earth may have accreted its core before the mantle, thereby forming with roughly its present structure. Such a process would yield an Earth with neither sulfur nor silicon in its core, and presumably with an FeO-free lower mantle. The volatile elements were added last as a thick veneer containing CI or other volatile-rich material, amounting to perhaps 2% of the mass of the Earth (the amount needed to match the primordial rare gas abundances). This model achieves the high oxidation state of the crust and upper mantle by selectively accreting oxidized material last. Larimer (1971), based on the abundances of the elements in crustal and upper mantle rocks, concluded that, although a homogeneous Earth accreted near 500 K from the solar nebula would be reasonably consistent with his data, a higher-temperature origin for the bulk Earth and a veneering of 1–3% C chondrite material would also be acceptable. Wasson (1971), in a comparison of the abundances of volatiles on the Earth and the Moon, suggests that a late bombardment of volatile-rich material at low velocities would very strongly favor accretion of these volatiles by the more massive Earth. An obvious problem with this idea is that, at the high encounter velocities expected for cometary and CI material, the relative capture cross sections of Earth and the Moon would be close to their geometrical cross sections. Hence, the 81.3 times more massive Earth, accreting with only 15 times the cross section area, might reasonably be expected to be volatile-*poor* relative to the Moon. Both Hutchison (1974) and Schuiling (1975) propose compositional models in which relatively volatile-poor high-temperature condensates of ordinary chondritic or achondritic composition are veneered by volatile-rich materials. Schuiling suggests 15% of the Earth's mass was ~400 K condensate. Hutchison's model includes very severe boiling of the Earth by the early Sun in order to lose vast amounts of Si and alkali metals, followed by accretion of Fe^{3+} and volatiles as a thin veneer.

Bringing in the ferric iron and volatiles as a late-accreting "veneer" must meet two stringent requirements which are not easily met. First, Ringwood (1975) emphasized the evidence for a remarkable degree of homogeneity in the distribution of siderophiles in Earth's mantle, a uniformity very hard to achieve if the original distribution of oxidized siderophiles is so strikingly nonhomogeneous. Second, there is the question of whether the late-accreting material on the Earth could possibly be so thoroughly dominated by CI material from very distant parts of the Solar System. Wetherill (1975) and Hartmann (1976) have explored the partitioning of late-accreting material between the terrestrial planets, assuming that the planets had nearly their present masses during that epoch. Although the models employed in the dynamic calculations are greatly simplified, it is worthy of note that Earth's late-accreting material is found to be almost entirely (96%) derived from the zones of Earth and Venus, with a negligible amount of material from the zone of Mercury and 4% of the total from the vicinity of Mars. The mount of mass derived from the asteroid belt would be smaller yet. Integral accretion models (Cox *et al.*, 1978; Cox and Lewis, 1980) show a similar behavior for the entire history of accretion, with little tendency for the composition of infalling material to depend on time.

The most plausible source for an enhanced abundance of volatile-rich material in the early Earth would be planetesimals from the region beyond the outer edge of the asteroid belt, which may be scattered by Jupiter's gravity into orbits which penetrate the inner solar system (Kaula and Bigeleisen, 1975; Weidenschilling, 1975). Such bodies, which were not included in the accretion model of Cox and Lewis, should be at least as volatile-rich (and oxidized) as C1 chondrites, and may well be similar to cometary nuclei in composition. Weidenschilling emphasizes their importance as projectiles capable of eroding and disintegrating belt asteroids and depleting and supply of planet-making material at the orbit of Mars. By Weidenschilling's estimate of the ejection probability per encounter for "cometary" planetesimals from the region of the Jovian planets, we would expect the supply of such objects to the inner solar system to decay with a half-life of a few tens of millions of years. If this supply dies out on a time scale shorter than the accretion time, then shock-heated debris from comet collisions with planetesimals in the inner solar system will be rather homogeneously distributed in the accreted planets. If, however, the flux of cometary material decays on a time scale long compared to the accretion time of Mercury, but much shorter than the accretion time of Mars, then comets penetrating to the orbits of Mercury and Venus would strike nearly fully accreted planets, while near the orbits of Earth and Mars the collisions would instead occur with a swarm of large numbers of small bodies with a total cross-section area much larger than that of the fully accreted planet. We can safely assume that the cometary bodies are widely varied in orbital inclination, and thus that shielding of the Mercury–Venus region by the protoplanetary swarm in the Earth–Mars region is not a major factor. Under these circumstances, cometary volatiles could be most severely depleted in Mars-forming planetesimals. Violent collisions of bodies with negligible gravity in a vacuum produces very high shock temperatures, efficient outgassing, and irreversible gas loss. We also might expect that the volatiles in Earth and Mars would be more homogeneously distributed than those in Venus: The latter would have more the character of a late "veneer."

If such a late veneering flux occurred, it has not persisted to the present. The present meteorite recovery rate on Earth is ~1.5% carbonaceous chondrites (0.5% C1s) and ~98.5% metal-bearing (ordinary chondritic, iron, stony-iron, and achondritic) objects by number. By mass, the predominance of reduced material is even more striking. Further, the overwhelming majority of the total mass of carbonaceous chondrite material known is from the single Pueblito de Allende fall in Mexico—and Allende is a magnetite-free, metal-bearing C3 chondrite. It is merely an improbable assertion that there was any era in the history of the Earth when the influx of mass was dominated by C1 material. It is most probable that the fraction of C1 material was ≤1% of the influx rate at all times in history.

We find when we consider the oxidation state of CI + H mixtures that the amount of metal in H chondrite material is sufficient to convert all the magnetite in the most magnetite-rich C1 chondritic material to FeO if even as little as 20% of the mass of the mixture is H chondrite. For a C2-type volatile carrier, with only 2% magnetite, mixing of only 2% H chondrite with 98% C2 material would suffice to reduce all the magnetite and leave the system on the Fe–FeO buffer.

When it is considered that only about 2% of the mass of the Earth need be CI material in order to provide all the observed volatiles (Anders, 1968), then the likelihood of domination of the outer layers of the Earth by meteoritic ferric iron and H_2O seems remote. The solution to the problem of the oxidation state of the crust and upper mantle would seem to be the high-pressure disproportionation of FeO. Once this has been recognized, the volatile element budgets themselves may be taken as the most useful discriminants of different possible meteorite analogs of the source material.

Before embarking on this comparison of specific meteorite types with the Earth, it should be remarked that the discovery of near identity of the Earth's volatile content with that of any single class of meteorite would be remarkable indeed. Large (factor of 2) fractionations have affected many of the major elements, most notably iron, in chondrites. As we shall see in Section 4.7 on asteroids, there is good reason to believe that all the ordinary chondrites originated on parent bodies with characteristic dimensions of only a few kilometers. Accretion of mineral grains to form such small bodies will to a large degree depend upon the physical properties of individual grains, such as electric charge, dielectric constant, conductivity, magnetic moment, friability, stickiness, density, and particle size. The composition of small bodies accreted mainly through interactions of small grains can be expected to depend on temperature, gas density, and streaming velocity, and other locally varying environmental parameters. On the other hand, terminal accretion of a terrestrial planet occurs by gravitational attraction, which is independent of the chemical and physical properties of the accreting bodies, and also samples vast numbers ($\sim10^{12}$) of kilometer-sized planetesimals. It is therefore very likely that each planet is assembled from a very representative collection of the locally available material, and that its present composition is dominated (say, 80–90% of its mass) by this local material. Indeed, it is precisely this idea which underlies the equilibrium condensation model for the planets, which describes to first order how the bulk composition (again, ~80–90%) of each planet, and thus its density, should depend on heliocentric distance if large-scale radial mixing were unimportant in the solar nebula. In any event, such phenomena as metal–silicate fractionation in chondrites are clearly easy to achieve, whereas it is not at all obvious that these same fractionations would be evident in large, gravitationally accreted planets. Thus, the observation that the Earth has slightly more iron than the most iron-rich H chondrites is not disconcerting, nor is the existence of E chondrites with more iron than Earth, or the Netschaëvo chondrite (Bild and Wasson, 1977), which likewise is richer in iron than the Earth. In this picture, the Earth is a better guide to the cosmic abundances of condensed elements than any class of chondrites, but can be used as a standard only if the distributions of the elements between crust, mantle, and core are fully understood.

4.3.1.2. Volatile-Element Composition

Table 4.10 summarizes briefly the visible abundances of H, C, N, S, and the rare gases in various classes of chondrites and on Earth. The Earth data are limited by the known high solubility of several of these elements, especially C, N, and S, in the core. Sulfur is a poor discriminant, since the sulfur content of the Earth could be as low as that in ordinary chondrites if there is a few percent FeO in the outer core, or as high as in the C1 chondrites if the only important light element in the core is sulfur. Every class of primitive meteorite gives ample carbon and nitrogen, as well as labile oxygen (in Fe oxides), as is illustrated in Table 4.11.

The rare gases are of special interest, since they are chemically unreactive under all known geologically relevant conditions and readily released into the (observable) atmosphere during planetary outgassing. The ^{36}Ar abundance, which is the most useful single indicator of the abundance of the fractionated (planetary) rare gas component, is about 40 times as high in C1 chondrites as in the Earth; thus, any Earth model containing more than 2.5% C1 material would provide more than the observed amount of ^{36}Ar. If this amount of C1 material were present, then it would bring in 0.25% of an Earth mass of water, which is embarrassingly large compared to the 0.03% represented by the mass of the oceans. The C2 chondrites have the same problem, only to a somewhat lesser degree. Further, the water in C1 chondrites is deuterium-rich compared to the Earth (Table 4.10). To confound matters fully, the isotopic composition of terrestrial crustal carbon is found to be much heavier than that in the ordinary chondrites, but closely similar to that in the C1 and C2 chondrites, whereas that of nitrogen is closer to that of ordinary chondrites (Kung and Clayton, 1978). The rare gas evidence clearly shows that ordinary chondrites come quite close to the terrestrial abundances, except for a

TABLE 4.10
Weight Fractions of Volatile Elements in Chondritic Meteorites and the Earth

Element	C1	C2	C3	H	L	LL	E	Earth
H (%)	1	0.5	0.03	?	?	?	?	0.004
(δD, %)	(+25)	(−10)	(−12, C3O) (+18, C3V)	?	?	?	?	(−1)
C (%)	3.1	2.5	0.5	0.1	0.1	0.1	0.4	>0.002
(δ^{13}C, %)	(−8)	(−7)	(−18)	(−24)	—	—	—	(−6)
N (%)	0.3	0.2	0.01	0.004	0.004	0.01	0.03	>0.0001
(δ^{15}N, %)	(+46)	(+20–70)	(−25)	(+20)	(+5)	(+10)	(−30)	(0)
S (%)	5.9	3.4	2.2	2.1	2.2	2.3	5.7	~4
$^{20}Ne^a$	2^{-10}	1.5^{-10}	5^{-11}	9^{-13}	1^{-22}	—	—	9^{-12}
$^{36}Ar^a$	1.5^{-10}	1^{-9}	$5–25^{-10}$	4^{-11}	4^{-11}	—	—	4^{-11}
$^{84}Kr^a$	4^{-11}	3^{-11}	2^{-11}	1.5^{-12}	1^{-12}	—	—	1.5^{-12}
$^{132}Xe^a$	4^{-11}	3^{-11}	2^{-11}	1.5^{-12}	1^{-12}	—	—	1.5^{-12b}

[a] Rare gases are given as follows: 2^{-10} means 2×10^{-10} gm gm^{-1}.
[b] *Total* terrestrial Xe ($\approx 10 \times$ atmospheric Xe).

TABLE 4.11
Weight Fraction of Labile Oxygen in Meteorites and Earth

Compound	Cl[a]	C2	C3	H	L	LL	E	Earth
FeO (%)	10	25	24	10	15	18	0.1	≥6[b]
Fe$_2$O$_3$ (%)	20	1.4	0	0	0	0	0	0.5

[a]Cl chondrites also contain several weight percent of sulfates and carbonates, which are also carriers of labile oxygen.

[b]In Ringwood's (1977) oxygen-rich model of the core, the FeO content of the Earth can be as high as 17%. The figure given is the weight percent of the Earth contributed by mantle and crustal FeO.

deficiency of about a factor of 10 in meteoritic neon. This could conceivably be an artifact due to partial diffusive loss of neon from meteorites in space, or to an additional neon contribution to Earth by, say, C3 meteorites, comets, the solar wind, or interstellar matter. Unfortunately, however, the diffusive loss idea is strictly ad hoc, and no chondrite class can provide a Ne : Ar ratio as high as that on Earth. The only meteoritic component which could help out is the dark (gas-rich) phase of certain solar-wind-irradiated chondrites, which have a rare gas composition approaching that of the Sun. This solar (as opposed to planetary) component is characterized by very high helium and neon abundances, but, since helium escapes readily from Earth, it would be discerned chiefly as an enhancement of the neon abundance.

Ever since Aston's (1924) early discussion of the deficiency of rare gases on Earth, the elemental and isotopic composition of these gases has been a fruitful field for research. It was from the high abundance of ^{40}Ar in Earth's atmosphere that von Weizsäcker (1937) first inferred the radioactivity of ^{40}K. Brown's (1952) discussion of the origin of the rare gases on Earth and Signer and Suess's (1949) distinction between solar and planetary (fractionated) gas set the stage for more detailed studies. Damon and Kulp (1958) argued that present ^{4}He and ^{40}Ar production rates were far greater than their outgassing rates, and that both isotopes are presently accumulating in the crust. They argued in favor of early strong outgassing of the Earth to provide the present inventories. A number of provocative papers were published in the book *The Origin and Evolution of Atmospheres and Oceans* edited by Brancazio and Cameron (1964). In that volume, Wasserburg (1964) pointed out that all argon in natural gas has a ^{40}Ar : ^{36}Ar ratio of 296, the same as the atmospheric value, which strongly implies crustal contamination by recycling of atmospheric argon. He also showed that the ^{4}He and ^{40}Ar outgassing rates were compatible with steady-state release from a crust with a K : U ratio of 10^{4}, several times less than the chondritic value of ~6 × 10^{4}. The only "primordial" (not recycled) gases seen were ^{129}Xe from ^{129}I and fission Xe from decay of ^{238}U (Butler et al., 1963). In the same volume Turekian (1964) also pointed out that the atmospheric ^{40}Ar inventory were nearly a factor of 10 lower than complete outgassing of a chondritic Earth would provide after 4.5 × 10^{9} years. Turekian also commented on the low ^{87}Sr : ^{86}Sr ratio in the Earth, reflecting a severalfold lower Rb : Sr ratio in Earth than in meteorites.

Wasson (1969) emphasized the close similarity of terrestrial rare gas relative and absolute abundances to those in ordinary chondrites. The major points of difference discernible between the ordinary chondritic and terrestrial atmospheric elemental abundances were a severalfold higher Ne content and a severalfold lower Xe content on Earth on a gram per gram basis. The major *isotopic* differences involved, interestingly, neon and xenon. The latter exhibited a nearly linear mass-dependent fractionation. There is also a far smaller linear fractionation effect seen in the Kr isotopes with opposite *sign* to that for Xe. These effects are discussed in Chapter 2.

The isotopic composition of terrestrial atmospheric Xe is shown in Figure 4.13. Note the nearly linear mass fractionation trend (solid line) with a slope of +30‰/amu. The departures from this trend are due to the higher "special anomaly," ^{129}Xe from decay of ^{129}I, in the Earth and the presence of a small component of an apparent exotic fission component in the carbonaceous chondrite reference material. By way of comparison, the slope of the Kr fractionation line for Earth's atmosphere is −5‰/amu, relative to the same standard (Marti, 1967).

Kuroda and Manuel (1962) attributed about 10% of the ^{129}Xe to decay of the extinct nuclide ^{129}I within the Earth after accretion. Based on iodine contents of a variety of crustal rocks, Kuroda and Crouch (1962) estimated that the time elapsed from the end of nucleosynthesis to the cooling of condensate grains below the iodine retention temperature was about 100 million years. The trend of the other xenon isotopes has been attributed (Cameron, 1963a,c; Kuroda, 1971) to neutron

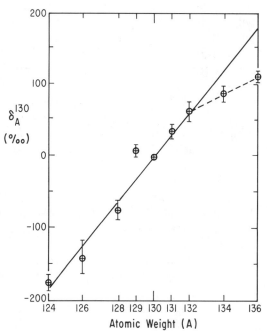

FIGURE 4.13. Isotopic composition of terrestrial atmospheric xenon relative to average carbonaceous chondrite (AVCC) xenon. The enrichment of each isotope of atomic weight A (amu) relative to the abundance standard is given by $10^3[n_A/n_{130})_0 - (n_A/n_{130})_{AVCC}]/(n_A/n_{130})_{AVCC} = \delta_A^{130}$ (‰). Note the departure of terrestrial xenon from AVCC xenon, even when linear mass-dependent fractionation is permitted (solid line). Earth is enriched in radiogenic ^{129}Xe and depleted in heavy Xe.

irradiation during the early deuterium-burning phase of the pre-MS evolution of the Sun, in which about half the terrestrial Xe inventory was irradiated and half shielded. Butler et al. (1963) identified two radiogenic xenon components in samples from natural gas wells: the ^{129}Xe special anomaly and a component of fission Xe from decay of ^{238}U. The presence of excess ^{129}Xe, and thus of continuing outgassing of ancient radiogenic (essentially primordial) heavy rare gases, is now well confirmed (Kaneoka and Takaoka, 1978; Phinney et al., 1978).

The interpretation of the abundance and isotopic composition of helium in Earth's atmosphere is closely tied to the relative rates of outgassing from the interior and loss by exospheric escape. MacDonald (1964) estimated that average rate of ^{4}He outgassing was about 2×10^6 cm^{-2} sec^{-1}, whereas the average ^{4}He escape rate is only 6×10^4. He also estimated a ^{3}He escape flux of 4 cm^{-2} sec^{-1}. Johnson and Axford (1969) similarly estimate the ^{3}He escape flux as 6 cm^{-2} sec^{-1}. Several sources of ^{3}He are in principle possible: Nuclear reactions in the Earth's interior, ^{3}He injection or decay of T(^{3}H) injected in the auroral zones, cosmic ray spallation, and release of primordial gases from the Earth's interior are the four most important sources. The geochemical setting for ^{3}He production by nuclear reactions in rock has been investigated by Morrison and Pine (1955) and by Gerling (1971). In their calculations, ^{3}He is made by (n, α) reactions of ^{238}U fission neutrons on ^{6}Li. The ratio of ^{3}He to ^{4}He, which is produced by the α, β decay chains of uranium and thorium, is calculated to be 10^{-7}, and the production rate of ^{3}He is estimated to be 0.1 cm^{-2} sec^{-1}. Craig and Lal (1961) calculate that high-altitude cosmic ray spallation in the atmosphere should produce ^{3}He at a mean rate of 0.2 cm^{-2} sec^{-1}. These are so much less than the escape rate that other sources must be sought. Johnson and Axford (1969) and Bühler et al. (1973) suggested a major source of ^{3}He from solar wind capture and auroral precipitation at high latitudes. Clarke et al. (1969) studied helium content of deep Pacific water and estimated a flux of 2 cm^{-2} sec^{-1} for ^{3}He diffusing through the ocean floor. Fairhall (1969) suggested that this ^{3}He, rather than being primordial helium diffusing out of the Earth's interior, was produced by decay of tritium that entered the lower atmosphere in the auroral zone. It appears, however, that this criticism was based on a misunderstanding (Craig and Clarke, 1970). Jenkins et al. (1972) have shown that, unlike the Pacific, Atlantic water does not display an anomalously high ^{3}He concentration relative to the amount expected for solubility equilibrium with the atmosphere.

High ^{3}He abundances have been found in gases from Kilauea on the island of Hawaii (Craig and Lupton, 1976; Jenkins et al., 1978), Red Sea hot brines (Lupton et al., 1977), and in the hot water jets emerging from the sea floor near the Galapagos Islands (Jenkins et al., 1978).

The escape dynamics of helium have been investigated by a number of authors. Nonthermal escape mechanisms suggested for helium include solar wind sweeping of the polar regions (Axford, 1968; Donahue, 1971), as well as possible ejection of hot He atoms by reaction of a highly excited He atom with atomic oxygen (Ferguson et al., 1965). The polar sweeping mechanism appears to be by far the most important route for helium escape from Earth. The picture is much more complex for hydrogen, for which much of the escape is by thermal (Jeans) processes, which is subject to large time-dependent variations caused by temperature variations at the exobase. Also, Jeans escape is modified by the depletion of the high-velocity tail of the Boltzmann distribution (Brinkman, 1970, 1971a; Chamberlain and Smith, 1971). The total escape of H from Earth is fixed at the limiting value of about 2.7 × 10^8 cm^{-2} sec^{-1} given by Hunten (1973b), whereas the proportion of the total flux carried by nonthermal mechanisms such as charge exchange ranges from about 60 to 90% (Liu and Donahue, 1974; Hunten, 1982).

Terrestrial neon, with a ^{20}Ne : ^{22}Ne ratio of 10, is intermediate between the solar wind ratios of about 13–14 and that in the "planetary" component of ordinary chondrite rare gases, for which ^{20}Ne : ^{22}Ne = 8. A modest contribution of solar-type rare gas to Earth would produce this result without having a noticeable effect on the abundance or isotopic composition of any of the other rare gases.

Fanale and Cannon (1971b) pointed out that the deficiency in the atmospheric Xe abundance might be explained by the adsorption of Xe on terrestrial shales; that in fact the *total* Xe content of Earth might be remarkably close to that in ordinary chondrites. Phinney (1972) showed that the adsorbed Xe and Kr in shales was indeed strongly linearly fractionated, with adsorbed Xe somewhat more strongly affected than adsorbed Kr (but recall that 90% of the Xe and only a few percent of the Kr would be adsorbed; thus the *gas*-phase fractionation would be far stronger for Xe). Also, the *sign* of the fractionation for Kr was complementary to that seen in the atmosphere. However, of the four Xe samples measured, half had the expected sign of fractionation trend and half had the opposite! The *magnitude* of the fractionation was easily adequate to explain the gas-phase trend. Podosek *et al.* (1980) surveyed the isotopic composition and abundance of xenon in terrestrial sediments. They do indeed find that the heaviest Xe isotopes are most strongly enriched, but doubt that there is enough adsorbed Xe to allow the Earth to have the planetary rare gas composition. We feel the issue is unresolved. Thus, we are left with the neon abundance and isotopic composition as the main clear point of difference between ordinary chondrite and terrestrial rare gases.

The most appealing mechanism capable of providing grossly solar-type neon to Earth (without affecting argon) in sufficient abundance to dominate the "chondritic" background is the interstellar cloud source proposed by Butler *et al.* (1978). Under some circumstances, the retention of neon by cometary solids may conceivably have been large enough so that comet impacts could have influenced the terrestrial rare gas inventory (Sill and Wilkening, 1978). However, the latter suggestion requires two caveats: The usual mechanism for rare-gas retention in an ice-rich body, formation of clathrate hydrates, is very unlikely to work for neon, and the availability of a sufficiently large cometary mass flux onto the Earth is not certain. The *present* cometary mass flux is grossly inadequate, and an intense early bombardment would be needed.

One puzzling phenomenon regarding the rare gas distribution in terrestrial rocks deserves mention. Pillow basalts, which are formed by the extrusion of magma directly into water, have been found to have greatly enhanced concentrations of radiogenic (Dalrymple and Moore, 1968) and nonradiogenic (Dymond and Hogan, 1973) rare gas isotopes. Dymond and Hogan found that the rare gases contained in the glassy rims of basalt pillows had an elemental abundance pattern more closely similar to solar rare gases than to the Earth's atmosphere. They, and later Fisher (1975), assumed the gas to be a valid sample of deep-mantle gases. The opposite approach has been taken by Ozima and Alexander (1976), who hold that some magmatic fractionation process may be responsible.

The nature of the storage reservoir for primordial rare gases in the deep interior of the Earth presents some very interesting questions. Certainly it does not appear that the Earth was subjected to prolonged and complete melting, with consequent complete outgassing and mantle homogenization. This does not rule out the idea of extensive, prompt outgassing of much of the volatile content of Earth during the accretionary era (Fanale, 1971b), nor does it lend credence to such a scenario.

Recent research on compositional variations along midocean ridges has led to the idea of a two-layer structure of the mantle. The upper ("depleted") portion has undergone extensive differentiation and loss of crustal and volatile materials, while the lower ("undepleted") portion still retains a much more primitive volatile-rich composition. "Plumes" (see Morgan, 1971) of volatile-rich material derived from the undepleted region are found along the midocean ridges, interspersed with typical midocean ridge basalts (MORBs) derived from depleted sources (Schilling et al., 1978). Calculated evolutionary histories of the Rb–Sr system suggest that roughly half the mass of the mantle remains undepleted (O'Nions et al., 1979). These complications are not considered in Ringwood's (1979) treatment of the origin of the Earth, and his repeated claim of the striking uniformity of the mantle can no longer be accepted (see also Sun and Nesbitt, 1977).

There is also the fascinating and difficult problem of *when* the rare gases were released from Earth's interior. Fanale (1971b) argued for outgassing of solids upon initial impact with the Earth, and almost all modern thermal history models of the Earth incorporate very early melting and differentiation. Extensive early outgassing of primordial Ar would leave a high present $^{40}Ar : {}^{36}Ar$ ratio in the mantle, whereas continuous first-order outgassing would provide a low $^{40}Ar : {}^{36}Ar$ ratio near 1000 (Ozima and Kudo, 1972; Ozima, 1973). Roddick and Farrar (1971) find a ratio of ~1220 in an ultramafic inclusion, whereas Schwartzman (1973) finds plausible ratios as high as 32,000, thus favoring rapid early outgassing. Perhaps the most novel approach is the suggestion by Hennecke and Manuel (1977) that the $^{40}Ar : {}^{36}Ar$ ratio in the Earth is *primordial*, essentially unaffected by ^{40}K decay. Fisher (1981) has shown convincingly that there is no evidence for ^{40}Ar-rich indigenous gas in an iron meteorite (chosen for the absence of K and, hence, of radiogenic ^{40}Ar).

Thus, after consideration of the abundances and isotopic composition of the volatile elements on Earth, we are left with no single meteoritic source that accounts for all observations. On the other hand, the ordinary chondrite model clearly comes closest to fitting satisfactorily. Perhaps the largest single problem with it is the carbon isotopic composition.

4.3.1.3. Accretional and Radiogenic Heating

The difficulties faced by one- and two-component meteorite-based planetary models, although not necessarily unsurmountable, nonetheless suggest that more complex mising models may be required. Such mixing of components, whether due to progressive chemical changes during temperature variations in the solar nebula at a particular heliocentric distance, to strong secondary heating events on moderate-sized planetesimals, or to radial mixing of solids by gravitational perturbations after the demise of the nebula, may in principle be extremely complex. An attempt to model such fractionation processes phenomenologically, with the number of components being held as low as possible in order to retain some sense of the chemistry, has been the main concern of the work of E. Anders and co-workers for the last decade. Although the full usefulness and detailed justification of the method can best be appreciated by reference to their extensive work on fractionation processes in meteorites, we shall here limit ourselves to a discussion of those aspects of their multicomponent models which apply to the Earth.

For these purposes, the two classes of high-temperature condensates, the refractory metals (Os, Ir, W, Pt, etc.) and the refractory oxide minerals (rich in Ca, Al, Ti, etc.), are lumped together in the same way that they are observed to associate in the high-temperature white inclusions from the Allende meteorite (Lord, 1965; Grossman, 1972). These constitute the first component.

The second component is dominated by magnesium silicates, and the third by iron, nickel, and associated elements, together making up by far the largest proportion of the total condensed mass.

The remaining minor elements are divided into two volatility classes, distinguished by their condensation temperatures from the solar nebula: those that condense between 600 and 1300 K, and those that condense between about 300 and 600 K. These five classes of material are those used in a recent discussion of the abundances of the volatile elements on Earth and Mars (Anders and Owen, 1977).

Mixing models for these five components have been used by Ganapathy and Anders (1974) to describe the compositions of the Earth and the Moon. The Earth is found by this model to be depleted in 600–1300 K volatiles by a factor of about 3 relative to the accepted cosmic abundances (which are, to a very large extent, based on C1 abundances), whereas the lower-temperature volatiles are deficient by about a factor of 40 relative to the same C1 standard. The fractionations between the refractory group, the iron-group metals, and the magnesium silicates are all small (a few tens of percent), so small in fact that these fractionations are all smaller than the uncertainties in the astrophysical sources of abundance data. It is only by assuming that C1 chondrites are unfractionated to this extent that such a conclusion may be reached. Alternatively, the Earth may be a better cosmic abundance standard than the C1 chondrites for the nonvolatile elements: The derived fractionation factors are then those for the formation of C1s from unfractionated material.

The reconstruction of the bulk composition of a planet mainly from crustal compositional data is necessarily dependent on a number of assumptions. Perhaps most crucial here are those concerning the bulk composition of the least accessible part of the Earth, the core, and the partitioning of crucial trace elements during geochemical differentiation of the Earth. Partitioning of potential siderophile and chalcophile elements into the core-forming Fe–FeS melt, with several percent Ni and important amounts of FeO, C, and P, is sufficiently complex so that we cannot claim to know all the relevant partition factors and activity coefficients even at a pressure of 1 bar, much less at hundreds of kilobars to 1.5 Mbar. But by an insidious fluke of nature, those elements which are "moderately volatile," having nebular condensation temperatures between 600 and 1300 K, are for the most part known chalcophiles or siderophiles. Further, there is the problem of what elements will be chalcophilic or siderophilic under conditions of first melting deep within the early Earth. There is ample room for sincere differences of opinion on this matter, and there will continue to be such differences until a large body of experience has accrued in the study of major-element phase relations and trace-element partitioning at pressures of 0.1 to ~1.5 Mbar. Equilibrium partitioning experiments at static pressures up to 1.7 Mbar are now possible (Mao and Bell, 1978). The crucial role of such partitioning studies becomes evident when it is realized that one of the moderately volatile elements which sometimes displays chalcophilic tendencies is potassium, the most important radioactive heat source in chondrites.

In view of this great diversity of opinions regarding the bulk composition of the Earth, it is not at all surprising that discussions of the thermal history of the Earth, which depend sensitively upon the bulk composition, mineralogy, volatile content, radioactive element endowment, and accretion history of the planet, should also be very diverse.

As part of the task of summarizing these ideas, we should especially pay attention to the ways in which the nature and strength of the available heat sources depend on the assumptions of the chemical and accretionary model.

First of all, there is the heat source due to accretional energy. In every model of the Earth this heat source exists; what is debated is (a) whether the total magnitude of this source is large compared to the other sources and (b) the time scale over which accretion occurred. The evidence which has been adduced to defend rapid accretion of the Earth is, as we shall see, very weak and certainly not diagnostic. Similarly, arguments purporting to require very rapid accretion of the Moon to achieve early melting remain unsupported by any plausible suggested mechanism for achieving this end. Finally, as we shall see later, several asteroids show convincing evidence of being igneous differentiates, and several classes of meteorites bear evidence that these differentiation episodes occurred at a very early date. But these asteroids are far too small for accretional heating to be significant. There therefore must have been an early heat source available to melt the igneous meteorite parent bodies at the same time that the Moon was melted. We are sure that, at least in the asteroidal case, this heat source could not have been accretion energy. Economy of postulate requires that we attribute the early lunar heating to the same energy source, in the absence of evidence to the contrary.

The most complete discussion of the evidence for rapid accretion of the Earth was that by Hanks and Anderson (1969). They stated that the presence of 3.5 billion-year-old rocks with apparently normal remanent magnetism demonstrates that the Earth's magnetic field was present at latest only 1 billion years after the accretion of the Earth (see Nagata, 1970). They presented model calculations on a chondritic Earth accreted cold ($\sim$300 K) and heated only by adiabatic compression and decay of long-lived radionuclides. It was found that such an Earth model does not warm to the melting point (defined as the olivine or iron melting curve) for almost 2 billion years. In order to get more rapid differentiation, one may not simply increase the radionuclide abundances arbitrarily, because the present terrestrial heat flux is essentially identical to that calculated for a chondritic Earth in steady state. Hanks and Anderson proposed instead that the Earth must have accreted in a time much shorter than 10^6 years in order that heat stored by the burial of strongly heated infalling debris may be sufficient to produce melting in $<10^9$ years.

However, Murthy and Hall (1970) have pointed out that the treatment of melting behavior in the Hanks and Anderson model is unrealistic for chondritic parent material, since the first melt would be the very dense Fe–Ni–S–O eutectic liquid, produced at low temperatures ($\sim$1180 K) and at a temperature which, unlike the melting points of pure substances, is relatively insensitive to pressure. Furthermore, the appropriate formation temperature for the Earth is near 500 or 600 K, not 300 K (Lewis, 1972b). It thus becomes likely that "warm" accretion of the Earth plus adiabatic compression of the interior would lead to eutectic melting of the Earth before it was fully accreted, regardless of the accretion time scale. The outer core would then contain 10–15 wt. % sulfur (King and Ahrens, 1973).

It must also be emphasized that the treatment of accretion heating by Hanks and Anderson was wholly phenomenological. They made no attempt to study the physics of accretion, and did not attempt to defend the feasibility of actually assembling the Earth in so short a period of time. However, as we discussed earlier, the time scale for terminal accretion of the Earth and the other terrestrial planets was on the order of 10^8, not 10^5, years, if the accretion models developed by Safronov, Weidenschilling, Wetherill, Cox, and others are at all relevant.

Another possible early intense source, first postulated by Urey (1955), is the short-lived radionuclide ^{26}Al. If this nuclide were the heat source, then meteorites formed with different initial Al : Mg ratios would, after decay of ^{26}Al (by β-emission to ^{26}Mg; $t_{\frac{1}{2}} = 7.2 \times 10^5$ years), contain an anomalously large amount of ^{26}Mg relative to the other stable Mg isotopes. Comparison of magnesium isotopic composition between different coexisting minerals with different Al : Mg ratios should then show a correlation of excess ^{26}Mg with the total Al content of the grain.

Such an effect has been searched for and not found in bulk meteoritic material (Schramm et al., 1970), although Lee and Papanastassiou (1974) and Lee et al.

(1976) have found that the white high-temperature refractory inclusions in the Allende meteorite (which, of course, have very high Al:Mg ratios) contain a large amount of radiogenic ^{26}Mg, nicely correlated with the total Al content of the grains. The deduced ^{26}Al content of the grains is so large that, if their aluminum isotopic composition is assumed to be typical of all solar system aluminum, any body which accreted to a size of a few kilometers in radius within 2 million years would have become hot enough to melt within another 1 or 2 million years. The expected result from accretion studies is that a very large fraction of the mass of the planets should have been accreted to kilometer-size bodies in much less than 10^6 years and therefore should have been thoroughly melted. Perusal of the asteroid belt, however, which is rich in much larger, very ancient bodies with apparent carbonaceous chondrite surfaces and bulk densities, suggests that this could not have happened. If these very ancient carbonaceous bodies were not generally affected by such heating, it is hard to imagine that a planet which takes 100 million years to accrete will be melted by the same source. Of course, the possibility remains that the ^{26}Al was injected very inhomogeneously into the solar nebula, and that only a minuscule fraction of solar system aluminum was of the type seen in the Allende meteorite inclusions. In that case, melting of bulk planetary material by this source would be most improbable, but a planet may accrete material from the general vicinity of the Allende parent body (location unknown) which, even in lumps a few kilometers in size, may already have been melted and differentiated at the time of accretion by the planet.

One other possible early heat source is that due to the early T-Tauri phase of the evolution of the Sun (Sonett et al., 1968). According to the analysis by Kuhi (1964), a young MS star expels roughly 10% of its mass over a time of about 10 million years as a very dense solar wind. According to Ulich (1976) the T-Tauri phenomenon is an infalling flux with a similar mass and energy content. In the Sonett model, the dense solar wind induces electric currents in the interior of large bodies, with maximum heating rates occurring at depths on the order of 100 km. This heat source, like ^{26}Al, cannot much affect the thermal state of a terrestrial planet, which has not yet accreted when the T-Tauri phase has ended. The magnitude of the heating effect is, unfortunately, dependent upon a number of variables, such as the electrical conductivity of the material, the radius of the body, its distance from the Sun, and the degree of reliance that may be placed on the general qualitative traits of the T-Tauri phenomenon.

In the absence of short-lived heat sources such as rapid accretion, ^{26}Al, or the T-Tauri solar wind, the dominant heat sources would be the familiar long-lived radionuclides ^{40}K, ^{232}Th, ^{235}U, and ^{238}U. The dominant heat-producing nuclide in solar material, ^{40}K, will produce melting on a time scale of 1 or 2 billion years or not at all.

Because the T-Tauri and ^{26}Al heat sources have lifetimes short compared to the usual estimates of the accretion time scale, extensive melting and outgassing of planetesimals may occur on small bodies prior to the terminal accretion of the terrestrial planets. Indeed, this mechanism is an essential part of the model of atmospheric origins proposed by McElroy and Prather (1981). They would, however, accrete Mars material most rapidly, and Mercury and Venus material least rapidly, contrary to the consensus of the students of planetesimal accretion. Such small bodies would have conductive cooling times comparable to their accretion times ($\sim 10^7$ years for a 30-km body; $\sim 10^8$ years for a 100-km body), and may well bring a large proportion of their heat into the planets which accrete them.

4.3.1.4. Differentiation and Outgassing

Weighing all of these factors, it is easy to accept the plausibility of catastrophic early degassing of the Earth, as suggested by Fanale (1971b). Since both primordial and radiogenic rare gas isotopes are available, it is possible in principle to distinguish the case of complete, very early outgassing from that of slow, steady release of volatiles (Damon and Kulp, 1958). As we discussed earlier, rapid outgassing is consistent with the evidence on the evolution of the $^{40}Ar:\,^{36}Ar$ ratio, although not proved by it.

Early differentiation and outgassing also have implications for the composition of the earliest atmosphere of the Earth.

The gases released by heating chondritic solids may span a wide range of compositions and oxidation states. The oxygen fugacity can range from well below the Fe–FeO buffer, for enstatite chondrite composition, to above the sulfur–sulfate buffer in C1 chondrites. Gas release temperatures can range from ~ 350 K (for volatile-rich carbonaceous chondrites) up to the liquidus temperature, which may be as high as ~ 1830 K. The former may suffice for extensive water loss from fine-grained phyllosilicates, whereas the latter may be necessary to expose the full carbon, nitrogen, sulfur, phosphorus, and chlorine content of metallic iron–nickel alloy to reaction with gases such as H_2 and H_2O. In addition, the composition of the gas phase is affected by pressure, not only through shifts in the equilibria between gases which involve a change in the number of moles of gas, but also in the buffer reactions by which the condensed phases and the gas interact. A final complication is the likelihood that the primordial preplanetary solids were a mixture of materials originating at different temperatures and different distances from the Sun.

For the first gases released during heating of a young planet, the most important gas release is likely to occur at temperatures above the solidus but well below the liquidus. Since, for H, L, LL, E, and many C3 chondrites the solidus is fixed by the Fe–FeS–FeO ternary eutectic temperature, a temperature near 1200 K would be expected at the time of first melting. A maximum temperature near 1300 K would also be reasonable.

We shall briefly consider the equilibrium gas pressure for a mixture of 98% H + 2% C1 chondrites. In order to avoid the tedium of a full discussion of the chemistry of such mixtures as a function of the percentage of C1 material, P, and T, we shall consider just this one composition along a single (P, T) profile for a planetary body. Figure 4.14 shows the initial gas composition for this composition model. Real-solution treatments of olivine, pyroxene, feldspar, and metal alloy (Fe, Ni, C, N)

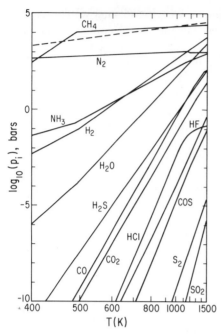

FIGURE 4.14. Gases released from an initially homogeneous mixture of 98% H5 chondrite with 2% C1 chondrite material along a geotherm. Gas pressures given are the equilibrium values for local P, T conditions, and do not include any consideration of differentiation, outgassing history, or Fe^{2+} disproportionation. See text for discussion. The dashed line gives the lithostatic pressure profile for one particular planetary model, of roughly the mass of the Earth, with a 10-fold larger radioactive heat source. (After Bukvic and Lewis, 1981.)

are used, including consideration of C saturation in the metal (Bukvic and Lewis, 1981). Results for temperatures below ~800 K become increasingly difficult to attain due to kinetic limitations, especially diffusion times for species such as C and N through the metal phase. Reequilibration of escaping gases along a geotherm will cause the rising gas composition to approach the low-temperature (<400K) values given in Figure 4.14. Eruption of a 1200–1500 K magma with 98% H + 2% C1 parental composition into a low-pressure environment would begin with a CH_4–N_2–H_2–NH_3–H_2O mixture with minor H_2S, CO, and CO_2. Eventual full outgassing would drop N_2 and NH_3 to the level of minor constituents, and comparable amounts of CH_4, CO, CO_2, H_2O, and H_2 would be released.

The surface mineralogy during this early era would, of course, begin with that of the partially equilibrated H + C1 mixture and evolve toward oxidation of metal, serpentinization of the mafic silicates, and eventually to water condensation. Water vapor, N_2 and NH_3, CH_4, CO_2 and CO, and ferrous phyllosilicates would be present at the surface at all times.

These volatile transport phenomena would transpire simultaneously with the large-scale radial redistribution of radiogenic heat sources in the Earth. The upward migration of U, Th, and a large fraction of the K in the Earth would lead to both a decreased heating rate in the deep interior and a shortened heat loss time scale after the initial differentiation event (Lee, 1968). From the similar upward partitioning of Sr and Pb parent nuclides (^{90}Sr, [238]U, [237]Th), Vidal and Dosso (1978) infer an end to Fe–S segregation from the upper mantle about 2 billion years ago. A schematic history of the Earth is sketched in Figure 4.15.

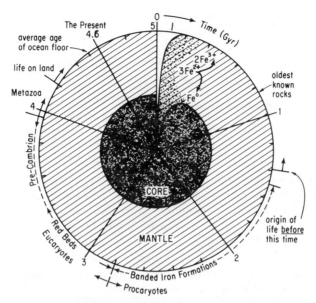

FIGURE 4.15. Schematic rendering of accretion, Fe^{2+} disproportionation, and other major events in the history of the Earth. The time scale for 90% accretion of the Earth is taken as $\sim 10^8$ yr, and the time scale for exhaustion of metal in the near-surface regions by upward-migrating Fe^{3+} is indicated to be a few times longer than this. Each 2×10^8 yr interval is equal to the galactic rotation period. The beginning of differentiation is assumed to occur as a result of eutectic melting of material heated by ^{26}Al decay and/or the T-Tauri phase solar wind when the Earth had only a few percent of its present mass. The arrow near 10^8 yr is a reminder that violent accretionary events involving the sweeping up of bodies with masses up to 5–20% of the mass of the Earth are likely to occur near this time, thus preserving high abundances of primitive metalliferous solids at shallow depths long after core formation has begun. A number of important times in the geological and biological evolution of the Earth indicated here are treated in more detail in Section 4.3.3.

The survival of undepleted mantle material, if correct, leads us toward a slow-accretion scenario for the Earth, strikingly similar to that advocated by Urey 30 years ago. The large volatile content of the undepleted lower mantle is not compatible with hot-accretion histories, including most nonhomogeneous accretion schemes, in which the deep interior is refractory. It also points toward absence of any significant heating by short-lived radionuclides in the early Earth.

At the present state of our ignorance of the initial differentiation process, we cannot be sure about the depth, temperature, and pressure at the time of the first extensive melting within the Earth. It is quite possible that the first differentiation event involved the descent of a dense Fe–Ni–S–O liquid to initiate core formation, followed by the generation of a low-density silica-rich magma which extracts volatiles, small-ion lithophiles (such as Li and Na), and large-ion lithophiles (such as the rare earths) from the siliceous interior of the Earth to initiate formation of the crust.

For smaller bodies, with low internal pressures and appreciable pore volumes, slow heating may outgas water at modest temperatures ($\sim$500–700 K). Release of CO, N_2, SO_2, etc. may take place only at temperatures close to the Fe–S–O ternary eutectic temperature, where diffusion rates for C and N in the metal, O in the FeO-bearing silicates, and S in FeS first become important. This scenario has important implications for smaller bodies (Lupo and Lewis, 1981), many of which were accreted by the growing Earth.

4.3.2. Primitive Terrestrial Atmospheres

Having established that the earth's atmosphere has been outgassed from its interior, we now turn to attempts to model the evolutionary aspects of the outgassed volatile species. Our discussion of this problem is gainfully prefaced by reference to a classical paper on this subject by Rubey (1951). In this paper he attempted to define the principal constraints imposed by geochemical, geological, and biological evidences on the problem of the evolution of the oceans and atmosphere.

In Table 4.12 we present Rubey's estimates of volatile materials now present at or near the earth's surface. From these estimates Rubey subtracted the quantity of each volatile compound which has been supplied by simple weathering of crystalline rocks rather than by degassing of the interior. One is then left with the quantities which must be provided by degassing or for that matter by comets and meteorites. This inventory will, of course, not include any H_2 which may have escaped in very appreciable quantities over geologic time. Note also that the carbon and oxygen inventories are combined as a CO_2 inventory: Since the CO_2 abundance (as sedimentary carbonate) far exceeds both the abundance of C in organic matter and the abundance of O_2 not combined permanently in rock oxides and water, then this

TABLE 4.12

Estimated Quantities (in Units of 10^{20} gm) of Volatile Compounds Now Present at or near the Earth's Surface[a,b]

	H_2O	CO_2	Cl	N	S	Others
In present atmosphere, hydrosphere, and biosphere	14,600	1.5	276	39	13	1.7
Buried in ancient sedimentary rocks	2,100	920	30	4	15	15
Total	16,700	921	306	43	28	16.7
Supplied by weathering of crystalline rocks	130	11	5	0.6	6	3.5
Excess volatiles to be supplied by outgassing, comets, etc.	16,570	910	301	42	22	13

[a]Also given are estimates of the quantities of these compounds which have been supplied by simple weathering of crystalline rocks. The remainder, usually referred to as "excess volatiles," must be supplied by mechanisms other than weathering.

[b]After Rubey (1951).

provides a reasonable approximation. The important point to make from Table 4.12 is that most of the volatile input must have been in the form of H_2O and CO_2 or in suitable precursor compounds. In this respect Rubey also argues that the expected input from meteorite and cometary bombardment falls far short of producing the observed large amounts of H_2O and CO_2. Therefore, these species must have been provided by crustal outgassing.

There are two viewpoints on the way outgassing proceeded, as we mentioned earlier. Rubey argues that a catastrophic outgassing at the time of the formation of the Earth is less likely than a more gradual accumulation of the atmosphere and oceans over a long period of time. His main point is that outgassed CO_2 must be principally removed by dissolution in the oceans followed by sedimentation as $CaCO_3$. This process can proceed only as fast as Ca^{2+} ions are provided to the ocean by weathering of crystalline rocks. Catastrophic early degassing would presumably lead to a predominantly CO_2 atmosphere and a very acidic ocean over a significant period of time. Acidic oceans would then presumably lead to rapid dissolution of Ca^{2+} and Mg^{2+} from crystalline rocks and the subsequent deposition of vast amounts of calcite and dolomite very early in the Earth's history. But there is no evidence of such an epoch in the sedimentary record dating back about 3 billion years and he concludes that the CO_2 content of the atmosphere has never differed markedly from present-day conditions. However, if one takes into account the modern view that crustal subduction and remelting leads to recycling of most sediments on the time scale of a few hundred million years and the fact that we have *no* rocks with ages exceeding about 3.7 billion years, it becomes very difficult to use these arguments to rule out catastrophic early degassing as a viable hypothesis.

In attempting to deduce the way in which atmospheric volatiles evolved, it is also instructive to look at the present-day degassing from active volcanoes, igneous rocks, and hot springs. Rubey's estimates are given in Table 4.13 in the form of percentages by weight of the total outgassed material for each of the important volatiles. These are compared with similar percentages calculated for the "excess volatiles" presently on the surface as given in Table 4.12. Agreement between present outgassing and the present volatile inventory is qualitative at best, but one cannot rule out the possibility that both the hydrosphere and atmosphere could have totally evolved from a suitable mix of these recognized outgassing modes. In doing a comparison of this type there is always, however, the problem of trying to assess just what fraction of the observed outgassing is pristine and what is derived from recycling of volatiles which had entered oceanic sediments. It has been shown by Anderson (1975) that present-day rates of release of such recycled volatiles by island arc volcanism could provide the entire Rubey inventory over the age of the Earth. Island arc volcanoes are located directly over subduction zones in which oceanic crust, with its superficial burden of sediments, is being subducted into the hot upper mantle.

Gases released in midplate volcanism, such as in Hawaii, clearly show the imprint of chemical equilibration between the gases and oceanic basalt: Computations of the oxidation state of fluids in equilibrium with basalts at or above the

TABLE 4.13
Percentage Composition of Principal Gases from Volcanoes, Various Igneous Rocks, and Hot Springs Are Compared with the Composition of "Excess Volatiles" Now Present at or near the Earth's Surface[a]

	Kilauea and Mauna Loa	Basalt and diabase	Obsidian, andesite, and granite	Fumaroles, steam wells, and geysers	Excess volatiles now present
H_2O	57.8	69.1	85.6	99.4	92.8
CO_2	23.5	16.8	5.7	0.33	5.1
S_2[b]	12.6	3.3	0.7	0.03	0.13
N_2	5.7	2.6	1.7	0.05	0.24
Ar^{40}	0.3	—	—	—	—
Cl_2[b]	0.1	1.5	1.9	0.12	1.7
F_2[b]	—	6.6	4.4	0.04	—
H_2	0.04	0.1	0.04	0.05	0.07

[a] After Rubey (1951).

[b] All compounds of S, Cl, and F (H_2S, COS, SO_2, HCl, and HF especially) are reported as S_2, Cl_2, and F_2 because the elemental assays are more reliable than the detailed molecular analyses.

solidus temperature yield $H_2O:H_2$ and $CO_2:CO$ ratios which are in reasonable agreement with Hawaiian volcanic gases (Muan and Osborne, 1956; Holland, 1964).

There is, however, an important (but purely conjectural) possible difference between the composition of the "crust" at the time of accretion and its composition in subsequent epochs (Holland, 1964). This difference, namely the presence of free iron and a much lower $Fe^{3+}:Fe^{2+}$ ratio, leads us to deduce that the initial outgassed volatiles may have been in a much less oxidized form than we presently observe. The relevant equilibrium reactions in the presence of free iron are

$$Fe + H_2O \rightleftarrows FeO + H_2, \tag{4.45}$$

$$H_2 + CO_2 \rightleftarrows H_2O + CO, \tag{4.46}$$

which H_2 slightly over H_2O and CO slightly over CO_2 at temperatures over ~ 1600 K (Fanale, 1971b). Similarly, in the absence of free iron the relevant equilibria are

$$2FeO + H_2O \rightleftarrows Fe_2O_3 + H_2, \tag{4.47}$$

$$H_2 + CO \rightleftarrows H_2O + CO, \tag{4.48}$$

which favor H_2O over H_2 and CO_2 over CO at high temperatures (Fanale, 1971b). Thus, for example, Abelson (1966) argues that CO and H_2 are expected to be major constituents in the Earth's initial atmosphere. Urey (1952c), on the other hand, preferred CH_4 over CO because the equilibrium

$$CO + 3H_2 \rightleftarrows CH_4 + H_4O \tag{4.49}$$

favors CH_4 at the low temperatures expected on the Earth's surface. However, the rate of CO to CH_4 conversion is very slow at room temperature, which leads one to favor CO over CH_4.

At this point it is useful to consider the customary division of the atmospheric evolutionary sequence into three distinct epochs. A Stage I outgassed atmosphere, as we have just discussed, would be by definition distinctly reducing in nature. As Fe was oxidized successively to Fe^{2+} and Fe^{3+} in the crust, and as H_2 escaped from the top of the atmosphere, the outgassing and, thus, the subsequent atmosphere became chemically more neutral. This second atmosphere, which we will see is largely composed of N_2, is designated Stage II. Of course, a source of Fe^{3+} from FeO disproportionation would allow Stage I to be very brief or even to be omitted, so that the atmosphere would begin in Stage II. Finally, the evolution of aerobic photosynthetic organisms gave rise to the present-day distinctly oxidized atmosphere which is designated as Stage III.

It is difficult to be quantitative in our description of the Stage I atmosphere. Holland (1962, 1964, 1972) prefers to consider it as mainly composed of CH_4 and H_2 with minor or trace amounts of H_2O, N_2, H_2S, NH_4, ^{40}Ar, and ^4He. Most of the outgassed H_2O condensed into the oceans, which then became a sink for HCl, NH_3, H_2S, and other soluble gases. Except for adding CO as a probable major atmospheric constituent along with or in place of CH_4 in this compositional listing, our present ideas about Stage I atmospheric composition remain essentially the same as those of Holland.

The organic chemistry which would be expected to proceed readily in a Stage I atmosphere has generated considerable excitement. This interest began with the classical experiment by Miller (1955), who sent a spark discharge through a mixture of CH_4, NH_3, H_2O, and H_2 and produced the amino acids glycine and alanine. Later Abelson (1965) repeated these experiments using H_2, CH_4, CO, CO_2, NH_3, N_2, H_2O, and O_2 in varying amounts. Ferris and Chen (1975a) carried out similar experiments on N_2, CH_4, and H_2O mixtures. As long as this mixture was reducing (i.e., contained large amounts of H_2, CH_4, CO, or NH_3), then amino acids were produced. Later researchers successfully repeated these experiments using UV radiation of wavelengths <1000 Å. Miller and Urey (1959) argued that such radiation was probably the dominant form of energy for dissociating chemical bonds in the primitive atmosphere. It was apparent that amino acids, the basic building blocks of polypeptides, are easy to produce in the Stage I atmosphere. Although many crucial questions remained unanswered, the idea that life could have evolved from such simple chemical compounds gained considerable support. Indeed, it is fair to say that the strongest evidence for a Stage I atmosphere on Earth was the inference that such a stage was essential to the origin of life.

Later experiments were designed to understand how, for example, amino acids could be condensed into polypeptides and how simple bases, sugars, and phosphates could be condensed into mono- and polynucleotides. These experiments have been recently reviewed by Horowitz and Hubbard (1974). Additional energy sources were also investigated. Bar-Nun et al. (1970) sent shock waves through mixtures of CH_4,

C_2H_6, NH_3, H_2O, and Ar and produced glycine, alanine, valine, and leucine. They argued that shock waves originating in thunderstorms and during atmospheric entry by comets and meteorites could have provided a significant energy source for bond dissociation (see also Bar-Nun and Shaviv, 1975). Sagan and Khare (1971a) and Khare *et al.* (1979) showed that addition of H_2S to the reducing mixture enabled the hot H atoms from H_2S dissociation below 2700 Å to be effective in dissociating CH_4 and C_2H_6. This would have enabled amino acid formation to be induced by the much more abundant long-wavelength solar photons in the 2000–2700 Å range. Whether H_2S concentrations in the primitive atmosphere were ever high enough to enable this long-wavelength production of organic material to be very effective remains a problem. Also, hot H atoms would be very rapidly quenched in an H_2-rich atmosphere (Lewis and Fegley, 1979).

The question of the efficiency of production of complex organic compounds is also of interest. Although some workers have attempted to put numbers on total production rates of organic matter, too many *ad hoc* assumptions have to be made to make them believable. There is also the question of the relative fractions of amino acids and simple hydrocarbons produced in the ancient atmosphere. To give an idea of the disparities in estimates that have been made, Lasaga *et al.* (1971) conclude that a 10-m-thick layer of total hydrocarbon material would be produced from a pure CH_4 atmosphere in 10^7 yr, whereas Sagan and Khare (1971a) would produce a 14-m-thick layer of just the amino acids in the same period of time.

The possible catalytic role in organic syntheses of minerals such as magnetite, hydrated silicates, shales, and clays has also been considered. Studier *et al.* (1968) consider that Fischer–Tropsch reactions between CO and H_2 in the presence of NH_3 may have played a role in producing organic material on the primitive Earth. Some experiments by Hubbard *et al.* (1971) suggest that organic synthesis could also have proceeded without the atmosphere being highly reducing (i.e., containing significant amounts of CH_4 or H_2). They irradiated mixtures of CO, H_2O, and CO_2 in the presence of silicates and were able to produce formic acid, formaldehyde, acetaldehyde, and glycolic acid. The silicate surface must act as a catalyst in bond cleavage since UV wavelengths as long as 3000 Å were found to be effective. These reactions may still be relevant in the contemporary Martian atmosphere.

Since the organic chemistry in this primitive atmosphere had profound implications for the origin of life, it is important to know just how long a Stage I atmosphere could have lasted. Its existence obviously depended on the presence of significant amounts of hydrogen, and we must therefore examine the principal source and principal sink for this element; namely, outgassing and atmospheric escape. McGovern (1969) has studied the escape problem for a CH_4–NH_3 atmosphere. If the exospheric temperature T_e were as high as its present-day maximum value of around 2000 K, he concludes that a CH_4–NH_3 atmosphere would have a lifetime of only 10^5–10^6 years even taking into account substantial outgassing from the crust. Recognizing the importance of an accurate knowledge of T_e, he carried out computations of thermospheric energy balance in his primitive atmosphere and concluded $500\ \mathrm{K} \lesssim T_e \lesssim 1000\ \mathrm{K}$, with the actual value depending on the H_2

content of the thermosphere. Since CH_4 and its photodissociation products are good radiators and H_2 is a poor radiator, the highest exospheric temperatures were obtained for the highest predicted H_2 mixing ratio; namely, ~ 0.15. As a baseline for his calculations, McGovern used the average hydrogen outgassing rate of 3×10^{14} gm per year estimated by Holland (1962). As the outgassing rate for H_2 decreased, the exospheric temperature also decreased until it reached its minimum value around 500 K. Initially, when T_e exceeded 700 K, H_2 was the main source for escaping H atoms. However, once the hydrogen outgassing rate had decreased to less than 3×10^{13} gm per year and $T_e < 700$ K, then CH_4 became the principal source. The important hydrogen-containing constituents above the turbopause were predicted to be in diffusive steady state; the escape flux was limited by the Jeans escape rate. Solar wind interactions were probably of negligible import as a sink or source of H_2 (Wasson and Junge, 1962).

Jeans escape, of course, will lead to preferential loss of ^1H and enrichment of the Earth in D. Thus, hydrogen source materials *richer* in D than the *present* oceans are clearly improbable. Those sources which are too rich in deuterium are the C1 and C3V chondrites.

Following the general sequence outlined by Holland (1964), we find that as the H_2 is depleted by escape, CH_4 is converted into either CO_2, which dissolves in the oceans, or into organic matter, and NH_3 is converted to N_2. Thus, a Stage I atmosphere would evolve into the predominantly N_2 Stage II atmosphere.

Shimizu (1976) has argued that very rapid eddy diffusion, driven by gravity waves excited by motions in the lower atmosphere, homogenizes the upper atmosphere and causes it to become a good absorber of solar energy. Hydrogen mole fractions of $\sim 0.15\%$ or higher lead to exospheric temperatures over 1000 K, which causes rapid blowoff of H_2 on a time scale of less than a million years.

A newly evolved Stage II atmosphere would have contained minor or trace amounts of H_2O, CO_2, CO, SO_2, ^{40}Ar, and ^4He. Although NH_3 played no major role in escape from the Stage I atmosphere, McGovern suggested that it would have sufficient concentrations, despite its solubility in water, to shield tropospheric H_2O from dissociating radiation. Only once NH_3 was converted to N_2 would it apparently become possible to produce significant amounts of O_2 and H_2 by photodissociation of water vapor below the stratospheric cold trap.

The implications of H_2O photodissociation assume much greater importance as we enter the Stage II epoch. The question generating most interest involves the possible role of oxygen from this source for oxidation of crustal rocks and volcanic gases and for building up O_2 in the atmosphere. Urey (1952c, 1959) argued that because O_2 absorbs at longer wavelengths than H_2O, the levels of atmospheric O_2 which could originate from this mechanism were limited. Berkner and Marshall (1965) concluded that this "self-regulation" mechanism meant that H_2O photodissociation could proceed at significant rates only until the O_2 reached 0.001 of its present atmospheric level (PAL). This conclusion was later challenged by Brinkman (1969), who argued that the penetration of UV radiation between lines in the Schumann–Runge bands of O_2 made O_2 a much less efficient shield for H_2O. He made the reasonable assumption that H_2O mixing ratios in the upper Stage II atmosphere were probably not significantly different from those today and concluded that H_2O photodissociation could have built up atmospheric O_2 to 0.25PAL or more. The method used for computing H_2 escape rates and hence O_2 accumulation rates was, however, very approximate (see also Brinkman, 1971a; Van Valen, 1971).

In this type of model the very large amount of oxygen needed to produce the presently observed high oxidation state of the crust and upper mantle places severe demands on the oxidation mechanism. For an upper mantle presently containing 8% FeO and 0.4% Fe_2O_3 (the Ringwood pyrolite model), an injection of 2.4×10^{23} gm of oxygen is required. This requires the destruction of 50 kg of water per square centimeter of surface area of the Earth. Additionally, about 20 kg of oxygen would be required to produce the carbonates in the crust by oxidation of CO, or 40 kg to make the same mass of CO_2 by oxidation of CH_4 (with H_2 escape assumed). Formation of sulfates would require yet more O_2, for a grand total of 70–100 kg cm^{-2}. If 3 km^3 of pristine, wholly unoxidized mantle material were exposed each year, then the amount of Fe_2O_3 presently in the mantle could form in 10^{11} yr. Such a surficial source of oxidant has the great disadvantage that it would not, in the normal course of events, come into contact with metal. Thus, Ni would not readily become oxidized and would be essentially absent from the mantle today.

But when we consider that the Stage I atmosphere is unstable against rapid hydrogen escape on a time scale of only a million years, while the time scale for terminal accretion of the Earth is about 100 million years, it becomes very unclear whether a recognizable reducing atmosphere was ever actually present. Add to this the continuous upward flux of ferric iron from ferrous iron disproportionation in the mantle, which is almost certainly occurring during terminal accretion, and we have the strong likelihood that late-accreting ordinary chondritic material may have been quickly oxidized to the point of exhaustion of metal. Any organic matter synthesized during this phase in the history of the Earth would almost certainly be destroyed by the violent impact events attending the last stages of accretion, buried by accreted material or impact ejecta, or oxidized to CO_2, H_2O, and N_2 by ferric iron. It should be kept in mind that we are here discussing events that took place long before the formation of the earliest surviving crustal rocks.

Several authors have commented on the fact that all current theories of stellar evolution ascribe a luminosity about 25% lower than the present value to the early Sun, immediately after the end of the T-Tauri phase (see, e.g., Sagan and Mullen, 1972). This would have placed the mean temperature of the Earth below the freezing point of water for half of the life of the planet. Sagan and Mullen proposed that a mole fraction of ammonia of 1–10 ppm would have produced enough IR opacity so that the greenhouse effect would have kept the surface temperature above the freezing point. But the photochemical lifetime of ammonia is extremely short, and it is very unlikely that a steady ammonia source would have been available to maintain this very unstable state. The only argument of any weight for the prolonged presence of ammonia in the early atmosphere is that suggested by Bada and

Miller (1968). They pointed out that one essential amino acid, aspartic acid, is unstable against decomposition in water:

$$
\begin{array}{ccc}
\text{CH}_2\text{—COO}^- & & \text{CH—COO}^- \\
| & \rightleftharpoons \text{NH}_4^+ + \| & \\
^+\text{NH}_3\text{—CH—COO}^- & & \text{CH—COO}^- \\
\text{aspartate ion} & & \text{fumarate ion}
\end{array}
\tag{4.50}
$$

In order to maintain aspartate at reasonable levels of concentration (that is, to maintain aspartate more abundant than fumarate), a concentration of about 3×10^{-3} M is required for NH_4^+ ion. This, in turn, requires a significant NH_3 partial pressure. The NH_3 pressure, however, cannot be calculated without knowledge of the pH of the solution.

Recognizing the photochemical instability of NH_3, Owen et al. (1979) have suggested an alternative scenario in which small amounts of CO_2, not NH_3, assist in providing the necessary greenhouse effect. This leaves the question of both the production and survival of amino acids unanswered.

Let us then pose the following problem: Can any biologically interesting chemistry occur in a system in which the gas phase is composed of CO_2, H_2O, and N_2, and in which liquid water is in contact with weathered basalts? Ferrous-iron-bearing phyllosilicates would be formed as weathering products of primary mafic igneous rocks under these conditions, and would presumably be abundant and easily accessible on the surface.

Several factors combine to make this a most interesting system. First, as originally pointed out by Bernal (1951), phyllosilicates provide an excellent substrate for adsorption of amino acids, nucleotide bases, and other building blocks of macromolecules. Adsorption on clays greatly concentrates these species and, thus, strongly encourages polymerization (Ferris et al., 1979; Lahav and White, 1980). In a paper developing Bernal's initial suggestion, Cairns-Smith (1965) proposed that the adsorption properties of the various monomers were different at chemically distinct lattice sites: The ions exposed at the surfaces of clay grains could serve as selective templates for adsorption and polymerization of these monomers. Thus, the information stored in the cation layers could be replicated and transferred to an organic "recording tape," such as a polypeptide. Cairns-Smith further suggested that the clay layers are themselves self-replicating, in that growth of a clay grain by deposition of ions from solution should faithfully follow the patterns laid down in the previous layer. A flaw in the replication process would be the logical equivalent of a mutation.

Paecht-Horowitz et al. (1970) found that activated amino acids in water will polymerize on montmorillonite clay particles, and Lahav et al. (1978) and Lawless and Levi (1979) have demonstrated the polymerization of unactivated amino acids on clay. All this, of course, requires the prior presence of amino acids, which are not stable in a mildly oxidizing CO_2–H_2O–H_2 environment.

There is, however, a strong reducing agent that is ubiquitous in igneous rocks: ferrous ion. Getoff (1962) showed that CO_2 and H_2O in the presence of 10^{-3} M

$FeSO_4$, when irradiated with long-wave UV (260 nm) radiation, produces formaldehyde and acetaldehyde with a quantum yield of about 0.01. Cannizzaro reactions of these aldehydes can then yield acids, alcohols, and esters:

$$RCHO + R'CHO \xrightarrow{H_2O} RCH_2OH + R'COOH \xrightarrow{-H_2O} R'COOCH_2R \qquad (4.51)$$

A recent theoretical treatment of the energetics of Fe^{2+}-driven organic synthesis by Baur (1978) suggests that the essential reducing agent in the early Earth, once thought to be H_2, could easily have been Fe^{2+} instead. The ferrous ion has several nonneglible virtues: It certainly was present; it cannot escape rapidly from the Earth(!); it is very reactive at ordinary temperatures; it is not limited by the availability of rare, very short-wave UV light. Hartman (1975) proposed that fixation of CO_2, H_2O, and N_2 to form organic molecules could occur on the surfaces of clay grains under the influence of long-wave UV light or heat. Photochemical reduction of N_2 on mineral (TiO_2) grains has since been demonstrated (Henderson-Sellers and Schwartz, 1980). These ideas have a strong unifying effect on theories of the origin of biological materials. When combined with the fact that a variety of organic molecules are capable of catalyzing the synthesis of clays (Siffert, 1978), we see emerging a remarkably simple, efficient, and nearly universal synthetic route for complex organic matter in which neither a strongly reducing atmosphere nor a concentrated solution of organic matter ("chicken soup oceans") is needed. This scenario can be briefly summarized: Ferrous clays, universally produced from the action of H_2O on mafic silicates, catalyze the formation of simple organic acids from CO_2 and H_2O. These acids are strongly adsorbed onto the surfaces of clay particles, where they catalyze the growth of the clay particles. The overall process of organic synthesis is fairly described as autocatalytic. The adsorbed simple molecules are then catalytically polymerized by the clays, assuming structures regulated by the structure of the clay surface template. Complex biopolymers may then desorb and exist autonomously.

A close connection between clay geochemistry and the origin of life is probable. The compositional, structural, and evolutionary study of living organisms will hereafter benefit from increased interaction with geochemical studies of planetary environments (Banin and Narrot, 1975).

Prior to the development of this line of research, there was only a single piece of evidence for the presence of a true Stage I atmosphere on Earth. This evidence was certainly not geochemical or geological, since the oldest rocks show no evidence of exposure to such a highly reducing environment: It was biological. Since the only known processes for making abundant organic matter on the early Earth were the Miller–Urey lightning or UV synthesis (Miller, 1955; Miller and Urey, 1959) or lightning–thunder synthesis (Bar-Nun et al., 1970) in a strongly reducing atmosphere, the presence of life on Earth has often been used to infer the prior existence of a Stage I atmosphere. This inference is no longer valid. A geochemically plausible Stage II atmosphere is fully capable of satisfying the biological requirements in an atmosphere dominated by N_2, with small amounts of CO_2 and H_2O, negligible free oxygen, and H_2 mole fractions of 10^{-2} or less.

The escape physics for the early Stage II atmosphere was considered in some detail by Hunten (1973b). Assuming an H_2 mixing ratio of about 10^{-2} in the lower atmosphere, he concluded that the H escape flux will be limited by diffusion through the turbopause for an eddy-diffusion coefficient of 10^6 cm^2 sec^{-1} and for exospheric temperatures of 1000–2000 K. The computed H escape flux was 4.2×10^{11} atoms cm^{-2} sec^{-1}. If the source of escaping H atoms was H_2O photodissociation, then the O_2 evolved over a period of about 10^9 yr would have been sufficient to have provided all the oxygen now present in CO_2 (in carbonates) and O_2 (in the atmosphere) and also all the oxygen needed to oxidize FeO to Fe_2O_3. If a good deal of the total carbon outgassing had been in the form of CO_2, the demand for oxygen would have been up to 6 times less.

Hunten justified exospheric temperatures in the range of 1000–2000 K as being reasonable for a N_2–H_2 atmosphere; he computed a rough estimate of 1300 K using a scaling formula. When the exospheric temperature was assumed to be as low as 750 K, the computed H escape flux was only about 5×10^9 atoms cm^{-2} sec^{-1}. This lower flux was controlled by the Jeans escape rate provided the eddy diffusion coefficient was not substantially less than 10^6 cm^2 sec^{-1}.

Once O_2 begins to accumulate in the atmosphere, these conclusions concerning high exospheric temperatures have to be altered. Visconti (1975) computed the dependence of this temperature on X_{O_2} and obtained values of 506, 483, and 607 K for O_2 levels of 0.001, 0.01, and 0.1PAL, respectively. Apparently T_c beins at the very high value envisaged by Hunten when essentially no O_2 is present, then steadily decreases to a minimum value when O_2 is at about 0.01PAL, and then steadily rises again to its present high value. Visconti (1975, 1977) points out that escape fluxes are therefore significantly less than the diffusion-limited fluxes while O_2 evolves from 0.001 to 0.01PAL. He also argues that $X_{H_2} \lesssim 10^{-4}$ during this epoch and that it would thus take $\gtrsim 1.3 \times 10^{10}$ years to evolve from 0.01 to 0.1PAL. Even allowing $T_c = 1000$ K, it would still take $> 3 \times 10^9$ years to reach the 0.1PAL level. These arguments would tend to rule out the contention (Brinkman, 1971a; Towe, 1978) that atmospheric O_2 evolved principally from photodissociation of H_2O accompanied by escape of H_2. It does not, however, preclude the possibility that the oxidation of FeO and volcanic gases may have been at least partially accomplished using oxygen from this source. In his computations Visconti implicitly assumed that these two latter oxidations proceeded proportionately with the buildup of atmospheric O_2, and this may not have been so. These oxidations could conceivably have proceeded with O_2 at $\lesssim 0.01$PAL; that is, prior to oxygen becoming a major atmospheric constituent.

At this point we appear to require a method for producing large amounts of O_2 before evolution from the Stage II to the Stage III atmosphere can proceed. The simple physics and chemistry of the situation does not lead us to conclude that evolution of the Stage III atmosphere was either probable or inevitable. A remarkable event, the appearance of living organisms, and in particular, photosynthesizing organisms, apparently provided the mechanism for further evolution of the atmosphere. This event and its implications will now be discussed in some detail.

4.3.3. Evolution of the First Life Forms and the Photosynthetic Production of O_2 from CO_2 and H_2O

Throughout the existence of any primitive reducing or Stage I atmospheric ubiquitous photochemical production of organic matter, together with the existence of a chemically ideal environment in the oceans, would favor chemical evolution leading toward the first forms of life. Whether the first life forms arose within the lifetime of a Stage I atmosphere or under the more chemically neutral conditions in the Stage II atmosphere is a matter of debate. There now exists, however, a considerable literature on the nature and subsequent evolution of the earliest life forms. Notable contributions to this field include, among others, Calvin (1965), Glaessner (1966), Abelson (1966), Barghoorn and Schopf (1966), Cloud (1968, 1972), Banks (1970a,b), Margulis (1970), and Schopf (1970). We will not undertake herein the burden of a comprehensive review of this work, but instead refer the reader to the review by Schopf (1970) and particularly to the book *Origin of Eukaryotic Cells* by Margulis (1970).

The earliest life forms were presumably very simple (prokaryotic) single-celled organisms. Their metabolism was probably anaerobic and initially heterotrophic; that is, dependent on the fermentation of photochemically produced organic matter for energy and growth. Examples of such organisms are the lactic acid bacterium *Lactobacillus* and certain sulfur bacteria such as *Desulphovibrio* which also reduce sulfate to H_2S or FeS during their metabolic cycle. Then autotrophic prokaryotic cells evolved. These were able to synthesize their own carbohydrates, presumably using the CO_2 dissolved in their aqueous environment as a source for carbon. Some of these cells may have been chemoautotrophs which use specific chemical reactions to provide the energy for synthesis. For example, the sulfur bacterium *Thiobacillus* oxidizes H_2S. Alternatively, the comparative biological complexity of the chemoautotrophs may suggest that the first autotrophs were simple photoautotrophs which manufactured organic matter using solar radiation as the energy source. These earliest photosynthesizers were still anaerobic. Some, like the bacterium *Chlorobium thiosulphatophilum*, used H_2S as a source of hydrogen and precipitated sulfur or sulfates. The bacterium *Rhodospirillum* extracted hydrogen from simple organic compounds. Some early photosynthesizers may even have used H_2O but, instead of releasing oxygen, used it to oxidize FeO to Fe_2O_3 or Fe_3O_4.

Before oxygen-producing photosynthesis (and thus an aerobic metabolism) could evolve, it is apparent that certain complex enzymes must have been synthesized by the primitive organism to modulate or prevent the oxidation of carbohydrates by the free oxygen. Otherwise the organisms would literally burn up. The first oxygen-producing organisms were probably the prokaryotic blue-green algae, for example,

Anacystis. With the evolution of these organisms, the buildup of atmospheric oxygen could begin. Further evolutionary steps resulted in the essentially aerobic eukaryotic cell. These advanced cells contained nuclei, and also mitochondria wherein enzyme-controlled burning of carbohydrates (respiration) could occur. The development of distinct single-celled plants and animals then followed.

The sequence of evolutionary events which we have briefly outlined above can be inferred from biological arguments alone. In order to establish its feasibility and to frame it in a suitable time scale, a detailed analysis of both the fossil and geologic records is necessary. In Figure 4.16 we have summarized the fossil evidence for life in the Precambrian age (the period from Earth's formation to 600 million years ago at which time the development of phosphate and calcareous shells produced an enormous explosion in the number of buried fossils). It is apparent that prokaryotic organisms probably existed at least 3.2×10^9 yr ago. Oxygen-evolving blue-green algae may have existed even 2.8 billion years ago, although they appear to have become common only about 2.2 billion years ago in the Early Proterozoic age. In the Late Proterozoic age the eukaryotic cell had definitely evolved.

The geologic evidence appears to substantiate the time scale deduced from the fossils. Precambrian geology has been discussed extensively by Holland (1972), Cloud (1972), Rubey (1951), Berkner and Marshall (1965), and Rutten (1966), among others. So-called "red-beds," which are sediments containing several percent Fe^{3+} as hematite or limonite, are common only in formations younger than 1.9 billion years. The high oxidation state of iron in these beds implies the existence of some atmospheric oxygen. Also, sediments of $CaSO_4$, which must be formed by oxidation (by bacteria or atmospheric ozone) of H_2S or SO_2, occur only in Paleozoic and younger sediments, implying large amounts of atmospheric oxygen at least 600 million years ago. It has also been suggested that the marked increases in dolomite $[CaMg(CO_3)_2]$ and calcite $(CaCO_3)$ deposition toward the end of the Precambrian are due to increases in atmospheric O_2 about this time. These increases presumably caused greater weathering rates of silicates and enabled higher oceanic concentrations of Mg^{2+} and Ca^{2+}. Finally, there is the presence of special "banded iron formations" in sediments between 1.9 and 3.2 billion years ago, but not younger. These extensive interleaved layers contain significant amounts of partially oxidized iron (FeO and some Fe_2O_3), and it has been theorized that they were formed by fluctuating colonies of primitive anaerobic organisms using Fe^{2+} in the oceans as a sink for O_2 during photosynthesis. These organisms became extinct when atmospheric O_2 appeared.

In summary, it appears that oxygen-producing organisms became prevalent roughly 2 billion years ago and, as carbon became buried in organic sediments, an equivalent amount of oxygen appeared in the atmosphere. Certainly by the dawn of the Paleozoic, some 600 million years ago, oxygen was an important constituent in the Earth's atmosphere and all life was either aerobic or dependent on aerobic organisms.

Before we can accept this proposed evolutionary sequence, one particular problem must still be resolved. During this sequence the Earth was being continuously

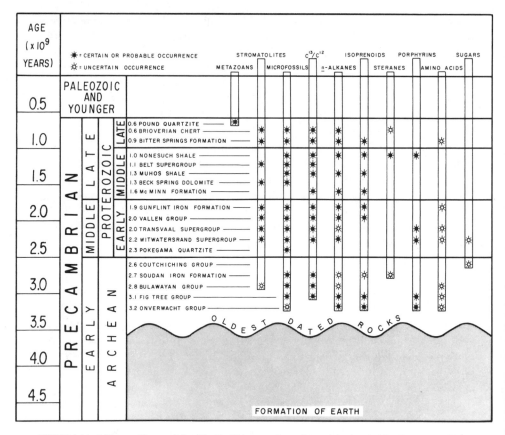

FIGURE 4.16. Evidence of Precambrian life. The histogram shows the organic chemical fossils (amino acids, *n*-alkanes, isoprenoids, porphyrins, steranes, sugars) and true fossils (stromatolites, microfossils, metazoans) which have been identified in known Precambrian deposits. Amino acids and *n*-alkanes could be derived from abiotic photochemistry, but isoprenoids constitute chemical evidence for biosynthesis by primitive autotrophic or heterotrophic prokaryotic cells. The porphyrins are contained in chlorophylls and provide possible evidence for the existence of autotrophic photosynthetic prokaryotic cells (anaerobic bacteria or perhaps aerobic oxygen-producing blue-green algae). Those sediments in which the $^{12}C : ^{13}C$ ratio is the same as that observed in organic matter produced in oxygen-evolving photosynthesis may also provide evidence for primitive photosynthesis. Stromatolites are laminated structures formed by blue-green algal communities. Microfossils are unicellular prokaryotic organisms, including blue-green algae in the sediments younger than 2×10^9 yr. Steranes and sugars may be evidence for organisms with an aerobic metabolism. The metazoan fossils prove the existence of aerobic eukaryotic multicelled organisms by the end of the Precambrian age.

bombarded by solar UV radiation, and these energetic photons are generally lethal for living organisms. In our present atmosphere N_2 and O_2 absorb the radiation below 2000 Å, while the ozone layer in the stratosphere absorbs the radiation in the 2000–3000 Å region. This ozone layer is, however, susceptible to depletion by anthropogenic pollutants, and we will discuss this possibility in more detail later. This latter possibility has, not surprisingly, resulted in a number of studies on the sensitivity of both plants and animals to UV radiation. The results of this research are directly relevant to our discussion here.

There are two important effects of radiation on living cells: the first is alteration of genetic material (DNA and RNA) leading to mutations and cancer; the second is actual destruction of cell material (protein etc.) ultimately leading to death of the cell. Experiments on algae such as *Chlamydomonas* and *Euglena*, on blue-green algae such as *Anabaena*, and on simple animal organisms such as *Paramecium* have shown that simulated solar radiation in the 2900–3200 Å wavelength region can kill those species on the time scale of an hour or less (CIAP, 1975; National Academy of Sciences, 1975). Wavelengths below 2900 Å are about 20 times more lethal than those above 2900 Å. The typical protein molecule begins to photodissociate at and below 2600 Å. Higher plants and animals show a similar sensitivity to UV light. A typical response curve for DNA molecules to UV light, together with that for erythema or sunburn in human skin, is shown in Figure 4.17 (National Academy of Sciences, 1975). These various studies lead us to the conclusion that the evolution of life undoubtedly required shielding of living organisms from UV wavelengths below 3200 Å and particularly below 2600 Å.

The only gas which absorbs beyond 2300 Å in the primitive Stage I atmosphere and which could conceivably have significant concentrations in H_2S. This particular gas absorbs out to roughly 2700 Å with each absorbed photon, leading to the dissociation reaction

$$H_2S \rightarrow H + SH. \tag{4.52}$$

Let us make the reasonable assumption that wavelengths less than 2200 Å were mainly absorbed by other gases such as CH_4, NH_3, and H_2O in the Stage I atmosphere. The photodissociation lifetime for H_2S obtained by convolving the H_2S absorption spectrum with the observed solar spectrum in the wavelength region

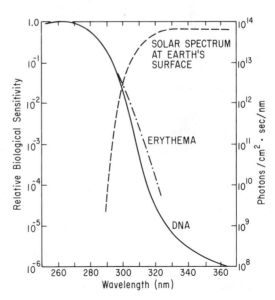

FIGURE 4.17. **Effects of ultraviolet light on DNA. The solid curve is the biological sensitivity spectrum for DNA, and the dashed curve is a typical Earth-surface solar UV spectrum. The response of human skin to sunlight ("sunburn") is shown as the dot–dash curve labeled erythema. (After National Academy of Sciences, 1975.)**

2200–2700 Å and dividing by 4 for a planetary average is then about 0.8 day. Once dissociated, H_2S can be reformed by

$$SH + SH \rightarrow H_2S + S, \tag{4.53}$$

$$SH + H + M \rightarrow H_2S + M, \tag{4.54}$$

or it may ultimately yield elemental sulfur particles through reaction (4.53) and

$$SH + H \rightarrow S + H_2, \tag{4.55}$$

$$SH + S \rightarrow S_2 + H, \tag{4.56}$$

$$SH + SH \rightarrow S_2 + H_2, \tag{4.57}$$

or it may lead to even further destruction via

$$H + H_2S \rightarrow H_2 + SH. \tag{4.58}$$

At a pressure of 1 atm, reactions (4.53) and (4.55) are more rapid than (4.54), implying that at least 50% of the dissociated H_2S molecules are converted irreversibly to particulate elemental sulfur which settles to the surface. If we refer now to Figure 4.18, where we have illustrated the H_2S absorption spectrum, we see that extinction of the wavelengths below 2600 Å requires $X_{H_2S} \gtrsim 10^{-5}$ or an H_2S column abundance exceeding 2×10^{20} molecules cm^{-2}. It is difficult to contemplate nonbiological mechanisms for recycling the elemental sulfur which has settled to the surface directly back to H_2S. The sole source of this latter gas will therefore be outgassing. The required rate is $\gtrsim 2 \times 10^{20}/1.6 = 1.25 \times 10^{20}$ molecules cm^{-2} day^{-1} or about 10^{19} gm per year of sulfur on a global basis. Such a large outgassing is simply not feasible; in $\lesssim 10^5$ yr the mass of outgassed sulfur would equal the mass of water in the oceans. The entire Rubey inventory of S would be released in only 280 years!

In Figure 4.19 we illustrate the absorption spectrum of solid elemental sulfur, which is clearly another potential UV absorber in the Stage I atmosphere. Let us

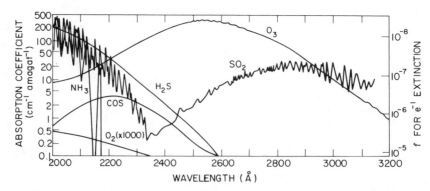

FIGURE 4.18. UV absorption spectra of sulfur compounds, NH_3, O_2, and O_3. The right-hand scale gives the mole fraction f of each gas which would be able to attenuate the incident solar flux by a factor of e.

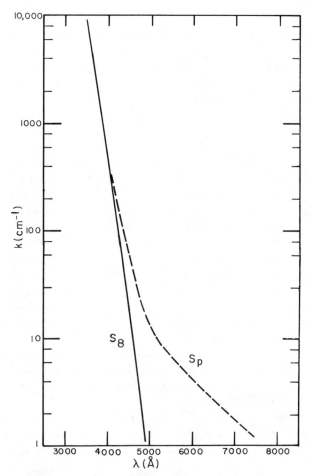

FIGURE 4.19. **UV absorption spectrum of solid sulfur. The solid curve is for cyclic octatomic sulfur crystallized in the rhombic system; the dashed curve is for "red" polymeric sulfur.**

take the H_2S outgassing rate as equal to the total mass of sulfur in the crust (1.2×10^{22} gm; Ronov and Yaroshevskii, 1967) divided by the age of the Solar System (5×10^9 years). Studies of the physics of photochemically produced clouds (Prinn, 1975) imply that the mixing ratios of sulfur in the atmosphere cannot exceed that of its precursor H_2S. Therefore, the maximum amount of sulfur in the atmosphere is roughly $(1.2 \times 10^{22} \times 1.6)/(365 \times 5 \times 10^9) = 10^{10}$ gm corresponding to a 10^{-9}-cm column of yellow rhombic sulfur. The required absorption coefficient is therefore 10^9 cm^{-1}, which is about 100 times larger than that observed (Figure 4.18). We must conclude that a Stage I atmosphere could not provide an adequate shield for UV radiation.

In the neutral Stage II atmosphere, the most probable gaseous form for sulfur is SO_2, and this is a very efficient absorber for wavelengths <3200 Å (Figure 4.18).

Unlike the case for H_2S, UV photons between 2280 and 3200 Å do not dissociate SO_2 but produce stable excited states (Warneck et al., 1964). We therefore expect SO_2 to have a long photodissociation lifetime. It is, however, very soluble in water and is readily oxidized to SO_3 by OH, HO_2, and H_2O_2, all of which are expected products of water photodissociation. Of these removal mechanisms, rainout is the fastest and would result in an SO_2 lifetime of about 1 month. Using the sulfur outgassing rate derived in the previous paragraph, we therefore expect an atmospheric SO_2 content of $(1.2 \times 10^{22} \times 1 \times 64)/(12 \times 5 \times 10^9 \times 32) = 4 \times 10^{11}$ gm corresponding to $X_{SO_2} \approx 3 \times 10^{-11}$. Figure 4.18 implies that $X_{SO_2} \sim 10^{-6}$ is required to absorb UV radiation efficiently. Since there are no strong UV absorbers other than SO_2 in the Stage II atmosphere, it is apparent that this second stage also lacked an adequate UV shield. Apparently only after life had evolved, and O_2 (and thence the ozone layer) had become an important component of the atmosphere, was such a shield provided by atmospheric constituents.

Another feasible UV absorber, at least for aquatic organisms, is the surface layer of oceans and lakes. In Figure 4.20 we show the attenuation of 2900–3200 Å radiation in typical freshwater (CIAP, 1975). A depth of a few tens of centimeters of water would appear adequate for protection. This absorption would, of course, be provided by impurities in the water, since pure water does not absorb significantly beyond 2000 Å. Some of these impurities may be biospheric and therefore not applicable to the prebiospheric Earth. We can speculate that colloidal sulfur produced by autoreduction of SO_2 in the oceans might be an efficient UV-absorbing impurity in prebiological oceans. Such arguments are not particularly convincing, and the problem of UV shielding during the Stage I and Stage II epochs can hardly be considered resolved at this time.

Once life has become sufficiently well established so that photosynthetic oxygen is a major constituent, the photochemically produced ozone layer will, of course, provide excellent UV shielding short of ~3000 Å. This case of biological interven-

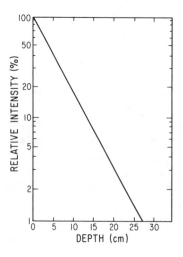

FIGURE 4.20. **Attenuation of 2900–3200 Å UV radiation versus depth in freshwater. (After Grobecker et al., 1974.)**

tion in planetary affairs is by no means unique. In Section 4.3.5 we briefly summarize some of the most important ways in which life participates in the regulation of the abundances of atmospheric gases. A very interesting theory for the spontaneous creation of such regulatory networks has been presented (Lovelock and Margulis, 1974; Margulis and Lovelock, 1974), which describes the resulting control of the global environment by life as a kind of planetary-scale homeostasis. If such biological control of planetary conditions has in fact been effective since the end of the Precambrian (or, in other forms, even longer), then both secular climatic changes due to continental drift (Smith, 1977; Whyte, 1977) and resultant sea-level changes (Clark and Lingle, 1977) and to secular and temporary changes in the brightness of the Sun (Sagan and Mullen, 1972; Hartmann, 1974a; Newman and Rodd, 1975) may have led to systematic changes in atmospheric composition.

Monitoring the global balances of sulfur and carbon and the oxidation state of the biosphere by stable isotopic studies on C and S in ancient sediments holds the promise of unraveling much of the complexity of our early atmospheric evolution. Detailed models of the sulfur isotope history of marine sulfates in ancient evaporites (Rees, 1970; Perry *et al.*, 1971; Holland, 1973) have been developed. Holland concludes that the sulfur isotopic evidence indicates a gain in atmospheric O_2 (sulfate reduction) in the lower and middle Paleozoic, and a loss in the Permian. The effect of the oxidation–reduction cycle for carbon is, however, even more important as a regulator of O_2, and the actual behavior of the atmosphere thus remains uncertain. Becker and Clayton (1972) find from stable isotope studies of sedimentary reduced carbon and carbonates that the oceanic C content has probably been close to constant for the last 3 billion years, but they are unable to prove constancy of the O_2 pressure. A detailed review of the application of stable isotope studies to the geochemical and biological evolution of the Earth by Kaplan (1975) is recommended.

4.3.4. Influence of Catastrophic Events on the Evolution of the Biosphere

Two centuries ago there were two major schools of thought in historical geology. The first of these, catastrophism, viewed the history of the geological record as long periods of utter blandness, in which nothing significant happened, punctuated by global catastrophes of unimaginable extent. In these catastrophes all the important work of geology, such as the extinction of ancient flora and fauna, the leveling of ancient mountains, the construction of new generations of mountains, and the deposition of immensely thick layers of sediment, were carried out in rare, isolated, extremely brief events. This view was challenged in 1785 by James Hutton's *Theory of the Earth*, which laid out the uniformitarian viewpoint. In Hutton's theory, the entire geological record is assumed to have been laid down by processes that we can observe at work on Earth today, acting slowly and continuously over vast extents of time.

In order to provide a sensible description of planetary history as it is presently understood, however, it is necessary to go beyond the most narrow interpretations of uniformitarian evolutionary theory to include the class of naturally occurring, statistically predictable, and undeniably catastrophic events arising from physical causes exterior to the Earth. Violent cosmic events, the effects of which are so obvious in the photographs of Mercury, Mars, and the Moon, were also instrumental in the evolution of the Earth. We remained ignorant of these events until quite recently, in part because of the rapid obscuration of the evidence for them by erosive processes, and in part because the enormous scale of many of these structures makes them more obvious to an observer in space than to one on the ground.

A specific type of catastrophic intervention of cosmic physics in terrestrial affairs was suggested briefly in the classic text *Astronomy* (Russell *et al.*, 1945). They showed that a comet should strike the Earth with an average frequency of one every 80 million years. They attempted no calculation of the consequences of such an impact. The first quantitative assessment of the effect of a comet collision on Earth was by Lin (1966), who described the basic physics of extremely energetic detonations on the surface of the Earth with tektite formation in mind. Urey (1973) explicitly proposed that cometary collisions caused the major discontinuities in the sedimentary record, including causing mass extinctions of species. Urey especially mentioned the extinction of the dinosaurs at the end of the Cretaceous era, about 65 million years ago. McLaren (1970) and D. A. Russell (1979) have both drawn attention to the possible importance of major impact events in causing extinctions.

The statistics on the sizes of small solar system bodies given by Dohnanyi (1972) imply that about 10 bodies with masses of 10^{18} gm or larger will impact with the Earth over the age of the Solar System, even at the *present* flux level. (The same statistics imply that 10^{11}-gm bodies should fall with a frequency of about one per 2000 years, and 10^{12}-gm bodies should fall at the rate of about one per 12,000 years.) Each 10^{18}-gm body, impacting with an average energy of about 10^{31} erg, would be the equivalent of 250 million megatons of TNT. As emphasized by Whipple (1976), it is entirely likely that cometary debris was much more abundant in the early Solar System than it is now, due to the perturbation of icy planetesimals from the region of the outer planets into Earth-crossing orbits. It is therefore impressive to find that even at present fluxes, the number of bodies hitting Earth over 4.6 billion years with sufficient energy to blow off part of the atmosphere ($\sim 10^{13}$ gm; $> 10^{25}$ erg) is about 10^5, assuming an asteroid-like mass distribution:

$$N(m) = Am^{-0.8}. \tag{4.59}$$

However, the evidence from studies of the brightness distribution of comets (Hughes and Daniels, 1980) suggests an even shallower mass distribution:

$$N(m) = Am^{-0.6}, \tag{4.60}$$

which makes very large bodies even more probable. In either case, the two or three largest impacting bodies would contribute most of the mass and energy flux.

The mean recent rate of energy input to Earth by catastrophic impacts is nearly 1 Mton per year (Napier and Clube, 1979). There are, on the average, two events per million years with energies sufficient to excavate craters with diameters of 20 km or greater. The diameter D of a crater (in kilometers) is related to the impact energy E (Mton) by $D = 0.747E^{0.3}$, a rule good for craters with a diameter >14 km. It could also blow off a mass of the Earth's atmosphere at least equal to the mass of the impacting comet.

Cometary or asteroidal bodies with sufficient strength to reach the surface intact will produce large impact craters. The large majority of the explosive impact craters known on Earth were produced by bodies of unknown composition, and hence it is not possible to tell from available data on these craters what proportion of them were produced by active comets rather than by Earth-crossing asteroids. Shoemaker (1977) estimates that the number of bodies with diameters of 100 m or larger in Earth-crossing asteroidal-type orbits is close to 10^5. He estimates that the present production rate for craters 10 km or more in diameter is $(1.2 \pm 0.6) \times 10^{-14}\,\mathrm{km}^{-2}$ per year. According to Wetherill (1976), a majority of the Earth-crossing asteroids are the outgassed remnants of short-period comet nuclei.

The nature of the surface scar left by an impact depends sensitively on the density and strength of the impacting body. Large solid bodies impacting at hypersonic speeds on the solid Earth would excavate craters with a depth which increases with the projectile density. A large metallic asteroid, with a density three times that of typical sedimentary rocks, would penetrate deeply and leave a most impressive crater, whereas a long-period comet nucleus, made of dirty ice of low density, would excavate only a very shallow crater. In the very likely event of an impact in water, a comet may well leave no discernible crater in the sea floor if the water depth is more than two or three times the diameter of the nucleus.

A number of physical and chemical consequences of the impact of a large comet are worthy of investigation. It is clearly of interest to discover whether cometary molecular species, such as CO_2, CO, and HCN, could survive the explosion. The synthesis of new molecular species in the cooling fireball may be of interest and, of course, the production of high-temperature species such as CN, HCN, and nitrogen oxides by passage of the shock front through the Earth's atmosphere must also be considered. Another point of interest would be the deposition of neon-poor but otherwise solar-type rare gases in the atmosphere by destruction of their carrier phase, water ice. Finally, large impacts, although rare, are capable of physically removing and ejecting at escape velocity a portion of the atmosphere above the impact site.

There is a high likelihood that since the origin of life, some 25–100 major cometary and asteroidal collisions may have occurred on Earth, and the need for careful study of the immediate physical and chemical consequences of impact events becomes clear. By "major" we mean an impact energy in excess of 10^8 Mton of TNT for each event.

We will here sketch out only briefly the basic physics of a major cometary impact event, following first the simple treatment of Lin (1966).

A comet which is sufficiently massive and sufficiently strong so that most of its mass reaches the surface of a terrestrial planet intact would strike the surface carrying roughly 10^{12}–3×10^{13} erg gm^{-1}. We assume here that massive, long-period comets are in highly eccentric (nearly parabolic) orbits with a random orientation of their lines of apsides over the celestial sphere. The impact kinetic energy of the nucleus is expended in a spherical wave which displaces the atmosphere outward until the work done in excavating the cavity is equal to the input kinetic energy. The cavity, filled with extremely hot shocked gas (largely water vapor from the comet nucleus), then rises buoyantly and radiates off energy as the shock wave continues to propagate through the planetary atmosphere. The radius of the plasma cavity is approximately

$$R_0 = (3E/4\pi P_s)^{1/3}, \tag{4.61}$$

where E is the impact energy and P_s is the atmospheric pressure at the surface of the planet. The fireball is approximately spherical if R_0 is small compared to the scale height of the atmosphere, H. For an explosion on the surface of the Earth to produce a fireball of radius H (about 8 km), an energy of 2×10^{24} erg, or about 50 Mton equivalent of TNT, is required. For an impact energy density of 10^{13} erg gm^{-1}, this would require a comet nucleus of mass at least equal to 2×10^{11} gm. For a spherical nucleus with a solar-proportion mixture of ices and rock-forming minerals, a density close to 1.3 gm cm^{-3} would be expected, giving a radius of 34 m for the nucleus. Defining E^* as the explosion energy for which $R_0 = H$, we can expect that explosions with $E \geq E^*$ would generate moderately nonsperical cavities, whereas for $E \gg E^*$ it is possible for the fireball to blow off a substantial proportion of the overlying atmosphere. Lin shows that, for explosion energies of 2×10^{30} erg, it becomes energetically possible for the explosion to blow off and remove to infinity the entire mass of atmosphere present above a plane tangent to the Earth's surface at the point of impact. This would result from impact of a spherical nucleus with a radius of about 3.4 km.

To place impact events in the perspective of recent research, let us turn to the available evidence regarding the behavior and effects of actual impact events on Earth.

The largest observed fall in modern times occurred on 30 June 1908 in central Siberia near the Tunguska River. The energy dissipated in the explosion has been estimated as 4×10^{23} erg (Hunt et al., 1960). The impacting body has been tentatively identified as a small comet nucleus (Whipple, 1930). One of the most striking features of this event was the fact that no impact crater was formed. The detonation apparently occurred at an altitude of about 10 km (Fesenkov, 1968). Violent high-altitude detonations are apparently not rare (Shoemaker and Lowery, 1967). That so massive a body should detonate at such a high altitude (about $1.2H$) is strongly suggestive of a low density and a low crushing strength. No macroscopic meteorite fragments were recovered; however, Fesenkov (1949) has shown that the increase in the turbidity of the atmosphere immediately after the Tunguska event implies the injection of a vast mass of dust, perhaps as much as 10^{12} gm, into the upper atmosphere.

Since cometary impacts bring in matter with a kinetic energy density of 10^{12} to $\sim 3 \times 10^{13}$ erg gm^{-1}, the observed detonation energy suggests a mass of about 10^{10}

to $\sim 4 \times 10^{11}$ gm involved in the final explosion. Debris shed along the entry path prior to the terminal explosion would, of course, not contribute to the explosion energy. These considerations suggest that the entry speed was toward the lower end of the allowed range; let us estimate 15 km sec^{-1} at the point of breakup. The dynamic pressure on a spheroidal body due to aerodynamic drag at hypersonic speeds (here nearly Mach 50) is given by Öpik (1958) as

$$p = 0.6\rho_g v^2, \tag{4.62}$$

where ρ_g is the gas density and v is the speed of the body. This gives minimum dynamic pressure of about 300 bar at the 10 km altitude level in Earth's atmosphere. This is comparable to the crushing strengths of weakly consolidated sedimentary rocks or of ice–dirt mixtures.

Park (1978) and Turco $et\ al.$ (1981) have shown that dissipation of the entry kinetic energy of the Tunguska object must produce nitric oxide (NO) with high efficiency. They also present evidence that the NO produced by this event may have caused a detectable depletion of the ozone layer. Thus, a small (roughly 10 Mton) explosion is already marginally capable of disturbing Earth's main shield against solar UV light.

The high efficiency of nitrogen oxide production by the Tunguska object can be attributed to the fact that the detonation occurred at high altitude, so that all of the kinetic energy of the incoming body was dissipated as a strong shock wave in the atmosphere. Any weak body would behave as the Tunguska object did: It would disintegrate at high altitudes, liberating its kinetic energy as a violent explosion. The Tunguska explosion stripped the limbs from trees within about 5 km of ground zero, leaving the charred trunks standing. From about 5 km out to as far as 30 km the trees were knocked down in a radial pattern. The shock wave from the explosion was recorded by microbarograph stations over the entire planet. Yet no crater was formed, and the largest specimens of extraterrestrial material found in the region are micron-sized droplets of nickeliferous magnetite. Within two or three centuries all visual evidence of the explosion will have disappeared. Almost anywhere else on Earth the biological damage would be even more quickly repaired, and all evidence of the event would probably be lost in a few decades.

Impact events much more severe than Tunguska are, of course, much rarer. The expected severity of their effects is to a large degree offset by the degradation of their remains by the passage of millions of years. Nonetheless, there is extremely strong evidence for several major impact events that seem to have been responsible for the termination of some of the major eras in the Earth's geological history. The evidence does not consist of huge, deep craters: The tectonic activity of the Earth and the activity of erosive processes reworks the crust rapidly and efficiently. Nearly three quarters of the surface area of our planet is covered by water, and virtually all of the ocean floor is less than 150 million years old. The tape upon which the planetary impact history is recorded has been extensively erased and rerecorded. The best evidence for ancient giant impacts comes from geochemistry, and the most crucial evidence for understanding the effects of these impacts comes from paleontology, which records a number of severe biological extinction events in the fossil record, such as that at the end of the Cretaceous era.

Direct experimental evidence for an anomalously large concentration of siderophile elements of extraterrestrial origin has recently been reported (Alvarez et al., 1980; Canapathy, 1980; Hsü, 1980; Kyte et al., 1980; Smit and Hertogen, 1980) for widely separated samples taken from a thin green clay layer at the Cretaceous–Tertiary boundary.

The mechanism by which massive extinctions were caused is a matter of considerable debate. Alvarez et al. favor the idea that a vast quantity of fine-grained dust was injected into the atmosphere by the impact and that, as a result of the obscuration of the Sun by this dust mantle, photosynthesis was shut down over the entire Earth for a period of several years. The food chain then, of course, must have collapsed.

McLean (1978) proposed that the extinctions were due to an enhanced CO_2 greenhouse effect, causing a sudden increase in mean global temperature, melting of the icecaps, release of vast amounts of CO_2 from the deep oceans, and consequent dissolution of the shells of calcareous marine organisms by the increased CO_2 and HCO_3^- content of surface waters. Although McLean did not mention an extraterrestrial source of CO_2, he does clearly show that either an abrupt global warming or an injection of CO_2 can have profound effects on the ecological stability of many of the genera that were in fact annihilated at the end of the Cretaceous. Hsü proposed that the impacting body was cometary and that the extinctions were caused by the injection of HCN. This mechanism, however, like that of Alvarez et al., lacks the necessary selectivity to explain why some genera disappeared altogether, whereas other genera in similar ecological niches were largely unaffected.

A few examples of the biological evidence regarding the Cretaceous extinction, drawn from the review by Russell (1977), suffice to demonstrate the selectivity of the event. The number of genera of freshwater fishes, amphibians, and reptiles changed from 36 to 35 across the Cretaceous–Tertiary boundary, whereas ammonites, with 34 genera in the Upper Cretaceous, became wholly extinct. Coccoliths, which have calcareous tests, were reduced from 43 to 4 genera, whereas the ecologically similar radiolarians, which have siliceous tests, maintained 63 genera across the boundary, with no evidence of any significant extinction. Bottom-dwelling species such as sponges and corals suffered a loss of 50–70% of their genera. Among the free-swimming (nektonic) organisms, the least affected were the cartilagenous fishes. Land plants suffered only about a 10% loss of genera, and amphibians and land mammals actually benefited from the decreased reptile population by undergoing rapid diversification immediately after the "catastrophe." Russell (1977) suggests that as many as 75% of all the genera present in the Upper Cretaceous became extinct.

The timing of the most famous part of these extinctions, that of the dinosaurs, is somewhat unclear. Both the dinosaurs and the ammonites seemed to have faded away over a finite period of time, during the earliest million or so years of the Tertiary (that is, during the early Paleocene), not instantly. A symposium dedicated to the paleontological and geological study of the Cretaceous–Tertiary boundary was held in 1979 (Birkelund and Bromley, 1979; Christensen and Birkelund, 1979), during which several useful generalizations emerged. One of these is that the most profound population changes occurred in the near-surface (pelagic) layer in the tropical oceans. The extinction of the dinosaurs is attributed by one author (Schopf, 1981) to sea-level changes that drained their swampy North American habitat. In this view, the demise of the last 20 or so dinosaur species becomes "a rather trivial question."

The crucial issue seems to be whether large-impact events can have effects that are severe, global, and yet selective, so that the extinctions of certain genera and the survival of others can be explained in causal terms.

All large impacts dissipate at least 0.1% of their incident kinetic energy in the form of atmospheric shock waves. The case of least interaction would be for vertical entry of a strong, dense, metallic asteroidal core bearing a high iridium content. The incident mass could then be as low as 10^{17} gm and still deliver the observed siderophile element abundances. With a minimum entry energy density of about 10^{12} erg gm^{-1}, and with only 10^{-3} of its energy deposited directly in the atmosphere, we find that $\sim 10^{26}$ erg is the minimum amount of shock energy put into the atmosphere. A fragile, ice-rich body with the same siderophile content, which enters at high speed and at a shallow entry angle, and which therefore deposits almost all of its energy in the atmosphere, could inject as much as 10^{32} erg. This energy can shock the atmosphere out to a distance of several hundred kilometers to a peak temperature of at least 2000 K. Lewis et $al.$ (1982) have shown that large impacts, such as the Cretaceous bolide, can synthesize NO with an efficiency up to 2×10^{10} NO molecules per erg of kinetic energy. A typical large long-period comet ($\sim 10^{18}$ gm; 10^{13} erg gm^{-1}) may produce over 10^{41} NO molecules. However, the shocked portion of Earth's atmosphere, even in the case of much smaller explosions, will convert NO rapidly to NO_2 via thermochemical reaction with oxygen. Nitrogen dioxide rapidly dimerizes to N_2O_4, which readily dissolves in water and reacts with water vapor to make nitric acid (HNO_3) and nitrous acid (HNO_2). The latter spontaneously disproportionates to make more HNO_3 and NO. The HNO_3 is made on a time scale of days, and is lost from the atmosphere dominantly by rainout on a time scale of weeks. This leads to severe acidification of the near-surface layer of the ocean to a depth of tens of meters. The entire top 75 m of the Earth's oceans can be acidified by ~ 0.5 pH units and brought to a state of calcite undersaturation by addition of 2×10^{40} HNO_3 molecules.

The result of such acidification is to destroy calcareous-shelled organisms at shallow depths in the ocean, while leaving benthic calcareous species and siliceous shallow-water species unaffected. This is precisely the nature of the selectivity observed in the marine extinctions at the end of the Cretaceous. Other lethal effects might be due to the toxic, mutagenic, and carcinogenic effects of intermediates such as NO, NO_2, and HNO_2, by climatological effects caused by absorption of visible sunlight by NO_2, and by the direct effects of the powerful explosion blast wave.

A very stimulating collection of 48 papers on the astronomical, geological, paleontological, and geochemical aspects of the Cretaceous extinctions has recently been published (Silver and Schultz, 1982). This volume is strongly recommended to readers interested in the stability of the biosphere against large disturbances.

It is becoming clear that small impacts bring in volatiles of asteroidal and cometary origin, while the largest impacts can cause explosive blowoff and severe loss of volatiles. The net effect of prolonged exposure to a realistic mass and energy distribution of impacting bodies is not yet known, and the effects of such a bombardment history on Venus and Mars are only now being investigated for the first time.

The bombardment of the Earth was certainly more severe in early Precambrian times, prior to the formation of the most ancient known terrestrial rocks. If, as has been argued by several authors, outgassing of the early Earth was substantially effected during accretion and primary differentiation, then the last 1% of the mass of the planet was brought in through a primitive atmosphere comparable in mass to the present atmosphere. Severe shock processing of the gas is unavoidable, and explosive blowoff of much of the gas is possible. Furthermore, if, as is often supposed, life originated well before the oldest known rocks, then the mass and energy flux from comet and asteroid infall may have influenced the course of early evolution many times. More recent catastrophic interventions in terrestrial affairs, such as the event at the close of the Cretaceous, can at least leave evidence of their occurrence in the surviving part of the geological record. However, imperfectly this record may attest to such events, it is nonetheless our best available source of data, fully worthy of the sophisticated techniques for detection of extraterrestrial material which have recently been brought to bear. There is real reason to suppose that at least this one event may have left sufficient evidence of the geological, climatological, and ecological consequences of a major impact to permit the formulation of detailed theories of such events.

Other types of catastrophic events of cosmic origin have also been proposed as influences in biological and geological evolution of the Earth. These include the explosion of a nearby supernova (Hunt, 1978) and the passage of the Solar System through a dense interstellar cloud (McKay and Thomas, 1978). Both of these events may occur in a quasi-periodic fashion, since the presence of dense interstellar clouds, rapid star formation, large abundances of O and B stars, and supernova explosions are all closely correlated in space and, of course, causally related. Such a rich and potentially dangerous environment should be experienced fairly regularly with a frequency near the mean orbital frequency of the Sun about the galactic core, or roughly once every 100–200 million years. As predicted by Urey (1966), a rich variety of complex organic molecules characteristic of dense interstellar clouds may be acquired by Earth.

4.3.5. Contemporary Atmospheric Cycles and Anthropogenic Perturbations

At this point we appear to have a plausible scenario for the evolution of our present (Stage III) atmosphere. Two other important questions which should be asked both concern the future: (1) Is the atmosphere still evolving? (2) Even more important, are the industrial, agricultural, and social activities of man liable to cause catastrophic deviations from this natural evolutionary path? In order to answer these questions we must make a detailed study of the chemical cycles in which the various constituents in our contemporary atmosphere are involved.

In formulating atmospheric cycles, it is clear that we must consider both the overall budgets on a global scale and the kinetics of the chemical and biochemical reactions on the microscale. The quantitative study of these cycles is unfortunately still in its infancy. Some of the details in the cycles which we will outline are therefore subject to change. This present lack of definition does not, of course, detract from their importance, particularly as the basis for studies of human influences on the future evolution of the atmosphere.

The atmospheric cycle which is of primary significance to life on Earth is that of carbon, which is illustrated in Figure 4.21. The CO_2 content of the present atmosphere is controlled to a large extent by evaporation from and dissolution into the oceans. Release of CO_2 into the atmosphere over tropical oceans and uptake by polar oceans results in an atmospheric CO_2 residence time of about 5 years (Bolin, 1960; Junge, 1963). The observed CO_2 content of the oceans is about 50–60 times that of the atmosphere. This content is largely controlled by the temperature and

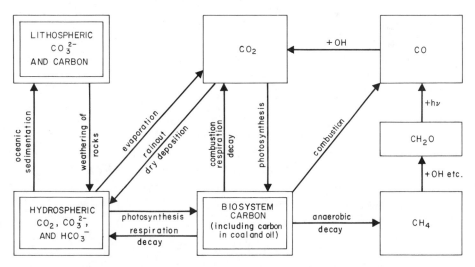

FIGURE 4.21. **The carbon cycle on Earth. Single boxes indicate gaseous species; double boxes indicate condensed and dissolved (lithospheric, hydrospheric, and biological) materials. For explanations, see the text.**

acidity of the seawater, and by the biological production rate: The CO_2 pressure thus varies seasonally (Wilkens, 1961) and is higher in surface water than in the terrestrial atmosphere (Takahashi, 1961). Carbon dioxide is cycled through the biosphere at a somewhat slower rate. The biospheric processes of photosynthesis, respiration, combustion, and decay result in atmospheric CO_2 being cycled through living material only once in about 20–30 years. The total carbon content of the biosphere is about 20 times that of the atmosphere, but only about 3% of this carbon is present in living material. About 71% is present as coal and oil deposits and 26% as dead organic matter (Bolin, 1960).

In contrast to the above cycling processes, the cycling of atmospheric CO_2 into coal and oil deposits has been an extremely long-term process. It is particularly significant that, through combustion of these fossil fuel deposits, modern man will be releasing this stored carbon (so painstakingly accumulated over a time span of several hundred million years) in the relatively short period of a few centuries.

The bulk of the carbon on the Earth is, of course, present in inorganic crustal compounds. Ronov and Yaroshevskii (1967) estimate that about 91% of crustal carbon is contained as carbonates in sedimentary (e.g., limestone, dolomite, marble, chalk), granitic, and basaltic rocks, and about 9% as free carbon. This crustal reservoir contains about 3×10^4 times the quantity of carbon present in the atmosphere. Cycling of atmospheric CO_2 through this substantial reservoir is accomplished through the opposing processes of oceanic sedimentation and continental weathering of rocks, but the time periods involved are extremely long—perhaps several hundred million years. It is interesting to note that if all the CO_2 buried in the crust were released into the atmosphere, a massive CO_2 atmosphere similar to that on the planet Venus would result. The role of the oceans in preventing such an occurrence is to provide a medium for formation of HCO_3^- ions, which can react to form carbonate minerals even at low (~ 300 K) temperatures.

In addition to CO_2 in the atmosphere, there are minor amounts of CH_4, CO, and CH_2O which also play a role in the carbon cycle. Methane, produced by anaerobic decay of organic matter (e.g., by *methanobacterium*), is oxidized in the atmosphere to formaldehyde and then to carbon monozide (Levy, 1971; McConnell *et al.*, 1971; Wofsy *et al.*, 1972; Ehhalt, 1974). The resultant lifetime for methane is about 2 years (Levy, 1971). The CO produced is itself then oxidized to CO_2 with a time constant of a few months. Carbon monoxide is also produced by incomplete combustion of fossil fuels, particularly in automobile engines, and measurements suggest that on both the local and global scales this source may be more important than the natural methane oxidation source (Seiler, 1974).

Combustion of fossil fuel deposits appears to be the most important human activity affecting the natural carbon cycle in the atmosphere. There is good evidence that CO_2 concentrations in the lower atmosphere have been steadily increasing during the present century, presumably as a natural consequence of coal and oil combustion (Revelle and Suess, 1957; Callendar, 1958). An expected result of increased CO_2 concentrations is to increase the IR opacity of the atmosphere. As we discussed in Chapter 3, increases in IR opacity should lead to higher atmospheric and surface temperatures as a result of the greenhouse effect (Möller, 1963). At present, the main source of IR opacity in the atmosphere is H_2O, which absorbs over a wide range of wavelengths. The important 15 μm band of CO_2, although strong, is not particularly broad at 1 bar of total pressure. As a result, Rasool and Schneider (1971) predict temperature increases of only a few degrees even for an increase in CO_2 by a factor of 10. Conditions on Venus show us that if sufficient CO_2 is added to begin to increase the total pressure, then surface temperatures exceeding the boiling point of water could evolve. However, Rasool and de Bergh (1970) point out that such large increases in CO_2 on Earth are probably prevented by the presence of the oceans; carbon dioxide will dissolve in seawater and, if sufficient concentrations of Mg^{2+} and Ca^{2+} ions are continuously provided by weathering of crystalline rocks, this dissolved CO_2 can then be removed from the oceans as carbonate sediments. This problem has been reviewed by Siegenthaler and Oeschger (1978).

Although large alterations in atmospheric temperatures appear unlikely, it is certainly conceivable that a doubling of the atmospheric CO_2 content can give rise to a 2–4 K increase in global mean temperature (Wetherald and Manabe, 1981; Hansen *et al.*, 1981). This warming may appear insignificant but the sensitivity of our present climatic regime to even very small perturbations is not yet understood. The Earth's climate is governed by a complex balance of radiative, diffusive, and dynamic energy fluxes in the ocean and atmosphere. Feedback mechanisms between composition changes, temperature changes, cloudiness, ice cover, vegetation cover, and the general circulation are only now beginning to be examined in detail (e.g., Budyko, 1971; Sellers, 1973; Stone, 1973; Schneider and Dickinson, 1974; Charney, 1975; Ramanathan, 1981). The possibility that continued fossil fuel combustion will lead to a substantial melting of the icecaps is an unproven but certainly not impossible hypothesis.

The nitrogen cycle, like that of carbon, plays a significant role in the biosphere. The principal elements in this cycle are illustrated in Figure 4.22. It appears that the fastest cycles in which atmospheric nitrogen compounds participate involve some of the simplest of living organisms (see, e.g., Burns and Hardy, 1975). Denitrifying bacteria (e.g., *Pseudomonas, Micrococcus, Achromobacter, Bacillus, Thiobacillus*) operating under usually anaerobic conditions in poorly aerated soil or in ocean water convert soil nitrate to N_2 (about 90% on a global scale) and N_2O (about 10% on a global scale). Most of the N_2O is photodissociated in the upper atmosphere to form N_2 and has a resultant atmospheric lifetime of about 50 years. However, a small fraction of the N_2O is converted to NO and NO_2 in the stratosphere. The latter two gases play a significant role in governing ozone concentrations as we will see shortly. NO_2 converts into HNO_3 at lower altitudes (Harris, 1978), and is then removed by rainout in the troposphere and destroyed by reactions in soil (Abeles, 1971). Some NO_2 is also produced from N_2 in high-temperature combustion engines and by lightning, from organic nitrogen by combustion of fossil fuels, and possibly by atmospheric oxidation of NH_3 (Gopala Rao, 1931; McConnell, 1973a).

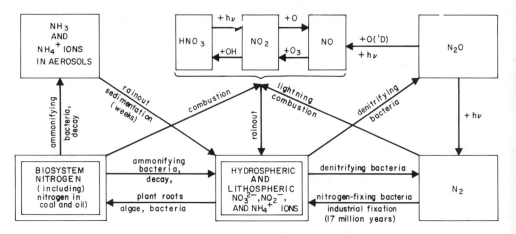

FIGURE 4.22. **The nitrogen cycle on Earth. Gaseous or predominantly airborne species are in single boxes; condensed hydrospheric, lithospheric, and biological species are in double boxes. See the text for further explanations.**

Most of the N_2 added to the atmosphere by microbial action appears to be recycled back to the biosphere–lithosphere–hydrosphere system by the action of nitrogen-fixing bacteria and algae. These bacteria and algae convert atmospheric N_2 to nitrate ion. Many of these organisms (e.g., *Rhizobium*, which lives in the root nodules of legumes) are symbiotic, but some (e.g., *Azotobacter* and *Clostridium* in soil and the blue-green algae *Trichodesmium* in tropical ocean waters) are free living. Certain species (e.g., *Clostridium*) prefer an anaerobic environment, whereas others (e.g., *Azotobacter*) prefer aerobic conditions. Nitrogen is also removed from the atmosphere for production of artificial fertilizers (e.g., the Haber–Bosch process, which converts N_2 to NH_3) and thus returned to the soil. Hardy and Havelka (1975) estimate that nitrogen-fixing organisms presently account for about 67% of the total nitrogen removed from the atmosphere, industrial fixation for about 15%, combustion in engines for about 8%, and lightning and other phenomena for about 10%. Overall, the processes of denitrification and nitrogen fixation appear to be in a rough balance. These opposing processes result in an atmospheric N_2 lifetime or recycling time of about 17 million years (Burns and Hardy, 1975).

Nitrogen is also added to the atmosphere as NH_3, which is produced during decay of organic matter by aerobic bacteria and fungi. This process of ammonification converts organic nitrogen to amino acids and thence to NH_3. The amount of nitrogen added to the atmosphere in this manner appears to be comparable to the quantity added as N_2 by denitrifying bacteria. Some of this ammonia is converted in the atmosphere to ammonium ion in $(NH_4)_2SO_4$, NH_4Cl, and NH_4NO_3 aerosols. All these ammonia compounds are very water-soluble and are rapidly rained out in the lower troposphere: The atmospheric lifetime for NH_3 or NH_4^+ ions is probably less than 1 month. Once removed from the atmosphere, some of the NH_3 is consumed directly by living organisms and some is recycled back to nitrite and thence to nitrate ions by aerobic nitrifying bacteria (e.g., *Nitrosomonas, Nitrosococcus, Nitrobacter*). Nitrogen is found in sedimentary rocks largely as nitrates (Chalk and Keeney, 1971) and in igneous rocks as NH_4^+ in substitution for K^+ ion (Stevenson, 1962).

Two portions of the nitrogen cycle are particularly sensitive to anthropogenic influence. This sensitivity is a direct consequence of the importance of nitrogen oxides in the stratospheric ozone layer. As we will discuss shortly, the principal natural source for these nitrogen oxides in the stratosphere is N_2O. It has been conjectured that increased use of nitrate fertilizers may lead to increased N_2O production by microbes in agricultural soil. The magnitude of the increase is, however, controversial (Crutzen, 1976; Liu *et al.*, 1976; McElroy *et al.*, 1976b). The most difficult part of the problem is to assess the fraction of the present global N_2O production resulting from denitrification in fertilized farm soils. To answer this we will need to know much more about the role of the oceans as a source, sink, or modulator for N_2O. In this respect, since N_2O and CO_2 have similar solubilities in water, we might expect warm tropical waters to appear as N_2O sources and cold polar waters to appear as sinks, in analogy to the CO_2 cycle.

A more direct perturbation to the ozone layer could result from supersonic aircraft designed to fly in the stagnant lower stratosphere, including the Anglo–French Concorde, the Russian Tupolev-144, and the proposed but canceled American Boeing 2707. These produce NO and NO_2 in their engines by thermal decomposition of atmospheric N_2 and O_2. A fleet of about 500 of the now-canceled American aircraft were expected to be flying by the year 2000. Such a fleet would inject NO and NO_2 into the stratosphere to a level some 3 times greater than that from the present N_2O sources. Computations using two- and three-dimensional dynamical–chemical models imply that the above fleet flying 8 hr day^{-1} at 20 km altitude and 45°N latitude would lead to ozone reductions of 8% and 16% in the Southern and Northern hemispheres, respectively (Prinn *et al.*, 1978; Alyea *et al.*, 1975). Similar ozone reductions would result from fleets of roughly 3000 of the smaller, lower-flying Concordes (Cunnold *et al.*, 1975). International restrictions on fleet sizes for stratospheric aircraft are clearly required. In the past, atmospheric tests of thermonuclear weapons have also resulted in significant deposition of nitrogen oxides in the lower stratosphere. It is hoped that such tests will never again be necessary.

Another portion of the nitrogen cycle is disturbed during the formation of photochemical smog in many urban areas. Nitrogen oxides are an important constituent of automobile exhaust gases and high NO_2 concentrations lead to high urban ozone concentrations through the reactions

$$NO_2 + h\nu \rightarrow NO + O \ (\lambda < 3950 \ \text{Å}),$$
$$O + O_2 + M \rightarrow O_3 + M. \tag{4.63}$$

Here $h\nu$ is a solar photon, λ is the wavelength, and M is any other molecule. This locally produced O_3 is a key component in smog development. Nitrogen dioxide also leads to production of another important component of many smogs, namely the toxic eye irritant peroxyacetyl nitrate.

The atmospheric sulfur cycle (Junge, 1960) is of interest both because sulfur is an important component of proteins and because sulfuric acid has become a particularly noxious pollutant in modern times. The basic elements of the cycle are illustrated in Figure 4.23. Robinson and Robbins (1970) estimate that on a global scale the majority of the gaseous sulfur present in air results from anaerobic decay of proteins, producing H_2S. This latter gas is also produced from inorganic sulfates by sulfur bacteria (e.g., *Desulfovibrio*) under anaerobic conditions.

Some of the H_2S is converted back to sulfates by other sulfur bacteria (e.g., *Thiobacillus*) operating under aerobic conditions. However, once in the air the bulk of the H_2S is rapidly oxidized; the principal reactions are probably

$$H_2S + OH \rightarrow H_2O + HS,$$
$$HS + O_2 \rightarrow OH + SO,$$
$$SO + O_2 \rightarrow SO_2 + O. \tag{4.64}$$

The resultant H_2S lifetime ranges from a few hours in polluted urban air to a few days in clean air. This biological H_2S accounts for about 60% of the SO_2 added to the atmosphere. There may be an additional as yet unquantified biological source of atmospheric sulfur in the form of $(CH_3)_2S$ which would also undergo atmospheric oxidation to produce SO_2. The remaining 40% is produced by combustion of fossil fuels (particularly coal) and this anthropogenic source is steadily increasing with time.

Some of the atmospheric SO_2 is apparently directly utilized by plants, but most of it appears to be removed by rainout and by oxidation to sulfuric acid. The phenomenon of acid rain—which is due to both H_2SO_4 and HNO_3 production

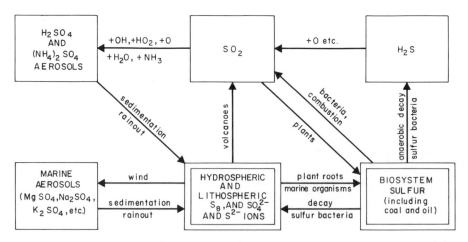

FIGURE 4.23. **The sulfur cycle on Earth. Gaseous and aerosol species are in single boxes; condensed surficial materials are in double boxes. See explanations in the text.**

from anthropogenic NO_2 and SO_2 emissions—is now a widely publicized regional pollution problem in Northern Europe and the Northeastern United States. In photochemical smogs, oxidation of SO_2 is rapidly accelerated; the resultant SO_2 lifetime may be reduced to only a few minutes. This rapid oxidation may be caused by organic peroxy free radicals which can result from the action of sunlight on atmospheric NO_2, O_3, and hydrocarbons. Both hydrocarbons and, as mentioned earlier, nitrogen oxides are present in automobile exhaust gases. The burning of high-sulfur fuels produces anomalously high local SO_2 and thus H_2SO_4 concentrations. Such burning is therefore carefully regulated along with automobile exhaust emissions in many urban areas.

An addition to the sulfur cycle has resulted from the discovery of COS in the troposphere (Hanst *et al.*, 1975). This latter gas is produced mainly from coal combustion, but unlike SO_2 it is relatively insoluble in water and chemically rather inert in the troposphere. It can thus be mixed upward into the stratosphere where it is destroyed at and above an altitude of 25 km by photodissociation and other reactions leading to SO_2 and H_2SO_4 (Crutzen, 1976). It is even now making a significant contribution to the Junge layer. Increased coal combustion can therefore lead directly to increases in the optical depth of this lower stratospheric cloud. It is unlikely that clouds as thick as those on Venus (Prinn, 1973) could ever develop, but some researchers (Naill *et al.*, 1975) have predicted at least a fivefold increase in coal combustion rates will be necessary to meet the energy needs of the industrialized nations toward the end of this century.

The COS is not, however, a significant source of SO_2 and H_2SO_4 on a global scale. The global cycle is dominated by reactions in the lower troposphere. Of the total sulfate appearing in the atmosphere, Robinson and Robbins (1970) estimate that about 80% is attributable to oxidation of biological H_2S and anthropogenic SO_2. The remaining 20% results from the formation of sea-salt particles over the oceans. These contain the sulfates of Mg, Na, K, and other metals. Once added to the air, such involatile particles are, of course, simply removed again by sedimentation and rainout.

Table 4.8 given earlier (p. 198) indicates the presence of several chlorine-containing gases in trace amounts in the atmosphere. The main natural source for atmospheric chlorine is undoubtedly acidification of chloride aerosols producing HCl near the surface (Eriksson, 1959). The chlorine cycle (pictured in Figure 4.24) has received considerable attention because any significant concentrations of Cl and ClO in the stratosphere will lead to depletion of ozone, as we will discuss later. Because the HCl produced at the ground is severely depleted by rainout, it does not give rise to significant stratospheric chlorine concentrations. However, the chlorocarbons CCl_4, $CFCl_3$, CF_2Cl_2, and CH_3Cl are relatively insoluble and are not rained out. They appear to decompose in the stratosphere and thus provide a source for Cl and ClO. An unchecked buildup of these compounds could therefore lead to significant ozone depletion. The chlorofluoromethanes CF_2Cl_2 and $CFCl_3$ are manufactured for use in refrigerators and air conditioners, as aerosol propellants, and for manufacture of plastic foams. Methyl chloride is produced naturally by

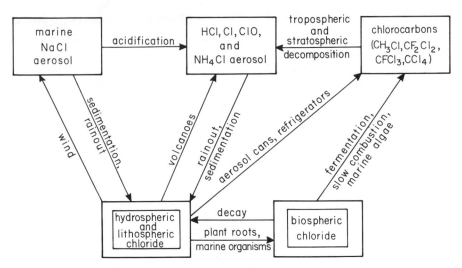

FIGURE 4.24. **The chlorine cycle on Earth. Single boxes indicate gaseous species; double boxes, condensed species. See text for detailed explanation.**

microbial fermentation and by slow combustion of vegetation, and CCl_4 is probably derived from both natural and industrial sources.

The oxygen cycle in the atmosphere deserves considerable discussion although its principal components are simply photosynthetic production balanced by removal by respiration, combustion, and decay. The resultant atmospheric lifetime for O_2 is about 18,000 years. There is also a relatively slow production of O_2 by photodissociation of H_2O and this is limited by the H_2 escape rate as we discussed earlier. Some atmospheric O_2 is also very slowly used in oxidation of crustal Fe^{2+} and volcanic gases such as H_2, CO, H_2S, and SO_2. From an evolutionary viewpoint we can affect atmospheric O_2 levels in a number of ways. One is, of course, to deplete photosynthetic organisms, such as oceanic algae, catastrophically. Another is to increase the rates of combustion of fossil fuels enormously. However, because the atmosphere contains about 700 times more O_2 than CO_2, we would expect the much larger fractional increases in CO_2 resulting from increased combustion to be felt much earlier than the O_2 decreases.

In addition to its obvious importance for respiration to all aerobic life forms, atmospheric oxygen, or more specifically oxygen in the form of its trace allotrope ozone, plays a critical protective role to which we alluded earlier. The ozone layer in the lower stratosphere (15–30 km altitude) is responsible for absorbing UV photons with wavelengths λ between 2400 and 3200 Å which would otherwise be transmitted to the Earth's surface. This radiation is lethal to simple unicellular organisms (algae, bacteria, protozoa), and to the surface cells of higher plants and animals. It also damages the genetic material of cells (DNA) and is responsible for sunburn in human skin. In addition, the incidence of skin cancer has been statistically correlated to the observed surface intensities of the UV wavelengths between 2900 and 3200 Å which are not totally absorbed by the ozone layer.

Ozone also plays an important role in heating the upper atmosphere by absorbing solar UV and visible radiation ($\lambda < 7100$ Å) and thermal IR radiation ($\lambda \simeq 9.6$ μm). As a consequence, the temperature increases steadily from about 220 K at the tropopause (8–16 km altitude) to about 280 K at the stratopause (50 km altitude). This ozone heating provides the major energy source for driving the circulation of the upper stratosphere and mesosphere.

Almost all of the ozone in the atmosphere is formed above the 30 km altitude where O_2 is dissociated during the daytime by UV photons:

$$O_2 + h\nu \rightarrow O + O \quad (\lambda < 2400 \text{ Å}). \tag{4.65}$$

The oxygen atoms produced then form ozone by the reaction

$$O + O_2 + M \rightarrow O_3 + M, \tag{4.66}$$

where M is any other molecule. Ozone has a short lifetime during the day due to photodissociation:

$$O_3 + h\nu \rightarrow O_2 + O \quad (\lambda < 7100 \text{ Å}). \tag{4.67}$$

However, except above 90 km where O_2 begins to become a minor component of the atmosphere, reaction (4.67) does not lead to a net destruction of ozone. Instead, the O is almost exclusively converted back to O_3 by reaction (4.66). If we define the *odd oxygen* concentration as the sum of the O_3 and O concentrations, then odd oxygen is produced by reaction (4.65) and in the earliest theories of the ozone layer (Chapman, 1930) is removed by

$$O + O_3 \rightarrow O_2 + O_2. \tag{4.68}$$

We see that reactions (4.66) and (4.67) do not affect the odd oxygen concentrations but merely define the ratio of O to O_3. Because the rate of reaction (4.66) decreases with altitude whereas that for reaction (4.67) increases, most of the odd oxygen below 60 km is in the form of O_3 whereas above 60 km it is in the form of 0.

More recent research (Crutzen, 1971; Johnson, 1971) has disclosed that reaction (4.68) is responsible for only about 18% of the odd oxygen removal rate. The bulk of the removal (about 70%) is due to the trace gases nitric oxide (NO) and nitrogen dioxide (NO_2) which serve to catalyze reaction (4.68) by

$$NO + O_3 \rightarrow NO_2 + O_2, \tag{4.69}$$

$$NO_2 + O \rightarrow NO + O_2. \tag{4.70}$$

This catalytic destruction cycle is partially short-circuited in the daytime because reaction (4.69) can be followed by the photodissociation:

$$NO_2 + h\nu \rightarrow NO + O \quad (\lambda < 3950 \text{ Å}), \tag{4.71}$$

which regenerates odd oxygen. The gases NO and NO_2 constitute only about 3 ppb of the air in the ozone layer. They are produced naturally in this layer by the decomposition of atmospheric nitrous oxide (N_2O) with excited oxygen atoms [$O(^1D)$] derived from ozone

$$O_3 + h\nu \rightarrow O_2 + O(^1D) \quad (\lambda < 3100 \text{ Å}), \tag{4.72}$$

$$O(^1D) + N_2O \rightarrow NO + NO. \tag{4.73}$$

The NO_2 is removed by reaction with hydroxyl radicals (OH) to form nitric acid (HNO_3), and is reformed by photodissociation of the HNO_3:

$$OH + NO_2 + M \rightarrow HNO_3 + M, \tag{4.74}$$

$$HNO_3 + h\nu \rightarrow OH + NO_2 \quad (\lambda < 3450 \text{ Å}). \tag{4.75}$$

If we define *odd nitrogen* as NO, NO_2, and HNO_3, about half of the stratospheric odd nitrogen is present as HNO_3, and downward transport followed by rainout of this water-soluble HNO_3 is probably the main stratospheric odd nitrogen removal mechanism.

Another catalytic cycle which is responsible for about 11% of the total odd oxygen removal rate involves H atoms and OH and HO_2 (hydroperoxyl) radicals. These species are produced in the upper atmosphere by dissociation of water vapor (H_2O), methane (CH_4), and hydrogen peroxide (H_2O_2).

There has also been considerable concern over the past decade that continued industrial production of $CFCl_3$ (Freon-11) and CF_2Cl (Freon-12) may lead to ozone depletion. Once released into the atmosphere, their only presently recognized removal mechanism involves photodissociation in the stratosphere (Molina and Rowland, 1974):

$$CFCl_3 + h\nu \rightarrow CFCl_2 + Cl \quad (\lambda < 2260 \text{ Å}), \tag{4.76}$$

$$CF_2Cl_2 + h\nu \rightarrow CF_2Cl + Cl \quad (\lambda < 2140 \text{ Å}). \tag{4.77}$$

The chlorine atoms released in these and subsequent reactions can catalytically destroy ozone by conversion to chlorine oxide radicals (ClO). The following reactions (Stolarski and Cicerone, 1974) play roles similar to reactions (4.69)–(4.71), (4.74), and (4.75)

$$
\begin{aligned}
Cl + O_3 &\rightarrow ClO + O_2, \\
ClO + O &\rightarrow Cl + O_2, \\
ClO + h\nu &\rightarrow Cl + O \quad (\lambda < 3035 \text{ Å}), \\
Cl + CH_4 &\rightarrow HCl + CH_3, \\
OH + HCl &\rightarrow Cl + H_2O.
\end{aligned}
\tag{4.78}
$$

Computations in one-dimensional atmospheric chemical models (Cicerone *et al.*, 1974; Crutzen, 1974; Wofsy *et al.*, 1975) imply that the chlorofluoromethanes are presently responsible for at most a 1% reduction in stratospheric ozone. However, when injection of $CFCl_3$ and CF_2Cl_2 into the atmosphere is continued at current rates the models predict about a 10% decrease in ozone by the year 2050. If reactions (4.76) and (4.77) are indeed the only atmospheric removal mechanisms for $CFCl_3$ and CF_2Cl_2, then their atmospheric lifetimes are about 45 and 70 yr, respectively. Therefore, even if chlorofluoromethane production were halted, the effects would still be felt for several decades thereafter. Unfortunately, stratospheric chemistry is still not as fully understood as needed to reliably assess the role of chlorine. For example, observations and calculations of ClO have yet to be reconciled. A comprehensive and urgent program of laboratory and atmospheric measurements is presently being conducted in order to shed light on these and related problems. The reader is referred to some recent reviews (Prinn *et al.*, 1978; World Meteorological Organization, 1981) for more detailed discussions.

In a purely chemical model of stratospheric ozone with no transport we would obtain the maximum ozone concentrations at altitudes and latitudes where reaction (4.65) is fastest; that is, in equatorial regions above 30 km. However, the observed maximum ozone concentrations occur at about 18 km altitude in polar regions. A substantial poleward and downward transport of ozone must therefore occur in the lower stratosphere. This poleward transport results in the column abundance of ozone being about 50% greater at the poles than at the equator with a maximum in polar ozone in the spring. Recent computations of ozone circulation and chemistry utilizing detailed two- and three-dimensional chemical–dynamical models have succeeded in simulating these observed seasonal and latitudinal ozone variations (Alyea *et al.*, 1975; Cunnold *et al.*, 1975; Prinn *et al.*, 1975).

Although the horizontal winds in the stratosphere can be quite large, the increase in temperature with altitude in this region makes it much more stable to vertical motions than the troposphere. Consequently, once destructive species such as Cl, ClO, NO and NO_2 are introduced into the ozone layer it takes roughly 3 years before they can be removed again by mixing down to the ground. This long stratospheric residence time, combined with the catalytic aspect of the ozone destruction caused by these species, makes the ozone layer particularly sensitive to the perturbations in nitrogen and chlorine oxide concentrations which we discussed earlier.

Of all the anthropogenic gases which may influence the ozone layer, only one appears to be benign: CO_2, although chemically unreactive with O_3, tends to cool the stratosphere, which decreases the rate of O_3 destruction (Groves *et al.*, 1978). Finally, we should mention some additional ozone-destroying processes that have been identified. Solid fueled rockets using ammonium perchlorate NH_4ClO_4, as the oxidizer (Minutemen missiles, NASA Space Shuttle booster) inject small amounts of hydrogen chloride (HCl) into the stratosphere (Stolarski and Cicerone, 1974); expanded production of methyl bromide (CH_3Br), an agricultural fumigant, may result in significant stratospheric bromine concentrations (Wofsy *et al.*, 1975b). Methyl chloroform (CH_3CCl_3) may also contribute to the ozone depletion problem (McConnell and Schiff, 1978). Catalytic odd oxygen destruction by bromine is similar to but faster than that by chlorine. Fortunately the global measurements of ozone which are required to ascertain the reality of all these possible long-term depletions have recently become possible from satellites. Such continuous ozone monitoring is clearly mandatory for the future. In addition, an extensive program of stratospheric measurements of species such as ClO, Cl, OH, O, NO, NO_2 and several fluorocarbons has been underway for several years and will continue. Measurements to date provide at least qualitative support for the chemistry outlined briefly in this section.

One message should come through rather clearly from our discussions in this section: Our own atmosphere may not be as resilient to pollution as has sometimes been considered in the past. It is clear that a long-term program of research into the chemistry of the atmosphere will be necessary before we can adequately assess the true role of human activities in its future evolution.

4.4. MARS

For Mars, more than any other planet, the disparity between our preconceptions based on Earth-based observations and our present knowledge, derived from several highly successful spacecraft missions, is immense. As recently as the early 1960s the literature on Mars focused on vehement partisan debates concerning a few central subjects. There was the famous "wave of darkening"—a phenomenon reported by many observers in which contrast increased and dark areas became much darker during local springtime (see, e.g., Focas, 1962). There were the even more famous "canali" of Schiaparelli, popularized by Percival Lowell, which decorated the planet from polar cap to polar cap (and which, according to Lowell, were also present on Venus). Much debate concerned the nature of the coloring agents responsible for the "red" bright areas and the "green" dark areas. This debate was based almost exclusively on visual color impressions as opposed to photometry, but a majority of observers were convinced that the dark areas were truly green. The most plausible interpretation of vast green regions which intensified in color in the spring seemed to be that Mars was heavily vegetated. This speculation, unfettered by hard evidence, quickly took on a life of its own. Thus, Mars quickly became endowed in the public imagination with enough greenery to keep a planet full of (itinerant) herbivores happy. By simple extension, nothing could be more reasonable than to postulate the existence of such ecologically appropriate herbivores. It is then a simple matter to suppose the existence of carnivores and, not surprisingly, even intelligent beings. Lowell's attribution of the "canals" to a planetary engineering program of meltwater distribution and irrigation is familiar to all.

Other areas of debate included the "violet haze" which made contrast vanish at the blue end of the visible spectrum and in the UV, the nature and composition of the condensates responsible for forming the clouds sometimes observed from Earth, and the mass and composition of the atmosphere. Explanations were also sought for the mysterious disappearance of UV haze whenever Mars was in opposition to the Sun, the famous "opposition effect." It was almost universally believed, on the basis of studies of the spectroscopy of the atmosphere and the polarization of reflected sunlight, that the atmosphere had a pressure of 100–250 mbar and was dominantly composed of nitrogen, with a small fraction of carbon dioxide. Water vapor (and liquid water) was assumed to be present, nicely meeting the needs of the inhabitants.

Sinton (1957) claimed the detection of features in the spectrum of Mars that came to be interpreted as due to organic matter, presumably present in the green areas.

Three theories regarding the red color of most of the planet were debated: In one of these, the red color was iron oxides formed by weathering of iron-bearing surface rocks. Two other theories based on the photochemistry of the atmosphere attributed the red colors to nitrogen tetroxide (N_2O_4) or carbon suboxide (C_3O_2).

Because of the widely held belief that the planets form an evolutionary sequence, with Mars older than Earth and Venus younger than either, the prior existence of oceans, oxygen, and flourishing life was widely assumed. The loss of vast amounts of hydrogen to space, with concomitant absorption of an equivalent amount of oxygen by surface rocks, was assumed to account for the absence of oceans and the redness of the surface at the present time.

Over the years since 1964, great advances in both spectroscopic techniques and spacecraft observations have led to a wholly new appreciation of Mars. In the present view, the furious debates of a few years ago seem strangely irrelevant.

The first major blow to the adversary structure pictured above was the discovery by Kaplan et al. (1964) that close analysis of high-resolution spectra could not be reconciled with high surface pressures. From their studies of the intensities of intrinsically weak (and therefore unsaturated) lines in the CO_2 IR spectrum, they were able to show that the surface pressure was less than 10 mbar, not near the long-accepted values of 100 or 200 mbar which had been the consensus of prior observers. This meant that it was not clear whether there could be any place on Mars where the atmospheric pressure was high enough to permit the presence of liquid water, which will boil away at such low pressures even at 0°C.

The second, even more shattering, blow was delivered by the Mariner 4 flyby in 1965, which returned 28 photographs of a heavily cratered, starkly lunar landscape. Although only a very small portion of the surface area of Mars was well covered, these areas seemed typical on the basis of Earth-based observations. No canals were seen in the Mariner 4 pictures. Mariner 4 also confirmed that the surface pressure was indeed 5–10 mbar, confirming the conclusions of Kaplan et al. (1964).

Four years later, Mariners 6 and 7 were sent on flyby missions to Mars to extend the coverage provided by Mariner 4. Some 200 pictures were returned, covering several percent of the planet at moderate resolution and virtually all of it at low resolution. The widespread occurrence of craters was confirmed, and new types of terrain were discovered. The Hellas region was found to be utterly featureless, and an area of "chaotic" terrain was photographed near the equator. The low surface pressure and high CO_2 content were confirmed, and the south polar cap was found to be cold enough for precipitation of solid CO_2.

Meanwhile, numerous Earth-based spectroscopic studies of Mars had been carried out with a great deal of success and many interesting conclusions, which we shall review in detail below.

By 1970, therefore, it was well established that a large portion of the surface of Mars was extremely ancient, geologically inactive terrain, as densely populated with large impact craters as the lunar highlands. A mean surface pressure below the vapor pressure of water at its triple point assured the absence of pure liquid water anywhere on the present surface. Further, none of the Mariner 4, 6, or 7 pictures showed evidence of the presence of "canals."

It was toward this dismally uninteresting planet that four terrestrial spacecraft were launched in 1971. The American Mariner 8 and 9 were to orbit Mars with very different orbital inclinations to map the surface with high resolution. A variety of other instruments were also flown, including experiments to measure the water content and thermal structure of the atmosphere. At the same time, the Soviet Union launched the Mars 2 and 3 vehicles, which were intended to orbit Mars and drop landing capsules to the surface. The chronologies of these and other Mars missions through 1981 are given in Table 4.14.

Of the four announced launches in 1971, Mariner 8 failed to achieve orbit, whereas Mars 2 and 3 survived the trip to Mars and orbited that planet, although their landing capsules failed. Mariner 9, which operated successfully in Mars orbit for well over a year, again completely revolutionized our understanding of the planet. The initial Mariner 9 results (Masursky et al., 1972) showed that dramatic scene of a dust-shrouded Mars, wrapped in the midst of an intense dust storm at the time of arrival of Mariner 9, Mars 2, and Mars 3. As the atmosphere slowly cleared, the few "spots" visible above the dust blanket slowly emerged as immense volcanic constructs far larger in volume than the greatest volcanoes on Earth. The Mars 2 and 3 orbiters, with insufficient lifetime to wait out the end of the dust storm, dropped their landers into the storm, where one crashed and the other survived less than 2 min. The most intense dust storms occur when Mars is near perihelion; however, the most economical transfer orbits from Earth to Mars also occur when Mars is nearest the Sun. Thus, terrestrial spacecraft launched under optimum conditions tend to encounter Mars under the worst possible viewing and entry conditions.

After the dust storm had subsided, Mariner 9 revealed a stunningly complex and paradoxical planetary geology. Part of the surface was indeed Moonlike in its austerity, densely cratered, and very ancient. However, much of the planet (missed by the limited coverage of the Mariner 4, 6, and 7 flybys) was of extraordinary interest. Aside from the huge volcanoes, numerous features indicative of the presence and withdrawal of huge volumes of ice were found, along with many valleys apparently carved out by flash floods with flow volumes at least the equal of any terrestrial river. Large areas of "chaotic" terrain were found, almost all in close conjunction with evidence of ancient water flow. The plans for the Mariner 8 and 9 missions prior to launch were explained by Steinbacher and Gunter (1970), and an overview of the first year of operation of Mariner 9 was given by Leighton and Murray (1971). A short but instructive survey of the entire mission by Steinbacher and Haynes (1973),

TABLE 4.14
Announced Mars Probe Launchings

Launch date[a]	Spacecraft	Results
1 Nov. 1962	Mars 1	Communication failure 21 Mar. 1963 en route
5 Nov. 1964	Mariner C	Launch failure
28 Nov. 1964	Mariner 4	Mars flyby 15 July 1965
30 Nov. 1964	Zond 2	Communication failure 4 Mar. 1964
1 Feb. 1967	None	—
15 Mar. 1969	Mariner 6	Mars flyby
	Mariner 7	Mars flyby
9 May 1971	Mariner 8	Launch failure
19 May 1971	Mars 2	Orbited Mars 27 Nov. 1971; lander crashed
28 May 1971	Mars 3	Orbited Mars 1 Dec. 1971; lander failed
30 May 1971	Mariner 9	Orbited Mars 14 Nov. 1971
21 July 1973	Mars 4	Failed to orbit; flew by 10 Feb. 1974
25 July 1973	Mars 5	Orbited Mars 12 Feb. 1974
5 Aug. 1973	Mars 6	Communication failure just before landing, 12 May 1974
9 Aug. 1973	Mars 7	Failed to initiate landing sequence; flew by 9 Mar. 1974
20 Aug. 1975	Viking 1	Orbited Mars 19 June, 1976; landed 20 July, 1976
9 Sept. 1975	Viking 2	Orbited Mars 7 Aug., 1976; landed 3 Sept. 1976
12 Oct. 1977	None	—
1 Dec. 1979	None	—
20 Jan. 1982	None	—
11 Mar. 1984[b]		

[a]Launch windows are roughly 1 month long.

[b]Mars launch windows repeat every 2 yr, 1 month, and 19 days on the average. Because of the eccentricity of Mars' orbit, intervals vary about this mean.

and a special issue of the *Journal of Geophysical Research* (Vol. 78, pp. 4007–4440, 1973) provide a valuable perspective.

It was during the period of Mars 2 and 3 and Mariner 9 flight activity that the long-range American plans for Mars exploration were fully crystallized about the Viking orbiter and lander spacecraft (Soffen and Young, 1972). Mariner 9 photographic mapping results played a central role in selection of landing sites for Viking. Basically the philosophy of the Viking mission was rather similar to that employed in the Mars 2 and 3 landing attempts, and in their closely similar sister missions, Mars 4–7, in 1973. All of these missions involved paired orbiters and landers, with complementary roles, involving remote sensing and data relay by the orbiter, and *in situ* sensing by the lander.

The Mars 4 and 5 spacecraft were intended to be orbiters; Mars 6 and 7 were separately launched landers. Mars 4 failed to enter Mars orbit and flew by at high speed on 10 February 1974. Mars 5 orbited successfully on 12 February 1974 and eventually carried out a very successful mapping and reconnaissance mission. Mars 6 entered the atmosphere but failed prior to landing, and Mars 7 suffered an engine failure which caused it to fly by Mars, without entering, on 9 March 1979. The early results of these missions, most notably, the Mars 5 orbiter, are given by Moroz (1975), and a special issue of *Cosmic Research* (Vol. 13, p. 1, 1975) describes the detailed results of these missions. The most important single advance from this very ambitious set of missions was the extremely high-quality mapping of part of the Martian surface by the Mars 5 cameras, including some coverage superior in resolution to the comparable Mariner 9 images.

The equally ambitious, and far more successful, Viking missions, launched in 1975, were intended to place two spacecraft in Mars orbit and two heavily instrumented landers on the surface to search for evidence of life. The progress of these missions from launch to goal is followed by Soffen and Snyder (1976), Soffen (1976a,b, 1977), and Snyder (1977, 1979a,b). All four spacecraft performed virtually flawlessly, and Viking data were received through 1982, 7 years after launch.

Perhaps the most profound insight to arise from the Viking missions is the extreme hostility of the Mars surface environment to any even remotely Earthlike forms of life. Although the Viking data at times arose to ambiguity on the question, it is fair to say that there is no longer any shred of evidence for the present or former existence of life on Mars. The detailed results of the Viking mission are rich, varied, and voluminous, and we shall return to them repeatedly in this section. The five most important sources of Viking and post-Viking Mars data are the three special issues of *Science* (Vol. 193, pp. 759–815; Vol. 194, pp. 57–105; and Vol. 194, pp. 1274–1353, all in 1976) and two large special issues of the *Journal of Geophysical Research* (Vol. 82, pp. 3959–4681, 1977, and Vol. 84, pp. 7909–8519, 1979).

Post-Viking work carried out under the Mars Data Analysis Program was presented in the first two issues of Volume 45 of *Icarus* (January and February, 1981), and Mars's climate cycles are discussed in Volume 50 of *Icarus* (pp. 127–472, 1982).

For historically important scientific reviews of the nature of Mars and its atmosphere, see Brown (1950) and Urey (1950). Urey argued from the fact that the apparent iron content of the Sun was about a factor of 10 lower than its abundance in the terrestrial planets (expressed, for example, as the Fe:Si ratio), that extensive fractionation of metal from silicates had occurred during the early history of solid solar system material. Brown (and later Ringwood, 1966) argued that a similar Fe:Si ratio could be present in each of the terrestrial planets if the iron in Mars were more highly oxidized than in Earth and Venus. Early reviews with emphasis on the atmosphere include Koval' (1971), Hunten (1971), Meadows (1973), and Ingersoll and Leovy (1971). Numerous recent papers, dealing with aspects of the origin and evolution of the atmosphere, will be discussed in 4.4.4 and 4.4.6. Of the recent more general treatments of Mars, we recommend the excellent and broad reviews by Snyder (1979a,b).

4.4.1. Structure, Composition, and Thermal History

Among the very few facts known about the interior of Mars prior to the beginning of spacecraft exploration of the planets were the mass, radius, ellipticity, and rotational moment of inertia of the planet. Since all of these parameters have been improved by more recent investigations, there is little point to recounting in detail the evolution of the quantitative details of the models of Mars which were based on these early data. However, certain general conclusions of the early modeling attempts remain valid, and many of the interpretive ideas suggested in the prespacecraft era of Mars studies, though not completely tenable today, nonetheless had an important influence on the evolution of our ideas.

It was suggested by Lamar (1962) that the basalt–eclogite transition should lie far deeper in Mars than in Earth and that the less dense basalt layer is much thicker near the equator, thus contributing to the optical ellipticity of the disk. Lyttleton (1965, 1968) developed internal structural models in which the entire planet was composed of rocky (silicaceous) material, with no core present. Jeffreys (1937), Urey (1951), and MacDonald (1962, 1963) contended that Mars could be assembled of a mixture of Earthlike core and Earthlike mantle materials in proportions different from those found on Earth. Kovach and Anderson (1965), using 3306 km (the pre-Mariner 6 and 7 value) for the radius of Mars, found a metal abundance for Mars not very different from that for the Earth. Kliore *et al.* (1969) showed that the mean radius of Mars compatible with the Mariner 6 and 7 radio occultation measurements was in fact 3394 km, which led to a substantial revision in the mass of the core of Mars. Binder (1969b), using a density of 8 gm cm^{-3} for the zero-pressure density of Martian core material, found a core mass of only 2.7–4.9%. However, this is surely a substantial overestimate of the core density, and a value near that of FeS is likely, rather than Binder's value which is suggestive of Ni-rich Fe–Ni alloy.

The startling difference between most early models of Mars and contemporary Earth models was the far lower density of Mars, which in every case required that the core mass be proportionately far smaller in Mars than in the Earth. Urey (1951, 1952a,c) attributed this density difference to physical accretionary processes which discriminated between metal and silicate grains prior to the formation of planet-sized bodies. This model was proposed in the context of the unanimous opinion of astrophysicists and solar spectroscopists (prior to 1969) that the abundance of iron-group elements in the Sun (relative to, say, silicon) was about a factor of 10 less than in chondritic meteorites and the terrestrial planets. Thus, Urey was forced to invent a process for extreme *enrichment* of the metal phase in accreting solid bodies. Applying this hypothetical enrichment process to different degrees in different planets accounted for the apparent fractionation of metal relative to silicates found by comparing the uncompressed densities of Mercury (then poorly known), Venus, Earth, and Mars. As we have seen, Ringwood (1959, 1966) chose to disbelieve the spectroscopic iron abundance and proposed that all of the terrestrial planets originally formed with the same bulk composition. That composition was taken to be C1 carbonaceous chrondritic. The present large density differences between these planets were then assumed to be due to extreme heating, devolatilization, and atmospheric escape, which most severely depleted the volatiles from the largest planets. In this view, Earth has the most devolatilized interior and hence the lowest FeO content. The bulk composition of Mars is pictured as more FeO-rich than Earth's mantle, with a core composed of a high-pressure polymorph of magnetite (Ringwood and Clark, 1971).

Anderson (1972b) favored models containing an Fe–Ni–S core with about 12% of the planetary mass, with a mantle containing a high concentration of FeO or metal. A total Fe content of about 25% was preferred. Anderson described Mars as an "incompletely differentiated planet."

The theoretical treatment of the condensation of preplanetary material by Lewis (1972b) ascribes the density differences between the terrestrial planets mainly to the temperature gradient in the solar nebula at the time of last equilibration of grains and dust. By this model, Mercury, we will recall, should have an elevated density due to incomplete silicate condensation at the high temperatures that close to the Sun (see Section 4.1 for a critical discussion of this hypothesis). The other terrestrial planets are pictured as having very nearly the same relative abundances of rock-forming elements, which are fully condensed below about 1000 K. Venus, Earth, and Mars are thus expected to differ in composition and density only with respect to the volatile elements, especially water and sulfur. This model then predicts an Fe:Si ratio of about unity for all three of these planets, with the iron becoming progressively more oxidized with increasing heliocentric distance. Mars should then differ from Earth principally in having more FeO and less Fe metal, and thus a denser mantle and less-dense core than Earth. Core formation would then involve Fe–FeS eutectic melting, as has been modeled by Johnston et al. (1974).

Several other works on the figure and internal structure of Mars published prior to the Viking orbiter missions (Binder and Davis, 1973; Meshcheryakov, 1973, 1975; Petersons, 1975; Reasenberg et al., 1975; Standish, 1973) depend in varying degrees on older planetary data, and we shall turn directly to more recent work.

Tracking of spacecraft in orbit around Mars has resulted in a refinement of the value of the reduced moment of inertia, I/MR^2. Reasenberg (1977) and Kaula (1979), using very different assumptions about isostatic compensation of the Tharsis province, both find a reduced moment of inertia of 0.365.

Models of Mars constrained by these more recent estimates of the moment of inertia have been constructed by Johnston and Toksöz (1977), Okal and Anderson (1978), Arvidson et al. (1980), and Goettel (1981). Johnston and Toksöz find that models with a higher mantle density and lower core density than Earth can be fitted to Mars. They estimate an FeO:(FeO + MgO) ratio of 0.25, about twice the terrestrial mantle value, with an allowable range of mantle zero-pressure densities from ~3.44 to 3.54 gm cm^{-3}. McGetchin and Smyth (1978) and Morgan and Anders (1980) have proposed geochemical models for Mars which have mantle FeO contents similar to those models of Johnston and Toksöz which have FeS (metal-free) cores.

Goettel (1981) has reported a careful error analysis for interior models of Mars and finds a zero-pressure mantle density of 3.44 ± 0.06 gm cm^{-3}, with a density near 3.41 favored for FeS-core models. He finds that the density of Mars is compatible with the assumption that the planet contains the major rock-forming elements in their solar proportions, but differs from Earth in that some 30% of the total Fe content of Mars is tied up as FeO in mantle silicates. These models use a reduced moment of inertia of 0.365, in agreement with the recent work of Reasenberg and Kaula mentioned above, and further assume a bulk density of 3.933 ± 0.002 gm cm^{-3} for the planet (Bills and Ferrari, 1978). These conclusions are generally compatible with both the model of Ringwood and Clark (1971), who attribute the high degree of oxidation to the small mass of the planet, and that of Lewis (1972b, 1974a), which ascribes a higher FeO content to Mars because it accreted at greater heliocentric distance, and hence at lower temperatures, than Earth.

However, carefully interior structure models are generated, and however accurately the equations of state and phase stability relations of geochemically plausible materials are known, it is simply not possible to produce unique and reliable models of the internal composition and structure of a planet without additional, usually chemical, constraints. Further, especially in the case of a planet as small as Mars, extreme assumptions about the accretionary history and radioactive element content can generate very different predictions of the present thermal state and tectonic activity of the planet.

One source of information concerning the chemical composition of the crust of Mars, and hence a possible indicator of the degree of geochemical differentiation of the interior, is reflection spectroscopy of the surface. Prior to the Viking landing missions, these remote observations were the best available source of data on Martian geochemistry.

One of the earliest chemical and physical discussions of Mars (Wildt, 1934a) attributed the red color to oxidation of iron-bearing igneous rocks by photochemically produced O_2 and O_3.

Prior to the 1960s, very little information was available on the reflection spectrum of Mars. Adamcik (1963) suggested that the water abundance in the atmosphere of Mars was about what would be expected at the point of equilibrium with coexisting goethite and hematite. Draper et al. (1964) showed that a four-point IR spectrum in the $0.8-1.6$ μm region could be very approximately fitted to a laboratory sample of limonite [a mixture of goethite (FeOOH) and hematite (Fe$_2$O$_3$)]. Binder and Cruikshank (1964a) concurred with this identification and suggested that the limonite constituted a weathering crust produced by atmospheric oxidation of primary ferrous-iron-bearing minerals. Van Tassel and Salisbury (1964) pointed out that the large majority of the mass of the surface might be transparent or very fine-grained silicates which would contribute little or nothing to the spectrum, while a minuscule mass fraction of limonitic stain could provide a rather intense reddish coloration. Sagan et al. (1965a) likewise favored limonite, which they interpreted as the result of oxidation of the surface attendant upon the escape of an ocean equivalent of hydrogen. Thus, they were able to infer biologically benevolent conditions in the past from observations of a dead and hostile present, a true logical *tour de farce.*

Hovis (1965) claimed detection of a band near 3 μm which he attributed to water of hydration. Polarimetric evidence for ferric iron compounds was summarized by Rea and O'Leary (1965). Fish (1966) presented thermodynamic arguments suggesting that hematite was probably stable relative to goethite at temperatures above about 253 K on the Martian surface. Sagan (1966), referring to the failure of the Mariner 4 flyby to detect any trace of a planetary magnetic field and to its success in finding a high abundance of very large impact craters, concludes that Mars had not differentiated, and thus had never possessed oceans. He attributed the red coloring to limonite produced by photochemically oxidized iron-rich minerals.

O'Connor (1968a), using newly available data limiting the CO abundance on Mars, calculated equilibrium phase assemblages in the Ca–C–O–H, Mg–C–O–H, Al–O–H, and Fe–C–O–H systems, assuming that the equilibrium oxygen fugacity of the atmosphere (f_{O_2}) could be calculated from the CO:CO$_2$ ratio at temperatures well below 300 K. Despite this erroneous assumption (f_{O_2} is photochemically regulated and independent of equilibrium regulation by CO and CO$_2$ for purely kinetic reasons), many of O'Connor's results remain valid. He finds feldspars, corundum (Al_2O_3), and boehmite ($AlOOH$) to be plausible minerals, and finds the calcite + quartz, magnesite + quartz, and siderite + quartz assemblages to be stable. The latter is, however, in error because of the assumption that the CO:CO$_2$ ratio could be used to deduce an oxygen fugacity (about 10^{-80} bar!) by assuming chemical equilibrium at a temperature below 300 K. Clearly, however, the CO is photochemically produced, which means that O$_2$ is also photochemically produced in at least roughly comparable amounts. This must larger O$_2$ pressure easily stabilizes ferric iron, and goethite becomes stable. O'Connor (1968b) also criticized Sagan *et al.* (1965a) by maintaining that Mars must be in an early stage of development, and hence could never have had an extensive atmosphere or hydrosphere. This argument, however, proved ineffectual, since Sagan (1966) had already reversed polarity on this issue.

O'Leary and Rea (1968a,b) found that the absence of the 0.87 μm absorption band of ferric oxide in the Mars spectrum argued strongly against the presence of large amounts of limonite, and indicated that the ferric coloring agent was present as a stain with a rather small mass fraction.

Berner (1969), arguing from phase coexistence data on the stability of goethite in terrestrial settings, concluded that the P_{H_2O} corresponding to the goethite–hematite buffer at typical Martian surface temperatures should be several orders of magnitude larger than the usual P_{H_2O} near the surface of Mars, and hence that hematite should be the stable phase. Morrison *et al.* (1969) and Pollack and Sagan (1969) have interpreted the photometric and polarimetric properties of the surface in terms of a systematic particle size difference between bright and dark areas, with mean radii near 25 μm. The 3-μm feature found by Hovis (1965), although compatible with the presence of goethite, could also be explained by adsorbed surface water or other hydrous minerals. They also find that laboratory reflection spectra of limonite powders do not support the idea that a grain size variation alone suffices to fit the spectra of light and dark areas. They again suggest the presence of hematite-stained transparent mineral grains. This conclusion was supported by Salisbury and Hunt (1969). Egan (1969) concluded that a bimodal size distribution of limonite particles could fit the optical properties of the Martian surface, a conclusion disputed by O'Leary and Pollack (1969) and defended by Egan (1971).

McCord (1969) reported the interesting observation that the spectral change undergone by regions of the surface affected by the "wave of darkening" could be simply described: The albedo changed by the same factor at all wavelengths. Thus, the visual darkening is *not* accompanied by a discernible color change. This generalization, however, rapidly breaks down toward longer wavelengths.

Dollfus *et al.* (1969) presented a model for the spectroscopic and polarization properties of the surface, including seasonal variations, in terms of a limonite–goethite grain model. Pollack *et al.* (1970a,b) found in laboratory simulations that goethite should be unstable on the surface in the daytime, but should exist at nighttime temperatures and persist at depths great enough to escape the diurnal thermal wave, which penetrates ~4–7 cm (Moroz *et al.*, 1975).

Gibson (1970) proposed that the surface of Mars could be colored by the oxidation of infalling meteoritic material, and thus the color might not be indigenous to Mars. In the complete absence of tectonic activity, such a possibility would have to be regarded seriously. Of course, this would defeat our purpose of discovering clues to the bulk composition and evolution of the plant by studying the composition of its surface.

McCord *et al.* (1971) studied the spectral reflectivity of selected small areas of the surface over the 0.3–2.5 μm region and were able to confirm that the most pronounced "green" area on Mars, Syrtis Major, is in fact red rather than green or neutral in color. They also reported a faint 0.95 μm spectral feature due to ferrous iron in the spectra of several bright regions. Other narrow-band photometric spectra of selected small areas were also published by Bronshtén and Ibragimov (1971) and by McCord and Westphal (1971). Albedo maps for comparison with these results are also available (Aleksandrov and Lupishko, 1972).

After the Mariner 6 and 7 flybys, new data on the photometric function of Mars at phase angles never before measured became available. Pang and Hord (1971) described these data, and Binder and Colin Jones (1972) developed an interpretation of the surface in terms of two fundamental types of material: transparent mineral grains and limonite stain.

The IR spectroscopy experiment on the Mariner 9 orbiter found from the thermal IR spectrum of the surface that the best single-component fit to the silicate bands was for an intermediate igneous rock, quite incompatible with ultramafic composition and indicative of extensive differentiation of the crust (Hanel *et al.*, 1972a,b).

Infrared and microwave radiometry of the surface from Earth (Kuz'min *et al.*, 1971; Cuzzi and Muhleman, 1972) and from the Mariner 9 spacecraft (Kieffer *et al.*, 1973) reveal a dusty surface with a low dielectric constant, which curiously showed no correlation between thermal inertia and radiometric albedo; contrary to more recent findings (Kieffer *et al.*, 1977).

Hunt *et al.* (1973), in a reanalysis of the IR spectrum, claimed that the clay mineral montmorillonite could be the principal dust component raised during dust storms. This is in reasonable accord with the results of IR spectroscopic studies by Houck *et al.* (1973), who found an absorption band due to chemically bound water, probably in phyllosilicates, in the reflection spectrum of the surface. The strength of the absorption suggests about 1% water content. The presence of a significant mass of goethite and clay weathering products is a reasonable expectation based on those physicochemical processes known to exist on the Martian surface (Huguenin, 1974). The nature of these weathering processes will be discussed in Section 4.4.2.

Seasonal and regional differences in reflection spectra are probably due to several different factors, among which particle size, degree of oxidation, and aeolian transport of dust are clearly important. In general, high-albedo areas, although active and variable in albedo (Capen, 1976), are spectrally very similar to dust clouds observed during the great dust storms (McCord *et al.*, 1977). Although some of the lighter areas contain detectable traces of ferrous iron (McCord and Westphal, 1971; McCord *et al.*, 1971), the general rule is that ferrous iron absorption near 0.95 μm is strongest, and ferric absorption is weakest, in darker surface units (Singer *et al.*, 1979). The band position and shape suggest that the Fe^{2+} resides in olivine and pyroxene, probably in the form of a primary basaltic rock.

The silicate reststrahlen band near 9 μm, discovered by Hanel *et al.* (1972b) and attributed by Hunt *et al.* (1973) to montmorillonite, is also consistent with other silica-rich minerals, such as feldspars or micas, present in major amounts in the mobile dust (Aronson and Emslie, 1975). Toon *et al.* (1977) concluded that the dust could not be pure montmorillonite, and may be dominated by mixed intermediate and acidic rock dust, containing no more than a few percent of limonite, carbonates, and nitrates. Moroz (1976c), on the other hand, favors a mixture of igneous rock dust with goethite as the material of the light areas, while agreeing that the dark regions are more basic than the light dust.

We conclude that the evidence from spectrophotometric and polarimetric studies strongly indicates the presence of silica-rich dust. The simplest explanation, that these silica-rich materials are produced by extensive geochemical differentiation and require density-dependent fractionation of core, mantle, and crust, may well be correct. However, it is clearly important to consider the nature of photochemically driven weathering reactions on the surface to see whether they are capable of generating such a dust by direct attack of water and oxidizing agents on primary mafic or ultramafic rocks.

Another crucial line of evidence, not available prior to the spacecraft missions of recent years, concerns the surface morphology of the planet. One of the principal contributions of the Mariner and Viking missions has been the compilation of a vast library of photographs and maps, which reveal a great wealth of detail concerning the tectonic activity of Mars.

The first great surprise generated by spacecraft studies of Mars was the heavily cratered appearance of the provinces seen by Mariner 4. Crater counts quickly revealed that the region seen must be extremely ancient (Marcus, 1968; Chapman et al., 1969; Hartmann, 1971a), and that degradation of these craters by wind and thermal stresses caused them to have more subdued profiles than lunar craters of comparable diameter (Sharp, 1968; Hartmann, 1971b). Most of the features found on earlier Mars maps, which had been based largely on drawings made by visual observers with moderate-sized telescopes, could not be identified in the enormously superior flyby photographs. However, certain of the "canali" did seem to be associated with apparent tectonic features (Smith and Robinson, 1968). Katterfeld and Hédervári (1968) identified some of the largest craters as "volcano-tectonic ring structures"—a view which would sound very modern if it were not for the fact that the region they studied did not include the presently known volcanic structures.

The much wider but still very unrepresentative coverage of the Mariner 6 and 7 missions (Leighton et al., 1969a,b,c; Dunne et al., 1971) revealed, in addition to the heavily cratered terrain, other regions of "chaotic" and essentially featureless appearance. The Hellas basin was found to be almost completely devoid of discernible features, possibly due to erosional obliteration, while the chaotic terrain seemed to have been produced by slumping and collapse on a vast scale, possibly related to geothermal activity (Sharp et al., 1971a). It was generally accepted that the cratered regions were representative of Mars, and the craters were studied intensively (Cordell et al., 1971; Murray et al., 1971) in addition to the crater count studies mentioned previously. Several large craters were identified and studied by Earth-based radar (Pettengill et al., 1971).

One interesting issue is the relationship between albedo and elevation. Sagan and Pollack (1968) and Ksanfomaliti et al. (1975a,b) claimed that the dark areas tend to lie higher than the bright areas. Wells (1969, 1972), measuring the surface relief by means of the horizontal variations in the CO_2 column abundance above the surface, found no correlation between albedo and height. Radar observations (Pettengill et al., 1969; Zachs and Fung, 1969) tended to confirm this generalization. Robinson (1969) found that the radar reflectivity varied between bright and dark regions, suggesting that the bright "desert" regions were rougher. Binder (1969a), exactly reversing the conclusions of Sagan and Pollack (1968), claimed that the bright desert areas were high, that the dark "maria" were low or on slopes, that canali occurred in broad and deep valleys, and that the dark areas were probably of biological origin. More extensive IR CO_2 abundance mapping, both from the Mariner 6 and 7 IR spectrometer and from Earth-based IR studies (Herr et al., 1970; Belton and Hunten, 1971), led to the conclusion that the bright areas are generally high, whereas low areas can be dark, light, or mixed. The dark Syrtis Major region was found to have a large mean slope (Belton and Hunten, 1971). Moroz et al. (1971) found by the same technique that dark areas could be either low or high, whereas Inge and Baum (1973), Cutts et al. (1974), and Frey (1974) concluded that no correlation between albedos and topography could be found!

One of the most tantalizing results of these early flyby missions was the hint that the Tharsis region, not well observed at close range, might depart dramatically from the character of the regions seen most clearly (Leighton *et al.*, 1969c).

The thorough mapping of the Mars surface by the very successful Mariner 9 orbiter mission revealed a rich variety of previously unknown or poorly characterized phenomena. One of the most striking developments was the mapping of a long chain of extremely massive volcanoes, several of which were larger than the largest volcanic constructs on Earth. Further developments of crater-count dating, made possible by the very high resolution and nearly complete areal coverage of the Mariner 9 photography, permitted the determination of the relative ages of different parts of the surface, and, with certain assumptions about the calibration of the flux of asteroidal and cometary bodies at the orbit of Mars, absolute ages could be estimated. A smooth decay of the cratering flux and a constant degradation rate for the impact craters (McGill and Wise, 1972) seem to fit the data well, and a brief, severe past episode of crater degradation, as proposed by Jones (1974), does not seem necessary. The first analysis of the Mariner 9 crater density data suggested that the ancient heavily cratered terrain was about 3–4 billion years old, whereas the large volcanic constructs were only 300 million years old (Hartmann, 1973a). The most ancient areas were found to be fully as heavily cratered as the lunar highlands (Cordell *et al.*, 1974; Wilhelms, 1974). Degradation processes, initially largely due to the bombardment process itself, later became dominated by weathering and wind erosion (Arvidson, 1974; Chapman, 1974a,b; Soderblom *et al.*, 1974). Studies of the effects of degradation on crater morphology have been reported by Burt *et al.* (1976), Cintala *et al.* (1976) and Smith (1976).

From the time of the first tentative identification of volcanic features (Steinbacher *et al.*, 1972), the age and nature of the volcanic activity on Mars have been of great interest. Carr (1973) reported finding remnants of a number of very ancient large volcanoes lying along several lineaments. Several huge, fresh (nearly uncratered) volcanic piles are seen along the Tharsis ridge, the largest and youngest of which, Olympus Mons, has a volume of at least $8.5 \times 10^5 \text{ km}^3$, comparable to the total volume of all the volcanoes in the Hawaiian–Emperor seamount chain. The great height of the rim of the caldera, ~24 km (Davies, 1974; Whitehead, 1974), presents a severe challenge for sources of magma. Either a very low-density magma was extruded, or the base of the magma column must have extended to very great depths within the planet. The low viscosity required for such long surface flows suggests a basaltic, rather dense magma. Morphologically, the volcanic systems seen by Mariner 9 seem more analogous to intraplate volcanism on Earth than to activity above crustal subduction zones, and qualitatively suggest the presence of a very few extremely intense "hot spots" (following the terrestrial model described by Morgan, 1971). Carr (1974a) relates the Tharsis ridge to a vast, domelike expression of the Martian crust, covering over 20% of the surface area of the planet and centered somewhat to the southeast of the line of volcanoes, which seem to have formed at a later time than the bulge.

The Tharsis bulge is seen as both a topographic and gravitational excess (Jordan and Lorell, 1975). The youngest volcanic constructs are only partially compensated at depths on the order of 100 km (Phillips and Saunders, 1975), and may possibly be more fully compensated by low-density roots at greater depth. This tends to somewhat erode the reliance which can be placed on the uncompensated model for Tharsis used by Reasenberg (1977) in deriving a reduced moment of inertia of 0.365.

One interesting feature of the structure of Olympus Mons is the high scarp which surrounds it. King and Riehle (1974) interpret the scarp in terms of ash flow emplacement mechanisms which cause the deposited ash to become less dense and physically weaker when deposited (at lower temperatures) far from the vent. Alternatively, erosion of a soft older substrate by winds may be responsible for the scarp (Head et al., 1976). Pyroclastic activity and formation of ash flows is apparently a widespread phenomenon (West, 1974). Narrow lava channels seen on the slopes of Olympus Mons attest to the ability of the lava extruded from the summit to flow for distances of hundreds of kilometers. The flow behavior deduced from study of these features departs from that of the closest terrestrial analog, Hawaii, in that the Martian lava was apparently more silicic and its eruption rate was much larger (Hulme, 1976).

Neukum and Wise (1976) initially found ages of about 2.5 billion years for Olympus Mons based on the crater density, but Blasius (1976), Hartmann (1977), and Carr et al. (1977a) have found extensive areas near the summit of Olympus Mons and in the calderas of that volcano and of Arsia Mons, indicating an age on the order of 100 million years. Wise et al. (1979) conclude that Tharsis "must be regarded as at least a warm corpse." It is likely that active volcanism has spanned most of the history of Mars (Carr, 1973).

The hypsometric curve for Mars is, like that of Venus, unimodal (Neiman, 1975), with one enormous topographic high associated with the Tharsis volcanic complex and bulge, and two low areas, the great Coprates canyon system and the Hellas basin. The large majority of the surface of the planet has elevations which are distributed fairly symmetrically about the mean.

The simplicity and unimodal character of the hypsometric curve do not, however, imply a simple or uninteresting geology at intermediate elevations. Thorough high-resolution coverage of the planet by the imaging experiments on Mariner 9 led to the development of photometrically calibrated mapping (Batson, 1973; Carr et al., 1973; Wu et al., 1973), and thence to geological interpretation of planetary-scale geological processes (McCauley et al., 1972; Masursky, 1973). These results greatly extended the meager evidence for active tectonism from the Mariner 6 and 7 photography (Elston and Smith, 1973). Extremely high-resolution imaging by the Mars 5 orbiter provided improved knowledge of limited regions of the surface (Florenskii et al., 1975; Selivanov et al., 1975), and the best coverage to date was provided by the Viking orbiters. Because of the wealth of information from the Mariner 9 and Viking orbiter missions, much of the most recent data remains to be digested. Most cartographic products to date rely principally on Mariner 9 data. The reader interested in the equipment and processing techniques used in gathering imaging data should consult the description of the imaging methods employed on the Viking orbiters (Carr *et al.*, 1972). A broad discussion of the general results from the imaging experiment was given by Carr *et al.* (1976), and papers describing the colorimetric (Soderblom, 1976), photometric (Thorpe, 1977), and photogrammetric (Wu, 1979) aspects of imaging data interpretation are available.

Only a few of the detailed geological studies can be discussed here, and only those which have some evident relationship to the overall tectonic history of Mars or to the release, storage, or cycling of volatiles will be mentioned.

Pre-Viking studies of the origin of large craters and of their subsequent evolution led to the early suggestion that extensive isostatic adjustment after the Hellas impact had modified the terrain extensively (Thorman and Goles, 1972), more than in the comparable-sized craters on the moon (Wilhelms, 1973). The extreme flatness and featurelessness of the floor of the Hellas basin, which lies some 6 km below its western rim (Kliore *et al.*, 1972), appears to be due to such adjustments. Schultz and Glicken (1979) have documented a number of related degradational and igneous processes which are widely associated with large Martian craters. Extensive fracturing of crater floors, often with intrusive dike activity, attests to volcanic activity stimulated by impacts. Other internal modifications appear to be due to sapping of meltwater from mineral ice in the preimpact rock. Sinuous channels are often found associated with large impact craters, especially at low latitudes and in the Coprates region. External lobate flow structures (Figure 4.25) are also commonly found around craters in the Coprates province (Carr *et al.*, 1977b; Mouginis-Mark, 1978), suggesting flow of the fluidized substrate derived from shock heating of volatile-rich crustal materials (Mutch and Woronow, 1980).

An even more striking type of evidence for the existence of massive deposits of volatiles (presumably water ice) in the Martian crust is the prevalence of large areas of jumbled "chaotic" (Sharp, 1973b) and deeply depressed troughed terrain (Sharp, 1973a). The volume of ice which must have been withdrawn from beneath the surface in order to produce the observed depressions is roughly 1.6×10^6 km^3. The greatest puzzle attaches to the greatest valley of all, the Valles Marineris complex, of which Coprates is a major component. Geological interpretation of the valley (Blasius *et al.*, 1977) shows no plausible mechanism for excavating the vast amount of missing material. The enormous landslides along the walls of the valley appear to have had the character of mud flows, originating in the release of water from subsurface aquifers at the heads of the slide regions (Lucchitta, 1979). No such features are known on either the Moon or Mercury. A portion of the great canyon complex is shown in Figure 4.26.

By far the most exciting evidence for the presence of water is, however, the channeled terrain, which is laced with large dendritic valley systems (Masursky, 1974; Masursky *et al.*, 1977). Some of the sinuous, branched channels can be traced for 2000 km, and channel widths range up to tens of kilometers (Baker and Kochel, 1979). The only reasonable interpretation of the morphology of these channels seems to be that they were cut by catastrophic flooding by liquid water (Baker and Milton, 1974; Theilig and Greeley, 1979). The widespread presence of ground ice seems essential to provide the observed flows. There is little evidence for drainage of surface water or for precipitation on a large scale. The stream valleys are not mature, but seem to have been formed in one or perhaps a few major catastrophic episodes, perhaps lasting only a few days (Baker and Milton, 1974). These flooding episodes may have covered billions of years. It has been claimed that flooding may have persisted down to fairly recent times (Florenskii *et al.*, 1975), but the weight of the evidence is that the channels are very old (Malin, 1976). Some of the smaller Martian channels may have been formed by lava, but the most impres-

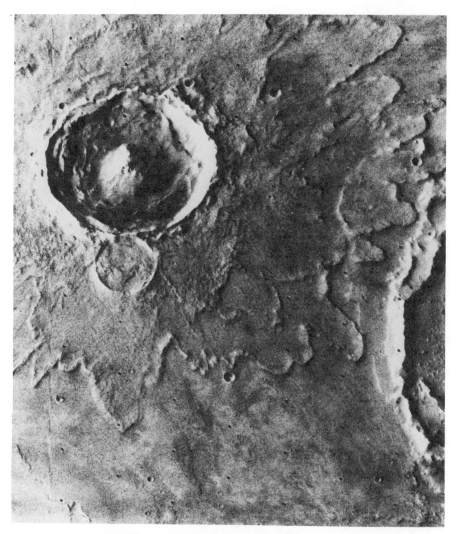

FIGURE 4.25. **Lobate "splashes" surrounding the crater Yuty. (Viking 1 image P-16848.) The rim-to-rim diameter of the crater is ~18 km, with thick, fluid ejecta deposits extending outward more than 20 km from the rim. Such crater morphology may be formed by impacts in a thick, ice-rich layer of sediments.**

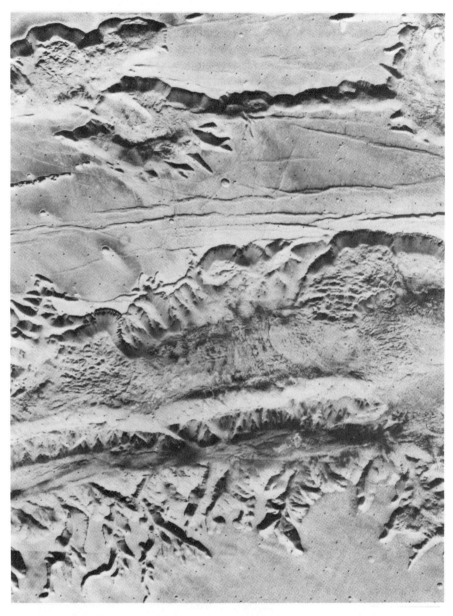

FIGURE 4.26. The Valles Marineris canyon, which spans over 3000 km, is marked by a central ridge, severely eroded gullies, and gigantic landslides. Terrain to the north of the canyon is broken by grabens which are of tensional origin. (Viking 1 image P-17872, courtesy of NASA.)

sive channels were not (Carr, 1974b). They were rather probably due to melting or evaporation of permafrost (Hartmann, 1974b). Some large channels are shown in Figure 4.27.

The antiquity of the channels is attested to by the fact that they are found in ancient cratered terrain (Pieri, 1976). Masursky *et al.* (1977) date the channels between 0.5 and 3.5 billion years old. Frey *et al.* (1979) have interpreted the numerous small (~1-km diameter) domes in the Cydonia region as due to volcanic eruptions through ice-saturated ground, analogous to similar features seen in Iceland.

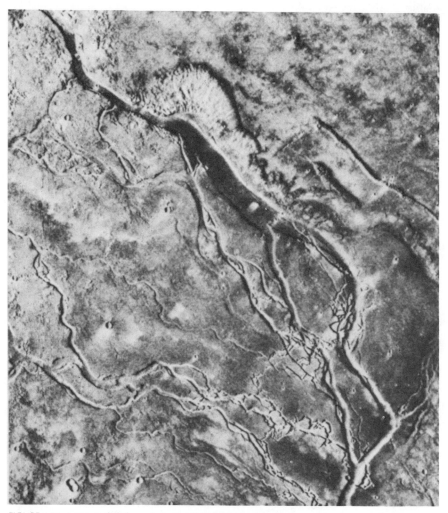

FIGURE 4.27. **Large eroded channels, such as this system near the Elysium Mons volcano, may originate from ground ice melted by volcanic activity. (Viking 1 image P-1867-004, courtesy of NASA.)**

The significance of these enormous surface expressions of the presence and mobility of water lies in the fact that present Mars surface conditions cannot support the presence of liquid water because the total atmospheric pressure is less than the vapor pressure of H_2O at the liquid–vapor–ice I triple point. Thus, the presence of large, open expanses of liquid water would require a much higher atmospheric pressure, surface temperature, and atmospheric IR opacity than now exists. It is of first importance to determine whether there has been a secular decline in the atmospheric mass, or whether periodic or random fluctuations occur by which the atmospheric mass is changed by a factor of several. We shall return to this question. In any event, the channels clearly attest to the presence in Mars of a reservoir of volatiles vastly larger than the atmospheric inventory, and show that more volatiles were exposed at the surface at times in the past than can presently be observed.

Another line of photogeological evidence for global-scale tectonic activity, possibly related to core formation or to changes in the orientation of the spin axis, is found in the widespread fracture systems in the crust. Binder and McCarthy (1972) identified sets of concentric and radial fractures associated with both the Hellas and south polar basins, as well as an apparent global-scale set of lineaments, in Mariner 4, 6, and 7 photography. Further evidence for the global lineament system was adduced by Schultz and Ingersol (1973), and Masson (1977) identified two different generations of lineaments in the Valles Marineris–Labyrinthus Noctis region. The older fractures trended along WSW–ENE directions, and are of tensional origin. The later, less well-defined, lineaments trend WNW–ESE and N–S, and appear to be associated with the opening of the Valles Marineris canyon system. Many of the largest channels may be initiated by tensional fracturing (Schum, 1974).

One of the ironic aspects of the discovery of the enormous chasms and apparently flood-cut valleys is that they remind us forcibly of the prespacecraft days when Mars was covered with canals. It is clear that there is essentially no correspondence between the features mapped by Lowell and those seen by Mariner 9 and the Viking orbiters (Sagan and Fox, 1975).

For useful synoptic reviews on Martian geological processes, we recommend Mutch and Saunders (1976), Wise et al. (1979), Arvidson et al. (1980), and Greeley and Spudis (1981). These papers serve as valuable guides to a literature too extensive (and in many cases too specialized) to be discussed in the present context.

These global investigations of the geology and tectonic evolution of Mars were given a wholly new dimension by the in situ surface observations of the Viking landers. The overall geological setting of the Viking 1 and 2 landing sites are described, respectively by Binder et al. (1977) and by Mutch et al. (1977). Viking 1 landed on volcanic terrain in the Chryse region. The terrain is broken and blocky as a consequence of both faulting and impacts, and is weathered by the formation of a ferric oxide stain and iron-bearing clays. A wide variety of rock types are suggested by the lander photography. The intrusion of dike complexes and the ballistic transport of impact ejecta from nearby craters may contribute to this diversity. Drifts of fine-grained dust are small but numerous (Mutch et al., 1976a; Shorthill et al., 1976b).

The Viking 2 landing site lies on the outskirts of an ejecta debris apron surrounding the crater Mie, in Utopia Planitia. The site is quite flat and rich in wind-transported sediments, and is covered moderately densely and uniformly with boulders with dimensions of about 1 m (Mutch et al., 1976b; Shorthill et al. 1976c). Most of the rocks seen are visibly vesiculated, although it is not clear whether these voids are caused by bubbles in volcanic extruded lava, or by solvent extraction of water-soluble salts from shock-lithified or compacted dirt, or even by differential mechanical and chemical weathering. Loosely cemented clods of dirt and thin sheets of apparent duricrust appear to be held together by a chemical sediment such as calcium sulfate.

The geological, geophysical, and geochemical experiments conducted by the Viking landers were the same at both landing sites. The first experiment is, of course, that implicit in the landing site descriptions given above: imaging. The instrumentation and mission plans for this experiment were described by Mutch et al. (1972), and the interpretive and operational procedures were described in detail in the proceedings of the first Mars conference (Huck et al., 1977; Levinthal et al., 1977a,b; Liebes and Schwartz, 1977; Mutch, 1977; Patterson et al., 1977).

The physical properties of the Martian surface were studied via the dynamics of the landing legs upon impact with the surface, the penetration of objects dropped by the lander, and the mechanical behavior of the surface sampling arm (Shorthill et al., 1972, 1976a; Moore et al., 1977). Magnets attached to the sampling arm head were used to search for and detect magnetic materials (Hargraves and Peterson, 1972; Hargraves et al., 1977, 1979).

One of the most important instruments on the Viking landers was an X-ray fluorescence (XRF) spectrometer for analysis of the surface and atmosphere (Clark and Baird, 1973; Toulmin et al., 1973). Preliminary results reported for the Viking 1 landing site (Toulmin et al., 1976) revealed that the soil contains a high amount of iron, with calcium, aluminum, silicon, and sulfur also easily discerned. Only a minor amount of titanium was detected, and a very high Ca:K ratio was indicated. Early data from the Viking 2 lander showed a distinct resemblance (Clark et al., 1976b): The sulfur concentration at both sites ranged from 10 to 100 times higher than in terrestrial crustal rocks, and the potassium abundance was clearly several times lower than would be expected by comparison to Earth's crust. Suspected samples of duricrust and cemented clods exhibited an even higher sulfur content, in keeping with the idea of cementing by a sulfate. Other general features of the analysis (Baird et al., 1976) suggest that the soil was produced by weathering of a mafic igneous rock. An approximate modal interpretation of the analyses gave about 80% iron-rich clays, 10% magnesium sulfate, and about 5% each of calcite and higher oxides of iron, such as hematite, goethite, or maghemite. The instrument, data interpretation techniques, and early results are summarized by Clark et al. (1977) and Baird et al. (1977). Interpretation of the XRF results is presented in detail by Toulmin et al. (1977) and by Clark and Baird (1979b).

Toulmin et al. (1977) emphasize that the Martian surface samples all exhibit low trace element, alkali, and alumina contents, and that no evidence for the presence of acidic rock types can be adduced. Clearly the weathering products are easily derived from mafic rocks, but not from diorites or granites. The soil compositions are all dominated by iron-rich smectite clays, carbonates, sulfates, and Fe_2O_3, as was found in the preliminary analyses mentioned above. Toulmin *et al.* (1977) hypothesize that the clays were formed by subsurface interaction of a mafic (basaltic?) magma with ground ice. The enrichment of sulfur in the loosely cemented duricrust and in clods suggests an upward transport of salts by migration of thin, highly concentrated brine films, so that soluble species leached out of the weathering basalt are transported upward and deposited in a thin, desiccated, surface layer.

One curious result of these analyses is that the surface samples closely resemble the Cl chondrites in elemental composition. The main expected difference, the water content, cannot be determined by the Viking XRF experiment. Further interpretations of the chemistry of the Martian regolith will be discussed in Section 4.4.2.

We come finally to the only geophysical experiment carried by the Viking landers—the seismometer (D. L. Anderson *et al.*, 1972).

The first 60 days of experience on the surface of Mars (Anderson *et al.*, 1976) did not detect a single seismic event. Furthermore, the Viking 1 seismometer was not successfully deployed, so that data were returned from only the Viking 2 landing site in Utopia Planitia. Over the lifetime of this experiment, many noise events were recorded which correlated with wind gusts, and only a single event which might be attributed to a Mars quake was recorded (Goins and Lazarewicz, 1979). It is likely that Mars exhibits much less seismic activity than Earth, and that success in studying the seismic activity and internal structure of the planet will have to await the deployment of much more sensitive instruments directly on the surface, and not attached to a large, complex, and active lander which can be easily agitated by the winds (Anderson *et al.*, 1977; Tittmann, 1979).

A very interesting and tantalizing piece of information regarding the thermal state and interal activity of Mars is the reported detection of a small but significant planetary magnetic field (Dolginov *et al.*, 1973; Marov and Petrov, 1973). The magnetometers on the Mars 2 and 3 landers reported maximum field strengths of about 60 γ (1 γ = 10^{-5} Γ) very close to the surface. If this field, which exceeds the nearby solar wind field in strength by nearly a factor of 10, is in fact due to an internal dipole field and not to some odd ionospheric currents or bow-shock interaction, then a planetary magnetic moment of about 2.5 × 10^{22} Γ cm^3 is indicated. The Mars 5 spacecraft confirmed the Mars 3 magnetometer measurements (Dolginov *et al.*, 1975). A model for a Martian dipole field has been presented by Dolginov *et al.* (1976), who postulate a 64-γ equatorial surface field with an axial tilt of only 15–20° and no offset of the dipole from the center of the planet.

Russell (1978a,b) has reexamined the Mars 3 and 5 data, and has suggested that field strengths as low as 5 γ might be acceptable. Intriligator and Smith (1979) have considered the standoff of the solar wind from Mars by both the ionospheric backpressure and the magnetic pressure, and conclude that the atmospheric contribution is insufficient by a factor of several to stop the solar wind at the observed altitude. They find that a Mars dipole field of about 20 γ is necessary. As previously pointed out by Johnstone (1978), Mars is a borderline case in which both Venuslike (ionospheric) and Earthlike (magnetospheric) interaction are important.

If there is indeed a present-day dipole field, then it is very likely that there is active convection in the core of Mars. This observation, then, places a potentially extremely important constraint on thermal models of the interior of the planet. It would require that, contrary to theoretical expectations (Young and Schubert, 1974), the core of Mars is liquid. It is particularly frustrating that the only spacecraft mission to date which might have measured directly the surface field strength, the Viking landers, did not carry magnetometers. Indeed, the only geochemical and geophysical experiments aboard the landers were added as an afterthought, in response to the criticism of the scientific community that the scope of the mission was too narrow, and that there was a substantial likelihood that life would not be detected, thereby leaving the Viking mission with no useful results whatsoever. It is clearly a great mistake to send a highly complex, expensive, and specialized space-craft into a hostile environment which has never been studied in a general way.

One further method of investigating the internal structure and mass distribution in a planet is by the mapping of its external gravitational field. This has been done by tracking the Mariner 9 and Viking 1 and 2 orbiters (see Lorell et al., 1972, 1973). We have already mentioned some of these results, such as the finding that the Tharsis region is characterized by a mass excess, with a thick and partially compensated crust (Phillips et al., 1973). A stress analysis of the interior based on the gravity field data has been done by Arkani-Hamed (1975). Tectonic scenarios have been generated to account for the present gravity field and surface morphol-ogy, such as the interpretation of the Valles Marineris–Coprates trough system as a tensional feature caused by polar wandering (McAdoo and Burns, 1975). Perhaps the most enthusiastic assessment of crustal mobility is that of Courtillot et al. (1975), who claimed that they had identified transform faults and even a possible triple plate junction. In light of their claims to have found a number of areas of crustal extension, the absence of any evidence for convergence is most striking. It is most reasonable to conclude that continental drift never really got started on Mars (Hartmann, 1973b).

We are still far from a consensus on all major aspects of the evolution of Mars. Nonetheless, a number of features seem to have achieved a rather wide acceptance. Mars is generally agreed to have accreted at about the same time as the other terrestrial planets, over a time scale of 100 to several hundred million years. The raw material out of which Mars accreted undoubtedly originated over a significant range of heliocentric distances (10% or more from at least 0.3 AU away), and therefore should have sampled a range of oxidation states and volatile-element contents. Only Ringwood seems to hold that the primordial oxidation state and volatile content of Earth and Mars were the same. Only Anders and co-workers seem convinced that the volatile-element inventory in Mars at the time of forma-tion was much smaller than in Earth. In order to refute Anders's arguments, however, it would be necessary to prove that the mass of volatiles lost (or *nonselec-*

tively buried) by Mars since formation is much greater than the presently visible inventory. Note that Anders assumes no massive volatile loss from little Mars, whereas Ringwood assumes massive volatile loss from the much more massive Earth.

The increase in oxidation state of iron with increasing distance from the Sun is nearly universally accepted. The evidence from the study of chondritic meteorites closely associates increasing oxidation of iron with increased volatile content. Then where is the water in Mars today? How much can be hidden in the polar regions, in permafrost, in clay minerals, and in groundwater? How much hydrogen might have escaped? It is clearly very important to assess the nature and magnitude of the chemical and physical interactions between the atmosphere and crust in order to attack these crucial questions.

The time scale for the beginning of melting and differentiation, and the time required for core formation and outgassing, are only poorly known, but the warm-up period almost certainly did not last longer than a billion years. Once an extensive region of the interior has differentiated, then the undifferentiated outer layers, which are no longer buoyant, are likely to founder both quickly and thoroughly, leaving the planet stably density-stratified. The most ancient cratered terrain on Mars today probably represents the primitive basaltic crust formed at this era. There is still no clear evidence for acidic rocks, and extensive reprocessing of Mars material may not have occurred. Lacking data on the regional variation of rock compositions and on the ages of rock units, it is most unlikely that any progress will be made in understanding these issues. Again, geochemical evolution may be related to, and obscured by, interactions between the atmosphere and the lithosphere.

For these reasons, we turn next to an analysis of atmosphere–surface interactions.

4.4.2. Atmosphere–Surface Interactions

One of the firmest conclusions arising from recent research on Mars has been the discovery that chemical and physical interactions between volatiles and crustal rocks have been nearly as important a factor there as on Earth. In the last few years there has been a great proliferation of ideas and evidence regarding polar caps, glaciation, permafrost, cratering in frozen mud, liquid water and eutectic brines, H_2O and CO_2 adsorption, chemical weathering, and wind transport of dust. Several of the underlying themes of this literature concern the question of what the actual volatile-element inventories on Mars are, whether they have demonstrably changed over geological time, and what the chemical and mineralogical composition of primary crustal igneous rocks might be. The latter issue, of course, bears directly on the degree of geochemical differentiation of the lithosphere—an issue which we have already addressed. For the present, we will consider it likely that the extrusive rocks on Mars are dominantly basaltic and concern ourselves with the ways in which such a magma, once extruded into the upper crust or onto the surface, can interact with the atmosphere and with ice.

The oldest evidence for the presence of massive amounts of nonatmospheric volatiles on Mars is the great body of telescopic observations of the polar caps. The caps can easily be seen to wax and wane with the local seasons. Because of the high orbital eccentricity of Mars (0.0933), the south polar insolation at the northern winter solstice and the north polar insolation at the local summer solstice are very different, and, due to the precession of the rotation axis, this hemispherical asymmetry changes periodically. Even longer-period temperature fluctuations are driven by cyclic changes in the orbital eccentricity and axial tilt (Ward, 1974). Presently perihelion occurs just before the southern summer solstice, so that southern polar summers are both more intense and briefer than summers at the north pole. The south polar cap sometimes disappears completely, but some small central portion of the north polar cap always survives summer. It is natural to wonder whether the caps are made of H_2O, which is easily condensed but rare in Mars's atmosphere, or CO_2, which is very abundant but harder to condense.

The Mariner 9 and Viking orbiter missions have provided a vast body of data on the temperature, altitude, atmospheric composition, reflection spectrum, areal extent, and total mass of the polar caps, as well as providing excellent photographic maps of the south polar regions (de Vaucouleurs et al., 1973). In addition, radio occultation measurements have helped map out altitude variations in the polar regions (Kliore et al., 1973).

The case for very CO_2-rich polar caps was first made by Leighton and Murray (1966). Mariner 9 IR spectra (Herr and Pimentel, 1969) initially seemed to show weak bands due to strange species such as NH_3 and CH_4, but upon close analysis these were found to be intrinsically weak features in the spectrum of solid CO_2. Larson and Fink (1972) identified a number of weak features in the polar cap spectrum as being due to solid CO_2. Ingersoll (1974) argued against the likelihood that the observed CO_2 pressure in the atmosphere could be maintained if the permanent caps were solid CO_2, but Kieffer (1979) has shown convincingly from Viking IR data that the *south* polar cap does in fact remain at the saturation temperature of CO_2 all year. This is in sharp contrast to the northern cap, which clearly does not keep CO_2 over the summer.

The polar temperatures are of crucial importance in determining what species may condense at the poles. Morrison et al. (1969) and Neugebauer, et al. (1969) estimated a polar temperature of about 145–150 K, which is essentially the saturation temperature of CO_2, and concluded that at least some dry ice should be present at the pole. Sharp et al. (1971b) showed that the condensation process is horizontally inhomogeneous, with condensation occurring preferentially in low areas. Cross (1971), in a theoretical study of the thermal physics of the polar regions, predicted that the hemispherical seasonal inequality would cause the southern cap to be not only larger, but also thicker than the northern cap. He further predicted that the total atmospheric pressure would vary by almost a factor of two in response to the seasonal deposition cycles of CO_2 frost, with the pressure a minimum at the northern autumnal equinox. Aleshin (1971) presents a somewhat similar analysis in which he predicts that as much as 17 gm cm^{-2} of condensate (over 80% of which is

CO_2) could accumulate at the north pole in a single winter. From Mariner 6 and 7 IR radiometer data, Neugebauer *et al.* (1971) find a south polar cap temperature of 148 K, in good agreement with the earlier estimate by Morrison *et al.* and again consistent with CO_2 condensation. This was confirmed by the IR radiometer experiment on the Viking orbiters (Kieffer *et al.*, 1976a). Kieffer *et al.* suggest that the very low polar night temperatures observed require either a high-altitude (20 km) CO_2 cloud layer or a large enrichment of noncondensible gases over the winter cap.

Spectroscopic evidence on the composition of the caps, based on laboratory data on the spectra H_2O CO_2, and mixed frosts (Kieffer, 1970a,b) and theoretical thermal models (Aleshin and Shchuko, 1972), agree that the caps must contain considerable H_2O, and the surviving north summer cap is essentially dirty water ice, free of solid CO_2 (Kieffer *et al.*, 1976b; Farmer *et al.*, 1976b). The transport of water vapor to and from the polar regions is easy to observe spectroscopically, since the H_2O partial pressure varies enormously with latitude and very high water vapor abundances are seen over the northern cap during local spring (Conrath *et al.*, 1973).

There are other condensates in addition to solid CO_2 and solid H_2O ice I which may be important: There is a well-characterized solid clathrate hydrate of CO_2 (Miller and Smythe, 1970) and a possible metastable phase of $H_2O \cdot CO_2$ stoichiometry (Harrison *et al.*, 1968) which may be relevant to Mars. Milton (1974) suggests that ground ice on Mars should largely be found as the clathrate $CO_2 \cdot 6H_2O$, and that dissociation of this hydrate could lead to formation of melt water, a phenomenon which would be useful in explaining the origin of the chaotic terrain. Dobrovolskis and Ingersoll (1975) point out that the summer polar temperatures are generally too high for solid CO_2 persistence, although the clathrate hydrate might persist and even buffer the partial pressure of CO_2. In fact, by the Gibbs phase rule, coexisting H_2O ice and $CO_2 \cdot 6H_2O$ can buffer *both* the CO_2 and H_2O pressures.

Another atmospheric component which interacts strongly with the polar caps is ozone, which is readily adsorbed and trapped by CO_2 frosts in the laboratory (Broida *et al.*, 1970). About 10^{-3} cm agt of O_3 is seen in the vicinity of the polar cap, with a maximum of 5.7×10^{-3} cm agt over the polar hood in local autumn (Barth and Hord, 1971; Barth *et al.*, 1973). Ozone formation is correlated with low temperatures, and hence with the formation of the clouds which constitute the polar hood (Barth and Dick, 1974).

There are very clear and widespread geomorphological features associated with the polar regions which suggest important interactions between the volatile caps and the crust. One of the most striking of these is the very subdued appearance of topography and albedo markings at high latitudes, which on Mars means any latitude above $\pm 30°$. Much of this region, which occupies half the surface area of the planet, has been buried under a deep mantle of condensates and dust. Close to the poles there are remarkable layered deposits (Figure 4.28) called laminated terrain, exhibiting alternating layers of dark and light materials, partly exposed and deflated by wind erosion, with total thicknesses of kilometers (Murray *et al.*, 1972; Cutts, 1973a). The much more extensive mantling deposits which extend down to

FIGURE 4.28. **Laminated terrain in the north polar region. The alternating layers of dark and light (dust-rich and ice-rich?) materials average about 20 m in thickness. A broad cross section of deposits is here exposed by wind erosion. (Viking image M2111-018, courtesy of NASA.)**

30° latitude may be largely derived from aeolian debris carried equatorward from the eroding laminated deposits (Soderblom *et al.*, 1973a,b) or from volcanic ash and volatiles (Sharp, 1973c). Topographic features are generally aligned with, and probably dominated by, the prevailing winds (Cutts, 1973a). The time required to accumulate the layered deposits is at least 500 million years, and the deposition rate is closely similar to the rate of deposition of H_2O ice. Thus, ice may be a major component of the laminae (Cutts, 1973b). The pressure is too low for the layered deposits to be rich in CO_2 (Dzurisin and Blasius, 1975), and the core of these deposits is seen to be covered with a layer of persistent H_2O ice (Cutts *et al.*, 1976). The exposed edges of the polar laminated terrain are girdled by a ring of dunes of surprisingly dark color (Cutts *et al.*, 1976), and the entire polar region is devoid of fresh-looking impact craters.

Numerous escarpments have been eroded into the polar laminae, apparently in one or a few brief episodes (Nash, 1974). Wind erosion is probably adequate for producing these features, but Sagan (1973b) has suggested that buried CO_2, heated by the internal planetary heat flux and compressed to pressures in excess of 5 kbar by overburden weight, may liquefy and permit very easy flow. Kane *et al.* (1973) have suggested that some of the valleys may have been cut by Alpine-type glaciation of water ice, and Clark and Mullin (1976) have presented the case for glaciation by solid CO_2 sheets in the polar regions as an alternative to aeolian erosion and redeposition of dust. Some of the channels at lower latitudes have a vague resemblance to glacier-cut valleys, and Belcher *et al.* (1971) have pointed out ridges which resemble glacial moraines. It is likely that glaciation as an agent of large-scale topographic change will continue to be investigated.

The widespread presence of permafrost was suggested by Leighton and Murray (1966) and by Smoluchowski (1967a), who associated such deposits with the dark regions. The seasonal color changes would then be due to water vapor from the surface, liberated and migrating upward in response to the penetration of the thermal wave in the spring, hydrating and darkening the surface. Wade and de Wys (1968) defended the likelihood of widespread permafrost, and Carr and Schaber (1977) have gleaned a large number of apparent permafrost-related features from the Voyager orbiter mapping photographs. A preponderance of these features is associated with the old cratered terrain, and some of them have scale sizes so large that they suggest ice-saturated ground many kilometers deep.

Mouginis-Mark (1979) examined the morphology of over 1500 fresh Martian craters, and he finds essentially no variation of the crater morphology with latitude. This suggests that fluidized craters, due to impact in volatile-rich target material, have nearly global distribution, and hence that permafrost or other volatile-rich mineral phases are likewise global in extent (Figure 4.29).

There are numerous volcanic features of unusual and distinctive morphology on Mars which bear a striking similarity to the products of volcanic eruption through thick ice sheets on Earth (Allen, 1979). Ice thicknesses of 0.1–1.2 km can be inferred from the Martian features. Hodges and Moore (1979) present a similar analysis of the origin of Olympus Mons, in which eruption through a layer of ice several kilometers thick is proposed.

Coradini and Flamini (1979) have shown how the annual thermal wave penetrates the Martian surface layers, and they demonstrate that an active permafrost zone should exist to about a depth of 100 m. Solid CO_2 or solid clathrate hydrate permafrost should be unstable down to a depth of at least a few hundred meters.

The past and present stability of liquid water on the surface of Mars has also been widely discussed. Frolov (1966) suggested liquid water as a possible surface agent, and Davydov (1971) ascribed the featureless filling of the Hellas basin to a local

FIGURE 4.29. **Severe mass-wasting and glacier-like flow have reshaped the landscape in the Nilosyrtis region. The freeze–thaw cycle in terrestrial permafrost regions produces similar features. (Viking 1 image P-18086, courtesy of NASA.)**

source of large amounts to liquid water, which caused a local episode of flooding, with no important source of atmospheric precipitation. He further suggested that large amounts of liquid water might be present today beneath a thick permafrost layer. Sagan and Veverka (1971) found that the decrease of the radio frequency brightness temperature toward shorter wavelengths, the opposite of the behavior expected for dry soil, could be modeled as due to a very thin ($\sim$30 μm) layer of liquid H_2O in the topmost few millimeters of the Martian regolith.

Ingersoll (1970), in a careful study of the stability of liquid H_2O under Martian conditions, argued that the very low atmospheric pressure, below even the vapor pressure of pure water at the ice I–liquid–vapor triple point, prevented pure liquid H_2O from being present. Warming of ice under Martian surface conditions would cause it to evaporate at a temperature well below 273 K, the normal melting point. At very low points on the surface, such as the bottom of Hellas, where atmospheric pressure might reach the triple-point vapor pressure of H_2O, very moist air pockets hugging the ground would be very unstable against convection, since the molecular weight of H_2O is less than half that of CO_2. "Bubbles" of moist air would rapidly rise, thereby reducing the surface partial pressure of H_2O far below the triple point. Ingersoll concluded that, for any liquid water to be present on Mars today, it would have to have a very high solute content, and hence both vapor-pressure depression and freezing-point depression relative to pure water. Thus, saturated brines, such as NaCl, Na_2SO_4, or $CaCl_2$ solutions, might be stable either exposed at the surface in the daytime or buried at a modest depth. A very fine-grained surface dust layer can also assist in stabilizing liquid H_2O or brines near the surface (Farmer, 1976). The dust acts as a barrier to water vapor loss by inhibiting diffusion of water vapor away from ice or brine, and should make it possible for liquid water to exist at depths on the order of a meter at temperate latitudes in local summer.

Malin (1974) has proposed that salt weathering, such as is important in the relatively Marslike dry Antarctic valleys, might also be active on Mars. Brass (1980) has examined carefully the question of brine stability on Mars, and we must conclude that both surface and subsurface brines may be widely distributed. Recent radar reflectivity measurements of Mars by Zisk and Mouginis-Mark (1980) show strong absorption in the Noachis–Hellespontis region and NW of Hellas. They find the easiest interpretation of their data to be that there is liquid H_2O or aqueous solutions close to the surface (50–100 cm) in that region. These regions are among those previously designated as "oases" by Huguenin and Clifford (1979).

We have earlier mentioned the existence of apparent stream-cut channels, which attest to the former stability of vast volumes of some liquid on the surface. We do not, of course, know whether the water, if it was indeed water, was exposed directly to the atmosphere, or whether the water flow took place beneath a thin ice crust. We have no evidence that any of the streams persisted for an appreciable amount of time. Rather, the stream morphology suggests that they were excavated in a few catastrophic, widely spaced episodes. Much evidence exists that the source of the temporary floods was groundwater. Milton (1973) describes the sequence from groundwater sapping to collapse and retreat of headwalls, stream bed cutting by the runoff, and formation of bars and braided channels. Sharp (1973a) gives critiques of the various suggested modes of origin of the great troughs in the Coprates region, whether by running water, salt dissolution, wind deflation, or deterioration of ground ice. He also (Sharp, 1973b) attributes the fretted and chaotic terrains to evaporation of ground ice or liquid water emergence, or both. These and the other treatments of riverlike features lead us to the conclusion that water is stored in some fashion in the ground over essentially all of the surface of Mars. It is clearly of interest to find out what other volatiles may be retained in quantity in the regolith and, if possible, to make some estimates of the total mass of outgassed volatiles. Until such an inventory can be constructed, it is obviously extremely rash to make sweeping generalizations about how thoroughly Mars has been outgassed, how much of a particular element may have been lost from the planet, or how the volatile-element composition may have resembled or differed from particular classes of meteorites.

Volatiles may be hidden from our view in the present atmosphere of Mars by any of several processes: incomplete outgassing, escape, condensation, adsorption, and chemical reaction with the regolith are the most important possiblities.

Prespacecraft spectroscopic studies had, as we have seen, already shown the presence of limonitic dust, which evidently betrays the action of oxidation and hydration of primary igneous minerals. Beer *et al.* (1971b) found evidence for a global distribution of minerals bearing water of hydration from their Earth-based interferometric IR spectroscopy. The Mariner 9 IR spectrometer experiment (Pimentel *et al.*, 1974) verified from Mars orbit that ice or minerals with water of hydration were universally present. The exchange of water vapor between atmosphere and permafrost or adsorbed water in the regolith has been modeled by Leovy (1973) and by Flasar and Goody (1976). The discovery of a major clay component in the dirt sampled by the Viking landers, as inferred from the elemental analyses carried out by the XRF experiment, served to connect these spectroscopic observations to a particular mineral species, and hence to a particular suite of geochemical weathering processes. Clark (1978) has given a useful discussion of the significance of the ubiquity of hydroxyl silicates and hydrated salts, including the role of smectite clays in suppressing ice formation by adsorbing water. He also shows how geothermal heat can distill bound water out of clays and hydrated salts to form ice lenses, which form buoyant domes and which may eventually break through to the surface.

Adsorption of atmospheric gases in the regolith was first suggested by Davis (1969), and Fanale and Cannon (1971) considered the simultaneous adsorption of H_2O and CO_2. They concluded that the atmospheric $CO_2:H_2O$ ratio was probably controlled by regolith adsorption with an overall $CO_2:H_2O$ ratio equal to that in Rubey's terrestrial inventory of "excess volatiles." The amount of water adsorbed in even a basalt regolith of only 10-m thickness would suffice, if released into the atmosphere by an upward temperature excursion, to stabilize liquid H_2O on the surface (Fanale and Cannon, 1974). Fanale (1976) proposed that the actual amount of volatiles released from the interior of Mars over its history was on the order of 100 times the present atmospheric and polar cap inventory. He proposed that the weathered oxide–clay regolith could easily be 2 m thick, in keeping with the evidence from cratering and chasm morphology studies, and could contain nearly the entire inventories of H_2O, CO_2, and Cl, S, and N compounds necessary to give Mars the same overall volatile-element composition as the Earth. Fanale *et al.* (1978) also considered the adsorption of rare gases on a clay-rich regolith and suggested that most of the Xe should be adsorbed at typical Mars surface temperatures. These factors will become important in Section 4.4.3, where we shall discuss the composition of the atmosphere.

The probability that the atmospheric pressure would vary seasonally was discussed prior to the Viking mission (Aleshin, 1973; Gierasch and Toon, 1973) and confirmed first by Mariner 9 (Woiceshyn, 1974) radio occultation measurements. The seasonal variation in CO_2 pressure as regulated by adsorption and desorption of CO_2 in the upper ~10 m of the regolith was studied by Dzurisin and Ingersoll (1975), and the climatological significance of astronomically driven long-term CO_2 variations was discussed by Ward *et al.* (1974). The latter find that the mass of CO_2 bound in the caps is a maximum at times when the axial obliquity is a maximum.

The exchange of CO_2 between atmosphere, caps, and regolith was modeled in some detail by Fanale and Cannon (1978), who also carried out new experiments on the adsorption of CO_2 on rock powder at Martian temperatures and CO_2 pressures. They show that the atmosphere–caps system is essentially buffered by a large mass of CO_2 adsorbed in the regolith. This buffer has sufficient capacity so that, if the atmosphere and caps were discarded, the regolith would rather quickly reconstitute them at their present mass. They further show that the mass of exchangeable CO_2 is so large that, unlike N_2, atmospheric escape of light isotopes would have no detectable effect on the isotopic composition of the oxygen in atmospheric CO_2. They attribute the laminated terrain to cyclic changes in the mass of the atmosphere and to the precipitation rate at high latitudes, and suggest that, although excursions of the surface CO_2 pressure up to tens of millibars are common, maximum pressures of a fraction of a bar may well have existed in the past. Based on the Viking XRF evidence for large clay abundances in the regolith, Fanale and Cannon (1979) updated their model and carried out new supporting experiments on CO_2 adsorption on nontronite clay. These new data make it virtually certain that an "ocean" of adsorbed CO_2 is indeed present adsorbed on clay mineral grains, especially at high latitudes. Periodic changes in the axial inclination of the planet, as proposed by Ward (1974), could cool the polar regions enough so that regolith buffering could lower the total CO_2 pressure to less than 1 mbar. This effect guarantees *at least* a factor of 10 variation in surface pressure even without considering the very large store of adsorbed CO_2 which could be released by *upward* temperature excursions.

Reactive atmospheric gases, which demonstrably are present adsorbed in the regolith, are in some cases clearly capable of undergoing spontaneous and irreversible chemical reactions with primary igneous minerals. For example, if we consider a reasonable range of mafic rock types, primary minerals such as olivine, pyroxene, and plagioclase must be common. We have seen that there is very strong evidence that the abundance of iron in the crust and mantle of Mars is high by terrestrial standards. It is easy to imagine a number of weathering reactions by which CO_2, H_2O, and small amounts of photochemically produced O_2, O_3, and O can react with ferrous silicates or with magnetite. Likely products include siderite ($FeCO_3$), goethite ($FeOOH$), hematite (Fe_2O_3), and related phases. We must recognize at least two major weathering environments which may be of widespread importance on Mars and may produce widely different products. The first of these is surface exposure, where the atmosphere, lithosphere, and solar UV light all coexist. The second is the site of igneous intrusion into volatile-rich terrain, especially aeolian sediments containing ice.

An experimental simulation of the surface conditions of Mars was carried out by Huguenin (1973a). He subjected magnetite and rock powders containing ferrous silicates to atmospheres containing known concentrations of CO_2, CO, N_2, Ar, and H_2O in the presence of an artificial Mars-surface solar UV spectrum. There was no detectable evidence of reaction between any of these photostimulated gas mixtures and ferrous silicates. On the other hand, magnetite grains rapidly and reproducibly altered to form hematite, possibly with some ancillary formation of maghemite (γ-Fe_2O_3). The dependence of the rate of this reaction on the abundances of the various gases and on the incident UV spectrum was also studied (Huguenin, 1973b). Initiation of the process apparently always depends on the presence of adsorbed H_2O. Desorption of H_2O, subsequent adsorption and dissociation of O_2, and photoionization of the surface by UV photons of wavelength less than 275 nm then follow. The resultant chemisorbed O^{2-} ions then coordinate with Fe^{3+} ions to form a very thin layer of hematite on the surface of each magnetite grain. Since the unit cells of magnetite and hematite are not commensurable, thin hematite scales detach themselves readily from the surface, exposing more magnetite for further reaction. In the presence of a moist atmosphere, goethite and clays will also form. Over a period of a billion years the accumulated weathering products would be sufficient for a layer of clays and oxides at least 1 m, and possibly as much as 1000 m, thick. Clearly, wind erosion is necessary as a means of cleaning igneous rock surfaces and maintaining their contact with UV light and with the atmosphere (Huguenin, 1974). Oxidation may occur on such a vast scale that the hematite formed would require the use of all the oxygen in a comparable mass of water. For a 1-km regolith containing 5 wt. % hematite, the equivalent of nearly 40 m of liquid water would have to be destroyed. Thus, oxidation by this mechanism is possibly of great importance in depleting the planetary inventory of H_2O (Huguenin, 1975, 1976).

The photostimulated oxidation mechanism has been criticized by Morris and Lauer (1980), who suggest that the process observed by Huguenin was in fact due to the elevation of the surface temperature of the magnetite samples. It is then not obvious why the rate versus UV flux curves reported by Huguenin should be smooth and linear, with rate proportional to flux $\phi^{3/2}$. If this were interpreted as a thermal effect, then the temperature $T \propto \phi^{1/4}$, and the rate would then be proportional to T^6. This would be a most peculiar kinetic behavior. Further, Schaefer (1981) has repeated the same experiment under conditions in which the temperature rise of the sample is negligible, and has found that photochemical oxidation nonetheless occurs. The mechanisms of oxidation of minerals under Mars surface conditions remain poorly understood, but their importance in governing the surface mineralogy and in constructing planetary abundance inventories of volatiles is almost certainly great.

It is not yet clear whether surface photostimulated reactions can produce significant amounts of carbonates and goethite. The most promising alternative to such an origin would be hydrothermal alteration of primary (basaltic?) magma with a regolith rich in adsorbed CO_2 and H_2O or solid ice and clathrate hydrate. It is not known whether there are abundant carbonates in the regolith.

Because of the efficient transport of dust and wind erosion attendant on major dust storms (McCauley, 1973; Sagan, 1973a), the regolith is quite mobile, and even deeply buried bedrock may be intermittently exposed by sediment deflation. Sagan's lower-limit estimate of the abrasion rate for exposed surface rocks provides a surface removal rate of 3 km per 10^6 yr. Although it is obvious from the several billion-year-old cratered terrain, with its vertical relief of only a few kilometers, that the actual transport is far less efficient than this, it is still reasonable that the total mass of material exposed to surface weathering over the last 4 billion years could be in excess of 1 km. There is no reason to doubt the importance of wind transport of dust as a major factor in the geology of Mars (Cutts and Smith, 1973).

4.4.3. Atmospheric Structure and Composition

The thermal structure of the Martian atmosphere was investigated at four points by the Mariner 6 and 7 radio occultations (Kliore *et al.*, 1969, 1970), and Earth-based high-resolution IR spectroscopy of the day side provided mean rotational temperatures over the day side, averaged in a complex way over altitude, latitude, and time of day (Young, 1969b; Verdet *et al.*, 1972). Earth-based radio astronomical brightness temperature measurements (Counselman, 1973), and remote sensing and *in situ* observations by the Mars 2, 4, 5, and 6 spacecraft (Avduevskii *et al.*, 1975; Kolosov *et al.*, 1975; Ksanfomaliti and Moroz, 1975; Kerzhanovich, 1977) were also available. Our pre-Viking knowledge of the surface temperatures and vertical temperature profiles in the atmosphere of Mars was reviewed by McElroy (1969a), Ingersoll and Leovy (1971), and Moroz (1976b).

During the Viking missions, thermal IR mapping of the surface (Kieffer, 1976; Kieffer *et al.*, 1976c) by the Viking orbiters was combined with radio occultation data (Fjeldbo *et al.*, 1977), probe entry deceleration measurements (Seiff and Kirk, 1976, 1977), and *in situ* surface temperature measurements by the lander meteorology experiment package (Hess *et al.*, 1976a,b,c, 1977) to permit synoptic study of the thermal structure and motions of the atmosphere. The radiometry results are discussed in the series of papers by Kieffer *et al.* (1976a,b,c, 1977) and by Zimbelman and Kieffer (1979). A range of surface brightness temperatures from as low as 130 K to as high as 290 K was observed. It appears that there is no simple argument by which the very low winter polar temperatures can be explained away: Not low emissivity, nor high-altitude H_2O clouds, nor high altitude can resolve the problem posed by such low apparent temperatures. Even at 140 K the vapor pressure of CO_2 is below 2 mbar, and it is very likely that the near-surface atmosphere near the winter pole has a composition which reflects the thermodynamic necessities: The CO_2 partial pressure must be very severely depressed by condensation, whereas the total atmospheric pressure must be roughly the same. Therefore, the mole fraction of inert, noncondensible gases at the pole must be in excess of 0.5.

At temperatures low enough to condense CO_2 (about 148 K at typical polar surface elevations), the vapor pressures of a number of potential atmospheric constituents are high enough so that they would remain wholly in the gas phase. At this temperature, the vapor pressures of some of these species are as follows: Xe, 240 mbar; SO_2, 10 μbar; C_3O_2, 20 μbar. Ar, CO, N_2, and O_2 would be completely vaporized. It is thus clear that any retention of "inert" gases such as N_2 or Ar in the polar regions would have to be due to adsorption or trapping, not condensation.

The Viking lander meteorological package has provided, in addition to the direct measurement of diurnal temperature variations from about 190 to 240 K, a direct measurement of long-term seasonal variations in the total atmospheric pressure (see Hess *et al.*, 1977). The atmospheric pressure must vary by 20% or more over a year, and thus the mole fractions of noncondensible gases must vary inversely by the same amount.

Recalling that CO_2 features in Mars's IR spectrum have been under study for decades, we should remind the reader that the deduced surface pressures were in the range 90–200 mbar until the early 1960s. The reinterpretation of the CO_2 spectrum by Kaplan *et al.* (1964) first suggested low surface pressures, and the radio occultation experiments on Mariners 4, 6, and 7 confirmed that surface pressures were generally in the range of 6–7 mbar at the mean surface level (Kliore *et al.*, 1969). These conclusions were further borne out by a host of Earth-based spectroscopic (Belton *et al.*, 1968a; Giver *et al.*, 1968; Carleton *et al.*, 1969; Kaplan *et al.*, 1969; Moroz and Cruikshank, 1971; Betz *et al.*, 1977) and polarimetric (Koval and Yanovitskii, 1969) studies. These measurements generally found 50–90 m agt of CO_2 and revealed no evidence for the presence of an abundant spectroscopically inert gas. This implied mean surface pressures of about 6 mbar. Had there been a discrepancy between the pressure which would be provided by the observed abundance of CO_2 and the pressure deduced from the CO_2 linewidths, it would have been necessary to postulate a large mole fraction of, say, N_2 or Ar.

After CO_2, the first molecular species to be sought in spectra of Mars were H_2O and O_2 (St. John and Adams, 1926). They tentatively identified both species, giving a column abundance of water vapor on Mars equal to 0.03 of that in the terrestrial atmosphere and an O_2 abundance equal to 0.16 of that in Earth's atmosphere. Both of these estimates are enormously greater than the amount actually present. Water vapor was repeatedly and unsuccessfully sought in the following decades [see, for example, the upper limit established by Spinrad and Richardson (1963)], and water vapor was finally marginally identified by Spinrad *et al.* (1963).

In order to understand the literature on the water vapor abundance, we must first face up to a nomenclatural oddity. Observations of water vapor are customarily reported in strange units which make direct comparison with observations of other gases unnecessarily difficult: It is usual to report how thick a layer of liquid H_2O would be formed by condensation of the observed amount of vapor. The vapor abundance is then reported in units of microns or centimeters of precipitable water, as, for example, "15 μm ppt H_2O." The fact that ppt is a common abbreviation for "parts per thousand" does not deter this usage, and it is perhaps overly optimistic to assume that the disapproval of the present authors will have much effect on this unfortunate convention. However, we are unable to merely review this conventional abomination without entering a plea for the use of units which are compatible with those of other gases, such as cm agt or mole fractions, in future observational reports. The units can be converted as follows: 1 μm ppt H_2O is equivalent to 10^{-4} g cm^{-2} or 0.124 cm agt. We further request that the vague "cm atm" and the correct but needlessly awkward "cm atm NTP" or "cm atm STP" be abandoned to avoid unnecessary confusion.

Water vapor was first convincingly demonstrated in the atmosphere of Mars by Kaplan et al. (1964), who estimated a water abundance of 14 ± 7 μm ppt H_2O (1.7 $\pm$ 0.9 cm agt). Schorn et al. (1967, 1969c) have shown that the H_2O vapor abundance over the entire disk of Mars is variable and sometimes undetectably small. Owen and Mason (1969) reported values of 35 ± 15 μm ppt H_2O (4.3 $\pm$ 2.0 cm agt); Tull (1970) and Barker et al. (1970) have seen real variations from about 1.6 to 6.0 cm agt. Schorn (1971) reviewed the variations in water abundance in detail, showing that seasonal variations caused the abundance to vary from below 1.2 cm agt to about 6.0 cm agt. These results have since been confirmed qualitatively by both Soviet and American spacecraft results. The Mars 3 orbiter found peak water vapor abundances near 2.5 cm agt at low latitudes in the Northern Hemisphere (Moroz and Ksanfomaliti, 1972; Moroz and Nadzhip, 1975a,b), less than half of the Southern Hemisphere summer figures cited above. Thus, typical planetary average water abundances are a few hundred parts per million of the CO_2 based on the assumption of uniform mixing of these gases. However, Hess (1976) finds the water to be strongly concentrated into the lowest 6–10 km, with even further enhancement likely in the lowest $\sim$2 km. Observations throughout the year-long Mariner 9 mission disclosed a maximum water vapor content of 2.5–3.7 cm agt during local spring over the north polar cap (Hanel et al., 1972b; Conrath et al., 1973). The Viking orbiters in 1976 found only 0–0.4 cm agt (Farmer et al., 1976a). The water vapor abundance was found to be a smooth function of latitude, ranging almost linearly from 0 near 60°S to 1 cm agt near 20°N (Farmer et al., 1976b). Recent reviews of the water vapor abundance and distribution have been published by Barker (1976) and by Farmer et al. (1977). The latter reports water abundances ranging up to 12.5 cm agt in the darkest, and therefore warmest, parts of the summer circumpolar region.

Carbon monoxide, which must be considered an obviously likely component of any CO_2 atmosphere exposed to solar UV irradiation, was first detected and measured by Kaplan et al. (1969), who found 5.6 cm agt, equivalent to a CO mole fraction of 8×10^{-4}. This estimate was revised upward to $\sim$19 cm agt by Young and Young (1977), equivalent to a mole fraction of $\sim 2.7 \times 10^{-3}$. Kakar et al.

(1977) have studied a single microwave line of CO near 115 GHz and have demonstrated the ability to sound the atmosphere from low altitudes up to as high as 50 km.

Photolysis of CO_2 will, of course, generate not only CO but also atomic O, which will rapidly recombine via three-body reactions to form molecular oxygen, O_2. The early claimed detection by St. John and Adams in 1926 was corrected by Adams and Dunham (1934), who set an upper limit of <0.1% of the O_2 column abundance found on Earth (i.e., <160 cm agt O_2 on Mars). Belton and Hunten (1968, 1969) reported a marginal detection of about 20 cm agt of oxygen through very difficult observations in the A band of O_2 near 7636 Å, but Margolis et al. (1971) were unable to verify this detection at their sensitivity level of 15 cm agt. Barker (1972) and Carleton and Traub (1972) have succeeded in establishing the presence of a small but significant amount of O_2: Barker gives 9.5 ± 1.0 cm agt, in very good agreement. The mole fraction of O_2 is then about 1.4×10^{-3}. In the photolysis of CO_2, the molar ratio O_2:CO in the products is 0.5. Ozone (Lane et al., 1973) is also present in detectable amounts in the polar regions, as we mentioned earlier in our discussion of the polar caps.

The search for nitrogen compounds has been under way for nearly 20 years. Numerous papers dealing with nitrogen dioxide have been published without a positive detection (Sinton, 1961b; Adamcik, 1962; Kiess et al., 1963; Spinrad, 1963; Warneck and Marmo, 1963; Sagan et al., 1965b; Gaiduk, 1971; Owen et al., 1975). The UV spectrometers carried by Mariners 6 and 7 failed to disclose any nitrogen-containing atmospheric species (Barth et al., 1969). Based on these data, Dalgarno and McElroy (1970) deduced a firm upper limit of about 5 mole % on N_2, and possibly less than 0.5% under certain circumstances. As we shall see, the only nitrogen compound detected to date in the troposphere of Mars is the molecular nitrogen found by the Viking analytical experiments.

Another lengthy debate surrounded the "Sinton bands" near 3.46 μm (Sinton, 1957). Sinton (1957, 1959) argued that they were produced by the stretching mode of the C—H bond. This proposed identification was later refined to the aldehyde group and then to acetaldehyde (Colthup, 1961; Sinton, 1961a). Shirk et al. (1965) reinterpreted the bands as due to phenomenal concentrations of HDO in the Martian atmosphere, a phenomenon so striking as to virtually unambiguously require the loss of oceanic masses of hydrogen from Mars with attendant deuterium enrichment. However, Rea et al. (1965) were able to show that the Sinton bands are in fact due to terrestrial atmospheric HDO in entirely unsurprising amounts.

Preliminary analysis of the data from the Mariner 6 and 7 IR spectrometer indicated significant amounts of ammonia and methane in the reflection spectrum of the polar caps (Herr and Pimentel, 1969), but the features are now agreed to be due to solid CO_2. Likewise, the reported detection of several absorption lines due to halogenated light hydrocarbons (Connes et al., 1966) is no longer supported by the original authors.

Two distinct theories attempted to explain the red coloration of Mars on the basis of photochemical reactions of atmospheric gases. A major feature of the model of

Kiess *et al.* (1963) was the attribution of the red color to the dimer of NO_2, nitrogen tetroxide. On the other hand, Plummer and Carson (1969) ascribed the red color to polymers of carbon suboxide (C_3O_2). However, Beer *et al.* (1971b) set an extremely low upper limit on the abundance of gaseous C_3O_2 monomer—only 0.02 cm agt. This is some 10^7 times less than the vapor pressure of C_3O_2 monomer in the regions of Mars surveyed. Perls (1971) defended the C_3O_2 hypothesis qualitatively, but later, upon further study, concluded that C_3O_2 could not be rapidly formed and did not contribute to the reddening observed during dust storms (Perls, 1973).

The first attempted *in situ* chemical analysis of the Martian atmosphere was by a mass spectrometer carried by the Mars 6 lander. The mass spectrometer unfortunately failed to operate, but the ion pump current was successfully telemetered back to Earth. The ion current was observed to be large and to decay slowly, suggestive of a large mole fraction of a poorly pumped inert gas (Istomin *et al.*, 1975). From that time until the Viking mission, discussions of the origin and evolution of the atmosphere were predicated upon the apparent necessity of explaining the presence of about 30% ^{40}Ar in the atmosphere (Levine, 1974; Istomin and Grechnev, 1976; Moroz, 1976a; Surkov and Fedoseev, 1976). The basic conclusion of these and other authors was that the high amount of outgassed ^{40}Ar required the release of proportionately large amounts of other volatiles from the interior. So much ^{40}Ar requires essentially complete outgassing of a chondritic Mars. If one assumes (not because it is obviously true, but because, in the absence of any other evidence, it seems like a plausible first guess) that the relative abundances of volatile elements in Mars are similar to those in the Earth, the masses of H_2O and CO_2 sufficient to produce kilometer-thick ice layers would have been released. If one allows for the lower formation temperature, higher oxidation state, and higher volatile content of preplanetary Mars material favored by many authors, then even more H_2O and CO_2 must have been released relative to ^{40}Ar. But such enormous masses of volatiles are hard to find on Mars today. Without the most extreme exertions, such a large amount of H_2O cannot be accommodated in permafrost and hydrous minerals. Hiding a comparable amount of CO_2 is an even more severe problem, requiring the presence of several kilometers of $CaCO_3$ over the entire surface of the planet. The existing data on the composition of the surface, as reviewed in the previous section, show clearly that carbonates cannot be more than a very minor surface constituent. The need thus arises for escape of vast amounts of CO_2 (molecular weight 44) with simultaneous complete retention of Ar (atomic weight 40). This is a most unpalatable state of affairs.

Fortunately, the Viking lander carried three experiments capable of detecting and measuring atmospheric argon. The XRF experiment, although having only sufficient sensitivity to detect argon down to the 2% mole fraction level, was able to demonstrate that the argon abundance was less than the detection limit (Clark *et al.*, 1976a). The much more sensitive entry mass spectrometer (EMS) experiment (Nier *et al.*, 1976a) quickly established that argon was in fact a minor constituent. Owen and Biemann (1976) found that the early gas chromatograph–mass spectrometer (GCMS) data gave a total argon abundance of 1–2%, and that the argon was overwhelmingly composed of ^{40}Ar. The ^{40}Ar:^{36}Ar ratio was found to be 2750 ± 500, about nine times higher than the terrestrial ratio of 296. Thus, there is about 20 times less radiogenic argon in the atmosphere of Mars than the Mars 6 data at first implied. At the same time, the very high ^{40}Ar:^{36}Ar ratio clearly demonstrates that ^{40}Ar is relatively more important on Mars than on Earth!

The Viking GCMS data on the nitrogen abundance further underline the dilemma: The early results (Owen and Biemann, 1976) bracketed the N_2 abundance between 2 and 3 vol. %. Thus, the ^{40}Ar:N_2 ratio on Mars is about 0.4, whereas in Earth's atmosphere the ratio is near 0.0125. On the other hand, the ^{36}Ar:N_2 ratios on Earth and Mars differ by less than a factor of 3.

The EMS data also provided a reliable measurement of the isotopic composition of nitrogen. The ^{15}N:^{14}N ratio was found to be about 60% higher than the terrestrial or chondritic meteoritic value (Nier et al., 1976b; Nier and McElroy, 1977). This fractionation effect is a logical consequence of both the mechanism for non-thermal escape of nitrogen atoms suggested by Brinkman (1971b), involving pre-dissociation of N_2, and the more rapid dissociative recombination mechanism suggested by McElroy (1972). (We shall return to these processes below.) McElroy et al. (1976a) have interpreted the observed enrichment of ^{15}N on Mars as requiring the escape of large quantities of N atoms, so that initial N_2 partial pressures (with presumed "normal" isotopic composition) of several millibars would be required. Yanagita and Imamura (1978) have proposed that the ^{15}N may have been produced during an early era of intense cosmic ray bombardment; however, it seems unnecessary to postulate an *ad hoc* process of this sort to explain circumstances which can already be accounted for with known processes.

The abundances of the rare gases display some interesting similarities and differences relative to those seen on Earth and Venus and in meteorites. Biemann et al. (1976b) placed restrictive upper limits on the abundances of neon, krypton, and xenon, and Owen et al. (1976) demonstrated the positive detection of both krypton and xenon. The detection and positive identification of Ne is difficult because of the masking of the $^{20}Ne^+$ ions by $^{40}Ar^{2+}$ at high ionizing electron voltages. However, Owen et al. (1977b) succeeded in measuring the neon abundance at the level of 2.5 ppm. The absolute primordial rare gas concentrations in the atmospheres of Venus, Earth, and Mars are compared in Figure 4.30. The elemental abundance pattern for Mars is strikingly similar to the terrestrial pattern. The large excess of ^{40}Ar relative to the primordial rare gases is, of course, not visible in such a display. Likewise, there is a large positive anomaly for ^{129}Xe in the Martian atmosphere. Thus, the two radiogenic nuclides which are not capable of thermal escape are both enriched severalfold on Mars relative to the nonradiogenic rare gas nuclides. The ^{20}Ne:^{22}Ne ratio on Mars was not measured, and the total Ne abundance was estimated by multiplying the ^{22}Ne abundance by 11.

The Viking GCMS was also used to analyze gases released by stepwise heating of surface dirt samples. By this means, extremely small concentrations of a very wide variety of organic molecules could have been detected. Detection limits for a large number of species ranged from 0.1 to 1.0 ppb. Surface samples heated to 500°C

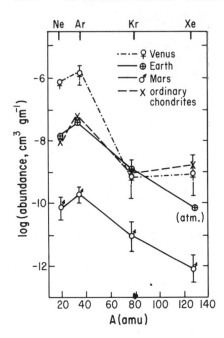

FIGURE 4.30. **Rare gas concentrations (cm³ NTP gm⁻¹) for Venus, Earth, and Mars. Data are from Pioneer Venus (Donahue et al., 1981), Venera 11 and 12 (Istomin et al., 1979), and Viking (Owen et al., 1977b). Only the primordial rare gases are shown: Radiogenic ⁴⁰Ar is omitted for all the bodies shown. The lines are for Venus (♀—·—·♀), Earth (⊕——⊕), Mars (♂——♂), and typical ordinary chondrite primordial rare gases (×---×).**

were observed to release abundant CO_2 and H_2O, as expected, but no evidence for the presence of any organic material on Mars was found (Biemann et al., 1976a).

We summarize our post-Viking knowledge of the atmospheric composition of Mars in Table 4.15. All species above the dashed line are known to be present and have been directly measured. Of these 10 species, 5 (nitrogen, neon, argon, krypton, and xenon) were discovered by the Viking mission.

Upper limits on a number of possible atmospheric constituents were set by Beer et al. (1971b), whose upper limit of C_3O_2 was mentioned. They also set an upper limit 3 cm agt on H_2S, and considered a number of other species such as formaldehyde ($<5 \times 10^{-3}$ cm agt) and formic acid ($>7 \times 10^{-3}$ cm agt). The best present upper limit on NO_2 is 8 μm agt (0.12 ppm) by Owen et al. (1975). Observational limits on a host of other plausible trace constituents have been reported by Horn et al. (1972). These results are summarized "below the line"in Table 4.15.

Further interpretation of these compositional data and their relationship to genetic and evolutionary models for the atmosphere will be given in Section 4.4.6.

4.4.4. Photochemical Stability and Atmospheric Escape

If we accept the hypothesis that the Martian atmosphere has always consisted largely of CO_2 with minor amounts of H_2O, we must then address the question of the stability of such an atmosphere to photochemical decomposition and to volatile element escape.

Concerning CO_2, we are confronted with the same problem here as on Venus. How do we explain the stability of CO_2 with respect to conversion to CO, O, and

TABLE 4.15
Composition of the Atmosphere of Mars

Species	Abundance (mole fraction)	Reference
CO_2	0.953	Owen et al., 1977[b]
N_2	0.027	Owen et al., 1977[b]
^{40}Ar	0.016	Nier et al., 1976[a]
O_2	0.13%	Barker, 1972
CO	0.08%	Kaplan et al., 1969
	0.27%	Young and Young, 1977
H_2O	(0.03%)[a]	
Ne	2.5 ppm	Owen et al., 1977[b]
^{36}Ar	0.5 ppm	Owen and Biemann, 1976
Kr	0.3 ppm	Owen et al., 1976
Xe	0.08 ppm	Owen et al., 1976
O_3	(0.03 ppm)[a]	Lane et al., 1973
	(0.003 ppm)[a]	Noxon et al., 1976

Species	Upper limit (ppm)	Reference
H_2S	<400	Beer et al., 1971[b]
C_2H_2, HCN, PH_3, etc.[b]	50	Horn et al., 1972
N_2O	18	Horn et al., 1972
C_2H_4, CS_2, C_2H_6, etc.[b]	6	Horn et al., 1972
CH_4	3.7	Horn et al., 1972
N_2O_4	3.3	Horn et al., 1972
SF_6, SiF_4, etc.[b]	1.0	Horn et al., 1972
HCOOH	0.9	Beer et al., 1971[b]
CH_2O	0.7	Beer et al., 1971[b]
NO	0.7	Horn et al., 1972
COS	0.6	Horn et al., 1972
SO_2	0.5	Horn et al., 1972
C_3O_2	0.4	Horn et al., 1972
NH_3	0.4	Horn et al., 1972
NO_2	0.2	Horn et al., 1972
HCl	0.1	Beer et al., 1971[b]
NO_2	0.1	Owen et al., 1975

[a]Very variable.
[b]A host of other exotic species are listed in the original paper, but are of no known planetological significance.

O_2? The oxygen problem is not quite as constrained here because the observed O_2 abundance is at least 2 or 3 orders of magnitude greater on Mars than on Venus. However, because there is much less CO_2, the time required for the Martian atmosphere to be converted to a predominantly $CO-O_2$ atmosphere in the absence of recombination mechanisms is very much shorter. McElroy and McConnell (1971) and McElroy and Donahue (1972) compute times as short as 3500 years. Even more telling is the fact that the observed amounts of CO and O_2 could be supplied in only 3 and 8 years respectively.

Initial attempts at answering this problem involved the formation of CO_3 (McElroy and Hunten, 1970) and a sequence requiring oxidation of CO by HO_2 (Reeves *et al.*, 1966). Subsequent laboratory measurements ruled out both of these possibilities.

Before discussing modern theories of Martian photochemistry, we will briefly review the principal constraints which must be applied to photochemical models of the Martian atmosphere. The reader is referred to an excellent review by McConnell (1973b) for many of the details. The constraints are as follows:

(*i*) From ground-based IR spectroscopic measurements the column abundances of CO and O_2 are $(1.5–5.4) \times 10^{20}$ and $(2.4–3.1) \times 10^{20}$ molecules cm^{-2}, respectively. These abundances correspond to average number mixing ratios of $(8–29) \times 10^{-4}$ and $(13–17) \times 10^{-4}$, respectively. These numbers strongly suggest but do not demand that the O_2 abundance exceed the CO abundance.

(*ii*) From rocket-borne UV spectrometer observations the column abundance of O_3 outside the polar regions is $<5.4 \times 10^{15}$ molecules cm^{-2}, corresponding to an average number mixing ratio $<2.9 \times 10^{-8}$. In polar regions the Mariner 7 and 9 UV spectrometers suggest ozone column abundances $\sim 2.7 \times 10^{16}$ molecules cm^{-2} or an average ozone mixing ratio of 1.5×10^{-7}.

(*iii*) From Mariner 6 and 7 UV airglow observations the CO number mixing ratio is $(0.3–1) \times 10^{-2}$ and the O atom number mixing ratio is $(0.5–1) \times 10^{-2}$ at ~ 135 km altitude. In the absence of *in situ* catalysis of CO and O recombination at high altitudes, this constraint necessitates very strong downward mixing in the upper Martian atmosphere. The upper atmospheric situation is therefore very similar to that on Venus.

In order to satisfy the first constraint, McElroy and Donahue (1972) have stressed the importance of the reaction

$$O + HO_2 \rightarrow OH + O_2 \tag{4.79}$$

in addition to

$$H + HO_2 \rightarrow OH + OH \tag{4.80}$$

for producing OH for CO oxidation and thus bypassing the very slow reaction between CO and HO_2. However, in order to obtain the high O atom concentrations at low altitudes which are necessary for efficient operation of reaction (4.79), they required a vertical eddy diffusion coefficient $K = 1.5 \times 10^8$ cm^2 sec^{-1} in the lower atmosphere. Such values are about 1000 times those in the terrestrial troposphere.

There have been a number of estimates of vertical mixing rates in the lower atmosphere of Mars. Gierasch and Goody (1968) have inferred the possibility of very strong turbulence near the Martian surface from theoretical considerations. In particular, they compute $K \sim 3 \times 10^8$ cm^2 sec^{-1} in a mixing layer below 12 km for early afternoon in the equatorial equinox. There are, however, very substantial diurnal, seasonal, latitudinal, and height variations in these computed K values.

For example, they are orders of magnitudes less at night and above the mixing layer, whereas during the midlatitude winter the mixing layer is confined to below 4 km. From these theoretical considerations it is certainly difficult to justify global- and time-averaged K values throughout the lower atmosphere $>10^7$ cm^2 sec^{-1}; this upper limit is probably substantially less above the 10 km level.

A second estimate of vertical mixing rates has been made by Hess (1976). He concludes that K cannot be much greater than 10^5 cm^2 sec^{-1} in the 0–20 km region if the extinction optical depth due to ice crystals in the troposphere is to be kept low enough to be consistent with observations. A third estimate has been made by Conrath (1975) utilizing the observed dissipation times for dust clouds during the 1971–1972 dust storm. He requires $K < 10^7$ cm^2 sec^{-1} in the 0–50 km region. Finally, a lower limit to K can be deduced if we accept the contention that the driving mechanisms for the Martian circulation are confined at or near the surface. In order for transient, vertically propagating waves to provide the large K values required at high altitudes, we then need $K \gtrsim 3 \times 10^5$ cm^2 sec^{-1} at the surface (Huguenin *et al.*, 1977).

Of course, all the above methods used for deducing K values must be regarded as providing, at best, order-of-magnitude estimates; the diversity of these estimates is therefore not surprising. The estimates of K by Gierasch and Goody (1968) do not include the planetary-scale circulation or the mechanical effects of the substantial Martian topography. It is therefore not clear why they are still so much larger than those deduced by Hess (1976). It is also difficult to equate the meteorological conditions during the 1971 great dust storm, analyzed by Conrath (1975), to those in the normal Martian atmosphere. In view of the many uncertainties, we feel that $3 \times 10^5 < K < 10^7$ cm^2 sec^{-1} provides a reasonable range for the expected vertical mixing rates in the region below 40 km on Mars. The K value of 1.8×10^8 cm^2 sec^{-1} required by McElroy and Donahue (1972) in the 0–60 km region certainly appears excessive.

In an alternative model designed to satisfy the first constraint, Parkinson and Hunten (1972) produced the OH necessary for CO oxidation by the reactions

$$HO_2 + HO_2 \rightarrow H_2O_2 + O_2,$$
$$H_2O_2 + h\nu \rightarrow OH + OH \quad (\lambda < 3700 \text{ Å}). \tag{4.81}$$

These reactions do not require large O atom sources at low altitudes, and thus much lower K values are mandated; Parkinson and Hunten (1972) utilized a value of 5×10^6 cm^2 sec^{-1} in the 0–40 km region. However, in order to obtain sufficiently high H_2O_2 concentrations for efficient operation of reaction (4.81), the total odd hydrogen (H, OH, and HO_2) required was almost four times greater than that found necessary in the McElroy and Donahue (1972) scheme. This large amount of odd hydrogen may result in H_2 production by the reaction

$$H + HO_2 \rightarrow H_2 + O_2, \tag{4.82}$$

exceeding the maximum H_2 removal rate possible by O(^{1}D), $CO_2{}^+$, and escape. The rate constant for reaction (4.82) is, however, not sufficiently well known for this conclusion to be definite.

A third model has recently been developed by Huguenin *et al.* (1977). In this model constraint (*i*) is satisfied by recombination of CO and O_2 adsorbed on ferrous minerals on the Martian surface. The chemical reactions involved are the following:

$$CO(g) + \tfrac{1}{2}O_2(g) + O^{2-} \quad \text{(in } Fe^{2+} \text{ crystal)}$$
$$\rightarrow CO_3^{2-} \text{ (adsorbed on } Fe^{2+} \text{ crystal surface)}, \tag{4.83}$$

$$CO_3^{2-} \text{ (adsorbed on } Fe^{2+} \text{ crystal surface)} + h\nu$$
$$\rightarrow CO_3^{2-} \text{ (adsorbed on } Fe^{3+} \text{ crystal surface)} + e^-, \tag{4.84}$$

$$CO_3^{2-} \text{ (adsorbed on } Fe^{3+} \text{ crystal surface)}$$
$$\rightarrow CO_2 + O^{2-} \text{ (in } Fe^{3+} \text{ crystal)}, \tag{4.85}$$

$$Fe^{3+} \text{ crystal} + e^- \rightarrow Fe^{2+} \text{ crystal}. \tag{4.86}$$

The rate-determining step is reaction (4.84); the basic process can therefore be regarded as a *photocatalytic* surface recombination mechanism. Predicted maximum rates for reaction (4.84) imply that surface recombination of CO and O_2 can easily compete with CO and O_2 production from CO_2 photodissociation.

The surface catalysis model is particularly appealing because it requires neither the very large $K(z)$ values necessary in the McElroy and Donahue (1972) model or the very large odd hydrogen concentrations present in the Parkinson and Hunten (1972) model. However, it should be emphasized that all three models lie within the present realms of possibility and indeed the actual mechanism for CO and O_2 recombination on Mars may involve contributions from all three processes.

The second observational constraint on Martian photochemistry involves ozone. Odd oxygen (O and O_3) is produced mainly by CO_2 photodissociation and in a dry Martian atmosphere is removed mainly by

$$O + O_3 \rightarrow O_2 + O_2 \tag{4.87}$$

or by surface reactions. Liu and Donahue (1976) and Huguenin *et al.* (1977) found that in order to keep O_3 concentrations sufficiently low in a model invoking surface removal of odd oxygen, we require $K(z)$ values in excess of 10^8 cm^2 sec^{-1} in the lower atmosphere. Earlier, Parkinson and Hunten (1972) had demonstrated that (4.87) acting alone cannot keep O_3 concentrations below observable levels outside of the polar regions. Both McElroy and Donahue (1972) and Parkinson and Hunten (1972) satisfied constraint (*ii*) by utilizing an odd hydrogen catalytic cycle familiar in our own upper atmosphere, namely:

$$H + O_3 \rightarrow OH + O_2,$$
$$O + OH \rightarrow O_2 + H. \tag{4.88}$$

The required odd hydrogen mixing ratio f_H may be simply estimated from computations presented in these latter two papers. For example, the Parkinson and Hunten (1972) run with $f_H = 5.6 \times 10^{-10}$ decreases the average ozone comfortably below its observed upper limit. The required f_H level is therefore a few parts in 10^{10} and this is easily provided by H_2O photodissociation as will be discussed shortly. Note that this level of odd hydrogen is not sufficient to catalyze recombination of O or O_2

with CO unless we invoke the large $K(z)$ values utilized by McElroy and Donahue (1972).

Recognizing that water vapor is the major source of odd hydrogen, we would expect that dry atmospheric regions near the Martian poles would contain much larger quantities of ozone. This is what is observed and indeed ozone levels at the poles appear to be close to those expected in an almost completely dry (i.e., no odd hydrogen) Martian atmosphere (Parkinson and Hunten, 1972).

Finally, in order to satisfy constraint (iii), we require only that $K(z)$ values be large in the upper atmosphere. McElroy and McConnell (1971) found a satisfactory fit for a constant $K(z)$ value of 5×10^8 cm^2 sec^{-1}. Huguenin et al. (1977) also found satisfactory fits for

$$K(z) = \begin{cases} 3 \times 10^5 \exp(z/2H) \\ 3 \times 10^6 \exp(z/3H) \end{cases} \tag{4.89}$$

clearly illustrating the need for large values only near the turbopause. As on Venus, this rapid vertical mixing is instrumental in removing O and CO from high levels down to low levels. There the O will be partially converted to O_2 and the O and/or O_2 then recombined with CO using the three mechanisms described above.

At this point the important role of H_2O as a source of odd hydrogen on Mars should be clear. Water photochemistry also plays a modulating role in the escape of hydrogen, which will be discussed shortly. A brief review of the water chemistry is therefore in order.

The water vapor concentrations observed in the Martian atmosphere are strongly dependent on the local temperature and, therefore, on altitude, latitude, and time of day. The bulk of atmospheric H_2O is probably confined near the surface, where it is destroyed by photodissociation (Hunten and McElroy, 1970):

$$H_2O + h\nu \rightarrow H + OH \qquad (\lambda < 2117 \text{ Å}), \tag{4.90}$$

and by $O(^1D)$ from ozone photolysis:

$$O(^1D) + H_2O \rightarrow OH + OH. \tag{4.91}$$

The odd hydrogen produced is rapidly cycled through H, OH, and HO_2 by reactions summarized in Figure 4.31. The odd hydrogen sinks are:

$$\begin{aligned} H + HO_2 &\rightarrow H_2O + O, \\ &\rightarrow H_2 + O_2, \\ HO_2 + HO_2 &\rightarrow H_2O_2, \\ HO_2 + OH &\rightarrow H_2O + O_2. \end{aligned} \tag{4.92}$$

The H_2O_2 produced here may conceivably freeze out in polar regions (Hunten, 1974a), but otherwise it will photodissociate [reaction (4.81)] to reform odd hydrogen. As mentioned earlier, spectroscopic observations suggest that the O_2 abundance exceeds the CO abundance on Mars and the O_2 generated here is the probable source of this additional oxygen; in a dry CO_2 atmosphere we expect the O_2 abundance to be one-half that of CO.

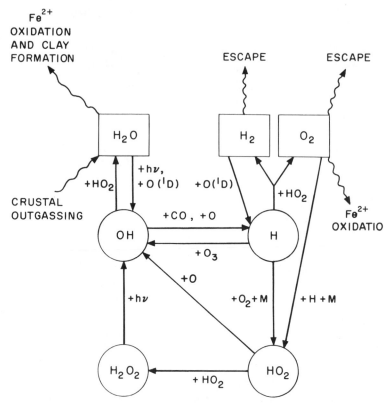

FIGURE 4.31. **Cycling and loss of water on Mars. The principal processes by which the atmospheric H_2O and O_2 contents are regulated are shown schematically.**

The next topic we must address concerns the escape of volatile elements from the atmosphere. Molecular hydrogen is generated by reaction of H and HO_2, which are both products of H_2O photodissociation in the lower atmosphere. Some of this H_2 is removed by

$$O(^1D) + H_2 \rightarrow OH + H \qquad (4.93)$$

near the ground to mainly reform H_2O and by

$$\begin{aligned} CO_2^+ + H_2 &\rightarrow CO_2H^+ + H \\ CO_2H^+ + e^- &\rightarrow CO_2 + H \\ H_2 + h\nu &\rightarrow H + H \quad (\lambda < 850 \text{ Å}) \end{aligned} \qquad (4.94)$$

in the ionosphere to produce H atoms. Molecular hydrogen is well mixed up to about the 80 km altitude, whereas above this altitude reactions (4.94) cause H atoms to become an increasingly important form of hydrogen. As a consequence, Hunten and McElroy (1970) estimate that at least 90% of the hydrogen escape flux is in the form of H atoms.

Anderson and Hord (1971) have analyzed the Mariner 6 and 7 measurements of H Lyman-α resonance emission from Mars and have deduced that the H escape level or exobase lies at an altitude of about 250 km, where the H number density is 3 $\times$ 10^4 (atoms cm^{-3}). The exospheric temperature is 350 K and the H escape rate, utilizing Jeans formula (3.78) is 1.8×10^8 atoms cm^{-2} sec^{-1}. For H atoms the net escape velocity $W_i = 53$ m sec^{-1}, whereas the limiting diffusion velocity w_i^* for $K^* = 5 \times 10^8$ cm^2 sec^{-1} is 6 m sec^{-1} (Hunten, 1973b). Thus, $w_i > w_i^*$ and escape should be roughly diffusion-controlled. At the turbopause H_2 is still the dominant form of hydrogen and the limiting H diffusion flux in the form a H_2 utilizing (3.79) is 1.2×10^8 atom cm^{-2} sec^{-1}. The observed escape flux is roughly equal to this value as expected.

In a steady-state model this outward flux of hydrogen must be balanced by the production of H_2 from H_2O. Hunten and McElroy (1970) were unable to estimate this production rate *a priori* due to poorly known rate constants and also to the unknown catalytic effects of the ground and atmospheric aerosols on a number of important reactions. One of their models where 30% of the dissociated H_2O produces H_2 does, however, yield an escape flux in satisfactory agreement with the observations.

It is instructive to consider the total amount of water vapor which would be destroyed on Mars if the present escape flux of H atoms were sustained over the lifetime of the planet (roughly 5 billion years). The amount is about one-thousandth of the water in the oceans on earth today (Rubey, 1951). In view of our earlier statements on the very large water content in hydrous minerals expected in the Martian lithosphere, the small water loss discussed here will certainly not have noticeably depleted Martian water resources. However, the other product of water photolysis, namely O_2, does present a problem. With the presently observed escape rate of H, we will produce the presently observed O_2 amount in just 100,000 years. To resolve this oxygen dilemma, we must search for ways to remove it from the atmosphere. Stewart (1972) has analyzed the Mariner 6 and 7 measurements of the $O(^3S) \rightarrow O(^3P)$ and $O(^5S) \rightarrow O(^3P)$ resonant emissions. He demonstrated the presence of sufficient O atom concentrations to enable the charge transfer reaction

$$O + CO_2^+ \rightarrow O_2^+ + CO$$

to cause O_2^+ to be the dominant ionospheric positive ion. McElroy (1972) pointed out that the two O atoms produced in dissociative recombination of O_2^+ and e^- have sufficient energy to escape the gravitational field of Mars even if one of these O atoms were produced in the ^1D state. Thus, the upward-traveling atom of the pair could escape (see Section 3.6.2). He noted a similar situation for recombination of N_2^+ and e^-, enabling escape of N atoms, and recombination of CO^+ and e^-, producing C atom escape. His computed escape fluxes for O, N, and C were 6×10^7, 3×10^5, and 4×10^5 atoms cm^{-2} sec^{-1}, respectively. We should add parenthetically that escape of volatile elements from Venus and Earth is not possible by this mechanism. The gravitational fields on these latter planets are considerably larger and the energies required for O, N, and C atoms to escape are roughly four times those required on Mars.

The above oxygen atom escape rate is in remarkable agreement with the production rate of O from H_2O photodissociation inferred from the present H escape rate. McElroy (1972) suggests that this balance is in fact expected from the buffering effect of molecular hydrogen. This effect has recently been confirmed and analyzed in considerable detail by Liu and Donahue (1976). If this escape rate remained constant over geologic time we can compare the total water lost by this mechanism to the total CO_2 now present in the atmosphere. McElroy (1972) thus infers that the $H_2O:CO_2$ ratio in the crustal outgassing is about 45:1. A comparison of the water in the oceans with the CO_2 in sedimentary carbonates on Earth actually yields a similar ratio (Rubey, 1951). However, as we discussed in some detail in Section 4.4.1, there are a number of cosmochemical reasons for expecting the $H_2O:CO_2$ ratio on Mars to considerably *exceed* that on Earth. If so, there must exist H_2O removal mechanisms much more potent than atmospheric escape. Some possibilities have already been discussed in the previous section.

The fact that the nonthermal O atom escape rate discussed above is roughly equal to one-half the H escape rate does not in fact mean that such escape is the dominant oxygen removal mechanism on Mars. First, the computed escape rates are model dependent. Second, we should also investigate possible surface sinks for oxygen since most of the total oxygen (O, O_2, O_3) is, of course, contained in the first few scale heights. In this respect the work of Adams and McCord (1969) has shown that the optical properties of the Martian surface are best simulated in the laboratory by olivine- and magnetite-bearing basalts which have been partially oxidized to the extent that they contain between 1% (dark regions on Mars) and 10% (bright regions on Mars) of Fe_2O_3. In view of our earlier discussion of the *primordial* mineralogy of the Martian surface, it is apparent that essentially all of the surface Fe should be in the Fe^{2+} oxidation state (FeO, FeS, etc). The question arises then as to the mechanism for producing the Fe^{3+} in Fe_2O_3 which is required to fit the observations.

Atmospheric O_2 would seem to be a plausible candidate for such oxidation, but O'Connor (1968b) pointed out that the present atmospheric and surface conditions on Mars are quite insufficient to oxidize Fe^{2+} to Fe^{3+} by *thermochemical* means. However, as we discussed earlier, more recent work by Huguenin (1973a,b) has demonstrated that the reaction between black magnetite (Fe_3O_4) and atmospheric O_2 to produce bright red hematite (Fe_2O_3) is powerfully stimulated by UV radiation with $\lambda > 2000$ Å. Since the thin Martian atmosphere transmits most of the radiation in this spectral region, this photostimulated oxidation of Fe^{2+} to Fe^{3+} offers both a suitable production method for Fe_2O_3 and a suitable sink for atmospheric O_2. Huguenin (1974) also suggested a similar mechanism for oxidation of Fe^{2+} in olivine and pyroxene to Fe_2O_3. The possible removal rate for O_2 was predicted to lie in the range $5 \times 10^7 - 5 \times 10^{10}$ molecules cm^{-2} sec^{-1}; even the smallest estimate exceeds that required to balance the removal of photochemically produced H_2 from the atmosphere by thermal escape. This surface sink for O_2 clearly has considerable appeal.

Oxidation of surface rocks may not have always proceeded by the above mechanism. Sagan (1971) and Sagan *et al.* (1973) have suggested that the north pole on Mars may contain significant quantities of water ice beneath the observable dry ice covering. The slow precession of the spin axis on Mars causes the permanent pole to change from north to south every 50,000 years. During this cycle substantial quantities of water vapor would be released into the atmosphere, perhaps enabling the temporary existence of rivers and lakes. Oxidation of Fe^{2+} proceeds readily in this environment and may provide yet another sink for oxygen and a production mechanism for Fe_2O_3.

The escape of H_2 operating together with the above mechanisms for removing O_2 from the atmosphere results in a net loss of H_2O. This loss rate is, of course, limited by the average H_2O photolysis rate which is 10^9 molecules cm^{-2} sec^{-1} (Hunten and McElroy, 1970). In the previous section we discussed geochemical mechanisms for removing H_2O from the atmosphere without prior photodissociation and we have given a number of reasons for suspecting that the major sink for Martian H_2O did not in fact involve atmospheric escape. Processes involving H_2O are summarized in Figure 4.31.

Nitrogen in the Martian atmosphere also deserves some attention. On Earth the $N_2:CO_2$ ratio is 0.03:1 (Rubey, 1951). We gave cosmochemical reasons in Section 4.4.1 for expecting the $CO_2:H_2O$ and $N_2:H_2O$ ratios to be significantly less on Mars than on Earth. It is difficult, however, to accurately assess the expected $N_2:CO_2$ ratio. If we assume this ratio to be the same as that on Earth, then the escape of N atoms by McElroy's (1972) mechanism keeps the N_2 mixing ratio $\leq 10^{-2}$, the actual value depending on the outgassing history and the position of the turbopause. In view of the results from Mariners 6 and 7 indicating that N_2 is $\lesssim 5\%$ of the atmosphere [and possibly as low as 0.1% according to Dalgarno and McElroy (1970)], it is very interesting to note the existence of this nitrogen loss mechanism. We might also expect N_2 to be decomposed in the Martian lower atmosphere by cosmic rays as it is in the Earth's lower stratosphere (Warneck, 1972). Reaction of the resultant N atoms with O_2 would lead to small amounts of NO_2 and HNO_3, which could then possibly be irreversibly removed from the atmosphere by formation of surface nitrites and nitrates.

We have already mentioned in Section 4.4.3 the large enrichment of ^{15}N in the atmosphere of Mars. McElroy's dissociative recombination mechanism can explain this fractionation if the original isotopic composition were similar to terrestrial nitrogen, and if escape has depleted the original nitrogen by at least a factor of 20.

As far as carbon is concerned, McElroy's (1972) escape mechanism has only a minor effect on the CO_2 abundance—perhaps a loss of a few tens of percent of the present atmospheric CO_2 content. This escape loss is probably secondary to the formation of surface carbonates mentioned in Section 4.4.2.

Before leaving this particular subject, we should mention that solar wind sweeping may also be making a contribution to escape from the Martian atmosphere. Cloutier *et al.* (1969) and McElroy (1972) have pointed out that near the planetary limb, specifically at latitudes higher than 70° and altitudes greater than 250 km, a number of ions are produced by photoionization and by charge exchange between solar wind protons and CO_2, O, and N_2. These ions which include O_2^+, CO_2^+, O^+, N_2^+, NO^+, and CO^+ can be swept up by the electric field of the impinging solar wind. Significant escape rates of CO_2 (5×10^5 molecules cm^{-2} sec^{-1}) and N_2 (2×10^5 molecules cm^{-2} sec^{-1}) may result.

4.4.5. Organic Matter and the Origin of Life

At the beginning of the era of spacecraft exploration of the planets it was widely, but by no means universally, accepted that Mars was inhabited by at least primitive life forms (Salisbury, 1962). The spirited debate concerning the possible biological significance of the IR "Sinton bands" (Sinton, 1957) was finally resolved in the negative by Rea et al. (1963a, 1965). The history of this debate was briefly recounted in Section 4.4.3. Laboratory experiments were carried out on the reflection spectra of organic and biological materials (Rea et al., 1963b) for comparison with Mars. The possibility of abiotic synthesis of organic matter on Mars was seriously discussed (Abelson, 1965), and the wave of darkening was still under active consideration as a biological phenomenon (Pollack et al., 1967).

Fanale (1971a) presented geochemical arguments, based on the presence of ferrous and ferric iron in the crust, which implied that the initial products of planetary outgassing may have provided a rather massive reducing atmosphere on Mars. This atmosphere is, of course, unstable against hydrogen escape, and Fanale emphasized in a series of papers that it was plausible that Mars outgassed rapidly and extensively very early in its history, presumably in response to severe accretional heating. Slow outgassing could not keep pace with H_2 escape, and the atmosphere would be oxidizing. Rapid outgassing could maintain a substantial reducing atmosphere for only a short time. Fanale also stressed that the present surface conditions on Mars are hostile to the formation and preservation of organic matter. Thus, if Mars ever had the ability to form organic matter and life, that chapter in its history must have been both brief and early in the evolution of the planet. Sagan (1971) offered the frank speculation that cyclic climatic excursions on Mars (specifically, those caused by the precessional cycle) would periodically provide benign conditions, and that life forms might survive in a freeze-dried state from one climate optimum to the next.

It was in this climate of opinion that the plans for the Viking landers were developed. Although the credibility of Martian life suffered greatly as a result of the Mariner 4, 6, and 7 flybys, which sampled dreary, ancient cratered terrain, the thorough mapping of the surface by the Mariner 9 orbiter raised new hopes. The photogeological evidence for river beds and flooding caused the freeze-dried hopes of exobiologists to germinate and grow with new vigor. Viking was planned. There was no question about the driving force behind Viking: It was not geophysical interest in whether Mars was an active planet, nor geochemical interest in the composition and history of the crust. There was no interest in whether Mars had a core or a magnetic field. Meteorology was found tolerable in small doses, but the addition of geophysical and geochemical experiments could only be carried out by the most determined interventions. Viking's main task would be to detect life on Mars.

The biological experiment package was designed to accommodate four experiments (Klein *et al.*, 1972). First, there was the carbon assimilation experiment (Horowitz *et al.*, 1972). In this experiment, ^{14}C-labeled CO and CO_2 are introduced into a "culture medium" which, by terrestrial standards, would resemble a torture chamber. Reasoning that, however inhospitable Mars seems to terrestrial organisms, Martian organisms (if any exist) must be adapted to it and can be expected to be very intolerant of the vastly different environments to which terrestrial microorganisms are accustomed. Accordingly, samples of Mars dirt are placed by the Viking lander sampling arm in a small chamber with a window which admits sunlight. Traces of labeled CO and CO_2 are admitted, and allowed to remain in contact with the dirt for several hours. Unreacted gases are then flushed from the system and the soil sample is heated to temperatures of about 850 K, sufficient to kill any microorganisms and thoroughly pyrolyze their organic remains. Radioactive gases released during pyrolysis are then cleaned of organic molecules, and the CO and CO_2 component of the released gas is then counted for its radiocarbon content. Hubbard (1976) described the philosophy, design, hardware, and execution of this experiment in some detail. Initial experience with this experiment on the Viking 1 lander mission (Horowitz *et al.*, 1976) showed that slow assimilation of CO and CO_2 took place, and preheating the surface sample or exposing it to moisture inhibited this process. The final report on this experiment (Horowitz *et al.*, 1977) shows more clearly how this activity depends on the severity of the preheating: 2 hr at 90°C has no effect, whereas 3 hr at 175°C reduces the rate of assimilation about 90%. Furthermore, the inhibition of assimilation by water vapor turned out not to be verifiable. The experimenters conclude that it is unlikely that the reactions observed, which apparently produced minuscule amounts of organic matter, were of biological origin.

The second major biological experiment (Levin, 1972), called the labeled release experiment, involved exposing dilute aqueous solutions of ^{14}C-labeled organic nutrient to a sample of Mars soil. The release of radioactive gases is taken as evidence for metabolic activity. The experiment is described in detail by Levin and Straat (1976a). Early Viking results showed a small response, which, like that in the carbon assimilation and pyrolytic release experiment described above, could be prevented by preheating to sterilizing temperatures. Both inorganic and organic processes in the Mars dirt could account for these results (Levin and Straat, 1976b). Further experience with this experiment on both Viking landers (Levin and Straat, 1977) has led to no further insights into the nature of the process.

The third life-detection experiment involved the use of a gas chromatograph for detection of gas exchange between Martian soils and the gases in contact with it (Oyama, 1972; Oyama *et al.*, 1976). All experiments performed on the Viking landers showed such composition changes (Oyama and Berdahl, 1977), which apparently are largely due to desorption of atmospheric gases from the soil sample combined with oxidation of organic matter in nutrient media which can be added to the sample. The oxidizing agent could, as suggested by simulations on Earth, be γ Fe_2O_3. Oxygen is evolved upon the addition of H_2O, possibly from the decomposition of photochemically produced superoxides in the surface dirt. Blackburn *et al.* (1979) find that MnO_2 produces this effect. No evidence requiring the presence of biological activity was found.

The fourth biology experiment originally planned for the Viking missions was the light-scattering experiment of Vishniac and Welty (1972). This experiment encountered severe problems during development and was deleted from the payload.

In a sense, one of the most powerful and general life-detection experiments aboard Viking was the GCMS experiment, which was designed to permit extremely sensitive searches for organic matter in surface samples (D. M. Anderson *et al.*, 1972). The results were, as we have seen, negative at the level of 0.1–1.0 ppb for a host of different organic species (Biemann *et al.*, 1976b, 1977). Nussinov *et al.* (1978) have claimed that the GCMS experiment results disprove the existence of strong oxidizing agents in the surface, but this is based on a misunderstanding of the function of that experiment and is incorrect (Biemann and Lavoie, 1979).

Sagan and Lederberg (1976) suggested that "large organisms, possibly detectable by the Viking lander cameras, are not only possible on Mars: they may be favored." This extreme view was subjected to experimental test by Levinthal *et al.* (1977b) and rejected. The latter authors generously considered a number of hypotheses which might permit life forms to be present but evade detection: They may be too small, too rare, too fast, etc. to be photographed by the Viking camera system. Curiously, two hypotheses evaded their attention: The organisms may be wholly transparent or they may be bashful. They did, however, conclude with the upbeat remark that "Martian photophobes could always be poised one scan line away, waiting for the reflected light from the nodding camera mirror to disappear." A critique of this suggestion is in preparation and will be published elsewhere.

General overviews of the results of the Viking search for life on Mars have reached cautiously dismal conclusions (Klein *et al.*, 1976; Klein, 1977). It is, of course, impossible to rule out the presence of life on Mars on the basis of the Viking results. Those who preferred to spin ethereal fantasies in the absence of diagnostic evidence are at liberty to continue their activities: However, we expect that most exobiologists will wisely seek employment elsewhere. Mars will for a time be left to the tender mercies of geochemists, geophysicists, aeronomers, and meteorologists. Life on Mars will be seen as a bad joke left over from the 1960s, and it will be well-nigh impossible to generate any new missions to Mars. The political climate will be poisoned by the failure of a billion dollar investment to prove the existence of little green men, and many of our most fundamental questions about Mars will remain unanswered for many years to come.

The fundamental questions regarding the organic and biological history of Mars remain unchanged. Was there ever a reducing atmosphere on Mars? If so, how long did it last and what did it produce? Yung and Pinto (1978) have shown that an assumed massive, mildly reducing atmosphere can generate vast amounts of hydrocarbons; in fact, they suggest that the fluvial features on Mars were produced by rivers of alkanes, floods of gasoline and lighter fluid which would dwarf the Amazon. Another fascinating question has recently come to the fore: Can organic matter be produced without a reducing atmosphere? The production of organic matter from CO_2 in the presence of UV light and ferrous iron was reported by Getoff (1962), and a thermodynamic rationale for the process was given by Baur (1978). Tseng and Chang (1974) found that UV light can produce formic acid from a mixture of CO_2 and H_2O at temperatures as low as 200 K. But if organic matter is formed, is it stable in the Martian regolith today? The Viking GCMS experiment tells us that organics are absent in the surface dirt, but does not tell us why. Chun *et al.* (1978) show that the powerful UV photochemical processes at the Martian surface will produce small amounts of organic matter and larger amounts of very strong oxidizing agents simultaneously. At steady state the concentrations of the interesting organic species are negligible, and buildup of organic products cannot occur because of attack by OH, O_2, O_3, and atomic O (see Hunten, 1974a). Sagan and Pollack (1974) proposed a "euphotic zone" at about 1 cm depth in the regolith, under the assumption that visible light is essential to life but UV is the main threat to microorganisms. The latter assumption is surely not correct, and strong oxidizing agents can readily diffuse to a far greater depth than visible light (whatever its utility) can penetrate. Enthusiastic attempts to sidestep these tedious arguments of stability, feasibility, and survival have been put forward, most recently by Abadi and Wickramasinghe (1977). They claim that a broad absorption feature in the UV spectrum of Martian dust is similar to that in a "wide class of organic molecules," and conclude that there is biology on Mars. We consider this claim to be proof of an assertion made in the previous paragraph.

However clouded the past history of life on Mars, we may not so readily dismiss the possibility of a glorious future. Burns and Harwit (1973) have proposed a particularly brute-force approach to ameliorating Martian climate by shifting around huge masses of material so as to increase the precession period and thus prolong (perhaps indefinitely) the coming "Martian spring." Sagan (1973d), less ambitiously, proposes transporting some 10^{10} tons of very dark dust to the polar regions, where it will absorb sunlight, evaporate the polar caps, raise the atmospheric pressure above the triple point of water, and thus permit a greatly enhanced greenhouse effect and a much warmer climatic regime. However, very much smaller masses of certain gaseous halocarbons may be capable of providing an enhanced greenhouse effect in a single step, without the danger of condensates covering the dark pigment or the wind burying it. Further thoughts along these lines may eventually lead to the discovery of a way to make the climate of Mars less forbidding.

4.4.6. Origin and Evolution of the Atmosphere

Our discussion of the bulk composition of Mars and the photochemical and aeronomical evolution of its atmosphere in Sections 4.4.1 and 4.4.2, and the treatment of volatile-element budgets and the geological evidence for massive deposits of mineral ice in Section 4.4.2 have covered most of the evidence and interpretation appropriate to this subject. We will therefore conclude with only a brief summary and prospectus for future work.

In Section 4.4.4 we surveyed the various mechanisms that have been suggested for depleting volatiles from the atmosphere of Mars. Thermal escape has been found to be satisfactory for helium and hydrogen, while nonthermal mechanisms such as dissociative recombination are capable of ejecting atoms of nitrogen, carbon, and oxygen. Solar wind sweeping is effective in removing any charged particles present in the ionosphere. Multiplying the present escape fluxes for these mechanisms times the age of the Solar System, we find that the amount of atmosphere that has been lost by these known processes is probably at least several times larger than the present atmospheric pressure.

The model for the formation of Mars advocated by Ringwood (1966) begins with C1 chondritic material, from which some gas loss has occurred. Since C1 material contains about 10 wt. % water, melting and differentiation of Mars would release enough water to cover the entire planet to a depth of 350 km. It is Ringwood's belief that wholesale loss of volatiles occurred on all the terrestrial planets very early in their history (Ringwood, 1979, p. 145), an idea which he supports by reference to McElroy et al.'s (1976a) interpretation of the nitrogen isotope anomaly found by the Viking landers. However, the latter effect is due to a slow, selective escape mechanism operating over the entire lifetime of the planet, not a massive, nonselective mass loss episode. The nitrogen isotope evidence is quite irrelevant to the issue of whether early, massive volatile loss occurred.

Ringwood (1979) accepts the idea that the high oxidation state of Mars requires low formation temperatures (Ringwood, 1966; Lewis, 1972b), but also believes that the XRF soil analyses prove that Mars is depleted in potassium relative to chondritic meteorites. He then argues that it is "reasonable to expect a corresponding depletion in sulphur, since the latter element is much more volatile than potassium . . . in the solar nebula." Thus, Ringwood (1979), like Anders and Owen (1977), believes that the surface dirt at the Viking landing sites reflects faithfully the overall composition of Mars. In that case, it is important to note that one of the most striking features of the Viking XRF data discussed in Section 4.4.1 is the remarkably high sulfur abundance in every sample. The significance of the high sulfur content is discussed in some detail by Clark and Baird (1979b), who explore the possibility that the surface dirt is produced by large-scale oxidation and disaggregation of primitive sulfur-rich material. Clearly, we cannot easily accept Ringwood's assertion that sulfur is severely depleted in Mars. Ringwood's conclusions that the core must be Fe_3O_4 and not FeS are based on the same logic, and thus do not follow from the observational evidence. Further, it is a well-known result of thermodynamic studies of condensation in the nebula that Fe oxides form at temperatures so low that both K and S are fully condensed (Lewis, 1972b; Grossman, 1972; Fegley and Lewis, 1980).

Ringwood then concludes that the low ^{36}Ar abundance in the atmosphere (a relative abundance of only 1% of the ^{36}Ar abundance in the Earth's atmosphere) requires either early massive atmosphere loss or very inefficient outgassing. We accept these conclusions, although we do not accept the arguments for *early* outgassing discussed above.

Ringwood mentions the fact that the ^{40}Ar:^{36}Ar and ^{129}Xe:^{132}Xe ratios are very high on Mars, but does not suggest an interpretation beyond the paradoxical remark that K is depleted in Mars. The latter, of course, would *decrease* the 40:36 ratio. We feel that this evidence demonstrates that radiogenic rate gases are relatively more abundant compared to primordial rare gases on Mars, which would be expected for volatiles released relatively *late* in the history of the planet. If Mars were to outgas its volatiles instantly upon accretion, with high efficiency (Fanale, 1971a,b), the initial atmosphere would be very poor in the radiogenic nuclides ^{40}Ar and ^{129}Xe. Further slow outgassing would add small amounts of a rare gas component which becomes progressively richer in radiogenic gases as time passes. Thus, early blowoff of a massive atmosphere formed at the time of planetary formation, followed by slow release of volatiles over a prolonged period of time, would be consistent with the observations. Massive early outgassing is clearly consistent with, but not absolutely proved by, the rare gas data.

The way in which the crustal elemental abundances are interpreted is of major importance in assessing bulk composition models for the planets. The XRF data refer to surface samples found at two widely separated points, both on relatively flat regions with a deep regolith. The dirt analyzed has certainly been subjected to planetary differentiation, possible remelting, possible hydrothermal alteration, extensive oxidation, wind sorting and transport, and probably at least some solvent leaching and deposition of water-soluble salts. The clay, oxide, and sulfate components may have different origins. The very existence of enormous amounts of clays and sulfates implies reactions destroying equally vast amounts of water. Nothing is known of the abundance of carbonates in the surface dirt, although they cannot be a major component.

An ambitious attempt at interpretation of the compositional data on the atmosphere and surface of Mars has been made by Anders and Owen (1977). They base their interpretation of the composition of Mars on a five-component model developed by Ganapathy and Anders (1974) to explain fractionation trends observed in chondritic meteorites. These five components are sensibly defined in terms of their geochemical behavior during condensation from the solar nebula. The five classes are as follows: (1) early condensates, (2) metal, (3) magnesium silicates, (4) moderately volatile elements (condensation temperatures below 1300 K but above 600 K), and (5) volatiles condensing below 600 K. Fractionation between the different classes of carbonaceous chondrites can be rather well described by such a scheme, and the differences between the classes of ordinary chondrites can be rather less satisfactorily explained. The fractionation process is envisioned as the selective accretion of condensate grains by small parent bodies, in which the physical properties of the grains, such as crushing strength, magnetic moment, and electrical conductivity, are more important than gravity in selecting the grains to be retained. We can thus picture small parent bodies accreting with different compositions because of these mechanisms. Such fractionation mechanisms may work for bodies up to 1 km in size, but it is hard to see how ensembles of 10^{12} such bodies, when accreted to form an Earth or Venus, could be other than perfectly democratic samples of all the solids available. Gravitation, which dominates accretion of bodies from the size of an average asteroid on up, does not distinguish between grains with different individual physical properties.

Two possible approaches for extending grain-by-grain physical fractionation processes to planetary-scale bodies have been suggested. First, there is the possibility that all solids in the early solar nebula were fully evaporated and then partially condensed during cooling (Grossman, 1972). In this case, radial sorting of grains by volatility can occur, leading to volatile-poor bodies close to the Sun (irrespective of size) and volatile-rich bodies far from the Sun (again, irrespective of size). This is also expected in the chemical equilibrium model of Lewis (1972b) wherein complete evaporation does not occur anywhere outside about 0.3 AU from the Sun. In either case, a fairly simple monotonic radial trend of Fe oxidation state and volatile content is expected.

The second mechanism for effecting composition differences between planet-sized bodies is aerodynamic sorting based on differences between the mean sizes and densities of grains in the solar nebula (Weidenschilling, 1978), which we discussed in connection with the origin of Mercury and the fractionation of metal from silicates in the early Solar System. Both this and the fractional condensation or equilibrium condensation approach predict a smooth and continuous (but not necessarily monotonic) variation of planetary bulk compositions with heliocentric distance.

The idea that the grain-related fractionation processes could shape the compositions of planets was also criticized by Ringwood (1979, pp. 161–162) on five grounds: First, the Ganapathy and Anders model is not obviously so successful on chondrites that it need be accepted. Second, it appears that the ordinary chondrites are derived from very small and quantitatively unimportant asteroid types whose places of origin are not known. Third, accretion of planets from ensembles of 10^{12} small bodies which are statistically sampled over a wide range of heliocentric distance would certainly wipe out the compositional variations of single bodies. Fourth and fifth, Ringwood finds that even the seven-component version of the model does not adequately describe the compositions of the Moon and the Earth.

Anders and Owen (1977), however, have applied this model to Mars, and Morgan and Anders (1980) have revised this approach for the purpose of comparing the composition of Mars to the other terrestrial planets. They assume strict coherence of the elements in each of the geochemical groups on Mars: An example is their assumption that the atmospheric inventory of ^{36}Ar fixes the planetary inventory of thallium and a number of other unobserved elements. The assumption is explicitly made that no large-scale loss of atmosphere has occurred. This assumption is defended by pointing out that the rare gas *composition* on Mars is very similar to that on Earth, and any thermal loss process would introduce very large (and quite unobserved) mass fractionation effects in, say, depleting ^{36}Ar by a factor of 10 while leaving the xenon abundance unchanged. This argument, of course, does not apply to catastrophic loss of an atmosphere by processes other than Jeans escape. Morgan and Anders explicitly claim: "Following Anders and Owen (1977), we are assuming that Mars never lost either ^{36}Ar or ^{40}Ar. First, both isotopes are too heavy to be lost from the present atmosphere. Second, there is no evidence whatsoever, from any planet, for a catastrophic loss of an early atmosphere." They then analyze the GCMS and XRF data on Mars and reach the expected conclusion that Mars is deficient in volatile (condensation temperatures below 600 K) elements by a factor of 40 relative to Earth. Moderately volatile (600–1300 K) elements are similarly depleted by a factor of only about 3, but that conclusion depends on guessing the efficiency of release of sulfur from the interior during degassing. They assume the same release efficiency for Mars as for Earth, despite the fact that the total abundance of sulfur-bearing gases in a mixed-volatile system is a sensitive function of both temperature and oxygen fugacity, whereas the sulfur gas composition depends on the water vapor fugacity as well. Such an assumption could easily be wrong by a factor of 100. For a discussion of the dependence of the abundance and composition of sulfur gases on T and f_{O_2}, see Lewis (1982).

Anders and Owen explicitly reject a more volatile-rich Mars by reference to the high ^{40}Ar:^{36}Ar ratio. Radiogenic argon is derived from moderately volatile (600–1300 K) potassium, whereas primordial argon is highly volatile (a "thallium-group" element condensing below 600 K). If equal release efficiency of the argon isotopes is assumed, then the Tl group could be raised in abundance by a factor of 3, but would still be depleted by a factor of about 12 relative to Earth. Of course, the abundance of the "Tl group" is based on the primordial rare gas abundances, which would have been most severely affected if large-scale blowoff had occurred.

Morgan and Anders (1980) argue that the volatile contents of planets increase with increasing planetary mass, whereas the FeO contents, which in meteorites are closely correlated with the abundances of volatiles, increase with heliocentric distance. They then point out that the parent body of the shergottite achondrites is even smaller than the smallest planet, yet has a higher volatile content! Morgan and Anders (1980) regard this as "no great calamity, because this trend has no obvious explanation anyway."

The entire argument upon which this analysis is constructed fails if massive, nonselective volatile loss can occur on small planets. If Mars, the Moon, and Mercury has suffered atmosphere blowoff without fractionation, then there is no such thing as a "thallium group" of elements in planets and the logical deductions made by assuming such a creature are invalid.

Although we have been assured by Anders and Owen that there is no evidence whatsoever for catastrophic loss of early atmospheres from planets, however small, it seems clear that a search for such a mechanism would be very much worthwhile. After all, whatever we believe about the primordial volatile contents of planets, it is obvious that escape is easier from the smallest ones. It is qualitatively appealing to interpret the FeO gradient in the inner solar system as requiring a smooth increase in volatile content with heliocentric distance in the raw planetary material. By this criterion, the Moon and Mars should have originally accreted with a higher volatile content than Earth (see Section 4.6).

We shall reject thermal escape and dissociative recombination as the agents of such an escape process; they are selective and would produce severe fractionation that is not seen. T-Tauri phase solar wind activity and tidal stripping by the early Sun are poorly understood at best, probably occurred too early in the accretion history of the planets to be relevant, and cannot be quantified.

The effect of large impact events is, however, of obvious interest. Mars, the Moon, and Mercury are all heavily cratered by objects which commonly left craters measuring tens of kilometers in diameter. Even small impact events are capable of blowing holes larger than a scale height in size in the lower atmosphere of Mars.

Such loss of volatiles by explosive blowoff would be highly nonselective, in that all uncondensable volatiles would be very easily dissipated into space, whereas only the condensed volatiles, H_2O and CO_2, would be retained.

A further consequence of such a scenario would be that the existence of a Stage I atmosphere on Mars may not persist through the end of the accretionary era. Let us suppose that a primitive atmosphere amounting to 1 ppm of the planetary mass was present after 99% of the mass of Mars had been accreted. If only the very conservative proportion of 10^{-4} of the impact energy of the late-accreting material were used to shock-heat the atmosphere, enough energy would be deposited to blow off the entire atmosphere. Clearly this possibility deserves intensive study.

In summary, the fundamental issue of the primordial volatile content of Mars thus remains unresolved by the available evidence. No one disputes that the present atmospheric inventory is small: What is at issue is whether Mars has outgassed thoroughly and whether massive loss of volatiles has taken place. The bulk density of the planet and its rotational moment of inertia reveal a high degree of oxidation, a property which, in chondrites, is always correlated with a high volatile content. For this reason we favor compositional models for the early Mars in which the proportions of water, carbon, etc. were higher than in the early Earth.

If Fanale's arguments about early catastrophic outgassing, occurring within about 100 million years of the dissipation of the nebula, are correct, then the accretional energy of the planet was available to dissipate any primary (Stage I) atmosphere. This phenomenon has never been modeled, although recent studies on the atmospheric effects of the Cretaceous impacting body should provide the necessary modeling techniques. Such dissipation would be of much less importance on planets with more massive atmospheres and deeper gravitational wells, such as Earth and especially Venus.

The suitability of Mars for the origin of life during its earliest history is far from obvious, and the impossibility of producing and accumulating organic matter in the present highly oxidizing photochemical regime is manifest.

There remain a number of urgent needs in the study of the evolution of Mars and its atmosphere. First and foremost among these, and at the top of the list in order of expense, is the dating of Martian geological and tectonic processes. Valuable insight into the chemical evolution of the planet may also be gained by compositional mapping of the surface, a goal which may be achieved at far lower cost than a Mars sample return mission.

Our discussions of Venus, Earth, and Mars have brought out the power of comparative study of the planets, and have emphasized how poorly we were able to comprehend planetary formation and evolution from our parochial studies of Earth alone. While we cannot yet claim to understand all the complexities we observe, we nonetheless have greatly widened our perspectives and can, on the basis of our new factual knowledge of the planets, greatly narrow the range of defensible models. Further, the clear differences between the terrestrial planets bear testimony to condensation and fractionation processes that depend critically on heliocentric distance. It is of great interest to us to compare these systematics to trends within the satellite systems of Jupiter, Saturn, and Uranus, where the formation of regular systems of satellites has imitated the formation of the planetary system itself. Thus, we now turn our attention to the outer solar system.

4.5. THE JOVIAN PLANETS

The massive Jovian planets, Jupiter, Saturn, Uranus, and Neptune, stand in sharp contrast to the small and rocky terrestrial planets. While the masses of the atmospheres of the terrestrial planets are in every case less than 10^{-4} of the total planetary mass, the Jovian planets are dominated by volatiles and are characterized by enormously massive hydrogen-rich fluid envelopes. An excellent and detailed review of our knowledge of the Jovian planets at the beginning of spacecraft exploration is given by Newburn and Gulkis (1973), and a review of the chemical aspects of the outer planets is given by Lewis (1973a).

The Jovian planets, because of their great distances from the Sun and their very deep gravitational potential wells, are not subject to escape of light gases, and their rates of capture of solar wind and cometary material are far too small to affect their volatile-element budgets even over several billion years. Thus, to an excellent approximation the Jovian planets evolve as closed systems.

This is not to suggest that there is no evolution of these planets: Indeed, there is excellent evidence for slow collapse with the gravitational potential energy change due to collapse being radiated off into space as excess thermal emission in the middle IR (Orton and Ingersoll, 1976). Thus, the radii and "surface" temperatures of these planets evolve with time (Grossman et al., 1972; Graboske et al., 1975; Bodenheimer, 1976). As we shall see, the effects of such an evolutionary change are not nearly as profound as those which we have already encountered in our discussion of the terrestrial planets. Interestingly, the present atmospheres of the Jovian planets allow us a glimpse of the kinds of changes attendant upon such a cooling trend through the comparative study of four atmospheres whose temperatures differ because of their different distances from the Sun.

Although the evolutionary changes occurring on the planetary scale are slow and not particularly profound, the processes for effecting chemical change in these atmospheres are active and of very fundamental interest. Consider, for example, a deep, massive atmosphere containing hydrogen, helium, methane, ammonia, water vapor, and a wide range of other chemically reduced species on a slowly collapsing planet. The process of collapse releases a great deal of heat in the deep interior of the planet. We have already commented on the high opacity of dense hydrogen due to collision-induced dipole absorption throughout the thermal IR. Because of this high opacity, heat transported from the deep interior to the tenuous, transparent outer layers of the atmosphere must be carried by convection.

This deep convection, essential to the transport of heat, also, of course, has the effect of lifting parcels of material from great depths up to the top of the convective region of the atmosphere. Thus, chemical species present at equilibrium at temperatures above 1000 K and pressures above 1 kbar may be transported up to levels where they are not in equilibrium, and where they may be detected by remote spectroscopic observations. Conversely, distinctive chemical species produced by solar UV photolysis at high altitudes will be carried down into the lower atmosphere by the strong vertical mixing, and will be destroyed there by chemical equilibration.

Further, a variety of other disequilibrating processes can be imagined, each with potential to contribute to the chemical diversity of the atmosphere in distinctive ways.

On a Jovian planet, there is no cool surface upon which interesting molecular species, such as complex organic molecules, may accumulate. Instead, the entire atmosphere is in approximate chemical steady state, in which the average rates of synthesis and destruction for each species are equal. One cannot expect that the steady-state concentrations of complex species will be high, but with highly sensitive modern analytical techniques, there is rational reason to hope that the steady-state concentrations of a number of species of interest will be high enough to be observable. In fact, as we shall see, there is very strong evidence that several of the minor species already observed in the atmosphere of Jupiter are produced by just such disequilibrating processes, and are therefore useful as global-scale tracers of vertical mixing of the atmosphere.

Models for the origin and evolution of the atmospheres of the terrestrial planets often are postulated to pass through a Stage I, in which the atmosphere is highly reduced and at a temperature of a few hundred degrees Kelvin. Although such environments could not persist long on any of the inner planets, their importance as reaction media for the production and chemical evolution of organic material is obvious. Such environments do, however, exist today within the atmospheres of the Jovian planets. On Jupiter alone, the mass of gas between 100 and 1000 K is roughly 10^4 times the total atmosperic mass of the Earth, or about 0.02 Earth mass. Thus, a vast natural laboratory for the study of fundamental evolutionary processes is available for our inspection and edification. Study of this environment will quickly tell us far more about the origin of organic matter than we can ever hope to deduce from study of the geological record on Earth, where billions of years of active tectonic evolution has utterly erased all evidence of the first billion years.

Our review of the Jovian planets scarcely begins to touch the great volume of recent literature, beginning with the Pioneer 10 and 11 Jupiter flybys (*Science*, Vol. 183, pp. 303–324, 1974; Vol. 207, pp. 400–430, 1980). We strongly suggest the excellent book *Jupiter*, edited by T. Gehrels (1976), as well as the recent avalanche

TABLE 4.16
Spacecraft Missions to Jupiter and Saturn

Launch date[a]	Spacecraft	Encounters
2 Mar. 1972	Pioneer 10	Jupiter, 4 Dec. 1973
		Crossed Saturn's orbit
		Crossed Uranus' orbit
5 Apr. 1973	Pioneer 11	Jupiter, 3 Dec. 1974
		Saturn, 1 Sept. 1979
20 Aug. 1977	Voyager 1	Jupiter, 5 Mar. 1979
		Saturn, 12 Nov. 1980
5 Sept. 1977	Voyager 2	Jupiter, 9 July 1979
		Saturn, 27 Aug. 1981
		Uranus, Feb. 1986
		Neptune, ?

[a]Jupiter launch dates recur at 13-month intervals.

of reports from the Voyager 1 and 2 encounters with Jupiter and Saturn described in *Science* (Vol. 204, pp. 945–1008, 1979; Vol. 208, pp. 925–985, 1979; Vol. 212, pp. 159–243, 1981), *Nature* (Vol. 280, pp. 725–806, 1979), and numerous other articles in the past few years. The greatest advances from these encounters have, however, concerned the magnetospheres and satellite systems of these planets, and have little impact on our discussion of disequilibrating processes and evolutionary trends on Jupiter and Saturn themselves. The short history of spacecraft investigations of the outer planets is given in Table 4.16.

4.5.1. Jupiter

The majority of the mass of our planetary system is in Jupiter. With a mass of 1.90×10^{31} gm, Jupiter can be said to lie in a range intermediate between planets and stars. The composition of Jupiter's atmosphere is strikingly close to that of a low-temperature sample of solar material (Wildt, 1938; Peebles, 1964; Hubbard, 1973; Podolak and Cameron, 1974; Zharkov and Trubitsyn, 1974) in which the relative abundances of hydrogen, helium, methane, and ammonia are all nicely in accord with the solar abundances of their constituent elements (R. Wildt, 1934a,b; McElroy, 1969b; Owen, 1970; Teifel, 1970). One dissenting opinion (Wallace and Hunten, 1978) favors a C:H ratio several times the solar value.

An external observer sees Jupiter as an immense, noticeably oblate spheroid streaked with varicolored cloud bands paralleling the equator, with almost feature-less grey cloud hoods over both polar regions. The constant change and complex swirling morphology of the visible surface suggests at once that we are in fact observing the top side of an actively circulating atmosphere with dense cloud cover (Peek, 1958). From the mean thermal emission temperature of the planet and from the spectroscopically observed abundance of ammonia above the cloud tops, it can be shown that the topmost cloud layer is made of tiny crystals of solid ammonia (Kuiper, 1952). The IR spectrum of Jupiter, when integrated over all wavelengths, gives an emitted flux equal to that of a 125 K blackbody, virtually independent of latitude (see Orton and Ingersoll, 1976). The steady-state temperature for a grey planet with Jupiter's albedo at 5.2 AU from the Sun is only 102 K, and therefore the flux emitted by the planet is $(125/102)^4$, or about 2.25 times that due to reemission of absorbed sunlight. In the 5 μm spectral region there occurs a narrow spectral window between absorption bands due to methane and ammonia, where the atmosphere of Jupiter is so transparent that brightness temperatures of over 300 K are seen. This observation plumbs to a deep level of the atmosphere where the total pressure is several bars, far below the NH_3 cloud tops.

Observations of microwave emission from Jupiter show that, at wavelengths at which ammonia is a strong absorber (near 1 cm), the topmost (approximately 125 K) cloud layers are opaque. To longer wavelengths, where ammonia is more transparent, progressively higher temperatures are seen. The wavelength dependence of the opacity is easily explained if the ammonia mole fraction is governed by vapor pressure equilibrium in and above the top cloud layer, and is equal to the solar $NH_3:H_2$ ratio below these clouds (Gulkis and Poynter, 1972).

When the chemical behavior of a solar-composition atmosphere under the ranges of temperature and pressure on Jupiter is considered, it is found that the region of the Jovian troposphere between about 125 and 300 K should contain three major cloud layers. The topmost of these, solid NH_3, is due to simple condensation of NH_3 vapor in adiabatically cooling updrafts. Next, near 210 K, there should be a layer of crystalline NH_4SH clouds formed by chemical reaction between gaseous NH_3 and S_2S. Even lower, near the 280 K level, water vapor would condense to form a dense layer of water droplet and ice crystal clouds. The liquid droplets would contain dissolved NH_3 (Lewis, 1969a). A solar-composition, adiabatic model for the vertical distribution of major cloud condensates and of condensable gases is given in Figure 4.32, after Weidenschilling and Lewis (1973). The transparency of the atmosphere down to the 300 K level at 5 μm wavelength requires not only the absence of abundant gaseous species which absorb strongly in this wavelength region, but also the total absence of cloud cover at all three of these condensation levels. Since rising atmospheric parcels of warm, moist gas will generate copious clouds, whereas cool, subsiding stratospheric gas will be extremely dry and free from condensables, it is most reasonable to interpret the 5-μm clear regions, such as Jupiter's North Equatorial Belt (NEB), as regions of atmospheric subsidence. Curiously, all regions of Jupiter with effective temperatures of 200–300 K are noticeably dark to the eye, often tinted yellow, orange, or brown, and in the case of the spots at and above ~260 K, dark blue (due to Rayleigh scattering in a very deep, clear molecular atmosphere.)

Models of the deep interior of Jupiter reveal several facts of great significance for understanding the atmosphere. First, although the heavy (rock-forming) elements are evidently enriched by a factor of several in Jupiter relative to the Sun (Podolak and Cameron, 1975; Podolak, 1977; Grossman et al., 1980), the overall low bulk

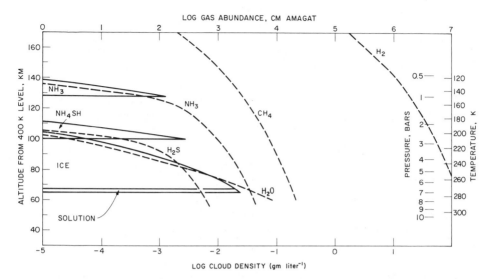

FIGURE 4.32. **Structure and cloud condensation in the upper Jovian troposphere. The main cloud layers, composed of aqueous NH$_3$ solution plus H$_2$O ice, solid NH$_4$SH, and solid NH$_3$, are formed by condensation of H$_2$O, NH$_3$, and H$_2$S, which are stable in the lower atmosphere. (After Weidenschilling and Lewis, 1973.)**

density of the planet, only 1.3 gm cm^{-3}, makes it clear that the planet is dominated by hydrogen and helium and is strikingly similar in overall composition to the Sun itself. Second, the large observed heat flux, which we have seen exceeds that attributable to reemission of absorbed sunlight by more than a factor of two, is only comprehensible if interpreted as revealing a substantial internal heat source. Jupiter's content of long-lived radioactive elements cannot be remotely large enough to produce the observed flux, and the production of heat by hydrogen fusion reactions in the core would not be of importance except for bodies with at least 60 times the mass of Jupiter. The most plausible heat source seems to be associated with leakage of heat from a hot, highly compressed fluid interior. Loss of thermal energy causes a decrease in interior pressures, which leads to slow shrinkage of the planet. The shrinkage converts gravitational potential energy into heat. In order to provide the observed heat flux, Jupiter must shrink in radius by only about 1 mm per year.

The heat produced by shrinking of the planet appears as compressional heating of the deep interior, and this heat must be efficiently transported upward at a sufficient rate to provide the observed flux. Radiative transport of heat through dense hydrogen is remarkably inefficient, because dense hydrogen is very opaque to thermal radiation. Hydrogen molecules have a pure rotational absorption line at about 17 μm wavelength, but direct absorption at this wavelength by isolated hydrogen molecules is symmetry-forbidden (H$_2$ is a symmetrical diatomic molecule with no dipole moment). When, during intermolecular collisions, the symmetry of the molecule is disturbed, it becomes a strong absorber. Because of this large opacity, radiative transport of the entire observed planetary heat flux through Jupiter's atmosphere would require very large vertical temperature gradients; so large, in fact, that the atmosphere would be very unstable against convective overturn. Thus, the entire outer fluid envelope of the planet, extending down to levels where the pressure is many millions of bars, is in a state of rapid convective overturn. If the global circulation regime of the atmosphere were known in detail (and it certainly is not), then we could trace out the trajectory in temperature–pressure space followed by a given parcel of atmosphere as it is carried about by the planetary-scale circulation. Since we are in fact unable to do this, we are forced to use a very much simpler description of the process of overturn and vertical mixing. In practice, this is usually done by regarding the vertical motions in the atmosphere as being a one-dimensional random walk.

The characteristic property of such a random-walk model is that the mean displacement of Δz of a parcel of gas from its altitude z_0 after time interval Δt is proportional to the square root of the time interval; in other words,

$$(\Delta z)^2/\Delta t = K, \qquad (4.95)$$

where K is a constant, called the eddy diffusion parameter, with dimensions of cm^2 sec^{-1}. Clearly, molecular diffusion, described by the constant D $(cm^2 \; sec^{-1})$ is a true random-walk process, whereas eddy diffusion is merely represented as one in the absence of a sufficiently detailed knowledge of the relevant fluid physics. The eddy diffusion constant for the Jovian troposphere can be estimated from the observed convectively transported heat flux to be near 10^8 cm^2 sec^{-1}. The characteristic vertical scale for a planetary atmosphere, the scale height H $(=RT/\mu g)$, is about the same as the spacing between adjacent cloud layers, and is about the altitude range over which convective cells maintain their coherence (Stone, 1976). A typical value of the scale height in the upper Jovian troposphere is about 30 km. We may then estimate that the characteristic mixing time over one scale height is $\tau \simeq H^2/K$, or 10^5 sec, whereas the characteristic vertical velocity, H/τ, is 30 cm sec^{-1}. Vertical mixing from a level 10 scale heights below the visible cloud tops, where the pressure is $e^{10} \simeq 10^4$ times higher and the temperature is over 2000 K, requires only $\sim 10^7$ sec. It is therefore not hard to imagine that certain exotic chemical species which are stable and present only in the hot lower atmosphere may be mixed upward to the cloud tops by the vigorous convection before they have time to be destroyed by chemical reactions along the way. Detailed theoretical studies of the chemistry of adiabatic, solar-composition atmospheres have been carried out by Lewis (1969a,b), by Barshay and Lewis (1978), and by Fegley and Lewis (1979) for the purpose of determining which chemical species ought to be observable if equilibrium were attained at all levels, and which species might serve as unambiguous tracers of deep atmospheric overturn if exact equilibrium is not attained on the time scale of the overturn process. An extension of the Jovian atmospheric model to $T \simeq 750$ K is given in Figure 4.33. The most general conclusion of these studies is that the troposphere of Jupiter is remarkably close to chemical equilibrium. The second is that certain observed trace components of the atmosphere, notably phosphine (PH_3), with a mole fraction of about 4×10^{-7}

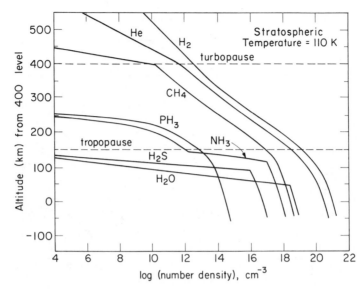

FIGURE 4.33. **Chemical structure of the atmosphere on Jupiter down to the ~750 K level. Stratospheric temperatures are near 110 K, and temperatures sufficient for rapid chemical equilibration occur only a few hundred kilometers below the visible cloud tops.**

(Ridgway, 1974a; Ridgway et al., 1976; Larson, et al., 1977), and carbon monoxide (Beer, 1975), with a mole fraction of about 10^{-9}, are almost certainly tracers of deep atmospheric overturn from the ~1000 K level (Prinn and Barshay, 1977; Barshay and Lewis, 1978). The equilibrium abundance profiles for carbon and phosphorus compounds along a Jovian adiabat are given in Figures 4.34 and 4.35 to illustrate the documentation available for these and a number of other elements. Other gaseous components of the deep atmosphere which may also serve as tracers of deep mixing (Fegley and Lewis, 1979) include germane (GeH_4) (Corice and Fox, 1972), arsine (AsH_3), thallium iodide (TlI), indium bromide (InBr), boric acid (H_3BO_3), hydrogen selenide (H_2Se), hydrofluoric acid (HF), arsenic trifluoride (AsF_3), and stibine (SbH_3). Of these, only germane has been detected (Fink et al., 1978), with an abundance requiring quenching at a temperature of ≥ 800 K.

Rapid vertical mixing is by no means the only mechanism by which the cloud-top atmosphere may be made to deviate from solar composition. Another major mechanism, photolysis, utilizes UV sunlight to break the bonds of the more abundant and stable hydrogen-containing species such as CH_4, NH_3, H_2S, and H_2 itself. In addition, the interloper PH_3 is readily destroyed by UV light.

Photolytic production of H and H_2 has, on most planets, a very important evolutionary role in that it makes these light species available for escape from the planet. On Jupiter, however, thermal escape is utterly negligible. Rather, it is the complementary aspect of photolysis, the production of heavier molecules, which gives interest to the study of the photochemistry of the Jovian planets.

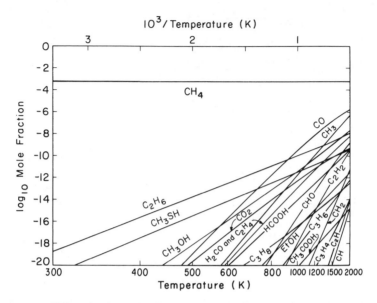

FIGURE 4.34. Equilibrium abundance of carbon compounds in the troposphere of Jupiter. Rapid vertical mixing from the ∞1100 K level is required to provide the amount of CO observed in the atmosphere near the cloud tops, a mole fraction $X_{CO} \simeq 10^{-9}$. The altitude scale gives depth below the 1 bar, ~175 K level. (After Prinn and Barshay, 1977, and after Barshay and Lewis, 1978.)

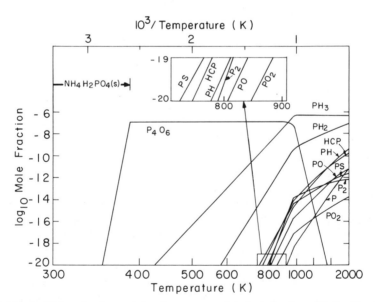

FIGURE 4.35. Equilibrium abundances of phosphorus compounds in the troposphere of Jupiter. The observed PH_3 abundance near the cloud tops is roughly solar, requiring rapid vertical mixing from below the ~800 K level. (After Barshay and Lewis, 1978.)

An incoming solar photon on Jupiter first encounters a diffuse but extensive upper atmosphere consisting almost entirely of hydrogen and helium, in which dissociation and photoionization reactions occur. Next, near the turbopause, the solar radiation first encounters methane, which is of negligible importance in the upper atmosphere due to diffusive sedimentation (eddy mixing cannot compete with diffusive unmixing when the mean free path of the gas molecules is very large). Little ionization of methane occurs, since most of the photons energetic enough to ionize methane are absorbed by the overlying layer of the enormously more abundant gases H_2 and He. Below the immediate region of the turbopause the main product from methane photolysis is the methylene radical CH_2, with minor methyl radical CH_3. These fragments can recombine (Strobel, 1969; Prinn, 1970) to produce simple hydrocarbon molecules. Among the most abundant species produced (Strobel, 1973a, 1974a) are ethane (C_2H_6) and acetylene (C_2H_2); ethylene (C_2H_4), which is an essential intermediate in the interconversion of these two species, is far less abundant in a photochemical steady state. Both acetylene and ethane, with mole fractions near 10^{-8} and 10^{-6}, respectively, are seen in emission in spectra of Jupiter (Combes *et al.*, 1974; Ridgway, 1974b; Aumann and Orton, 1976; Tokunaga *et al.*, 1976; Orton and Aumann, 1977), clearly indicating their presence in a warm layer, heated by absorption of UV and IR sunlight, above the cold stratosphere. The relevant reactions are summarized in Figure 4.36. Organic molecules more complex than ethane are unknown on Jupiter, although simple hydrocarbons containing three or four carbon atoms are surely produced in very low yields.

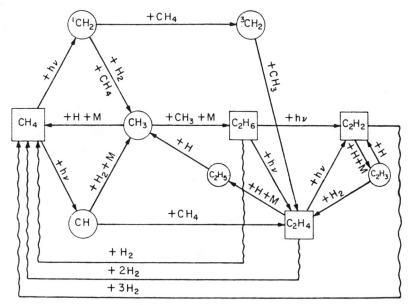

FIGURE 4.36. **Photochemistry and thermochemistry of methane on the Jovian planets. The main photolysis products from methane are ethane (C_2H_6) and acetylene (C_2H_2). (After Prinn and Owen, 1976.)**

Much debate has centered on the issue of whether photolysis can produce large amounts of more complex organic species in the atmosphere of Jupiter. Sagan *et al.* (1967) attempted to model the synthesis of organic molecules through UV excitation of the upper atmosphere by carrying out chemical equilibrium calculations for a very hot gas, a procedure entirely without theoretical or experimental justification. Numerous experiments have been carried out on "simulated Jovian atmospheres," in which vast UV or electrical discharge energy fluxes have been used to excite roughly equimolar mixtures of H_2O, CH_4, NH_3, and sometimes H_2, H_2S, C_2H_6, and PH_3 (Woeller and Ponnamperuma, 1969; Chadha *et al.*, 1971; Khare and Sagan, 1973; Molton and Ponnamperuma, 1974; Ferris and Chen, 1975b; Ponnamperuma, 1976; Khare *et al.*, 1978). Most such "simulations" are grossly dissimilar to Jupiter in composition, temperature, energy input, or all of these at once, and this genre has been criticized for its irrelevance to real processes on Jupiter by Lewis and Prinn (1971). Recent extensions of these simulations are considerably more relevant, and further exploration of at least moderately Jupiter-like conditions is under way (see Noy *et al.*, 1979; Raulin *et al.*, 1979).

Photolysis of ammonia is limited by the filtering of incident UV light by hydrogen, helium, and methane and by the condensation of most of the ammonia to form the topmost cloud layer. Accordingly, little ammonia lies above the cloud tops, and UV longward of the methane absorption cutoff at 1650 Å mostly penetrates into the region of the cloud tops before being absorbed. Phosphine, which is much less abundant than ammonia in the troposphere, is uncondensable and hence more abundant than ammonia in the stratosphere. The UV absorption spectra of ammonia and phosphine are very similar, and as a result phosphine photolyzes rather more rapidly than ammonia and helps to shield the ammonia vapor in the clouds from solar UV radiation. The studies of ammonia photolysis by Prinn and Lewis (1975) have been combined by Strobel (1977) to show that the photolysis of PH_3 is indeed the more important of the two. Further, radical-scavenging reactions between PH_2 and C_2H_2 cause an interesting local competition between these products (Vera Ruiz and Rowland, 1978). Since the available solar flux between the methane absorption cutoff at 1650 Å and the NH_3/PH_3 cutoff near 2350 Å is very large, the rate of NH_3/PH_3 photolysis will be about 200 times the rate of methane photolysis. The main product of ammonia photolysis is nitrogen, with condensable hydrazine (N_2H_4) an interesting second (Prasad and Capone, 1976). Atreya *et al.* (1977a) have shown that the steady-state nitrogen abundance in the cloud-top region, derived from NH_3 photolysis, is near a mole fraction of 10^{-10}. Nitrogen is, of course, essentially undetectable even at high concentrations, and this estimated abundance would tax even a sophisticated entry probe's analytical capabilities. The chemistry of ammonia is summarized in Figure 4.37.

The principal product from PH_3 photolysis has been shown by Prinn and Lewis (1975) to be solid particles of red phosphorus. The predicted steady-state abundance above the clouds, averaged over Jupiter, is too small to be visible, using the eddy diffusion constants estimated for the stratosphere based on studies of the photolysis of ammonia ($K \simeq 10^4 \ cm^2 \ sec^{-1}$). In regions of enhanced dynamical activity, such

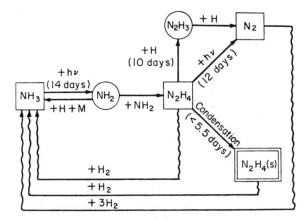

FIGURE 4.37. **Photochemistry and thermochemistry of ammonia on Jupiter and Saturn. The main photolysis products are N_2, which can also be made in the deep interior in trace quantities, and hydrazine (N_2H_4), which is condensable in the stratosphere (and also readily photolyzed to make N_2). (After Prinn and Owen, 1976.)**

as the Great Red Spot (Ingersoll, 1973), where the eddy diffusion constant may well exceed 10^6 in the lower stratosphere, the amount of solid red phosphorus suspended by the turbulent gas would be sufficient to give the area a distinct red coloration. Note that methane photolysis is far slower, occurs at such high altitudes that it is essentially uniform over the planet, and produces no known colored species. It is thus not a plausible source of colored organic matter to explain the appearance of the Great Red Spot. The chemistry of phosphorus is outlined in Figure 4.38.

In regions in which the ammonia clouds are sparse or absent, UV radiation longward of the NH_3/PH_3 absorption cutoff will encounter gaseous H_2S in the vicinity of the NH_4SH cloud layer (Wildt, 1937). The only important initial product is the HS radical, which recombines to make very photolabile hydrogen polysulfide chains, all of which combine to produce a series of yellow, orange, and brown tarry solids. Lewis and Prinn (1970) have explained the observed association of 5-μm temperatures in excess of 200 K with dark yellow to brown markings (Keay et al., 1973; Westphal et al., 1974) as being causally related, with the high temperatures and the photochemical production of sulfur chromophores both due directly to clearing of the ammonia clouds. Cyclic S_8 has been claimed (Khare and Sagan, 1975) and disclaimed (Khare and Sagan, 1977) as a major chromophore. Because of the ability of H_2S to absorb out to a wavelength of 2700 Å, where the solar flux is growing rapidly with increasing wavelength, the photolysis rate for H_2S in clear regions will be roughly 2000 times that for methane. Both photolysis and condensation of NH_4SH contribute to the absence of H_2S (Owen et al., 1977a). Figure 4.39 summarizes the chemistry of H_2S.

The photochemical production of N_2, N_2H_4, P_4, S_x, C_2H_6, C_2H_2, and cross products such as CH_3NH_2 and CH_3SH at the top of the atmosphere must, in the steady state, be balanced by reformation of CH_4, NH_3, PH_3, and H_2S deep within

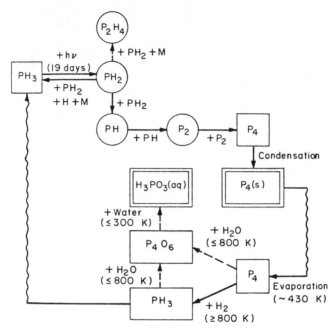

FIGURE 4.38. Chemistry of phosphine in the atmospheres of the Jovian planets. The only important product is the red chromophore P_4. (After Prinn and Owen, 1976.)

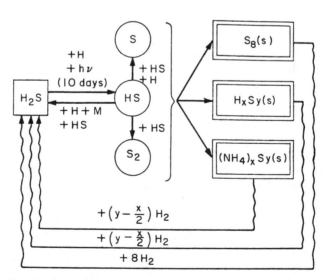

FIGURE 4.39. Photochemistry and thermochemistry of H_2S in the Jovian atmosphere. The most abundant photolysis products are yellow, orange, and brown polymeric sulfur species. (After Prinn and Owen, 1976.)

the planet, where the temperatures and pressures are so high that these are the equilibrium species. For PH_3, this requires temperatures and pressures near or above 1000 and 1 kbar. Thus, thermochemical reactions again are found to play a crucial role in atmospheric cycles. The relative abundances of photochemical products do not depend on time; their absolute abundances can be read from Figure 4.40 as a function of the vertical mixing time scale.

The only known constituents of Jupiter's atmosphere which are disequilibrium species not readily made from CH_4, NH_3, H_2S, and PH_3 photolysis are CO, PH_3, and GeH_4. All of these can be provided by deep mixing from the 1000 K level in the troposphere. CO has also been attributed to high-altitude photochemical processing of oxygen injected into the upper atmosphere by meteorite infall or captured gases from the Galilean satellites (Prather *et al.*, 1978; Strobel and Yung, 1979).

In the absence of a cool surface upon which chromophores and organic matter might accumulate, they are fated to come rapidly to a steady state in which their rates of production and destruction are equal.

Other disequilibrating processes are far more speculative and almost certainly of far inferior quantitative significance. For example, if the efficiency of generation of lightning discharges by the atmospheric heat flux is the same on Jupiter as on Earth (and if the efficiency of production of complex organic compounds is the same as in laboratory experiments on pure methane), then the weight fraction of total organics in the Jovian atmosphere would be near 10^{-13} (Lewis, 1976).

Lightning has been suggested as a possible source of acetylene and chromophores

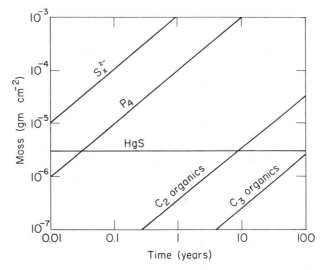

FIGURE 4.40. **Rates of production of photolysis products in a solar composition Jovian atmosphere. For a given accumulation time, the steady-state column densities of these photolysis products may be read from this graph. The only known colored products are S_x and P_4; they are also the two most abundant products. The accumulation time in the (sluggish) stratosphere, where $K \simeq 10^5$ cm² sec⁻¹, is ~1 yr, while tropospheric mixing time scales are much shorter. (After Lewis, 1976.)**

on Jupiter by Bar-Nun (1975, 1979). The formation process would invoke a very high frequency (10^2–10^4 times the terrestrial rate) of large lightning discharges at altitudes near the troposphere, causing shock-wave (thunder) heating of nearby atmosphere. Shock-wave synthesis in a solar-composition gas, such as would be found in and below the water clouds, forms CO as the dominant product (Lewis, 1980a). If the H_2O content of the lower atmosphere is in fact enriched above solar composition by a factor of about 10, as many interior models of Jupiter suggest, this conclusion is further strengthened. Only shock-processing of a very dry H_2–CH_4 mixture gives appreciable yields of C_2H_2 and HCN.

Several lines of evidence regarding the location, frequency, and energy dissipation rates of Jovian lightning were provided by the Voyager 1 Jupiter encounter. First, the imaging experiment photographed 20 flashes in a 192-sec time exposure of the night side, covering an area in excess of 10^9 km^2 (Cook et al., 1979). The individual flashes were observed to dissipate about 10^{17} erg each, for an overall energy dissipation rate of 10^{-3} erg cm^{-2} sec^{-1}. The upward total energy flux carried by convection is 0.9×10^4 erg cm^{-2} sec^{-1}, so the efficiency of conversion of convective energy into visible light is about 10^{-7} (Lewis, 1980b). The light emission from terrestrial lightning is about 10% of the total energy dissipated (Hill, 1979), and a luminous efficiency of 10–20% would be reasonable for Jupiter's atmosphere, which is more transparent in the near UV. Thus, the proportion of the convective energy flux which goes into visible Jovian lightning flashes is about 5×10^{-7}. The equivalent figure on Earth is about 10^{-4}.

A second detection of Jovian lightning was reported by the plasma wave experiment (Gurnett et al., 1979; Scarf et al., 1979), who recorded frequency-dispersed radio pulses (whistlers) presumably arising from Jovian lightning. A simple analysis of the whistler rate data (Lewis, 1980b) suggests that the total lightning stroke rate is comparable to the rate of visible flashes seen by the imaging experiment. Since the upper clouds are of optical depth ~1 (Hunt, 1973) and the water clouds are optically thick, all lightning in the region depleted of water (and hence capable of shock synthesis of C_2H_2 and HCN) should have been photographed. Whistlers can originate at any depth in the cloud structure: Whistlers in excess of the number needed to explain the observed flashes would have to originate deep in the water clouds; otherwise we would see them. Scarf et al. (1981) have calculated an upper limit on the total lightning rate and found that deep (optically unobserved) discharges could be up to ~100 times as frequent as the high-altitude visible flashes without violating the data. Such lightning would produce CO, not organics (Lewis, 1980a).

An energy budget for a number of disequilibrating processes is given on Table 4.17. We conclude that lightning is a very minor source of colored matter and organic polymers on Jupiter.

Readers interested in a detailed and general discussion of the photochemistry of the Jovian atmosphere should consult the review by Prinn and Owen (1976). The entire range of disequilibrating processes in the upper troposphere is addressed by Lewis (1976), while the chemical effects of rapid vertical mixing are discussed by

TABLE 4.17
Energy Available for Disequilibrating Processes on Jupiter[a]

Process	Principal products	Available flux ($erg\ cm^{-2}\ sec^{-1}$)	Reference
CH_4 + UV ($\lambda < 1600$ Å)	C_2H_6, C_2H_2	0.2	Strobel, 1973a
NH_3 + UV ($\lambda < 2300$ Å)	N_2, N_2H_4	10	Strobel, 1973b
PH_3 + UV ($\lambda < 2300$ Å)	P_4 (red)	15	Prinn and Lewis, 1975
H_2S + UV ($\lambda < 2700$ Å)	S_x (yellow → brown)	100	Lewis and Prinn, 1970
Lightning	CO, C_2H_2, HCN	<0.5	Lewis, 1980a,b

[a]After Lewis (1980b).

Barshay and Lewis (1978). Atreya *et al.* (1977b) present Lyman-α intensity data which show enormous ($100\ erg\ cm^{-2}\ sec^{-1}$) energy dumping in the auroral zone, which may have some chemical significance.

There is clearly no dearth of dynamic processes in the Jovian atmosphere. However, because Jupiter is a closed system to an excellent degree of approximation, we find that the only true evolutionary trend discernable is the slow increase in density of the planet with its concomitant lowering of the Jovian adiabat to ever more frigid (and chemically less interesting) temperatures. Thus, the distant future state of Jupiter near the end of the Sun's MS lifetime will lie intermediate between the present states of Jupiter and Saturn. Given nearly identical compositions for these two planets, present-day observations of both can effectively span several billion years of the evolutionary history of a single such planet.

With this idea in mind, let us turn our attention to a brief description of Saturn.

4.5.2. Saturn

Much of what is presently known about Saturn is readily conveyed by the simple statement that it is very similar to Jupiter, except a little smaller and a little farther from the Sun. Much of our ignorance about Saturn relative to Jupiter is also attributable to exactly these same two conditons.

Saturn, like Jupiter, is a net emitter of heat, with an internal heat source larger than the amount of absorbed solar energy. The atmosphere, though it has historically been observed at much lower resolution and brightness level, is banded and apparently turbulent and complex. The Pioneer 11 and Voyager Saturn flybys (*Science*, Vol. 207, p. 400, 1980; Vol. 212, p. 159, 1981) have recently provided a very impressive improvement in our knowledge of Saturn, but without any new knowledge of the origin and evolution of Saturn's atmosphere.

As with Jupiter, the IR and radio thermal emission spectrum is that of a body with a deep, hot atmosphere with an adiabatic temperature–pressure profile, dropping to about 100 K at the level of the Saturnian cloud tops, where the pressure is near 1 bar. The IR and red end of the visible spectrum are filled with extremely strong methane absorption bands. Intrinsically weak hydrogen absorption features are found, but with greater intensity than on Jupiter. Helium was detected by the IR radiometer spectrometer (IRIS) on Voyager 1 (Hanel *et al.*, 1981) in concentrations significantly below the solar He:H ratio. This helium depletion is evidently due to the downward concentration of helium into Saturn's core predicted by Smoluchowski (1967b). Ammonia is also present, but, with a cloud-top temperature of only 100 K, the vapor pressure is so low that it is barely possible to detect NH_3 in the spectrum. Detection of NH_3 may, in fact, require the temporary partial clearing of the ammonia crystal cloud layer, permitting visual access to lower, warmer levels where ammonia has not been depleted by condensation. A Saturn atmosphere model, in which cosmic abundances of volatiles are assumed, is given in Figure 4.41. The ammonia abundance below the saturation level can be estimated from the wavelength dependence of the brightness temperature in the vicinity of the ~1.3 cm inversion band of NH_3. Model atmospheres with different NH_3 mole fractions yield predicted brightness temperature spectra which can be fit to the Saturn data for an ammonia mole fraction of $(0.3–1.0) \times 10^{-4}$ (Kuz'min *et al.*, 1972; Ohring and Lacser, 1976). The main cloud deck has been generally accepted to be solid NH_3 (Bugaenko *et al.*, 1975; Caldwell, 1977).

The methane abundance has been estimated many times: Abundances of 50–100 m agt are usual (Bergstralh, 1973; Macy, 1976), suggesting a C:H ratio 1.5–3 times the solar ratio. The isotopically substituted CH_3D molecule has been

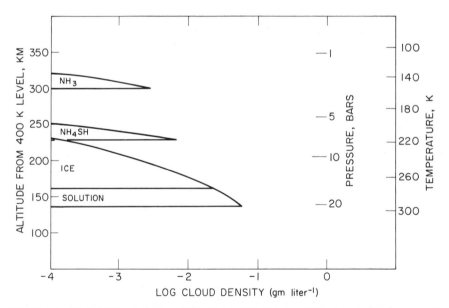

FIGURE 4.41. Atmospheric and cloud model of Saturn. Note the close similarity to Jupiter. Pressures are slightly higher and the liquid NH_3/H_2O clouds are more prevalent on Saturn. Solar composition is assumed for the condensable gases. (After Weidenschilling and Lewis, 1973.)

observed (Fink and Larson, 1978), implying a D:H ratio of about 2×10^{-5}. Beer *et al.* (1972) first detected CH_3D on Jupiter, and Beer and Taylor (1973) found a D:H ratio lower than the terrestrial value. Trauger *et al.* (1973) observed HD directly and gave a Jovian D:H ratio of 2×10^{-5} ($\pm 20\%$). D:H on Uranus is $<4 \times 10^{-4}$ (Lutz and Owen, 1974).

Phosphine (PH_3) was tentatively identified by Gillett and Forrest (1974), and measured to have a mole fraction of about 2×10^{-7} by Bregman *et al.* (1975). This is essentially the solar abundance. As on Jupiter, this requires rapid upward mixing from levels where the temperature is near 1000 K. Also, as on Jupiter, PH_3 may compete with NH_3 for UV photons between 1600 and 2300 Å. An even larger proportion of these photons will be used in PH_3 photolysis on Saturn because of the very low saturation vapor pressure of NH_3 at stratospheric temperatures.

Ethane band emission has been detected in the spectrum of Saturn, but estimation of its abundance requires assumptions regarding the thermal structure of the upper atmosphere which cannot presently be tested (Gillett and Orton, 1975; Tokunaga *et al.*, 1975). Larson *et al.* (1980) and Encrenaz *et al.* (1975) have also tentatively identified C_2H_2 and C_2H_4.

The Voyager 1 IRIS team (Hanel *et al.*, 1981) give mole fractions of 1×10^{-6} for PH_3 (5 times the solar abundance), 5×10^{-6} for ethane, and 2×10^{-8} for acetylene (both photochemical products of methane). They tentatively add methyl acetylene ($CH_3-C{\equiv}CH$) and propane (C_3H_8). Upper limits on several other gases are given by Cruikshank and Binder (1969) and by Treffers *et al.* (1978).

In light of our discussion of Jupiter, these results seem easily comprehensible. All the major processes observed or deduced for Jupiter seem likely to be present on Saturn as well. Certain differences due to the lower temperature of Saturn should perhaps be mentioned. First, we would expect that the ammonia abundance will be found to be nonuniform across the disk, with an enhanced NH_3 abundance likely to be found in visually darker bands which also exhibit 5-μm brightness temperatures noticeably higher than the planetary average, possibly over 200 K. Second, PH_3 photolysis will provide a thin planet-wide haze which will only be thick enough to make Jupiter-like red spots if regions of very high dynamical activity are present. It is possible that the elevation of the ammonia cloud tops by the high turbulence in such areas may decrease substantially the column density of methane visible from above, and thus the reddest regions may also be regions of abnormally low methane absorption.

The evidence regarding the bulk composition of Saturn derived from theoretical equations of state for mixtures of cosmically abundant materials suggests that Saturn is slightly more enriched in the heavy elements than is Jupiter. Whether the traces of such chemical distinctions suggested by the spectroscopic data are reliable is unclear at the moment: We are inclined to be cautiously positive. Unfortunately, verification would require an atmospheric entry probe to penetrate to levels where all the volatiles and even some of the rock-forming elements are present for inspection. This requires impressive technological breakthroughs and no little luck.

4.5.3. Uranus and Neptune

When we consider the general desirability of comparative planetary studies, we are reminded that the comparison of Jupiter with Saturn by no means exhausts the richness of variety provided by the outer planets. Both Uranus and Neptune are denser than either Jupiter or Saturn, although far smaller in mass and radius, and thus lower in internal pressure. Uranus and Neptune, therefore, must be composed of intrinsically denser material than either of their larger brothers, meaning either a proportionately greater content of rock-forming elements or an enhancement of both rocky and ice material relative to hydrogen and helium (see, e.g., Podolak, 1976).

If the differences between these planets are due solely to different contents of rock-forming elements (and different heliocentric distances), then we must expect the externally observable features of these planets to continue the trend which we have so far observed: That lower temperatures and greater distances will further shorten the list of uncondensed gaseous species and essentially obliterate all information about the meteorology of the planets. The regions of the atmosphere most directly affected by variations in the abundances of rock-forming elements will be even more deeply buried and hidden from external view, and even simple hydrocarbons produced by methane photolysis will saturate and condense out at very low concentration levels. If, however, both the rocky and icy components are enhanced in Uranus and Neptune, then the atmospheric abundances of H_2O, NH_3, and even CH_4 may be enhanced, depending on how completely the icy fraction was condensed in the nearby nebula at the time of formation. Hydrogen sulfide plays an interesting role in this process, since substantial sulfur may be retained as FeS in the rocky component, while any leftover H_2S after iron metal exhaustion would condense with the icy component as NH_4SH.

Given this general background, then, it is most interesting to recount the debate currently surrounding the interpretation of the spectra of Uranus and Neptune. Such vast quantities of both methane and hydrogen are present in the atmospheres of both planets that all moderately strong methane bands are badly saturated, and only relatively poorly understood intrinsically weak bands are available for abundance determinations. The quantity of hydrogen in view is so immense that the short-wavelength wing of the 17-μm collision-induced H_2 rotational band is encroaching into the red end of the visible region. Pressures of several bars are reached in the accessible portion of the atmosphere. The atmospheric models in Figures 4.42 and 4.43 assume a tenfold environment of condensables relative to H_2 and He.

Spectroscopic abundance determinations on hydrogen (Herzberg, 1952) and methane (Teifel and Kharitonova, 1970; Belton and Hayes, 1975; Bergstralh, 1975) have failed to resolve the composition, and $CH_4:H_2$ ratios ranging from solar up to 100 times solar (Danielson, 1977) have been proposed for Uranus in recent years. Belton *et al.* (1971) proposed a semi-infinite cloudless molecular atmosphere, whereas Sinton (1972) observed limb brightening in a methane band, implying a high-altitude methane haze. Prinn and Lewis (1973) and Trafton (1975a) proposed

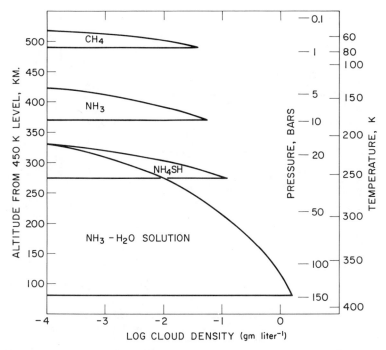

FIGURE 4.42. **Atmospheric and cloud model of Uranus. The ice-forming volatiles H₂O, NH₃, CH₄ and the rock-forming volatile H₂S are assumed to be enhanced by a factor of 10 above solar composition, relative to hydrogen and helium. See text for a discussion of the problems presented by the apparent absence of an NH₃ cloud layer. (After Weidenschilling and Lewis, 1973.)**

an interpretation of the Uranus spectrum using a high-altitude CH_4 haze layer overlying an extensive clear region, with a dense cloud layer, presumably of solid NH_3, beneath. Upper atmosphere models by Wallace (1975), Macy and Sinton (1977), and Danielson (1977) give temperature minima near 55 K at 100 mbar, with a temperature inversion of about 80 K above that level. The thermal structures of Saturn, Uranus, and Neptune are compared in Figure 4.44. The Uranus and Neptune thermal inversions are due to solar heating in CH_4 absorption bands (Macy and Trafton, 1976c). Curiously, Uranus is no warmer than Neptune: Like Jupiter and Saturn, Neptune has an internal heat source comparable to its net solar heating flux (Murphy and Trafton, 1974). Because of the 97° axial tilt of Uranus, which causes the poles to point almost directly at the Sun in turn, alternating every 32 yr, there is no simple relationship between the temperature of the visible half of Uranus at any time and the true planetary mean temperature. The puzzle could, of course, be nicely solved by a flyby of a spacecraft carrying an IR radiometer, which could indeed determine the distribution of emitted heat over the planet, and thus permit the total emitted thermal flux to be calculated.

A second observational puzzle regarding the thermal structure of the atmosphere of Uranus is that arising from measurements of the thermal emission in the radio-

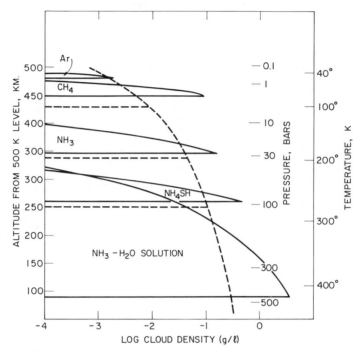

FIGURE 4.43. **Atmospheric and cloud model of Neptune. Condensables enhanced by a factor of 10 relative to hydrogen and helium compared to solar abundances. (After Weidenschilling and Lewis, 1973.)**

wavelength region. At wavelengths near 1 cm ammonia is a strong absorber, and it is no surprise to find that the centimeter-wavelength brightness temperatures of both Jupiter and Saturn are very similar, corresponding to the temperature of the solid ammonia cloud layer. In both cases, our view into the atmosphere is obstructed near the 130 K level because it is near this temperature that the vapor pressure of ammonia becomes large and the atmosphere becomes opaque. On Uranus, curiously, the entire microwave region is found to have a temperature 50 K or more higher than the condensation temperature of ammonia (Gulkis *et al.*, 1978). Thus, gaseous ammonia in vapor pressure equilibrium with solid ammonia is *not* a characteristic of the planetary-scale typical structure of Uranus. It is difficult to imagine a reason why Uranus, which appears to be enriched in ice-forming (and possibly rock-forming) elements relative to hydrogen and helium, should be greatly deficient in ammonia, a substance even easier to enhance in the icy phase than methane. Perhaps a more promising prospect is the possibility that ammonia, although present deep within the planet, may be chemically removed at a level far too warm for ammonia condensation. By far the most plausible means of effecting this "scrubbing" operation is to have the condensation of NH_4SH at the 210 K level result in complete removal of NH_3. This in turn requires that the abundance of H_2S must exceed that the NH_3 in the lower atmosphere, and strongly suggests that substantial amounts of H_2S gas may survive above the NH_4SH cloud tops to undergo condensation to solid H_2S and photolysis to solid polymeric sulfur species. In any case, if the real reason for the depletion of ammonia in the upper troposphere of Uranus were known, it would likely prove to be a very valuable hint regarding the origin and bulk composition of the plant. If the large enhancement of H_2S described above is in fact the real situation, then the most likely mechanism for effecting the enhancement is the capture of a large mass of rocky material during planetary accretion. The rocky component will have a very large excess of sulfur over nitrogen due to the presence of the abundant and stable sulfide troilite, whereas the icy component almost certainly would contain a large excess of ammonia over the cosmically less abundant H_2S.

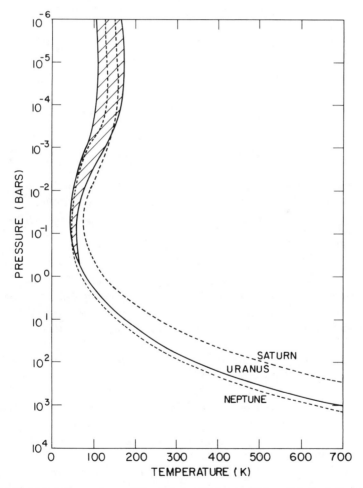

FIGURE 4.44. **Temperature–pressure profiles of Uranus, Neptune, and Saturn. Note the similar stratospheric temperatures of Uranus and Neptune.**

The identities of the stable condensates present in the outer solar system during the accretionary era are only poorly known, although many plausible statements can be made which do not conflict with the meager evidence. Most of this evidence derives from studies of the reflection spectra of the surfaces of the satellites of the outer planets and of the rings of Saturn, and from the available density data on asteroids, the Galilean satellites of Jupiter, and Saturn's largest satellite, Titan. The large lunar-sized satellites and the asteroids are also of interest with respect to their volatile-element contents and their potentials for producing and maintaining atmospheres, and are deserving of our attention.

For now, however, let us merely note that there is a vast range of bodies in the outer solar system whose cosmogonical histories are intimately interrelated. The Jovian satellite system and the satellite and ring systems of Saturn and Uranus reflect in microcosm the structure and compositional trends of the Solar System itself; it is almost as if Nature has provided us with several small examples of the formation of planetary systems within the confines of the main example, and we would be delinquent if we did not pursue the comparative study of these systems assiduously.

4.6. LUNAR-SIZED OBJECTS

The Solar System contains eight bodies of closely similar mass, intermediate in size between planets and asteroids. These eight—the Moon, Titan, Triton, the four Galilean satellites, and Pluto—have widely varying densities and bulk compositions, which makes their close coherence in mass even more startling.

The lunar-sized bodies are large enough to have interesting thermal histories, including the generation of magmas, volcanism, and active outgassing. In addition, they are marginally large enough to retain temporary atmospheres for an appreciable period of time. Most interesting of all, however, is the fact that several of these bodies are composed of comparable mass fractions of rocky and icy material, and that differentiation can lead to very deep convecting mantles and thin hard-ice crusts, with all rock material aggregated into a central muddy core (Lewis, 1971b,e). On the basis of present data, these satellites also can be seen to possess distinctive features which force us to consider them as individuals, not just as a set. A recent and very interesting survey of the natural satellites is the book *Planetary Satellites*, edited by J. A. Burns (1977). The Earth's Moon is not covered in that volume due to the magnitude of the present lunar literature, for which the primary source is the *Lunar* (an Planetary) *Science Conference Proceedings*, from 1969 to date, and the best single review is that by S. R. Taylor (1975). Burns' book also must be extensively supplemented by the Voyager encounter data on the Jupiter and Saturn systems, especially the articles collected in *Science* (Vol. 204, pp. 945–1008, 1979; Vol. 206, pp. 925–990, 1979; Vol. 212, pp. 156–243, 1981), *Nature* (Vol. 280, pp. 725–806, 1979), and *Geophys. Res. Lett.* (Vol. 7, p. 1, 1980). An issue of *Icarus* (Vol. 44, pp. 225–547, 1980) is devoted to the satellites of Jupiter, and another issue of *Icarus* (Vol. 44, pp. 1–71, 1980) contains a set of papers on Pluto.

Spacecraft missions to the outer solar system have been listed in Table 4.16 and will not be repeated here. Because of the great size of the lunar literature, the large number of spacecraft missions to the Moon, and the absence of any significant atmosphere, we shall deal only briefly with that body and omit a list of lunar missions.

We shall begin by extending several brands of condensation theory to low (ice condensation) temperatures, in order to establish the approximate relationships of the large low-density bodies to the inner planets and other rocky bodies, and then look in more detail at the individual lunar-sized objects.

The bulk compositions of these bodies may be approximately deduced from their bulk densities. Figure 4.45 gives the equilibrium stability fields for condensates forming from the solar nebula below 300 K. The series of chemical reactions begun at high temperatures and detailed in Section 2.6 continues as follows (Lewis, 1972a):

(9) Condensation of H_2O ice.
(10) Formation of solid $NH_3 \cdot H_2O$
(11) Exhaustion of H_2O ice to make the solid clathrate hydrate of methane $(CH_4 \cdot 7H_2O)$.
(12/13) Condensation of solid CH_4 ice and Ar ice.
(14/15) Condensation of solid Ne and H_2.
(16) Condensation of He.

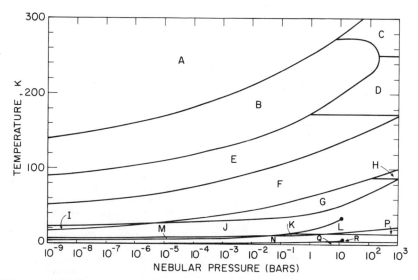

FIGURE 4.45. **Equilibrium stability fields for low-temperature condensation from the solar nebula. The condensation curves for solid H_2O, $NH_3 \cdot H_2O$, $CH_4 \cdot 7H_2O$, CH_4, Ar, and H_2 are given at low temperatures and pressures. The melting curve at high pressures running from 173 to 273 K is that for the NH_3/H_2O system, and the melt phase to the right of the curve is cold aqueous NH_3 solution. (After Lewis, 1972a.)**

The last step in this sequence requires temperatures below 1 K and is impossible in a universe with a background temperature of 2.7 K. Steps 14 and 15, the condensation of hydrogen and neon, may also have been impossible in the early solar system.

For this condensation sequence, the cumulative density of condensate varies widely, from ~5.5 gm cm^{-3} upon condensation of metallic Fe–Ni alloy down to ~1 gm cm^{-3} after condensation of solid methane. Figure 4.46 gives the dependence of density on formation temperature for the nebular $P-T$ profile of Cameron (1973), but a glance at Figures 2.14 and 4.45 will show that the condensation sequence is remarkably insensitive to the exact pressure profile selected, over a range of a factor of 10^2-10^3 in pressure. The detailed predictions of this model are given in two reviews (Lewis, 1973a,b). The placing of Ganymede, Callisto, and Titan on the temperature axis by density alone is not optimum, and a large uncertainty should be attached to that dimension. If we take seriously the JIII–JIV density difference and the presence of Titan's atmosphere, then a more precise assignment of formation temperatures can be made, as in Figure 4.47. Note also the totally irreconcilable entries made for the Moon at ~1800 and ~400 K. The former readily explains the high concentration of refractories and low concentration of volatiles in the lunar crust [indeed, in the models of lunar origin of Anderson (1972a) and Cameron (1972) the Moon *is* refractory]. The latter gives an easy explanation of the high FeO content of the lunar interior.

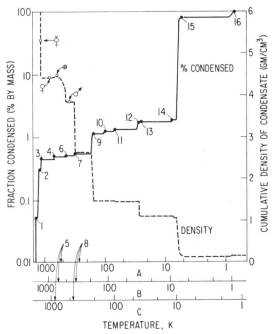

FIGURE 4.46. Cumulative density of equilibrium condensate in the solar nebula. Primitive, undifferentiated bodies condensed and accreted at a single temperature will have the dependence of uncompressed density on formation temperature given in this figure. These results do not directly pertain to bodies which are (a) differentiated as open systems, (b) produced by fission of differentiated bodies, or (c) inhomogeneously accreted. Note that an uncompressed density near 1.0 corresponds to complete methane condensation. The stages of condensation indicated are as follows: (1) refractory oxides; (2) metal; (3) Mg silicates; (4) FeS; (5) CO → CH$_4$ gas conversion; (6) Fe metal oxidation of FeO-bearing silicates; (7) serpentinization of Mg and Fe silicates; (8) N$_2$ → NH$_3$ gas conversion; (9) H$_2$O ice condensation; (10) NH$_3$·H$_2$O formation; (11) CH$_4$·6.75H$_2$O formation; (12) CH$_4$ ice condensation; (13) solid Ar; (14) solid Ne; (15) solid H$_2$; (16) solid He. (After Lewis, 1972a.)

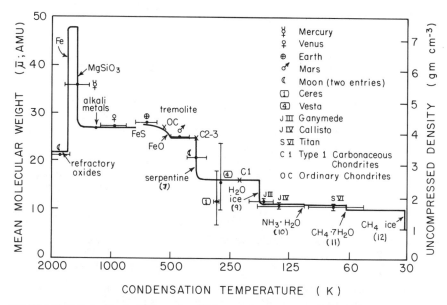

FIGURE 4.47. **Bulk density of satellites and equilibrium condensates versus nebular temperature. The nebular P–T profile of Cameron (1973) is assumed. Possible equilibration temperatures to produce bodies with the observed densities of a number of solar system members are indicated. See Figure 4.46 for further explanation. (After Lewis, 1972a.)**

The course of condensation at low temperatures can be dramatically altered if strict thermodynamic equilibrium is not attained. In Section 2.6 we saw that rapid accretion during cooling, which isolates newly condensed grains from further interaction with the gas, leaves metallic iron intact, unreacted with H_2O, H_2S, and silicates. Hydroxyl silicates likewise cannot form. Thus, such a nonhomogeneous (disequilibrium) process would leave abundant H_2S and moderately enhanced amounts of H_2O in the gas phase at lower temperatures. Cooling below the ice saturation temperature can then lead to NH_4SH condensation followed by pure solid NH_3 and then CH_4 (see Figure 4.48). This is the same sequence seen in Section 4.5 for cloud condensation on the Jovian planets.

A third possibility, particularly appropriate in a turbulent nebula with a strong radial temperature gradient, is that in which reduction of the high-temperature gases N_2 and CO is kinetically inhibited, so that only small amounts of NH_3, CO_2, and CH_4 are formed. In this case, NH_4HCO_3 may condense even above the H_2O saturation temperature. Ammonium carbamate (NH_4COONH_2) is also a possible condensate. Solid CO_2 condenses at lower temperatures, and finally a mixed-clathrate hydrate of CO, N_2, and CH_4 forms. The dependence of the condensation sequence on the quench temperature of gas-phase reactions was shown in Figure 2.19 (p. 65).

These three scenarios are compared in the flowcharts given in Figure 4.49.

It is important to understand that the equilibrium condensation theory relates the

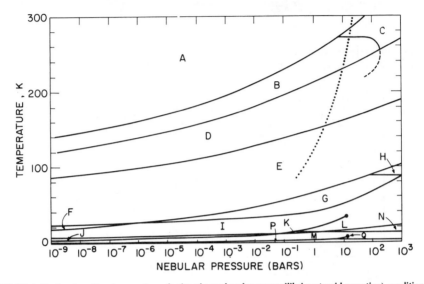

FIGURE 4.48. **Condensation temperatures for ices formed under nonequilibrium (rapid accretion) conditions. Condensates are deposited in layers as temperatures fall. In order of condensation, the species condensing are H_2O ice (B), aqueous NH_3 solution (C), solid NH_4SH (D), solid NH_3 (E), solid Ar (F), solid CH_4 (G), liquid CH_4 (H), solid Ar and CH_4 together (I), solid Ne (J), etc. The dotted line is an adiabat for the atmosphere of Jupiter; a solar nebula adiabat would cut the H_2O condensation curve near 10^{-6} bar. (After Lewis, 1972a.)**

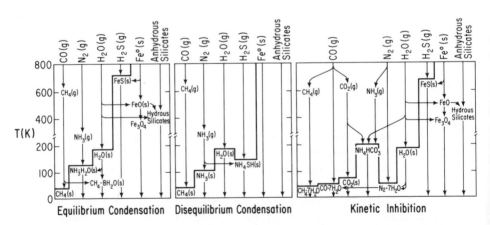

FIGURE 4.49. **Low-temperature condensation flowcharts for compounds of H, C, N, O, and S. The equilibrium and disequilibrium (rapid accretion) sequences are from Lewis (1972a) and from Barshay and Lewis (1976); the sequence for kinetically inhibited formation of NH_3 and CH_4 is adapted from Lewis and Prinn (1980). The heavy line separates gases (above) from condensates (below). The relative masses of the condensates cannot be read from these diagrams.**

composition of primitive condensate to its temperature of origin in the solar nebula. Objects which are derived from differentiated larger bodies, assembled by accumulation of material from very diverse sources, or formed at an unknown point in the Solar System cannot be so simply interpreted. Further, satellites formed in dense planetary subnebulae may be made from well-equilibrated materials, whereas similar-sized planets formed in the solar nebula at the same temperature would be very poor in ammonia and methane. This occurs as a consequence of the much greater rate of the reactions between H_2 and the high-temperature gases CO and N_2 in the high-density circumplanetary nebulae (Prinn and Fegley, 1981).

We shall now briefly examine present knowledge of the bulk composition, volatile content, atmospheric composition and structure, and evolutionary history of the lunar-sized bodies in the order of their heliocentric distance.

4.6.1. The Moon

Today, as for the last 20 years, the leading theories for the origin of the Moon are that it was (*a*) captured by the Earth from heliocentric orbit (Gerstenkorn, 1955, 1969; Goldreich, 1966; Kaula, 1971) by an improbable but not impossible encounter; (*b*) assembled from high- and low-temperature components of unrelated origin (Anderson, 1972a; Cameron, 1972); or (*c*) derived, by fission, from a portion of the Earth or of the hypothetical protoatmosphere of the Earth (Wise, 1963; Ringwood, 1970, 1972).

If we accept the burden of attempting to find a simple origin for the Moon, and if we believe the Moon to be a primitive body in the sense used above, then we must make do with either the 1800-K or the 400-K alternative merely to fit the observed density. High-temperature origin of the Moon requires a large additional source of iron oxides, which necessarily involves introduction of a second component, as in Anderson's model or Cameron's (1972) revision of it.

If, on the other hand, we derive the Moon from primitive material essentially devoid of metallic iron ($\sim$400 K), it will be rich in FeO but also well endowed with volatiles. The bulk density suggests an $\sim$C3 chondrite composition, whereas the FeO content appears to require an origin below 500 K. The extreme depletion of volatile elements observed in lunar surface rocks (Anders, 1976; Morgan *et al.*, 1973) requires either accretion at very high temperatures or intense volatilization after accretion. O'Hara and co-workers have argued for volatilization from an originally water- and alkali-rich magma upon extrusion into the hard vacuum on the lunar surface (O'Hara and Biggar, 1972). Other possibilities for depletion of volatiles and enrichment of refractories include prolonged cooking by the T-Tauri phase of the Sun, which apparently could melt the Moon to a depth of 200–300 km (Sonett *et al.*, 1970). Starting, then, with a low-temperature lunar origin (lower than or equal to that of the Earth), thorough outgassing by an external heating mechanism such as the T-Tauri phase or tidal dissipation of energy during capture by the Earth could drastically affect the outer portion of the Moon, while leaving the deep interior far less severely altered. Tatsumoto's (1973) discovery of indigenous highly nonradiogenic lead on the Moon is by such a model easily explicable, whereas a special hypothesis is needed to inject primitive lead (and FeO!) into a high-temperature moon. Indeed, several authors (Brett, 1973b; Taylor, 1973) have expressed preference for igneous differentiation and surface volatilization as the reasons for the high refractory and low volatile content of the lunar surface. Ringwood (1974, 1976) has favored homogeneous and roughly chondritic models with loss of volatiles and metal.

Nonetheless, it is clear that a "chondritic" model for the Moon is still too simple: The missing volatiles may indeed have been lost after formation, but what of the fact that the overall density of the Moon is only 3.34 gm cm^{-3}? This density, which is lower than that of any ordinary or C3 chondrite, cannot be attained by any object with the chondritic relative abundances of the rock-forming elements unless all metal is oxidized and several weight percent water is retained (see Figure 4.50).

This did not seem unreasonable prior to the Apollo missions. It was suspected and even hoped that, because of its small size, the Moon might have been spared the severe geological reprocessing undergone by Earth. The surface of the Moon might then be as ancient as the chondritic meteorites, and could in fact be the source of one or more classes of chondrites. Mild, ancient heating of the interior might have released volatiles such as water vapor, which could have accumulated in permanently shadowed regions near the lunar poles (Watson *et al.*, 1961a; Mukherjee and Siscoe, 1973). In addition, neutralization of incident solar wind ions could provide a sparse steady-state atmosphere derived from the heavier gases in the wind (Nakoda and Mihalov, 1962; Bernstein *et al.*, 1963).

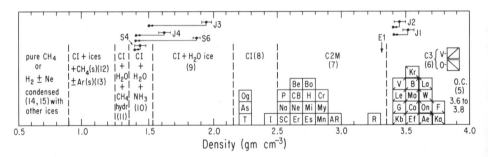

FIGURE 4.50. **Densities of carbonaceous meteorites and natural satellites. Meteorite data after Mason (1971); satellite densities generally after Morrison *et al.* (1977). The abbreviations for the meteorites are as follows: Orgueil (Og), Alais (As), Tonk (T), Ivuna (I), Pollen (P), Nawapali (Na), Santa Cruz (SC), Bells (Be), Cold Bokkeveld (CB), Nogoya (No), Erakot (Er), Boriskino (Bo), Haripura (H), Mighei (Mi), Essebi (Es), Crescent (Cr), Murray (My), Murchison (Mn), Al Rais (AR), Renazzo (R), Vigarano (V), Leoville (Le), Grosnaja (G), Kaba (Kb), Karoonda (Kr), Bali (B), Mokoia (Mo), Coolidge (Co), Efremovka (Ef), Lance (La), Warrenton (W), Ornans (On), Allende (Ae), Felix (F), and Kainsaz (Ka). The satellites are as follows: (J1) Io, (J2) Europa, (J3) Ganymede, (J4) Callisto, (S4) Dione, (E1) the Moon, and (S6) Titan. The range of allowable observed densities is given by the brackets, and the arrows below the brackets show how the *mean* observed densities would be reduced by reduction to 1 bar pressure. These corrections are important only for the massive ice-rich satellites J3, J4, and S6. The stage reached in the condensation sequence is given by the number of the last reaction involved (parenthesized).**

Analyses of the first returned Apollo samples from Mare Transquillitatis showed such severe depletion of volatiles (including alkali metals) that indigenous volatiles were hard to find. The first rare gas evidence for native volatiles (Manka and Michel, 1971) turned up small amounts of ^{40}Ar which had been outgassed, ionized, and implanted in the surface by solar wind fields. Wasson (1971) proposed that in fact the volatiles were a late addition, probably of cometary material, and that the very different present volatile contents could be accounted for by gravitational focusing by the Earth (which is not very significant at cometary speeds) and by the inability of the Moon to retain the hot gases from an impact fireball. Urey (1971) preferred the idea that the Moon had accreted cold and relatively volatile-rich, and had lost its gases later.

El Goresy et al. (1973) attributed the high abundances of Zn, Pb, Cl, and even FeOOH in one Apollo 16 sample to an impact of a volatile-rich carbonaceous chondrite or comet. Megrue (1973), on the other hand, proposed that solar wind implantation, impact heating, and consequent fractionation was the principal means by which volatiles were implanted in solid solar system bodies. Surely both processes contribute to the volatile inventory of the Moon; however, neither is of obvious importance on bodies capable of retaining outgassed volatiles. Further, we know from our experience with chondritic meteorites that FeO content and volatile content are positively correlated in primitive material. It is thus not surprising that many lunar igneous rocks contain abundant empty vugs and vesicles: An appreciable content of volatiles, carried by the magma at the time of extrusion, has been lost. The crucial issue on the Moon, as on Mars, is to establish the nature of the volatile-loss processes.

It is potentially instructive to try to relate the bulk, nonvolatile elemental composition of the Moon to meteorite analogs, such as the terrestrial planets, chondritic, and achondritic meteorites. A number of such genetic scenarios have been published and have later died or evolved markedly. There is now a rather wide consensus on the description of the bulk composition of the Moon (see, e.g., Ganapathy and Anders, 1974; Ringwood, 1979; Wänke and Dreibus, 1979; Wolf and Anders, 1980): The moon is about as similar to chondritic meteorites in composition as the various classes of chondrites are to each other.

Morgan and Anders (1980) have used five parameters (abundance of refractories, abundance of iron-group elements, abundance of moderately volatile elements, abundance of highly volatile elements, oxidation state of iron) to categorize the terrestrial planets, the Moon, the eucrite parent body, and chondrites. The extrapolations are very long for Venus and Mercury, whose data are sparsest, and moderately long for Mars, and, of course, the entire approach assumes a tight coherence between groups of elements which may be unaware of and uninfluenced by terrestrial theories of fractionation behavior. Nonetheless, the comparisons are of interest, especially that between the Earth and the Moon. The refractory elements are apparently twice as abundant (relative to Si) on the Moon as on Earth; the iron-group elements are depleted by about a factor of four in the Moon; the moderately volatile elements (600–1300 K condensation temperatures) are depleted about twofold in the Moon and highly volatile elements ($T_c < 600$ K) are depleted by about a factor of 50. At the same time, the FeO:(FeO+MgO) ratio, another measure of low-temperature volatile addition, is about three times higher in the Moon than on Earth. The Earth (and, less reliably, Venus) appear to be within the chondritic range for all these parameters, whereas the Moon lies far above the chondritic range of refractory content, far below the lowest chondritic content of Fe-group metals and of moderately volatile elements, and near the lowest extreme of the chondrites in volatile-element abundances. The oxidation state of the Moon appears to be nearly as high as the most recent and most carefully assessed value of the oxidation state of Mars (Goettel, 1981).

Conventional arguments about "volatile-element veneering" on the planets, which add the volatiles late in accretion and in extremely concentrated form, would have insuperable difficulties in reaching and oxidizing an amount of Fe sufficient to make, say, 10% of the Moon's mass into FeO. This would require a sizeable fraction of the mass of the Moon to be C1 or cometary material. A more modest approach to the FeO problem is to form the Moon out of condensates equilibrated in the solar nebula at about the same temperature as Mars-forming solids. The volatiles would then begin with abundances of only a few times their terrestrial values, and they must be lost by strong heating. The deficiency of iron-group elements would then have to be attributed to nebular fractionation processes (e.g., Wolf and Anders, 1980). Ringwood would have the Earth–Moon system undergo a complex (and safely unquantifiable) high-temperature fission, with the missing siderophiles lost to Earth's core prior to separation. Ringwood favors FeO contents for the Moon which are only slightly higher than Earth's.

Both of these theories address the data rather well, both contain some rather unsettling *ad hoc* elements, and neither has such an obvious excess of virtue as to compel our acceptance.

4.6.2. The Galilean Satellites

The four lunar-sized satellites of Jupiter—Io (J1), Europa (J2), Ganymede (J3), and Callisto (J4)—have been studied since their discovery by Galileo in 1610. The traditional astronomical background on the appearance and motions of these satellites is treated in the review by Harris (1961), which seems to have stimulated the application of a wide range of observational techniques to the study of planetary satellites.

Infrared radiometric and photometric studies of these satellites by Murray *et al.* (1964), Moroz (1966), Gillett *et al.* (1970), Johnson and McCord (1970, 1971), and Morrison *et al.* (1972) provided data on the surface temperatures of these bodies, determined their visible and near-IR reflection spectra at low resolution, and permitted the identification of water frost on the surfaces of Europa and Ganymede (Johnson and McCord, 1971). Improved thermal measurements on these satellites (Morrison and Cruikshank, 1973) revealed no evidence for significant atmospheres on any of them, whereas high-resolution IR spectra by Pilcher *et al.* (1972) and

Fink *et al.* (1973) confirmed the presence of ice on Europa and Ganymede, and set restrictive upper limits on the abundances of gaseous CH_4 and NH_3. No ice was seen on Io or Callisto. These conclusions were reinforced by further improvements in the laboratory data on the reflection spectrum of frosts and its dependence on particle size and temperature (Kieffer and Smythe, 1974; Fink and Larson, 1975).

The densities of the Galilean satellites were also approximately known, since the masses could be determined from their mutual perturbations, while the radii of their tiny discs (roughly 1″ arc) rendered accurate measurement of their diameters difficult. These data sufficed to show that the Jovian system was in a way a microcosm of the Solar System: These densities revealed evidence for the same kind of compositional trend with distance from Jupiter that we see among planets with distance from the Sun. The bulk densities of the satellites decrease in a smooth sequence from about 3.4 to 1.6 going outward from Jupiter. This has long been assumed to reflect a radial temperature gradient in the vicinity of Jupiter during condensation or accretion of the satellites. Grossman *et al.* (1972) have calculated collapse histories for Jupiter which show that Jupiter's luminosity should have prevented ice from forming at the distance of Io or Europa for millions of years (Pollack and Reynolds, 1974). Our explanation of these densities, based on the temperature dependence of the composition of condensed matter, is that Io and Europa should have had initial compositions near that of a C3 chondrite (possibly C2 for Europa), Ganymede should be of $C1-H_2O$ ice composition, and Callisto may or may not have retained ammonia hydrate in addition to H_2O. The reason for the absence of ice bands in Callisto's spectrum is not clear. In any case, these compositions correspond to zero-pressure densities of 3.5, 3.4, 1.6, and 1.5 for these bodies, respectively. The water contents of Io and Europa would easily suffice to permit formation of extensive surface deposits of ice consequent upon outgassing of their interior. Baking of their surfaces by the T-Tauri phase of the Sun is 27 times less severe at 5.2 AU than at 1 AU; hence melting of silicates may never have occurred near their surfaces. The density data on the satellites were compared to the densities of volatile-rich meteorites in Figure 4.49, in which the densities for equilibrium condensates are also shown.

We will describe in the discussion of thermal histories of asteroids in Section 4.7 how carbonaceous chondrites may, upon moderate heating, exude water containing quantities of soluble salts. Ammonium, alkali metal and alkaline earth carbonates, sulfates, and halides are most probable.

There have been numerous other fascinating observations of the Galilean satellites which relate to their volatile contents.

First, there are the measurements of the anomalous brightness of Io immediately upon emerging from eclipse by Jupiter (Binder and Cruikshank, 1964b, 1966; T. V. Johnson, 1969, 1971; Fallon and Murphy, 1971; Franz and Millis, 1971, 1974; O'Leary and Veverka, 1971; Cruikshank and Murphy, 1973; O'Leary and Miner, 1973; Millis *et al.*, 1974; Frey, 1975; Pavlovski, 1976). This effect was tentatively attributed by Lewis (1971b) to partial condensation of a trace atmosphere, of $P_s \simeq 10^{-7}$ bar, during eclipse. Sinton (1973) presented a model atmosphere, composed of NH_3, which, however, has a severe stability problem due to its short photochemical lifetime. An occultation of the star Beta Scorpio by Io failed to reveal an atmosphere with a pressure of $>10^{-7}$ bar (Bartholdi and Owen, 1972).

Second, Brown's detection of rather strong Na D-line emission in the spectrum of Io has opened a fascinating new chapter in the study of satellites (Brown and Chaffee, 1974). On some of the spectra, the emission spikes superimposed on the solar NaD absorption features have a higher bandwidth power than the neighboring reflected solar continuum. An intense observational effort was devoted to the study of these atomic emissions, and to theoretical interpretation of the observations in terms of sputtering of surface atoms by magnetospheric protons and excitation of surface atoms by resonant scattering of sunlight (Matson *et al.*, 1974; Trafton *et al.*, 1974). Both observations (Mekler and Eviatar, 1974; Bergstralh *et al.*, 1975; Trafton, 1975b; Trafton and Macy, 1975) and theories (McElroy *et al.*, 1974; Carlson *et al.*, 1975; Macy and Trafton, 1975a,b) proliferated. Mapping of the sodium distribution showed a disk of gas surrounding Io in its orbital plane, elongated along its orbit about Jupiter (Wehinger *et al.*, 1976; Matson *et al.*, 1978).

Several detailed atmospheric models were proposed in response to this flood of data, taking into account the Pioneer 10 radio occultation experiment results on Io (Kliore *et al.*, 1975) as well as the spectroscopic data reviewed above. Whitten *et al.* (1975) proposed an atmosphere in which the dominant neutral gas was neon (a species not observed to date, and one expected on cosmogonic grounds to be negligible). McElroy and Yung (1975) described a variety of models in which the surface number density of the neutral atmosphere lies in the approximate range from 10^{10} to 10^{12} cm^{-3}. The dominant atmospheric gases NH_3 and N_2, with the destruction of ammonia providing not only a large source of atomic hydrogen, but also a requirement for replenishment of the ammonia by outgassing of the surface of the interior. McElroy and Yung expressed some preference for the presence of solid NH_3, $NaNH_2$, or $NH_3 \cdot H_2O$ in the crust, but pointed out that other sources of ammonia are possible.

Magnetospheric protons may cause sputtering of sodium from exposed solid sodium compounds such as water-soluble salts. Sodium is supplied to high altitudes by convection of the lower atmosphere, with much of the energy to drive convection supplied by condensation of a tenuous haze of solid NH_3 crystals. Photoionization of sodium at high altitudes is postulated to be the source of the ionosphere detected by Pioneer 10, whereas collisional excitation of sodium by nitrogen molecules is the source of the D-line emission from the atmosphere of Io proper. The extended emission cloud found by Brown is powered by scattered D-line photons in this interpretation. Many of the features of this model are present in the discussion by Gross and Ramanathan (1976) as well.

The Pioneer 10 UV spectrometer detected an incomplete torus of atomic hydrogen about Io's orbit, but not extending all the way around the planet (Carlson and Judge, 1974). McElroy and Yung estimate that the atomic hydrogen escape rate from Io required to maintain this torus must be near 10^{11} cm^{-2} sec^{-1}, which requires a substantial proportion of the solar UV shortward of 2300 Å to be used to protolyze ammonia in Io's atmosphere. Loss of hydrogen from the torus must occur sufficiently rapidly to prevent the torus from reaching all the way around Jupiter. This means that the lifetime of neutral atomic hydrogen cannot much exceed 10^5 sec, which is so short that some efficient ionizing energy source must be at work. One plausible mechanism for H atom ionization is bombardment by large fluxes of low-energy protons trapped in the Jovian magnetosphere.

The suggestion by Matson *et al.* (1974) that sputtering of sodium from a salt-bearing surface layer could provide the sodium for the observed cloud has led to extensive laboratory work on this process. Nash *et al.* (1975) and Nash and Fanale (1977) have found that kiloelectron volt protons will sputter sodium atoms from solid salts such as NaCl with reasonable efficiencies. Fanale *et al.* (1979) proposed a model for the surface of Io in which outgassing of a carbonaceous-chondrite-composition Io provides an upward flow of volatiles, notably water and water-soluble materials. By analogy with the CI chondrites, which contain distinct veins of white, crystalline, water-soluble salts, the surface of Io can plausibly be imagined to be an evaporite, rich in halides, sulfates, and other oxysalts of sodium, potassium, calcium, magnesium, etc. This model nicely explains the reflection spectrum of Io in the visible and near-IR, where a very high Bond albedo combined with the inability of observers to detect absorption features due to H_2O ice had presented an interesting dilemma.

The discovery of atomic potassium in an extended cloud about Io by Trafton (1975c), combined with the continuing failure of observers to detect calcium, aluminum, and other atomic (Brown *et al.*, 1975; Trafton, 1976b) and molecular (Fink *et al.*, 1976) species helps to further constrain the chemistry of the surface layers. More recently, Kupo *et al.* (1976) have found an extended cloud of singly ionized sulfur about Io, forming a thin disk in the plane of the satellites' orbits. Sulfur, as elemental sulfur, had previously been proposed as a constituent of the surface by Wamsteker *et al.* (1974); as polysulfides, by Lebofsky and Fegley (1976); and as sulfates, by Fanale *et al.* (1979). The candidates for surface minerals are reviewed in detail by Nash and Fanale (1977), who postulate a mixture containing sulfur, bloedite [$Na_2Mg(SO_4)_2 \cdot xH_2O$], ferric sulfate [$Fe_2(SO_4)_3$], and traces of hematite (Fe_2O_3), with possible small amounts of halite (NaCl), sodium nitrate ($NaNO_3$), and the clay mineral montmorillonite. Further studies of sulfur-bearing mixtures have been carried out by Veverka *et al.* (1979).

Carlson *et al.* (1975) have suggested that the lifetime of sodium atoms near Io is limited by electron impact ionization by magnetospheric plasma electrons, and Johnson *et al.* (1976) have presented a revised model of Io's atmosphere in which the ionizing effect of these electrons on the atmosphere is included: Surface neutral particle densities as low as 10^9 cm^{-3} are found, and outgassing fluxes of H-bearing molecules of 10^{10} cm^{-2} sec^{-1} would suffice to provide the observed H cloud. The state of models for Io's atmosphere at the time of the Voyager 1 Jupiter encounter is indicated in Figure 4.51. Readers interested in more details regarding the atmo-

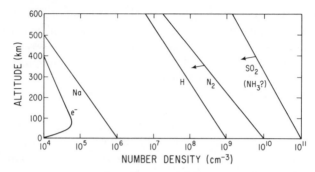

FIGURE 4.51. Io atmospheric models. This figure is based, in part, on the model by McElroy and Yung (1975), and incorporates the Pioneer 10 estimate of the electron density profile on the day side of Io (Kliore *et al.*, 1975). According to the models of Johnson *et al.* (1976), the total number density of neutral species at the surface could be as low as 10^9 cm^{-3} if electron impact is the dominant ionization mechanism at and near Io.

sphere of Io should consult the paper by McElroy and Yung (1975) and the reviews by Fanale *et al.* (1976) and Johnson (1978). All this early work on Io laid the groundwork for the Voyager 1 encounter with the Jupiter system.

On the eve of the encounter, a theoretical treatment of tidal dissipation in Io was published, in which "widespread and recurrent surface volcanism" was predicted for Io (Peale *et al.*, 1979). At the time of appearance of this paper, analyses of the first close-encounter photographs of Io (Morabito *et al.*, 1979) revealed several bright patches and plumes, some extending as high as 250 km above the surface. At the same time, the IR interferometer experiment (IRIS) found anomalously hot spots on the surface of Io, at temperatures perhaps several times the mean surface temperature (Hanel *et al.*, 1979). Extreme ultraviolet (EUV) spectra of the region of Io's plasma torus about Jupiter revealed ionized atomic oxygen and sulfur (Broadfoot *et al.*, 1979), which were also detected by the plasma experiment (Bridge *et al.*, 1979).

Consolmagno (1979) pointed out that most of the usual planetary volatiles would be rapidly lost from Io due to its low escape velocity and severe particles and fields environment, but that sulfur, with its low volatility and high molecular weight, might accumulate and dominate the surface. Then silicate magmatism, wherever it approaches the surface, would serve to drive violent sulfur volcanism. The accumulated sulfur deposits would be no more than a few kilometers thick, and the observed mountains are almost certainly silicates, not sulfur (Carr *et al.*, 1979; Masursky *et al.*, 1979). The prevalence of characteristic erosional scarps on Io similar to those produced by groundwater sapping and ice sublimation on Mars suggests that they are formed by sulfur melting or by vaporization of other sulfur-bearing volatiles such as SO_2 (McCauley *et al.*, 1979). The IRIS experiment successfully identified SO_2 as an important component of the volcanic gases in the plumes (Pearl *et al.*, 1979), and Smith *et al.* (1979) proposed that volcanic plumes reaching an altitude of as much as 280 km could be driven by SO_2 starting at a temperature as low as 400 K and a pressure near 40 bar.

Because of the powerful source, the dominant gas in the neutral atmosphere is probably SO_2, despite the rapid loss of atmosphere which occurs on Io (Kumar, 1979). SO_2 then ought to be in vapor pressure equilibrium with the surface and may be responsible for the posteclipse brightening: Daytime temperatures can provide an SO_2 vapor pressure in excess of 10^{-7} bar, which, by the model for the posteclipse brightening of Lewis (1971e), is sufficient to explain the extent and duration of the observed brightening.

The enormous eruption rates have other interesting consequences. For example, the observed mass flux from the volcanoes is sufficient to resurface Io at the rate of about 0.01–0.1 cm per year (Johnson et al., 1979). Thus, even a 3-km-thick sulfur layer could be fully recycled in as short a time as 10 million years. SO_2 and S ought to coat the entire exposed surface, and indeed the spectral signature of SO_2 frost is widespread over the surface (Fanale et al., 1979; Smythe et al., 1979). In addition to SO_2 and S, other sublimates may contribute to the near-IR reflection spectrum: Nash and Nelson (1979) make a case for the presence of NaHS, Na_2S, and K_2S, and possibly other sulfides as well. We should recall that hydrosulfides and polysulfides, suggested by Lebofsky and Fegley (1976), are easily formed in the presence of elemental sulfur.

Hapke (1979) proposed a "magmatic–volatile" model for Io in which the sulfur compounds at the surface are supplied by "dissociation of troilite brought up from the core." Io is assumed to have formed from metal-bearing chondritic solids, and to have segregated out a dense Fe–FeS core, which served to store sulfur during the outgassing of the rest of Io.

There are several fundamental problems with this model. First, the uncompressed density of Io is lower than the density of any metal-bearing chondrite. Second, if Io has been extensively outgassed (which seems quite unavoidable), then its previous density must have been even lower, within the range found for C2 chondrites (see Figure 4.50). Such material outgasses water if heated gently, and does not form a Fe–FeS eutectic melt because the oxygen fugacity is orders of magnitude too high. Third, a mechanism by which very dense core material could be carried upward for 1000 km by a hot, low-viscosity silicate magma in response to a heat source which lies above the core–mantle boundary is frankly incredible. Fourth, FeS does not dissociate thermally to give sulfur vapor under any possible Io surface conditions: Hapke evidently has referred to an isobaric 1-atm phase diagram in order to reach this conclusion. Lewis (1982) has shown that the behavior of the Fe–S system in equilibrium with its own vapor pressure does indeed allow sulfur vapor to be abundant, but only at sulfur fugacities far too high for FeS stability (Figure 4.52). Furthermore, the requirements that SO_2 be the major magmatic gas and that liquid sulfur be thermodynamically stable combine to place very tight restrictions on the source of these volatiles. The maximum SO_2 pressure is realized near the melting point of silicates (1200 K) at an oxygen fugacity of about 10^{-10} bar and a sulfur fugacity of nearly 6 bar, in equilibrium with hot liquid sulfur. Elemental sulfur is not stable in any solar-composition rock system at temperatures below 1200 K, nor at higher or lower oxygen fugacities: Higher f_{O_2} turns elemental sulfur

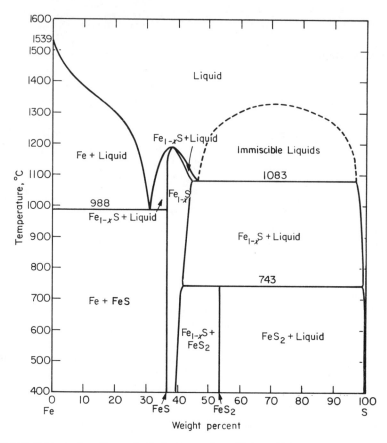

FIGURE 4.52. **Polybaric phase diagram for the Fe–S system in equilibrium with its own vapor. Note the "liquid sulfur" stability field to the right of the two-fluid dome. Incongruent melting of FeS$_2$ will drop dense solid pyrrhotite and liberate a low-density, low-viscosity melt of sulfur with a few weight percent Fe, which could be recycled on Io without further loss of iron sulfides. (After Lewis, 1982.)**

into sulfates; lower f_{O_2} makes pyrrhotite. Thus, the optimum conditions for both S and SO$_2$ magmatism lie on the sulfide–sulfate boundary at 1200 K. This point can readily be reached by heating C2 material which has been progressively warmed to first drive out water vapor. Alternatively, a sulfide-bearing assemblage with a trace of metal may have been oxidized during water outgassing to wholly destroy the metal and make some sulfates. It is interesting to note that this evolutionary path can be rather effectively blocked by H$_2$O retention rather than loss: On a body larger than Io, the same initial composition may have led to a planet similar in present state to Mars.

The SO$_2$ pressure as a function of the sulfur and oxygen fugacities is shown in Figure 4.53, superimposed on the stability of the main sulfur- and iron-bearing minerals at 1200 K. Diagrams for lower temperatures are given by Lewis (1982). If a

FIGURE 4.53. **Phase stability relations for Fe and S compounds and SO_2 pressures at 1200 K. The solid lines mark the major S_2 and O_2 buffers. Chondritic relative abundances of rock-forming elements are assumed. The ×s mark a sequence of unidirectional oxidation from E chondrite to ordinary chondrite to carbonaceous chondrite mineralogy. Note the appearance of liquid S at $\sim10^{-9.6}$ bar f_{O_2}. The diagonal dashed lines are contours of $\log_{10}$ of the SO_2 partial pressure in bars. (After Lewis, 1982.)**

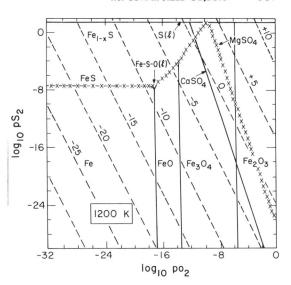

small amount of water vapor were present in this system, H_2 would be rather abundant even at the appearance point of liquid sulfur, whereas H_2S would be negligible (see Figure 4.54).

Several constraints on the energetics of the volcanic process on Io are available. First, thermal emission enhancements have been observed from Earth, bearing testimony to the eruption of a large area of hot magma (Sinton, 1980; Sinton *et al.*, 1980), and the heat flux from several hot spots has been measured (Morrison and Tedesco, 1980; Matson *et al.*, 1981; Sinton, 1981). The mean global heat flow observed is $(6 \pm 1) \times 10^{13}$ W, largely emanating from spots whose color temperatures are between 200 and 600 K. Second, Reynolds *et al.* (1980) have pointed out that the cold-SO_2 driving mechanism for the plumes is at best marginal, and argue for a 1200-K sulfur vapor source as the driver. They mention that SO_2 may be a very minor component of the plumes. However, the geochemical (iochemical?) considerations outlined above clearly make it easy for interaction of a silicate magma with liquid sulfur to be accompanied by very large SO_2 pressures, on the order of 1 kbar. These pressures greatly exceed the overburden pressure at any reasonable interaction depth.

Pollack and Witteborn (1980) have considered a wide variety of loss processes which may have been responsible for the evolution of Io's volatile inventory. It appears that even loss of N_2 can occur at sufficient rates to remove it quantitatively over the age of the Solar System. Of course, early outgassing of a hot, low-molecular-weight magmatic gas could assist in removing nitrogen by carrying NH_3 away by hydrodynamic blowoff. Also, the role of impact removal of volatiles has not yet been evaluated for Io.

For recent perspectives on the thermal and atmospheric history of Io, the articles of Consolmagno (1981) and of Kumar and Hunten (1982) should be consulted. It

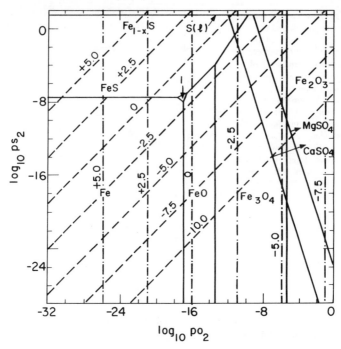

FIGURE 4.54. $H_2 : H_2O$ and $H_2S : H_2O$ ratios for the system in Figure 4.53. The vertical dot–dash lines are contours of $\log_{10}$ of the $H_2 : H_2O$ ratio. The diagonal dashed lines give $\log_{10}$ of the $H_2S : H_2O$ ratio. Note that if water were present at the point of appearance of liquid S, about 10% of the hydrogen would be found as H_2, 0.03% as H_2S, and the rest as H_2O.

seems at present that the most likely evolutionary path for Io involves a C2 or nearly metal-free C3 parent material, from which most hydrogen and hydrogen compounds were lost early in Io's history. The period of rapid outgassing of H_2O, NH_3, H_2S, CH_4, etc. may have produced a surface ice layer. Early heating of Io would involve not only the conventional long-lived radionuclide and accretion heat sources, but also other contributions such as the elevation of the steady-state temperature of Io by the thermal luminosity of the collapse phase of Jupiter, tidal interactions between the satellites, and the dissipation of the rotational energy of Io due to Jupiter's tidal force acting on a tidally produced equatorial bulge. Short-lived radionuclides may also have been present. The time scales for decay of these heat sources are about 10^5 yr for Jovian contraction, 7×10^5 yr for short-lived radionuclides, and roughly 10^6 yr for despinning (but this is very sensitive to unknown initial conditions). The time scale for accretion of the Galilean satellites has not been properly modeled, but is probably in the range of 10^5–10^6 yr. By about 10^5 yr after the beginning of the collapse of Jupiter, however, the contribution of Jupiter's luminosity to Io is down to about one-quarter of the incident solar flux, and thus ceases to be a very important effect (Bodenheimer et al., 1980).

Rapid early outgassing and strong internal heating may result in a thin ice crust surmounting an ocean of saline water; loss processes will rather rapidly remove ice, with possible salt deposition. It is therefore not possible to rule out an early evaporite stage for Io, of the sort expected when remote spectral observations suggested a surface dominated by salts and sulfur. However, the likelihood of any trace of such an early surface surviving to the present on hyperactive Io is extremely small. Such a history would presumably involve extensive photolysis of water vapor or surface ice, thus causing oxidation of the surface layer and assuring that oxysalts such as sulfates, phosphates, and carbonates would be stable. A second plausible consequence of such a history would be the oxidation of ammonia to nitrogen, with possible complete destruction of ammonium salts near the surface.

The CI chondrites, which contain the veins of water-soluble sulfates mentioned previously, also contain crystalline carbonates and phosphates. The sulfates, which contain a very unusual excess of the light sulfur isotopes, coexist with small amounts of elemental sulfur and sulfides, both of which are isotopically heavy. Since isotopic equilibration leads to the exact opposite effect, it is necessary to postulate that the water-soluble vein material was formed by a unidirectional oxidation process beginning with a solid sulfide such as FeS or possibly a water-soluble sulfide such as Na_2S or $(NH_4)_2S$. Laboratory oxidation experiments in which sulfides are reacted with strong oxidizing agents such as O_2 or HOOH (Lewis and Krouse, 1969) show isotopic fractionations indistinguishable from those seen in meteorites when oxidation is carried out near 0°C, with progressively closer approaches to isotopic equilibrium at higher temperatures. It is interesting in light of our recent studies of the Galilean satellites to recall that Lewis (1967) proposed that the CI parent body may have had not only aqueous solutions out of which sulfates and other salts were precipitated, but also a temporary trace atmosphere containing water vapor, in which oxygen and other strong oxidizing agents such as hydrogen peroxide were produced by solar UV irradiation. Such conditions may not be unique.

Europa, Ganymede, and Callisto all differ in several very important ways from Io. First, they all contain significant amounts of water ice. Second, they are all subject to much less heating by tidal dissipation than Io is. Third, none of them has a detectable atmosphere in excess of the very low vapor pressure of H_2O at their surface temperatures. The 1972 occultation of a faint star by Ganymede (Carlson *et al.*, 1973) seemed to indicate a trace atmosphere with a pressure near 10^{-6} bar. As we have seen, IR spectrophotometry by Pilcher *et al.* (1972) and high-resolution spectroscopy by Fink *et al.* (1973) have clearly demonstrated the presence of large amounts of water ice on Ganymede, although the albedo of Ganymede is lower than that of either Io or Europa. Yung and McElroy (1977) showed that photolysis of water vapor (or water ice) will lead to production of hydrogen, which will be very rapidly lost from Ganymede, and of oxygen, which will accumulate. If there were no loss mechanism for oxygen, then O_2 would collect up to the point where it provides an opaque UV screen and prevents further photolysis of water vapor and ice. The oxygen abundance which can be produced by this mechanism is about 10^{-6} bar, and the time required to reach this state is about 10,000 years at Jupiter's distance from the Sun. Any massive ice-bearing body in the Solar System would presumably reach the same state, although the time required to do so would vary with heliocentric distance. Yung and McElroy note that a torus of hydrogen and oxygen may be associated with Ganymede's orbit about Jupiter.

Unfortunately, however, the Voyager 1 UV experiment (Broadfoot et al., 1979) has placed an upper limit on the O_2 pressure on Ganymede of only 10^{-11} bar. Kumar and Hunten (1982) have shown that the $H_2O–O_2$ photochemical system on Ganymede can actually exist in either of two quite distinct states, which differ in their steady-state oxygen pressure by a factor of 10^5. Subtle changes in boundary conditions, such as solar flux or ground albedo, might cause a dramatic change of modes. Ganymede is presently in the low O_2 mode, but, within the confines of our very limited knowledge, Callisto might be in either mode (Kumar and Hunten, 1982).

If potassium is efficiently extracted into aqueous ammonia solution or pure water, then ^{40}Ar release may occur on Ganymede and Callisto. For the Galilean satellites, complete outgassing would provide $\sim 0.1–0.5$ mbar of ^{40}Ar, depending on the mass of the satellite, its acceleration of gravity, and the details of the compositional model. There is no evidence for any atmospheric mass larger than about 10^{-3} of this value on any of these satellites, so loss processes are clearly effective (Pollack and Witteborn, 1980).

Since the first thermal structure models of the icy Galilean satellites (Lewis, 1971b,e), considerable progress has been made in investigating their thermal and tectonic evolution. Conductive thermal models by Consolmagno and Lewis (1976a,b, 1978) and homogeneous, incompressible conductive models by Fanale et al. (1977) were made more realistic by use of accurate equations of state for both the rock and H_2O ice components (Lupo and Lewis, 1979) and by the very important inclusion of solid-state convection, which strongly depresses the internal temperature gradient and minimizes the extent of melting (Reynolds and Cassen, 1979; Thurber et al., 1980). The Voyager imaging data on these satellites (Smith et al., 1979) has spawned a new type of literature dealing with the geology of vast masses of ice (Parmentier and Head, 1979; Lucchitta, 1980; McKinnon and Melosh, 1980; Squyres, 1980). Recent studies of the interior thermal states of the icy Galilean satellites (Cassen et al., 1979, 1980) examine the question of the stability of liquid H_2O in their mantles. In the past few years, spectroscopic and photometric studies of the surface compositions of these bodies have also advanced. Lebofsky (1977) demonstrated that H_2O ice was in fact present on Callisto, and Clark (1980) has established that 30–90% of the surface material of Callisto is ice. Carbonaceous chondrite-like impurities in the ice serve to mask the ice absorption bands and darken the surface. McFadden et al. (1980) have studied the dependence of the reflection spectra of the Galilean satellites on orbital phase (longitude).

We cannot leave the Galilean satellites without emphasizing the value of comparative studies of these bodies, including Jupiter's innermost satellite, J5 Amalthea, as well (Gradie et al., 1980). The density trends seen in this system mirror those seen in the Solar System at large, and these bodies are by no means of negligible size: They have the dimensions of planets, and each can be expected to reflect a wide range of complex planetary-scale processes. The thermal histories of ice-rich bodies are, of course, radically different from those of equal-mass rocky bodies, due to the very low melting point of water ice and the even lower melting point of the H_2O-NH_3 eutectic ($-100°C$), which may be important in Saturn's satellite system.

4.6.3. Titan

Saturn's largest satellite, Titan, has been known for over 30 years to have methane in its atmosphere (Kuiper, 1944). Trafton (1972a,b) showed that the atmospheric pressure on Titan probably exceeds 100 mbar, and reported that weak hydrogen quadrupole absorption lines are marginally detectable in the spectrum of Titan. Veverka (1973a) and Zellner (1973a) presented polarimetric evidence, and Barker and Trafton (1973) and Trafton (1975a) give spectroscopic evidence that UV and visible light are scattered by a dense cloud layer. Lewis and Prinn (1973) tentatively identified this as a cloud layer composed of particles of solid methane near the triple point of CH_4. The cloud tops are so cold (70–80 K) that NH_3 is not spectroscopically detectable (Encrenaz and Owen, 1974).

The thermal IR emission from Titan is most strikingly different from that of a Planckian emitter, as can be seen in Figure 4.55 in which measurements by Gillett *et al.* (1973) and by Low and Rieke (1974) are given. Two very different classes of theories were developed to account for this phenomenon. Sagan (1973c) first proposed a greenhouse effect in a very massive H_2-rich atmosphere. Less hydrogen-rich models ($<50\%$ H_2) have been suggested briefly by Morrison *et al.* (1972), by Hunten (1973a), and by Lewis and Prinn (1973), and examined in considerable

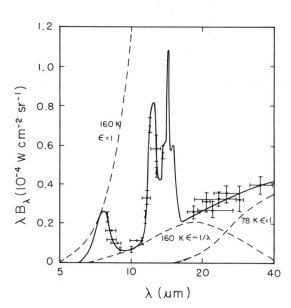

FIGURE 4.55. **Infrared emission spectrum of Titan. The data points due to the work of Low and Rieke (1974) and of Gillett *et al.* (1973) are plotted with a solid curve corresponding to the model of Danielson *et al.* (1973), as developed by Caldwell (1976). The model is discussed in the text.**

detail by Pollack (1973). In these models, optical penetration is limited by a dense CH_4 cloud layer, and surface pressures of several bars are possible. A very different model has been proposed by Danielson *et al.* (1973), who suggested that the high 8–13 μm temperatures observed by Gillett *et al.* (1973) and confirmed by Gillett (1975) might be due to molecular emission by photolysis products (notably C_2H_2 and C_2H_6) in a temperature inversion high above the surface of Titan. They also showed that some absorbing medium, such as photo-chemically produced "dust," is needed in order to explain the low UV albedo. This model pictures a bright, cold (<100°C), icy surface for Titan, whereas the greenhouse models permit surface temperatures of 180 K or even higher.

Prior to the Voyager Saturn encounter, it seemed that there were desirable and even convincing aspects of both models. The presence of a dark absorbing layer, presumably composed of tiny colored inorganic or polymeric organic particles—the "Axel dust" of Axel (1972), the "red clouds" of Khare and Sagan (1973), or the red phosphorus of Prinn and Lewis (1975)—at a temperature near 160 K, was viewed as very probable by all parties. On the other hand, there was general agreement by most authors (see Hunten, 1974b) that an optically thick, bright cloud layer was probably present well below this photochemical haze, and that the observed partial pressures of CH_4 and H_2 therefore must be regarded as lower limits on their true surface pressures. There is no clear evidence that the surface of Titan has ever been seen at visible wavelengths. An unfortunate feature of this combined model is that the true surface temperature and pressure remain undetermined. Hunten (1975) and Caldwell (1976) have given useful discussions of this issue.

Perhaps the most interesting issue regarding Titan is its suitability for the production and preservation of prebiotic organic material, especially light hydrocarbons. The greenhouse model for Titan envisioned a surface temperature $\gtrsim$140 K, at a pressure >500 mbar. Under these conditions, a significant trace of NH_3 could be present in the lower atmosphere. Further, if the surface temperature is $\leq$173 K, the icy crust may melt to yield the NH_3–H_2O eutectic liquid. This combination of circumstances need not be realized in entirety for the photolysis of ammonia and production of H–C–N compounds to become possible. Indeed, Hunten (1973a) has pointed out that the large hydrogen content suggested by Trafton's measurement requires the existence of a very large escape flux of H_2 from Titan, and that methane photolysis could not possibly supply enough H_2 to achieve a steady state. On the other hand, ammonia photodissociates out to a wavelength of ~2300 Å and can absorb a far larger proportion of the solar UV emission. If all incident sunlight of λ < 2300 Å goes into NH_3 photolysis, then the necessary H_2 flux could barely be maintained. Hunten (1973b) has also treated the general escape physics of hydrogen in detail, including blowoff and diffusion as well as thermal Jeans escape. Atreya *et al.* (1978) have treated NH_3 photolysis in detail, leading to the conclusion that up to 20 bar of N_2 could be formed.

One inescapable consequence of rapid hydrogen loss (about 10^{12} cm^{-2} sec^{-1}) is the injection of a vast flux of H_2 into space near Titan, where it would form a gas torus girdling Saturn, with highest density close to Titan's orbit. Indeed, the first suggestion of a gas torus associated with escape of a satellite atmosphere was that of Titan's H_2 ring by McDonough and Brice (1973a,b). Obviously, the interpretation of the very marginal Titan H_2 spectrum can affect such ideas profoundly.

All models of Titan begin with the presence of several tens of millibars of methane above the clouds (which, in Danielson-type models, are lying on the ground). The photochemical destruction rate for methane on Titan is about 1.6×10^9 cm^{-2} sec^{-1}, or 4×10^{-14} gm cm^{-2} sec^{-1}. The baseline methane abundance of about 20 mbar is equivalent to about 130 gm cm^{-2} of methane, which implies a lifetime of only $\sim 10^8$ years for total destruction of the methane. Strobel's (1974b) study of the photochemistry of Titan's atmosphere showed that the large amounts of ethane and acetylene required to produce strong emission features in the thermal IR are directly due to known features of the photolysis of methane. Yet all was not told regarding the chemistry of Titan's atmosphere.

Trafton (1974) reported strong features in the 1-μm region which are due to an as yet unidentified absorber. Studies of the ionic chemistry of the upper atmosphere, where photo-ionization and cosmic ray ionization are possible (Capone et al., 1976, 1980; Whitten et al., 1977), showed the possibility of small but interesting formation rates for a wide variety of hydrocarbons.

More recently, some of the problems associated with very hydrogen-rich atmospheric models were tempered by a more sensitive measurement of the spectrum in the vicinity of the (3, 0) S1 quadrupole line of H_2 by Münch et al. (1977). They fail to detect the feature, and place an upper limit on the hydrogen abundance of about 1 km agt, several times less than previous estimates. Thus, the presence of a large mole fraction of H_2 is thrown into serious doubt, and the magnitude of the associated hydrogen torus may be orders of magnitude less than at first anticipated. The lower hydrogen abundance and concomitant lower escape flux in turn make it very plausible that photolysis if methane alone may suffice as the source of hydrogen. Note that the production of hydrogen by methane photolysis is hardly open to doubt, since the methane is clearly seen. Adding more photolabile (and chemically plausible) species capable of shielding the methane from some of the incident solar UV radiation, such as ethane or phosphine, would merely produce hydrogen with the same or greater efficiency. The presence of ethane and acetylene prove the significance of methane photolysis, and thus of that source of hydrogen.

The fact that the atmosphere as seen from above contains far too little methane to survive for the age of the Solar System requires that there be either a much greater (about 1 bar or more) methane pressure at the surface of Titan than there is at the visible cloud level, or a massive surface source of methane. The latter would be plausibly provided by the presence of either solid methane clathrate hydrate or of methane itself as a major constituent of the surface, in vapor pressure equilibrium with the atmosphere. Among the preliminary compositional classes presented in Figure 4.50, models 11 and 12 meet these requirements. We have seen in Figure 4.50 that the accepted pre-Voyager radius and mass of Titan suggested an uncompressed density near 1 gm cm^{-3}, which would require the presence of methane in solar proportions. This interpretation is, however, predicated upon the depth of the atmosphere being small, since the observed radius is used as the surface radius. If, instead, the atmosphere is 300 km deep (as an example), then the uncompressed density of the planet becomes reconcilable with that of the model 11 of Figure 4.50, containing methane clathrate hydrate but not the full solar complement of methane. The latter model, although methane-poor by comparison, nonetheless would yield a surface atmospheric pressure of 1.5 kbar if fully outgassed. This would require that the surface lie 8.5 scale heights (about 300 km for T near 100 K) below the visible cloud level. The minimum mass of gaseous methane required in order for the photochemical lifetime to be longer than the age of the Solar System would place the surface at least 4 scale heights below the clouds: A condensed CH_4 reservoir on the surface would lower the required mass of atmosphere, and thus lower the density of "solid" Titan.

The main features of the two extreme types of models for Titan's atmosphere are sketched in Figure 4.56.

In the months prior to the Voyager 1 encounter, several workers (Bar-Nun and Podolak, 1979; Podolak and Bar-Nun, 1979; Podolak et al. 1979; Rages and Pollack, 1980) investigated the photochemistry of the C–H system on Titan and the formation and distribution of aerosols in its atmosphere. They predicted that C_2H_6 and C_3H_8 would accumulate in a methane-rich atmosphere, while C_2H_2 and C_2H_4 would rapidly reach a steady state (Bar-Nun and Podolak, 1979).

Recent IR spectroscopic and radiometric observations from the same time period include Fink and Larson (1979), Cruikshank and Morgan (1980), McCarthy et al. (1980), and Jaffe et al. (1979). The latter, observing at 6 cm wavelength with the very large array (VLA) radio telescope, found a surface temperature of 87 ± 9 K for Titan.

The Voyager 1 Saturn encounter included an approach to within 6490 km of Titan, thus permitting the entire powerful complement of experiments to be brought to bear simultaneously (Stone and Miner, 1981). The imaging experiment

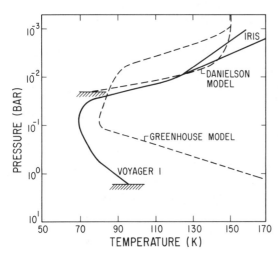

FIGURE 4.56. **Titan pressure–temperature profiles. The presence of a warm (150–160 K) inversion layer at high altitudes is accepted; pre-Voyager theoretical models, as reviewed by Hunten (1975) and Caldwell (1976) are shown as broken lines. A preliminary Voyager T, P profile from the radio occultation experiment (Tyler et al., 1981) is shown as the solid line at pressures above ~0.1 bar, and the IRIS temperature profile (Hanel et al., 1981) is shown at higher altitudes.**

covered Titan with very high resolution, but found it to be almost perfectly feature-less, covered with thick clouds tinged a dirty orange-brown. The northern hemisphere is slightly darker and redd' than the southern, and there are detached, diffuse high-altitude haze layers visible at the limb. The seasonal effects are presumably superficial, since the radiative time constant for Titan is well over 100 years. This causes seasonal changes to lag insolation changes by about 90° (Smith *et al.*, 1981). The IRIS experiment provided an exciting insight into the chemistry of Titan's atmosphere. In addition to CH_4, which has long been known to be present, C_2H_2 and C_2H_6 were known to be major photochemical products from CH_4 (Strobel, 1974b), C_2H_4 was known to be a minor photolysis product, and both C_2H_2 and C_2H_6 were seen in emission from the inversion layer (Caldwell, 1976). The IRIS experiment confirmed all of these species, detected HCN, and added tentative detections of C_3H_4 and C_3H_8 (Hanel *et al.*, 1981). Several prominent unidentified spectral features were found in emission at high latitudes.

The radio occultation experiment provided several crucial pieces of information about Titan's atmospheric and surface conditions. First, the occultation experiment succeeded in penetrating all the way to the surface, providing a measurement of the surface pressure of about 1.5–2 bar (Tyler *et al.*, 1981). The surface temperature was found to be about 93 K, as shown in Figure 4.56. The scale height of the atmosphere permitted a direct calculation of the mean molecular weight of the atmosphere, which was found to be about 28. This is entirely plausible if the theoretical calculations of N_2 photochemical production by Atreya *et al.* (1978) are considered. The positive location of the surface permitted the first accurate determination of the radius, and hence the density, of the "solid body" of Titan becomes 1.89 gm cm^{-3}. Corrected to zero pressure, this gives a density which, within the uncertainties in the cosmic abundances of the elements, is reconcilable with the composition classes 9, 10, and 11 of Figure 4.46: that is, water ice and rock with or \ ithout solid hydrates of ammonia and methane. Given the observed presence of methane, the necessity for a large surface reservoir of CH_4, and the presence of a massive nitrogen atmosphere, the most plausible conclusion is that these hydrates were in fact retained to some extent. However, aside from Titan's massive atmopshere, the bulk properties of Titan and Callisto (and also, to a slightly lesser degree, Ganymede) betray astonishingly similar zero-pressure densities and bulk compositions (Lupo, 1981).

The UV observations from Voyager 1 (Broadfoot *et al.*, 1981) found emission from both atomic and molecular nitrogen in the spectrum of Titan. A hydrogen torus fed by escape from Titan into planetocentric orbit was also seen.

Strobel (1981) has offered an early post-Voyager interpretation of the photochemistry and aeronomy of Titan. He describes the very important role of magnetospheric electrons in exciting, dissociating, and ionizing N_2, thereby providing N atoms and atomic ions which react very readily with a wide variety of hydrocarbon gases. These reactions make HCN with a production rate estimated at 1.4×10^9 cm^{-2} sec^{-1}. Polymers of HCN with C_2H_2 are apparently responsible for the red clouds. Because of the low surface temperatures, rainout of C_2 and heavier hydrocarbons should irreversibly deposit some 85 kg cm^{-2} of hydrocarbons on the surface of Titan over the age of the Solar System. The corresponding accumulated amount of HCN polymers is about 10 kg cm^{-2} (Strobel, 1981). This is sufficient to cover the entire surface of Titan to a depth of 1 km with organic matter. This conclusion, in addition to confirming what has long been suspected, places some interesting constraints on the nature of the massive methane reservoir which is required to replenish photolyzed CH$_4$.

Strobel also finds nitrogen escape rates of 3×10^{26} atoms sec^{-1}, which would cause the loss of 2.4 kg cm^{-2} over the age of Titan, or about 10% of the present atmospheric mass. Explosive blowoff of gases by impact events has not been modeled, but is probably not a very important factor because of the high surface denisty of Titan's atmosphere and the relatively low encounter velocity of colliding bodies.

Surface conditions on Titan remain obscure, but hydrocarbon oceans, N$_2$ precipitation at the poles, and large-scale formation of organic nitriles (R—C≡N) are all possible. Temperature contrasts on the surface are extremely small, of order 0.1 K (Leovy and Pollack, 1973; Golitsyn, 1975), so that surface condensation may depend more on the altitude of surface relief than latitude.

4.6.4. Triton and Pluto

The seventh of the lunar-sized bodies is Triton, the larger of Neptune's two satellites. It is unusual in being of large size, but in a retrograde orbit. The mass of Triton, determined by its effect on the apparent motion of Neptune, was found by Alden (1943) to be $(1.38 \pm 0.24) \times 10^{26}$ gm. The radius is most difficult to determine because of the very small angular size and the unknown degree of limb darkening. If the albedo is 0.5, then the visual brightness corresponds to a radius of 1600 km (Morrison et al., 1976). The most extreme albedos documented for solar system bodies are near 0.7 and 0.03, which allow radii ranging from as low as 1350 km to as high as 6530 km. Kuiper's (1950) estimate of the radius was 1900 ± 600 km, which favors higher values of the albedo and higher density for Triton. Even using the more restrictive radius estimate bu Kuiper, the bulk density can still range from 1.6 to 20 gm cm^{-3}, which clearly is of little value in determining the composition. With the full range of albedos presently regarded as possible, a density as low as 0.09 would be possible!

Until recently, there was very little observational evidence regarding the possibility of an atmosphere. Spinrad (1969) was unable to find any trace of methane in the spectrum. The very low temperature at Neptune's distance from the Sun would permit a pure methane atmosphere to condense. The radiative steady-state temperature of Triton, assuming it to be a rotating atmosphereless grey body with the same albedo as Jupiter, is only 43 K. The vapor pressure of methane at this temperature is near 10^{-6} bar, far too low to be detected. We should mention that the mean opposition magnitude of Triton is only 13.6, so that it provides a light flux 4000 times less intense than that of Ganymede at opposition, making spectroscopy extremely difficult.

Nonetheless, Benner *et al.* (1978) have claimed the marginal detection of a methane atmosphere on Triton. Because of Triton's large mass, atmospheric escape is not a problem, and the main limiting factor on the growth of atmospheric pressure resulting from outgassing is the low surface temperature. The spectroscopic evidence for the presence of both gaseous and solid methane has been confirmed and extended by Cruikshank and Silvaggio (1979) and by Cruikshank *et al.* (1979).

If we assume a chemically and cosmogonically plausible composition for Triton (composition class 12 in Figure 4.46 for equilibrium condensation, or condensation of H_2O, NH_4HCO_3, and CO_2 with or without CO and N_2 clathrate hydrates as in Figure 2.19 (p. 65), in the case of kinetically inhibited formation of CH_4 and NH_3 in the nebula out of which the solids formed), then the bulk density should be in the range of 1.0–1.5 (at zero pressure). Taking a geometric albedo in the range 0.35–0.70 as most likely, the radius would be near 1400–1900 km, and the mass should be $(1–4) \times 10^{25}$ gm. The surface gravity would then lie between about 40 and 80 cm sec^{-2}, and the escape velocity would be between 1.1 and 1.7 km sec^{-1}. Comparing the thermal speeds of plausible atmospheric gases such as CH_4, N_2, and Ar to this escape speed, as in Figure 3.6 (p. 108), shows that methane at 43 K should be bound for the entire age of the Solar System if the escape velocity is in the upper half of this range. Nitrogen and argon are bound down to about the lower limit on the escape velocity. Photolysis of a methane-rich or methane–nitrogen atmosphere on Triton should produce about 40 m of accumulated C_2 hydrocarbons over the age of the Solar System. For an optically thick atmosphere at wavelengths near 1600 Å, a methane photolysis rate of about 2×10^9 cm^{-2} sec^{-1} would be expected, and, assuming as a generous upper limit a column atmospheric CH_4 abundance of 20 m agt, we find that the photochemical lifetime of the atmosphere would be only 10^6 yr. Again, as on Titan, we would be forced to postulate a surface reservoir of CH_4 several orders of magnitude larger than the atmospheric inventory allowed by vapor pressure considerations. We should emphasize that, if this deduction is valid, then photochemical products should darken the clouds and surface of Triton, making the geometric albedo lower than the range assumed above, which in turn increases the radius, mass, acceleration of gravity, and escape velocity. For this reason, radii as large as ~2400 km cannot be ruled out.

Many of the same general considerations apply to Pluto. The discovery of Pluto, motivated by the discovery of unexplained apparent residuals in the position of Neptune, gave startling and unexpected results: The apparent mass of Pluto was found to be roughly 1 Earth mass, rather than the 5–10 Earth masses required to provide the necessary forces on Neptune (see, e.g., Mayall and Nicholson, 1931). Repeated quasi-independent estimates of the mass of Pluto between 1930 and 1966 gave generally concordant results. More sophisticated and complete analyses of outer solar system dynamics (see, e.g., Ash *et al.*, 1968; Seidelman *et al.*, 1971; Duncombe *et al.*, 1970) revised this estimate to 0.1–0.2 Earth masses. A dissenter was Brosche (1967), who favored a mass of only 0.011 Earth mass. The radiometric determination of the Bond albedo of Pluto by Cruikshank *et al.* (1976) combined with a reasonable density suggested a mass near 0.004 Earth mass (2.4×10^{25} gm).

The discovery of Pluto's satellite Charon (Christy and Harrington, 1978) led to a considerable improvement in the mass estimate to 0.0022 Earth mass (Christy and Harrington, 1980). A review of the history of Pluto mass determinations is given by Duncombe and Seidelmann (1980).

Radius determinations on Pluto have a similarly deflationary history. Kuiper (1950) estimated the radius to be about 3000 km, but the most recent measurements by Arnold et al. (1979) give a radius of 1500 ± 200 km for a Lambert disk. Previous to the most recent downward revisions in Pluto's mass, these radii implied a density of roughly 60 gm cm^{-3} (for one Earth mass) or 12 gm cm^{-3} (for 0.2 Earth mass). These fantastic results were generally disbelieved, although some authors speculated about an "iron-rich" Pluto denser than the Earth (Manning, 1971; Fix, 1972). The most probable density within the range of currently acceptable masses and radii is about 1.1 gm cm^{-3} (Lupo and Lewis, 1980a,b), with 1.5 gm cm^{-3} as a reasonable upper limit.

The presence of an atmosphere of neon on Pluto was suggested by Hart (1974); however, in the absence of any mechanism for retaining neon in the first place, and in view of the tiny partial pressure of neon on all other bodies studied, this conjecture seems without basis.

Benner et al. (1978) presented spectroscopic evidence for the presence of a trace of gaseous methane on Pluto, and the presence of solid methane is rather well established (Cruikshank et al., 1976; Lebofsky et al., 1979; Cruikshank and Silvaggio, 1980; Soifer et al., 1980). Barker et al. (1980) set an upper limit of about 3 m agt of gaseous CH_4, whereas Fink et al. (1980) estimate 30 m agt of methane based on the detection of seven visible and near-IR bands. Trafton has pointed out that, since Pluto is both smaller than Triton and closer to the Sun (until the end of the century!), there is a serious problem of atmospheric stability against blowoff. He proposes (Trafton, 1980) that the major constituents of Pluto's atmosphere are probably heavy rare gases, nitorgen, carbon monoxide, or oxygen. In order to make CO or O_2, some oxygen compound would have to be available for photolysis (H_2O?), and the most plausible source of N_2 on an icy body is NH_3. Both species, however, would have negligible vapor pressures at Pluto's temperature, and seem implausible. Primordial ^{36}Ar is probably the best candidate, but requires enough heating of Pluto's interior to release the argon from its ice clathrate. If formation of CH_4 and NH_3 is kinetically inhibited in the nebula, then rare gas concentrations in the ice can reach their solar abundances relative to oxygen. However, this entire debate has been shown by Hunten and Watson (1982) to be an artifact of the arbitrary (and incorrect) assumption that the exospheric temperature of Pluto must be roughly equal to the surface temperature. They show that, for reasonable exospheric temperatures, methane escape is quite slow.

In summary, the lunar-sized objects, when studied in parallel with the terrestrial planets, help to broaden our perspectives on the chance for volatile retention and loss, and, consequently, to enrich our understanding of the process for shaping the atmospheres of Venus, Mars, and Earth. A flyby of Uranus, Neptune, and Triton by one of the Voyager spacecraft is planned, but Pluto remains out of reach in the foreseeable future. Recent discoveries regarding the atmospheres of Titan, Io, and Ganymede will undoubtedly spur further observational efforts using Earth-based optical, radio, and radar techniques.

4.7. THE ASTEROIDS

We have seen in Section 2.2 that asteroids are far too small to capture any gas from the solar nebula. An appreciable atmospheric pressure, if ever present, could have been due only to outgassing of a secondary atmosphere from the interior.

It is worthwhile to review what is presently known regarding the composition of the asteroids, in order that we might have some grounds for speculation on their volatile content.

The most rudimentary type of compositional information for distant solid bodies is their bulk density. Unfortunately, there are approximate mass determinations available for only three asteroids—Ceres (Schubart, 1971), Vesta (Hertz, 1968), and Pallas (Schubart, 1975). The recent application of the polarimetric (Veverka, 1973; Veverka and Noland, 1973; Bowell and Zellner, 1974; Zellner et al., 1974b) and IR radiometric (Allen, 1971; Matson, 1971; Cruikshank and Morrison, 1973; Hansen, 1977) techniques to the determination of asteroid radii has resulted in the flood of data summarized in Table 4.18, which is based on the very useful reviews by Chapman (1976) and by Morrison and Chapman (1976). Both techniques generally give concordant radius and albedo estimates except for a few of the lowest albedo bodies. For these bodies, the errors of the IR determination are theoretically smaller, whereas the calibration of the polarimetric measurement techniques for bodies of such low albedo has been less convincing. We therefore have preferred the IR results for low-albedo objects whenever discordance between the two techniques is evident. Recently, Zellner et al. (1977) have recalibrated the polarimetric techniques for low-albedo objects and restored concord with the radiometric albedos.

For Ceres, we shall use a mass of $(1.2 \pm 0.2) \times 10^{24}$ gm (Schubart, 1971) and a radius of 477 ± 20 km (Chapman et al., 1975). For Vesta, we will use a mass of $(2.4 \pm 0.4) \times 10^{23}$ gm (Hertz, 1968) and Morrison's radius estimate of 252 ± 20 km, and for Pallas—a mass of $(2.3 \pm 0.4) \times 10^{23}$ gm (Schubart, 1975) and a radius of 279 ± 20 km.

The densities calculated from these data are ~ 2.6 gm cm^{-3} for Ceres, 2.6 gm cm^{-3} for Pallas, and 3.6 gm cm^{-3} for Vesta. The mass, radius, and density data for all asteroids known to have $r > 150$ km, including error estimates, are given in Table 4.19.

The deduced densities may profitably be compared with those of known meteorite classes as was done for the satellites of the planets in Figure 4.47. This comparison is given in Figure 4.57, whence it can be seen that the largest asteroid in the belt, 1 Ceres, is almost certainly a C1 or C2 chondrite. 4 Vesta, although

TABLE 4.18
Radii and Albedos of Asteroids[a]

	Asteroid	Type[b]	p_{vis}	r (km)
1	Ceres	C	0.062	477
2	Pallas	C	0.087	269[c]
3	Juno	S	0.190	113
4	Vesta	U	0.264	252
5	Astraea	S	0.177	58
6	Hebe	S	0.203	98
7	Iris	S	0.182	96
8	Flora	S	0.168	75
9	Metis	S	0.161	84
10	Hygiea	C[c]	0.054	191
11	Parthenope	S	0.130	74
12	Victoria	S	0.173	40
14	Irene	S	0.183	74
15	Eunomia	S	0.155	135
16	Psyche	U	0.091	127
17	Thetis	S	0.152	45
18	Melpomene	S	0.140	70
19	Fortuna	C	0.032	110
20	Massalia	S	0.214	67
21	Lutetia	C[c]	0.102	52
22	Kalliope	U	0.123	84
23	Thalia	S	0.172	51
27	Euterpe	S	0.121	56
29	Amphitrite	S	0.135	94
30	Urania	S	0.159	44
36	Atalanta	S	0.027	58
37	Fides	S	0.183	47
39	Laetitia	S	0.200	73
40	Harmonia	S	0.150	51
42	Isis	S	0.155	43
43	Ariadne	S	0.123	43
44	Nysa	U	0.113	67
51	Nemausa	C[c]	0.060	73
56	Melete	C	0.040	56
63	Ausonia	S	0.193	54
64	Angelina	U	0.290	31
65	Cybele	C	0.026	150
68	Leto	S	0.079	77
70	Panopaea	C	0.033	74
80	Sappho	S	0.123	41
88	Thisbe	C	0.039	103
89	Julia	S	0.119	77
93	Minerva	C	0.037	82
107	Camilla	C	0.042	103
116	Sirona	S	0.182	39
129	Antigone	S	0.179	57

(continued)

TABLE 4.18—*Continued*

Asteroid		Type[b]	p_{vis}	r (km)
131	Vala	S	0.121	17
139	Juewa	C	0.069	51
140	Siwa	C	0.065	45
141	Lumen	C	0.040	53
192	Nausikaa	S	0.200	46
208	Lacrimosa	S	0.129	20
230	Athamantis	S[c]	0.096	59
324	Bamberga	C	0.020	125
349	Dembowska	U	0.251	72
381	Myrrha	C	0.031	62
415	Palatia	C?	0.026	46
433	Eros	S	0.14	10
441	Bathilda	S	0.112	32
471	Papagena	S	0.159	66
511	Davida	C	0.038	172
532	Herculina	S	0.140	66
554	Peraga	C	0.059	49
558	Carmen	?	0.084	31
624	Hektor	U	0.028	105
654	Zelinda	C	0.04	48
707	Interamnia	S	0.118	57
747	Winchester	C	0.028	93
782	Montefiore	S	0.156	8
790	Pretoria	C	0.025	86
863	Benkoela	U	0.269	16
887	Alinda	S	0.146	2.2
1178	Irmela	C	0.057	10
1566	Icarus	S	0.178	0.7
1567	1941 HN	C	0.026	35
1620	Geographos	S	0.209	1.0
1685	Toro	S	0.148	3.0
—	1976 AA	S	0.18	0.9

[a]After Chapman *et al.* (1975) and Morrison and Chapman (1976).

[b]Types are designated as follows: C, carbonaceous; S, silicaceous, perhaps stony-iron; U, unclassified, including achondritic, metallic, and unidentified objects.

[c]Wasserman *et al.* (1979) occultation data.

probably similar in density to ordinary chondrites, achondrites, or C3 chondrites, has error bars wide enough to embrace the full range of C2 densities as well. The density of Pallas is even less diagnostic.

The most probable value of the density of Pallas is also typical of CI chondrites, although the upper limit on the density (3.7 gm cm^{-3}) would fit an ordinary chondrite. The probable *upper limit* on the density of Ceres, using both the upper limit on the mass and the lower limit on the radius, is only 2.8 gm cm^{-3}, a value

TABLE 4.19

Masses, Radii, Densities, and Escape Velocities of the Largest Asteroids ($r \geq 150$ km)[a]

Asteroid	Type	Radius (km)	Mass (gm)	Density (gm cm^{-3})	V_{esc} (km sec^{-1})
1 Ceres	C	477 ± 20	$(1.2 \pm 0.2) \times 10^{24}$	2.6 ± 0.7	0.58 ± 0.04
2 Pallas	C	279 ± 20	$(2.3 \pm 0.4) \times 10^{23}$	2.6 ± 1.0	0.33 ± 0.04
4 Vesta	S	252 ± 20	$(2.4 \pm 0.4) \times 10^{23}$	$3.6 \pm {}^{1.7}_{1.2}$	0.35 ± 0.04
10 Hygiea	C	191 ± 15	—	—	—
511 Davida	C	172 ± 25	—	—	—
65 Cybele	C	150 ± 20	—	—	—

[a]Radii from Chapman *et al.* (1975) and from Morrison and Chapman (1976); masses for Ceres, Pallas, and Vesta from Schubart (1971, 1975) and Hertz (1968), respectively.

more characteristic of carbonaceous meteorites than any other known class of extraterrestrial material. A density below 2.1 gm cm^{-3} would, if taken literally, mean a higher volatile (H$_2$O) content than CI chondrites.

These densities strongly suggest that Ceres is far more volatile-rich than any of the terrestrial planets, probably comparable to carbonaceous chondrites. A compositional model for planets and satellites which otherwise is in good agreement with observation (Lewis, 1972a,b) predicts a formation temperature near 300 K for Ceres, which would provide enough volatiles to give a density near 2.5 gm cm^{-3}.

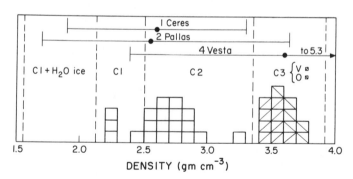

FIGURE 4.57. Densities of carbonaceous meteorites and asteroids. Asteroid data are those summarized in Table 4.19. The meteorite data are explained in the caption of Figure 4.47. The density of Ceres is considered to imply a C2 or C1 composition, since temperatures low enough to condense H$_2$O ice at Ceres distance from the Sun (a = 2.767 AU) are improbable for other reasons. The density of 2 Pallas is compatible with any composition from C1 to C3, and the formal error bars even reach the low-density end of the range of ordinary chondrite densities, roughly 3.63 gm cm^{-3}. For Vesta, any stony meteorite composition can be reconciled with the asteroid data, although C1 composition seems quite improbable. The error bars are for the *absolute* densities of these asteroids. Morrison (1976) has shown that the errors in the *relative* densities are smaller, so that the density contrast between Ceres and Vesta inferred from this figure is *more* reliable than the error bars on the absolute densities would suggest. Morrison concludes that $\rho_{Vesta} = (1.33 \pm 0.17) \times \rho_{Ceres}$.

A second type of suggestive compositional information can be found in the albedos derived for asteroids by the IR and polarimetric techniques. For Ceres, the preferred radius of 477 ± 20 km yields a visual Bond albedo of 0.062 ± 0.01 (Table 4.18). This low albedo is by no means unique, as can be seen from Table 4.18. Albedos of less than 0.09 have been found for 1 Ceres, 2 Pallas, 10 Hygiea, 19 Fortuna, 324 Bamberga, and 511 Davida, among many others. It is interesting that, solely on the basis of size, Davida and Bamberga should be among the "Big Ten," but they have such low albedos that they bear the unflattering catalog numbers of 324 and 511! No naturally occurring materials except carbonaceous chondrites are known to have albedos as low as 0.04.

Laboratory studies of the albedos and reflection spectra of rock, mineral, and meteorite samples by Gaffey (1976) and by Johnson and Fanale (1973), and comparison with astronomical spectra by Chapman (1976) led to the strong suspicion that most of the asteroid belt is dominated by carbonaceous chondrite material, and led to a great increase in the astronomical efforts to determine the reflection spectra of asteroids.

The early spectral reflectivity studies also revealed great variety and complexity in the inner part of the belt. McCord *et al.* (1970) have found that the reflection spectrum of Vesta bears a striking resemblance to that of the Nuevo Laredo basaltic achondrite. Further, 16 Psyche's reflection spectrum resembles that of an iron meteorite.

Detailed reviews by Gaffey and McCord (1978, 1979) discuss all available spectrophotometric and spectroscopic data on the asteroids, including many objects for which radius (and hence albedo) determinations are not yet available. The interpretation of the spectral properties of asteroid surfaces generally is limited to the detection and identification of major silicate phases (ferromagnesian silicates and plagioclase), opaques (metal, carbon, carbonaceous material), and hydroxyl silicates (clays, etc.). Experience with the laboratory study of the spectra of a large suite of meteorite and mineral samples then permits the interpretation of the asteroid spectra in terms of known meteorite classes and of mixtures of known minerals. The results of this effort are given in Table 4.20, except that the interpretation of spectral type F has been changed from C4 to C2 chondrites. Gaffey and McCord (1978) favored the C4 interpretation, but indicated that a second alternative would be C3/2. A recent extension of the IR spectrum of Ceres out to 3.5 μm by Lebofsky (1978) has led to the discovery of a strong hydroxyl band near 3.0 μm, in very good agreement with the C2 chondrites, but neither as deep nor as broad as the same feature in the spectrum of CI chondrites.

The overwhelming majority of the asteroids studied were found to be carbonaceous chondrites (35 of 59), whereas virtually all the others appear to be igneous differentiates—either achondrites, irons, or stony-irons (22 of 59). The two exceptions are 433 Eros and 1685 Toro, which appear to be H and L chondrites, respectively. The semimajor axes of the orbits of these asteroids are only 1.458 and 1.368 AU, far smaller than those of the belt asteroids. Zellner and Bowell (1977), summarizing the limited broad-band photometric and polarimetric data on 12

TABLE 4.20
Spectroscopic Classification of Asteroids[a]

Asteroid	Spectral Type[b]	Interpretation[c]
1 Ceres	F	None
2 Pallas	F	None
3 Juno	RA-1	Stony-iron: ol + px[d]
4 Vesta	A	Eucrite
6 Hebe	RA-2	Mesosiderite
7 Iris	RA-1	Stony-iron: ol + px
8 Flora	RA-2	Mesosiderite
9 Metis	RF	E chondrite or iron
10 Hygeia	TB	C1/C2
11 Parthenope	RF	E chondrite or iron
14 Irene	RA-3	Stony-iron: px
15 Eunomia	RA-1	Stony-iron: ol + px
16 Psyche	RR	E chondrite or iron
17 Thetis	RA-2	Mesosiderite
18 Melpomene	TE	C3
19 Fortuna	TA	C1/C2
25 Phocaea	RA-2	Stony-iron: px
27 Euterpe	RA-2	Stony-iron: px
28 Bellona	TE	C3
30 Urania	RF?	—
39 Laetitia	RA-1	Stony-iron: ol + px
40 Harmonia	RA-2	Mesosiderite
48 Doris	TA	C1/C2
51 Nemausa	TC	C1/C2
52 Europa	TA	C1/C2
58 Concordia	TABC	C1/C2
63 Ausonia	RA-3	Stony-iron: px
79 Eurynome	RA-2	Mesosiderite
80 Sappho	TD	C3
82 Alkmene	TE	C3
85 Io	F	?
88 Thisbe	TB	C1/C2
130 Elektra	TABC	C1/C2
139 Juewa	TB	C1/C2
140 Siwa	RR	E chondrite or iron
141 Lumen	TA	C1/C2
145 Adeona	TA	C1/C2
163 Erigone	TA	C1/C2
166 Rhodope	TC	C1/C2
176 Iduna	TA	C1/C2
192 Nausikaa	RA-2	Stony-iron: ol + px
194 Prokne	TC	C1/C2
210 Isabella	TABC	C1/C2
213 Lilaea	F	C(?)
221 Eos	TD	C3

(continued)

TABLE 4.20—*Continued*

Asteroid	Spectral Type[b]	Interpretation[c]
230 Athamantis	RF	E chondrite or iron
324 Bamberga	TABC	C1/C2
335 Roberta	F	C(?)
349 Dembowska	A	C1 achondrite
354 Eleonora	RA-1	Pallasite
433 Eros	—	HCh
462 Eriphyla	RF?	—
481 Emita	TABC	C1/C2
505 Cava	TA	C1/C2
511 Davida	TB	C1/C2
532 Herculina	TE	C3
554 Peraga	TA	C1/C2
654 Zelinda	TC	C1/C2
674 Rachele	RF?	—
704 Interamnia	F	C(?)
887 Alinda	TD	C3
1685 Toro	—	LCh(?)

[a]After Gaffey and McCord (1978).

[b]Spectral types are described by the visual appearance of the spectrum in the visible (first letter) and the near-IR (second letter). Flat spectra are denoted by F, spectra sloping upward toward longer (redder) wavelengths are denoted by R, and those showing a transition from R to F in the given spectra region are denoted by T. The subdivisions denoted by A, B, C, D, and E, while probably of objective compositional significance, are not of great importance for our present purposes.

[c]Interpretations are generally those given by Gaffey and McCord (1979), except for spectral type F (see text for explanation). The only ordinary chondrite bodies found are the H chondrite 433 Eros and the L chondrite 1685 Toro.

[d]The abbreviations are as follows: ol, olivine; px, pyroxene.

Apollo and Amor asteroids (which cross the orbits of one or more of the terrestrial planets), find that only one, the large carbonaceous object 887 Alinda, does not belong to the S classification. These S-type bodies include 1566 Icarus, 1620 Geographos, and 1864 Daedalus.

Because the asteroids are cataloged in order of discovery, selection effects favor attaching low catalog numbers to abnormally large, high-albedo bodies relatively close to the Sun and to Earth. Indeed, the list of all known asteroids is already strongly biased *against* carbonaceous material. Of asteroids with numbers up to 50, only 7 of 22 with known spectral type are carbonaceous, but of the 37 with larger catalog numbers, 28 are carbonaceous! Furthermore, directly relevant to the prevalence by *mass* of carbonaceous material in the belt, 5 of the 6 largest asteroids (Table 4.19) are carbonaceous. After 1 Ceres, 2 Pallas, and 4 Vesta, there are only 10 other asteroids known with $r > 100$ km (numbers 10, 511, 65, 15, 16, 324, 3, 19, 624, and 88), and the total mass of these 10 is only about equal to that of Vesta (20% of the mass of Ceres). Seven of the 10 are carbonaceous.

It is thus clear that the primitive material in the asteroid belt was carbonaceous.

This confirms the prediction of Lewis (1972b), who predicted this at a time when the only available compositional datum on any asteroid was McCord *et al.*'s (1970) interpretation of 4 Vesta as a basaltic achondrite. The consensus of meteoriticists at that time is aptly summarized in the review by Anders (1971): "The prevailing view at present is that most (or all) meteorites come from the asteroid belt. But a cometary origin of some types (e.g., carbonaceous chondrites) remains an intriguing possibility." Wetherill (1969) at that time favored a cometary origin for almost all meteorites, concluding on dynamic grounds that "it does not seem likely that any observed group of asteroids will serve as satisfactory sources."

Although all the asteroids in the belt which appear to be primitive (chondritic) bodies are carbonaceous, the abundance of C asteroids does vary in a systematic way with heliocentric distance: S asteroids are most common at the inner edge of the belt, with C asteroids becoming more common at larger distances. The composition of the asteroid belt versus distance from the Sun is given in Figure 4.58.

Spectroscopic data have also revealed a few very close matches between known classes of igneous meteorites and particular asteroids. The best-characterized possibility is that the basaltic achondrites come from 4 Vesta (McCord *et al.*, 1970; Larson and Fink, 1975) or possibly 349 Dembowska (Feierberg *et al.*, 1980). Spectral comparisons between Shergotty and Alan Hills 77-05, two unusual achondritics, and asteroids are given by Feierberg and Drake (1980). They suggest that both the eucrites and howardites derive from 4 Vesta, whereas Alan Hills 77005 may derive from 349 Dembowska.

Nonhomogeneous accretion scenarios have not been applied to the asteroid belt for the purpose of predicting observable compositions; however, the absence of detectable ordinary chondritic material in the belt is striking. Equally striking is the uniqueness of 4 Vesta, whose basaltic composition is more similar to that of lunar basalts than to any other known asteroid. If we assign the basaltic chondrites to an origin on Vesta, and assign carbonaceous chondrites, irons, stony-irons, and achondrites to plausible parent bodies in the belt, then the only important class of meteorites *not* accounted for is the most numerous of all—the ordinary chondrites on Earth. Merely assigning the ordinary chondrites to Earth-crossing asteroids is no

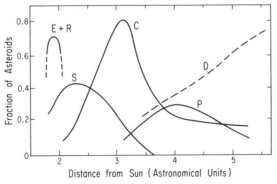

FIGURE 4.58. **Distribution of asteroid spectral types over heliocentric distance. These smoothed and bias-corrected data are taken from Zellner (1979) and from Gradie and Tedesco (1982). Mixing lengths for the S and C asteroids are about 0.5 AU. The P and D classes are apparently "super-carbonaceous," with even higher concentrations of volatiles than the C asteroids or C chondrites.**

solution to the problem, since such orbits have very short lifetimes: We cannot believe that these bodies have been in their present orbits for more than 10^7–10^8 years. Where were they for the rest of the age of the Solar System?

Mechanisms have been suggested for selective removal of asteroids from certain special locations in the belt (Kozai, 1962; Williams, 1969, 1973; Zimmerman and Wetherill, 1973), but it seems most improbable that, for example, ordinary chondritic material should be common near the Jovian orbital resonances but invisible to terrestrial astronomers who peruse the belt. A very general but less violent acceleration mechanism—the Yarkovsky effect—has been explored in detail by Peterson (1976). The radiation pressure anisotropy due to thermal emission on the afternoon side of rotating bodies provides a component of force tangential to the orbit, leading to outward evolution for prograde rotators and inward evolution for retrograde rotators. This mechanism is very effective for 1-km or smaller bodies, but may also move more massive bodies short distances into resonances, from which they will be violently (and possibly interestingly) perturbed by the mechanisms of the authors mentioned above. Again, however, no obvious reason for powerful selection in favor of ordinary chondrite material is evident. Perhaps the locus of origin of ordinary chondrites is closer to home than the asteroid belt.

One consequence of the apparent dominance of primitive carbonaceous chondrite material in the asteroid belt is that it requires an explanation of the apparent presence of a high-temperature igneous differentiate on Vesta (but not on Ceres). This is not a trivial problem, since some mechanism seems to have made, in the heart of the asteroid belt, basaltic achondritic material which is quite similar (but by no means identical) to that found on the Moon. Numerous authors have attributed early melting and differentiation of the lunar surface to a hypothetical very rapid accretion of the Moon—in some models within a period of 100 years. We shall briefly examine the success of this *ad hoc* postulate in explaining the essentially simultaneous formation of very similar material on Vesta. During the terminal stages of accretion of Vesta, each infalling element of mass will reach, upon impact with the surface, a kinetic energy per unit mass of

$$\frac{E}{m} = \int_{r_s}^{\infty} \frac{GM}{r^2} \, dr = -\frac{GM}{r_s} = 6.4 \times 10^8 \text{ erg gm}^{-1}.$$

All earlier-accreted material will bring in less energy per unit mass. The specific energy of the last-accreted debris is equivalent to the specific kinetic energy of material falling at the present escape velocity of Vesta:

$$E/m = (\tfrac{1}{2})v_{esc}^2; \qquad v_{esc} = 0.36 \text{ km sec}^{-1}.$$

If 100% of the gravitational potential energy liberated by accretion were stored in Vesta, then the temperature could possibly rise by as much as

$$\Delta T_{max} = \frac{E/m}{C_p/\mu} \simeq \frac{13 \text{ cal gm}^{-1}}{0.3 \text{ cal gm}^{-1}\text{K}^{-1}} \simeq 40 \text{ K}.$$

There is little to be gained by postulating that Vesta is a fragment of a larger body: If the entire mass of the asteroid belt were collected into a single body, a ΔT_{max} only slightly greater than 100 K would result. But then, of course, this superasteroid could not split up because it would have nothing to collide with! The surface of Vesta would, in addition, not have been warmed by even 40 K while the nebular gas was present, because a much smaller temperature rise would drive convection and result in rapid heat loss (see Section 2.2). Also, in the absence of the nebula, the 40-K heating could not occur if it took longer than ~20 yr for the surface 1-km layer to accrete, since radiation loss would otherwise cool the surface significantly in that time. Finally, Vesta is the largest differentiated asteroid, 6.5 times as massive as the next largest—15 Eunomia.

Radioactive heating is little better. The long-lived radionuclides will cause melting (or peak temperatures) one or more half-lives after the origin of the asteroids. The most potent radionuclide heat source, ^{40}K, would lead to differentiation 1–2 billion years after formation. Bogard et al. (1967) have found an igneous meteorite which melted about a billion after formation, nicely in keeping with the idea that melting was caused by long-lived radionuclide decay on a sizeable parent body. However, virtually all igneous meteorites seem to have differentiated very close to the time of origin of the Solar System. An appropriate heat source to effect rapid, strong heating must be found.

Lee and Papanastassiou (1974) and Lee et al. (1976) have shown the existence of a large ^{26}Mg anomaly in aluminum-rich mineral grains from high-temperature white inclusions in the Allende C3V metorite. This isotope is the decay product of the short-lived β^+ emitter ^{26}Al ($t_{\frac{1}{2}} = 7 \times 10^5$ years). Melting of a body made of these white inclusions would occur within about a million years for parent body sizes larger than a few kilometers in radius. The survival of bodies 10^6 times as massive as this in the belt without signs of melting (Ceres, Pallas) leads to a difficult dilemma. Either (a) ^{26}Al was never present in their material or (b) ^{26}Al was present in their material, but decayed before the asteroids had grown to kilometer size.

Alternative (a) requires that the isotopic abundance of ^{26}Al must have been very much smaller in C2 material than in the Allende inclusions. Alternative (b) requires a growth time scale $\geq 10^6$ years for subkilometer asteroids. Schramm et al. (1970) have failed to find ^{26}Mg anomalies in a suite of diverse bulk meteorite samples, which at least suggests that large ^{26}Al abundances were not common or widely distributed among meteorite classes. Studies of the accretion physics of small particles, such as the work of Goldreich and Ward (1973), suggest that 5-km-sized bodies should form on a time scale of ~10,000 years due to dust sedimentation into the central plane of the solar nebula. Hence the simplest conclusion that presents itself is to attribute the ^{26}Mg anomaly to high-temperature Allende inclusions only, and to form these grains within a few ^{26}Al half-lives (i.e., ~ 2×10^6 years) of the nucleosynthetic event which made ^{26}Al. In this case, ^{26}Al does not function as a useful heat source, since it may be wholly extinct before the refractory grains find themselves accreted into asteroid-sized bodies. One other possibility is the Sonett et al. (1970) mechanism for ohmic heating by a hypothetical dense solar wind during the T-Tauri phase of the early Sun. This mechanism can deposit large quantities of energy at a depth of ∼100 km in poorly conducting (chondritic) solid bodies. The T-Tauri phase, which is a commonly observed phenomenon in young clusters, is not to be confused with the hypothetical Hayashi (1961) phase of the Sun, in which the luminosity of a star of 1 solar mass may reach 10^6 times the present luminosity of the Sun before settling down as a well-behaved member of the Main Sequence. Such high luminosities are predicted only for the case of rapid collapse of a protostar with zero angular momentum. In the case of T-Tauri stars, enhancement of luminosity by only a factor of 2 or 3 is observed. Thus, there is some evidence for external heating of large asteroids in the belt. This heating mechanism could cause extensive melting and outgassing in water-rich bodies such as carbonaceous chondrite parent bodies. At the same time the surface temperatures of small asteroids need never be more than a factor of $\sim 3^{\frac{1}{4}}$ higher than at present. For a typical (220 K) surface temperature in the heart of the belt, a temperature increase of only 70 K would be produced, sufficient to melt surface ice in the daytime.

Because of the high surface : volume ratio of small asteroids, solar wind heating of their interiors will not accumulate, but rather diffuse out to the surface and be radiated away. Nonetheless, this may provide a mechanism for driving upward transport of volatiles such as water and ammonia. The surface of an asteroid, even a large one, is not conducive to the retention of an atmosphere at best, and both thermal escape and solar wind sweeping were more effective at this era in the history of the Solar System than they have ever been since. Interestingly, Black (1972a,b) has presented very good evidence that the regoliths of some asteroids were directly exposed to irradiation by an early intense solar wind.

Briggs (1976) and Brecher et al. (1975) have shown that the efficiency of T-Tauri phase heating is a very sensitive function of electrical conductivity of the body, as well as depending in more obvious ways on the size of the body and its distance from the Sun. Thus, they find that C3 chondrites would be heated at $\sim 10^6$ times the rate of CI bodies of the same size and same distance from the Sun. Thus, a very large C1 or C2 asteroid may remain unmelted when exposed to conditions capable of melting a small C3 body. This is clearly a most interesting conclusion, but the application of these calculations to the early Solar System presupposes a knowledge of the physics of the T-Tauri phase which we unfortunately cannot rely on. A trivial example suffices to illustrate this point: It is even uncertain whether the observed T-Tauri phase "solar wind" is heading outward from the star!

It is difficult to find in the available compositional data on asteroids more than a very few apparent trends. This is in part due to the fact that compositional gradients due to temperature differences during condensation are relatively subtle compared to the gross melting differentiation and high-grade metamorphism imposed by early intense heating. Chapman (1974c, 1976) has argued that the observed igneous asteroids are cores left behind by collisional erosion and fragmentation of originally larger (and more numerous) bodies. Because these achondritic, metallic, and stony-iron objects are, almost without exception, mechanically stronger than chondrites and far stronger than carbonaceous material, and because large collision rates are required to expose the cores of so many differentiated bodies in 4.6 billion years Chapman postulates a primordial population of asteroids about 300 times the present number, almost all of which were carbonaceous. Vesta, with its surface layer of basaltic, calcium-rich (crustal) achondritic material, is regarded as the sole asteroid of the original population of ~100 differentiated bodies to have survived more or less unscathed to the present. Chapman (1975) claims that the excess of stony-iron asteroids of radius 50–100 km is due to these bodies. In this model, many igneous asteroids may show compositional variations as they rotate.

There are good reasons to believe that the early mass density of the asteroid belt was far higher than it is now, based on reconstructions of the mass distribution such as that published by Weidenschilling (1977). A powerful mechanism for collisional disruption of asteroids and depletion of the belt has been found to be acceleration of planetesimals by Jupiter (Weidenschilling, 1975). This mechanism has the additional features that it reduces the mass density near the orbit of Mars and injects volatile-rich carbonaceous material into the inner solar system, especially to Mars. Such a process will to some degree cause radial mixing in the belt; indeed, some observational evidence for smooth compositional and albedo intergrading can be found in the astronomical studies of asteroids (Figure 4.58). This may be interpreted as radial mixing of materials over a scale length of 0.4–0.5 AU, similar to the degree of mixing expected from condensation–accretion modeling of the terrestrial planets (Cox and Lewis, 1980; Barshay and Lewis, 1981). In the model for the compositions of the carbonaceous chondrites favored by Anders and co-workers (see Section 2.5), the various types of C chondrites are composed of two components, one of which is the low-density, fine-grained black matrix material, rich in volatiles, and very similar to the C1 chondrites. The other component is made up of larger pieces of high-temperature minerals, including chondrites, fragments of chondrites and igneous polymineralic rocks, and white refractory inclusions, and is nearly devoid of volatiles. This model accounts in a simple way for the abundances of a number of moderately to highly volatile elements in the C1, C2, and C3 meteorites, which are found to vary coherently and systematically in abundance, *without* regard to differences in volatility, quite as one would expect from simple dilution of C1 matrix material by volatile-free high-temperature minerals. This two-component model cannot, as we have seen, be simply extended to ordinary chondrites without introducing several additional components of somewhat variable composition.

We may imagine a simple model for the origin of carbonaceous chondrites by considering an inner region of the solar nebula, populated by ordinary chondrites and igneous meteorite types, and an outer region of lower temperature in which the stable condensates are volatile-rich and dominated by hydroxyl silicates, low-temperature sulfides, magnetite, and water-soluble salts, similar to C1 chondrites. Radial mixing of the nebula by turbulence, and of solid material by dynamical processes such as the Yarkovsky effect and orbital resonances with Jupiter, will generate a wide transition region in which these materials are mixed. The mixing may be continuous, with all proportions of high- and low-temperature materials represented somewhere in the belt; however, our sampling of the belt is very incomplete and certainly not random: The discrete types C1(I), C2(M), C3V, and C3O probably represent merely four individual and discrete parent bodies, selected by an unknown algorithm from countless thousands of asteroids whose compositions span the entire C-chondrite range continuously and smoothly.

There is no reason to truncate this process arbitrarily at the inner edge of the belt: C chondrites are known because they presently fall on Earth! Collisionally produced debris from the destruction of a primordial asteroid population of mass $\sim 10^{27}$ gm will include vast quantities of bodies of subkilometer size, which will evolve out of the belt in both directions. Weidenschilling (1975) has demonstrated the importance of the flux of volatile-rich asteroidal matter upon the accretion of Mars, and it is not hard to imagine that a CI-type volatile influx of great magnitude has affected the planet's volatile-element inventory. Indeed, exactly this effect is found in the accretion sampling calculations of Cox and Lewis (1980).

As for the importance of the large volatile-element contents of the asteroids in generating possible temporary atmospheres or surficial ice deposits, the small size and low escape velocities of the asteroids offer little reason for optimism.

The escape velocities of Ceres and Vesta—0.58 and 0.35 km sec^{-1}, respectively (Table 4.19)—are of the same order as the thermal speeds of common molecules at asteroidal temperatures, as summarized in Figure 3.6. Thus, an atmosphere of even heavy gases such as CO_2 could escape almost instantly.

A trace steady-state atmosphere could be maintained either by outgassing from the interior or by evaporation of a deposit of some frozen substance on the surface. Watson *et al.* (1961b) have calculated the evaporation rates of several substances over a wide range of temperatures, and Lebofsky (1975) has extended these calculations to a number of additional materials for more detailed thermal models of satellite and asteroid surfaces. One example germane to the present discussion is that H_2O ice at 200 K would evaporate at the rate of 2.6×10^{-5} gm cm^{-2} sec^{-1}. A 1-km layer of ice at 200 K would thus evaporate in 120 years, providing an escape flux of 1.6×10^{10} molecules cm^{-2} sec^{-1}, if heat could be supplied rapidly enough through the escaping gas to maintain this evaporation rate. Ice should therefore not be present on the surfaces of belt asteroids.

We have already seen in a general way what the consequences of heating C1 material would be. In Section 2 we saw that C1 chondrites contain about 10% H_2O, 3.54% C, 0.25% N, and 6.0% S by weight. High-temperature equilibration will produce a mixture of H_2O, CO, H_2, and N_2, whereas at lower temperatures H_2O, H_2, CO_2, and N_2 or NH_3 will be found. A silicate magma rich in H_2O and CO_2 may be produced if confining pressures are adequate (>1 kbar), but the *central* pressure of Ceres is only ~ 1.3 kbar (0.7 kbar for Vesta). A more plausible possibility for such small bodies is an H_2O–CO_2 fluid bearing soluble materials such as NH_4^+, Na^+, K^+, and Ca^{2+} salts of CO_3^{2-}, SO_4^{2-}, and halide anions. It is interesting to note that C1 chondrites have veins of white, water-soluble salts, which appear to have been precipitated out of aqueous solution. The lowest temperature at which such a solution could remain liquid is not known. However, the binary H_2O–NH_3

eutectic temperature is $-100°C$ (173 K), and the binary $H_2O–NH_4Cl$ eutectic is at $-15.4°C$ (258 K). Ammonium chloride was reported as a constituent of the C1 chondrite Orgueil immediately after its fall (Pisani, 1864).

The dehydroxylation of minerals, the melting of ice, and the evaporation of water are very substantial heat sinks, especially for C1 parent material. Furthermore, the result of heating is to generate great quantities of electrically conducting aqueous salt solutions at quite modest temperatures. The course of thermal evolution of such material will be quite complex and very unlike that for even C2 or C3 meteorites. Our discussion of the history of Io in Section 4.6 will be found to have many fascinating reflections and distortions in the histories of the largest asteroids.

The reader interested in further information on asteroids is referred to the book *Comets, Asteroids and Meteorites*, edited by A. H. Delsemme (1977), and especially to the book *Asteroids*, edited by T. Gehrels (1979).

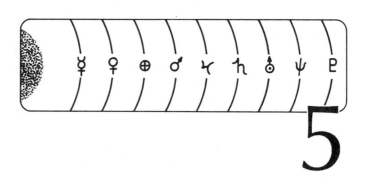

5

Conclusions

Now that we have considered in detail the evidence regarding the composition, abundance, origin, evolution, and loss of planetary volatiles, it is appropriate to step back to a broader perspective, assess the general state of our knowledge of these subjects, point out the most crucial needs for new data, and outline the prospects for securing these data.

5.1. ASSESSMENT OF COMPETING MODELS

It is now almost universally accepted that the terrestrial planets formed by slow (100 million years) accretion of small solid bodies, and that the major volatiles in their atmospheres were brought in by those captured solid bodies which originated at the largest heliocentric distance. No obvious evidence for a major component of primordial or captured solar-composition gas can be found on these planets. Even the heavy rare gases on Venus, which apparently display solar relative abundances, seem to have been strongly fractionated from the light rare gases during their incorporation into solid carrier materials such as ice or rock. Not one of the volatile-rich meteorite classes has a composition which satisfactorily fits the known volatile element inventories of any of the terrestrial planets. It is, of course, not certain whether most of the volatiles were accreted along with the bulk of the planetary material or instead accreted later as a "veneer"; however, studies of the celestial mechanics of accretion suggest only modest concentration of volatiles toward the end of the process.

The Jovian planets seem to have passed through a rather luminous collapse phase, during which they were surrounded by planetary subnebulae of gas and dust in much the same way that the early Sun was surrounded by the solar nebula. Formation of these planets by gravitational capture of gas and dust by solid planetesimals seems the most direct way of explaining what is known of their bulk compositions. Two-component (condensed solids plus uncondensable gases) models of these planets appear promising and need to be explored.

Direct derivation of any of the terrestrial planets from any single class of meteorites seems most unlikely, and of these models, derivation from C1 chondrites is the least satisfactory because of the numerous *ad hoc* assumptions which must be made. The absence of any quantitative predictions arising from this model is also an important factor in this judgment.

The concept of derivation of the terrestrial planets from mixtures of a few known classes of meteorites or of a few well-defined components made up geochemically coherent groups of elements has obvious heuristic value. The danger in this approach is that the known meteorite classes are highly quantized, clearly different samples from a handful of small bodies. The known classes of chondrites are all obviously affected by severe fractionation processes which are probably characteristic of kilometer-sized bodies accreting under conditions where magnetic, electrostatic, and other nongravitational forces were important. It is by no means obvious that such fractionation processes can provide compositional differences for gravitationally accreted bodies with radii of hundreds or thousands of kilometers. The concept of mixing of groups of geochemically coherent elements is at first sight more attractive, but the severe dynamical environment in which planets accrete almost certainly involved large-scale radial mixing of solids over several tenths of an astronomical unit, with collisional comminution and reaccretion acting as important processes.

Application of multicomponent mixing models to the planets is unfortunately entirely *post hoc*: No predictions can be made regarding the abundances of any elements on a planet which has not yet been visited. Once analyses are available for a few elements on a planet, then certain assumptions must be made about outgassing efficiency, the time history of gas release, the degree of differentiation, the partitioning of elements between core, mantle, and crust, and so on. Then predictions of the abundances of other elements may be made. Unfortunately, these assumptions are highly moot and likely to vary widely from planet to planet.

The equilibrium condensation concept, in which the local composition of solids in the solar nebula was determined by equilibration between dust grains and the solar-composition gas, is very obviously a gross oversimplification of the chemistry. Accretion of the solids to form planet-sized bodies must occur with substantial radial mixing of solids which formed at different places in the nebula, and terminal accretion apparently involves the capture of bodies with masses of 1% to over 10% of the final planetary mass. By the very nature of this accretion process, statistical fluctuations in the abundances of volatile elements must be large, while the relative abundances of the rock-forming elements should be quite close to the solar proportions.

The equilibrium condensation process is, of course, in principle completely quantifiable. Application of this model in the real Solar System, however, is hindered by the incompleteness and inaccuracy of the relevant thermodynamic data, by our possible complete ignorance concerning the very existence of relevant compounds, by the possibility that some elements, such as the primordial rare gases, may be retained by adsorption and implantation processes which transcend the limits of the model, and, of course, by kinetically caused departures from equilibrium. In addition, comparison of this (or any other) model with present-day reality requires that proper allowance be made for the acquisition and loss of volatiles since the end of the primary accretion era. In this area also there are numerous fascinating questions regarding the effects of impacts of volatile-rich bodies on the planets.

Finally, there is a very general shortage of crucial planetary data by which the existing theories may be tested. The planetary exploration program of the past 20 years has dramatically narrowed the range of tenable models, but, as in all exploratory ventures, new and wholly unexpected observations have also accrued. Further, because of the vast improvement in the quantity and quality of the data base, present theories are necessarily more complex than those of 10 or 20 years ago.

5.2. DECISIVE MEASUREMENTS

Let us briefly review some of the more crucial kinds of data which we still lack, planet by planet.

First, we have essentially no compositional data on Mercury. The total metal content is definitely very high, but we do not know whether the metal is nickel-rich, silicon-rich, or nearly pure iron; we do not know whether the core is fully solid or whether it contains sulfur. We know neither the abundances of the alkali metals nor the total magnitude of the planetary internal heat source.

Our analyses of the rare gases on Venus are mutually contradictory; the photochemistry of the atmosphere obscures the details of the crust–atmosphere interaction; the oxidation state and mineralogy of the surface are known only by inference; the tectonic history is equally poorly known. The evidence for a 100-fold enrichment of deuterium on Venus relative to Earth suggests that the abundance of water was once ≥ 100 times the present amount, or that Venus accreted a disproportionate share of D-rich water, possibly from polymeric organic matter. Among the most crucial necessities are the chemical analysis of the lower troposphere (the lowest 22 km, containing over 80% of the atmospheric mass, has not been analyzed), the elemental and mineralogical characterization of the crust, high-resolution geological mapping of the surface (necessarily by radar), resolution of the rare gas controversy, and careful accounting of the planetary hydrogen budget.

It is moderately tempting to embark here upon a lengthy itemization of the things we would like to know about the terrestrial piece in this puzzle. However, our knowledge of Earth is so incomparably superior to that of the other planets that we must attach higher priority to understanding the basics about the other planets: The main limitation in our ability to fit the terrestrial, Martian, and Venusian pieces together is clearly our ignorance of the general shape of the extraterrestrial pieces! Further, we are likely to learn at least as much about the general behavior of planets, the evolution of their atmospheres, and the process of planetary formation by broad investigation of the less well-known planets than from an equal-sized increment of effort devoted to Earth. We shall later return to specific illustrations of this contention.

Mars has received an inordinate amount of attention from biologists and only the most cursory geochemical and geophysical scrutiny. Regional compositional mapping is clearly feasible and of basic importance. The degree of seismicity and volcanic activity could be rather easily established. With a much higher level of effort, and at considerable additional expense, surface samples could be returned from Mars to be subjected to the vast range of experiments which require instrumentation too heavy, complex, and delicate to be sent to Mars. The global water inventory could be greatly refined, and the possible early existence of conditions suitable for the origin of life might be established. As with Venus, the role of major impact events in the evolution of atmospheric composition needs investigation.

The Jovian planets hide from outside observers almost all of the chemical evidence of their bulk compositions. Chemical analysis by deep-entry probes on all four of the Jovian planets would provide a broad sample of the spectrum of possibilities of the planetary formation process. The abundance of organic matter produced by disequilibrating processes and the nature of the coloring matter in the clouds could also be determined.

The satellites of the Jovian planets are in a number of ways more revealing of the origin of the terrestrial planets than are the Jovian planets themselves. The compact, regular satellite systems of Jupiter, Saturn, and Uranus have evolved into families of massive bodies by the same processes which produced the terrestrial planets, acting under different boundary conditions. Quite aside from the intrinsic interest of complex and fascinating bodies such as Io, Ganymede, Titan, and Iapetus, there is the broad comparative aspect of their study, whereby each body contributes to an overall appreciation of the general planet-forming process. The geochemistry and volcanic energetics of Io, the vast hydrocarbon deposits of Titan, the water-rich mantle of Callisto, and the tectonic history of Rhea all contribute to this understanding. Thorough mapping of the Galilean satellites by a Jupiter orbiter and probing of the atmosphere and surface of Titan by an entry vehicle are both clearly feasible.

The role of the small bodies in the Solar System should not be forgotten. There is an excellent chance that unaltered samples of the solids from a wide range of locations in the solar nebula may still be available on these bodies. The composition of the dusty and volatile components of comets may be essential to understanding the details of the volatile-element inventories on the terrestrial planets. The variation of the properties of asteroids with size and heliocentric distance will certainly provide a powerful insight into the nature of the short-lived heat sources present in the early Solar System, to say nothing of the practical advantages which may accrue to mankind from achieving the means to reach asteroids rich in scarce and valuable minerals. For this purpose, multiple-asteroid flyby missions carrying a diverse assortment of analytical instrumentation could be developed at once and flown within a few years.

Almost every area of fundamental importance in the above list can be addressed exclusively by spacecraft investigations. Carefully thought out strategies for the systematic exploration and eventual exploitation of the Solar System have been developed by the National Academy of Sciences, with annual expenditures well within the range of the average expenditures on planetary exploration over the past decade. It is interesting to review briefly these recommendations.

First, the main focus of the planetary exploration program is the comparative study of Earth's closest relatives in the family of planets, Venus and Mars. Radar mapping of the surface geology of Venus has been given high priority, as have the geochemical and geophysical characterization of the crust of Mars. Determination of the composition of the surface of Venus and detailed study of meteorology of Mars were also deemed important components of the program.

The goals of the recommended program for study of small bodies are strongly conditioned on the likelihood of finding extremely ancient material in comets and asteroids, thus extending our knowledge of the history of solar system materials back in time. Most of these bodies are expected to be unaltered and undifferentiated, so that sampling their surface composition will give us an excellent idea of their bulk properties. To this end, flyby missions, while interesting, do not suffice. It is clear that comet rendezvous and multiple-asteroid rendezvous missions will be necessary to document and understand the systematic variation of the composition of these primitive bodies with heliocentric distance and perhaps other unsuspected variables.

In the outer solar system, the immediate objectives are the detailed study of the Galilean satellites from orbit about Jupiter, and the probing of the atmosphere by a deep-entry probe. This approach can then be applied with little alteration to the study of Saturn and its satellites.

In the face of vigorous and growing scientific interest in planetary exploration, and in spite of the phenomenal degree of success and public interest associated with the planetary program to date, there are serious doubts about the willingness of the United States Government to continue with these programs. Some brief historical background and perspective on this erosion of will and its impact on future mission plans are in order.

5.3. PRESENT TRENDS AND FUTURE PROSPECTS

If an informed American were called upon to give a brief description of the history of space exploration, he or she very likely would respond somewhat as follows:

The first satellites were launched by the Soviet Union in 1957. The Soviets employed a booster related to their early ICBMs, and therefore could put up very heavy satellites from the start. The American program struggled frantically to catch up. The entire society was galvanized by the challenge, and American educational practices in the sciences and engineering were extensively reviewed and modernized. The first American satellites were tiny scientific packages launched by high-strung and temperamental small rockets. But a tremendous effort was expended on space and, galvanized by the first Russian probes of the Moon and Venus, the United States entered into the space arena with determination. After a few years of explosive growth in the 1960s, the American space program had leaped far ahead of the Russians'. While American space probes explored the inner solar system and American astronauts developed the techniques to fly to the Moon, the Soviet space program suffered from a prolonged series of disappointing failures of its planetary probes, and launched a long series of humdrum manned missions with no apparent purpose. This trend culminated with the Apollo 8 lunar flyby of December 1968 and the Apollo 11 lunar landing of July 1969. Since that time, the Soviet Union has played tortoise to the American hare, slowly building up its space program. The overall level of effort expended by the people of Earth on the use of space has been about constant for several years.

Although this popular understanding is correct in almost every detail, it gives a perspective on the present state and future prospects of the American space program that is woefully inadequate and seriously misleading (Figure 5.1). In fact, in 1978 the Soviet Union launched 119 satellites, whereas 36 were placed in orbit by American boosters. Of those 36, 5 were orbited on contract for other nations, 17 were military satellites, 4 were orbited for commercial interests or for other government agencies, and 10 were orbited by and for NASA. These 10 included the last two planetary probes, the Pioneer Venus orbiter and probe package.

In 1979, the Soviets launched 101 satellites, whereas American boosters placed 18 satellites in orbit. One of these was a British satellite, 10 were military, 2 were commercial communications satellites, and 1 was launched for NOAA. There were four NASA spacecraft launched (HEAO3, Magsat, Sage, Solwind), none of which were planetary missions. In 1980, while the USSR was launching 110 satellites, the American total fell to 13. Of these, 8 were military, 2 were commercial, 1 was launched for NOAA, and the civilian space research program ended up with 2 Earth-orbital missions (GOES and SMM), again with no planetary launches. This is the lowest level of American space activity since 1959, when of the 11 satellites orbited by the United States, 5 were NASA research missions.

It is the grim truth that the United States paid for and attained world leadership in space exploration, but then abdicated this position. In order to fund politically mandated programs such as the Apollo program and the Space Shuttle, virtually the

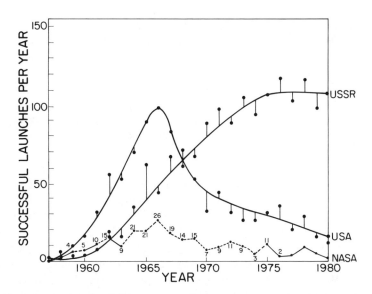

FIGURE 5.1. Spacecraft successfully placed in orbit by the United States and the USSR (1958–1980). The NASA flight program, which occupies a roughly constant proportion of the American effort, is shown at the bottom. The last NASA planetary launches were in 1978, and only one future planetary mission (Galileo) has been funded. Decomposition of the Soviet launch program into military, applied, and "civilian research" endeavors is rendered uncertain by the intentionally obfuscatory procedure of numbering almost all spacecraft in the Kosmos series. Our best estimate is that about 10 "research" missions were flown by the USSR in 1980, but this estimate counts all 6 manned missions as nonmilitary, and may err in the direction of generosity.

entire scientific exploration program of NASA has been sacrificed. Thus, a rich diversity of small, research-oriented missions have been sacrificed to fund a handful of manned spectaculars. The greatest loser among these programs is planetary exploration. It has now been 6 years since there was a budgetary new start in the planetary program. Only one future mission has been funded, the frequently de-layed and often-cancelled Galileo Jupiter orbiter and probe.

A vigorous and fruitful planetary exploration program with at least one new start per year could be purchased for a mere $300 million per year. The present starva-tion-level budget of $150 million is entirely used to maintain the ability of the planetary program to do missions; that is, supporting the most basic institutional needs. At this level, NASA cannot afford to fly any missions. At twice this level (but still lower than the peak funding era of the mid-1970s), the exploration of the Solar System could continue without interruption. The difference is about one-sixth the annual expenditures for bubble gum, several times less than the gate receipts of Star Wars, and roughly equal to 0.075% of the Department of Defense budget. The United States could again have a vigorous planetary exploration for an annual cost equal to the Defense Department's expenditures in the next six hours.

Now that we have established that the cost of such a program is an utterly negligible part of the national budget, let us for a moment ask what benefits are purchased by this investment in space research.

First, it is important to grasp that the entire modern scientific view of the Earth, whether it be geochemical, tectonic, economic, meteorological, or oceanographic, is global in perspective. As recently as the early 1960s, discussions of these processes were only rarely couched in global terms. The advent of the ability to view Earth holistically from space has entirely transformed these fields. Rather than randomly prospecting for minerals, or using purely regional criteria to identify potential oil and gas deposits, we now use the persepctive of global tectonics to predict where precious resources may be found. Gravity mapping of the Earth by tracking space-craft in low Earth orbits has led to the wholly unexpected discovery of apparent continental sutures deep within the crust in such unlikely areas as Wisconsin. With modern techniques for deep drilling, these potentially rich deposits are accessible for exploitation. It has become clear that many strategic minerals, such as copper and manganese, are governed in their terrestrial occurrence by global tectonic processes, and the search for new deposits is being managed accordingly.

The Earth, as we have seen, depends for its congeniality to life upon the tenuous ozone layer, which shields the surface from killing radiation. Among the life forms that are highly vulnerable to UV light, seedling wheat and rice are especially noteworthy. Our present understanding of the stability of the ozone layer, and certainly our appreciation of the most severe threats to its existence, derives largely from the NASA planetary program. The catalytic role of water vapor in the oxidation–reduction cycles which regulate ozone was first established by the study of the photochemistry of Mars, and the threat of destruction of the ozone layer by nitrogen oxides and water vapor from the exhaust of high-flying supersonic aircraft was averted by the timely acquisition of this knowledge.

Similarly, the catalytic destruction of ozone by chlorine compounds was identi-fied as a terrestrial problem on the basis of theoretical studies of the photochemistry of HCl in the atmosphere of Venus.

These are far from unique examples of the effects of human activities. Man's wastes are now a factor in many global balances, including the atmospheric in-ventories of aerosols and carbon dioxide. These materials alter both the visible and IR opacity of the atmosphere, with possible severe climatological consequences. A continuation of the present global warming trend could lead to the inundation of most of the major cities of the world as sea levels, augmented by the melting of the ice caps, rise substantially. If a large part of the temperature rise is due to the injection of aerosols and carbon dioxide from the burning of fossil fuels, then clearly a quantitative understanding of these effects is an essential prerequisite to intelligent energy policy planning.

One of the most distinctive features of anthropogenic perturbations on the bio-sphere is that they occur with a very short time constant, from a few years for halomethanes to a few decades for carbon dioxide. It is clearly unacceptable to wait passively until a global ecological disaster begins before taking interest in the prob-lem. By then, the backlog of the offending substances in the biosphere may be so large that, even if injection were stopped at once, the undesirable consequences would continue to cumulate for many years. It is remarkable that we have identified several such threats in recent years, and have successfully altered the course of our commercial and technical endeavors in time to avert the most serious consequences. If we were not engaged in a synoptic exploration of how planets work, we might already have made decisions which would have unacceptable future consequences. As the scale of human activities continues to grow, these problems become more acute, and an ever-greater premium is placed upon understanding. If we cease to pursue the fundamental laws of nature, the consequences will be predictable.

Of all the aspects of Earth essential to human survival, the atmosphere may well be the most delicate. The geological record testifies to us strongly of mass extinctions brought about by natural causes, which in several cases seem to have been very large explosions that have seriously perturbed the composition and turbidity of the atmosphere. It is within present human capability to bring about this form of catastrophe as well. Nuclear war, industrial pollution, and a terrestrial runaway greenhouse effect are avoidable, but only by global measures. It does not suffice for one nation to take unilateral measures to restrict its output of toxins, mutagens, and pollutants: Any sufficiently foolish nation could doom us all. Global problems demand global solutions.

It also does not suffice to express the wistful hope that our children or grandchildren will somehow clean up the messes we make. The time scale for the perturbations is too short for that. The issue lies squarely in the hands of those old enough and technically mature enough to read this book. We, the authors, see ample room for all varieties of involvement in this problem. But, among the many kinds of effort which will be required, we expect that the comparative study of planetary atmospheres will rank high. By this broad study, more than by any degree of detailed study of Earth, and within the very few years available, we may make major advances in understanding the fundamental laws governing the behavior of atmospheres. We hope that those who make and influence budgetary decisions on these matters will have the nerve and foresight to invest some of our money in space.

References

H. Abadi and N. C. Wickramasinghe, Pre-biotic molecules in Martian dust clouds. *Nature (London)* **267**, 687–688 (1977).

F. B. Abeles, Fate of air pollutants: Removal of ethylene, sulfur dioxide and nitrogen dioxide by soil. *Science* **173**, 914–916 (1971).

P. H. Abelson, Abiotic synthesis in the Martian environment. *Proc. Natl. Acad. Sci. U.S.A.* **54**, 1490–1497 (1965).

P. H. Abelson, Chemical events on the primitive Earth. *Proc. Natl. Acad. Sci. U.S.A.* **55**, 1365–1380 (1966).

J. A. Adamcik, On the formation of nitrogen oxides in planetary atmospheres. *Publ. Astron. Soc. Pac.* **74**, 328–329 (1962).

J. A. Adamcik, The water vapor content of the Martian atmosphere as a problem of chemical equilibrium. *Planet. Space Sci.* **11**, 355–358 (1963).

J. A. Adamcik and A. L. Draper, The temperature dependence of the Urey equilibrium and the problem of the carbon dioxide content of the atmosphere of Venus. *Planet. Space Sci.* **11**, 1303–1307 (1963).

J. B. Adams and T. B. McCord, Mars: Interpretation of spectral reflectivity of light and dark regions. *JGR, J. Geophys. Res.* **74**, 4851–4856 (1969).

J. B. Adams and T. B. McCord, Optical properties of mineral separates glass and anorthositic fragments from Apollo mare samples. *Geochim. Cosmochim. Acta, Suppl.* **2**, 2183–2195 (1971).

W. S. Adams and T. Dunham, Jr., Absorption bands in the infared spectrum of Venus. *Publ. Astron. Soc. Pac.* **44**, 243–247 (1932).

W. S. Adams and T. Dunham, Jr., The B band of oxygen in the spectrum of Mars. *Astrophys. J.* **79**, 308–316 (1934).

A. Adel, A determination of the amount of carbon dioxide above the reflecting layer in the atmosphere of the planet Venus. *Astrophys. J.* **85,** 345–361 (1937a).

A. Adel, Note on the temperature of Venus. *Astrophys. J.* **86,** 337–339 (1937b).

A. Adel, The importance of certain carbon dioxide bands in the temperature radiation from Venus. *Astrophys. J.* **93,** 397–400 (1941).

J. E. Ainsworth and J. R. Herman, Venus wind and temperature structure: The Venera 8 data. *JGR, J. Geophys. Res.* **80,** 173–179 (1975).

J. E. Ainsworth and J. R. Herman, On the reality of the Venus winds. *Icarus* **30,** 314–319 (1977).

H. L. Alden, Observations of the satellite of Neptune. *Astron. J.* **50,** 110 (1943).

Yu V. Aleksandrov and D. F. Lupishko, The distribution function of the normal albedo of the surface of Mars. *Sol. Syst. Res. (Engl. Transl.)* **6,** 7–9 (1972).

Yu. N. Aleksandrov, M. B. Vasi'ev, A. S. Vyshlov, G. G. Dolbezhev, V. M. Dubrovin, A. L. Zaitsev, M. A. Kolosov, G. M. Petrov, N. A. Savich, V. A. Samouol, L. N. Samoznaev, A. I. Siborenko, A. F. Khasyanov, and D. Ya. Shtern, Nighttime ionosphere of Venus from the results of two-frequency radioscopy from Venera 9 and Venera 10. *Cosmic Res. (Engl. Transl.)* **14,** 706–708 (1976).

V. I. Aleshin, Annual heat balance of the north polar cap of Mars. *Sov. Astron. (Engl. Transl.)* **15,** 462–464 (1971).

V. I. Aleshin, Variations of the CO_2 mass in the atmosphere and polar caps of Mars. *Sov. Astron. (Engl. Transl.)* **17,** 666–667 (1973).

V. I. Aleshin and O. B. Shchuko, Nature of the polar caps on Mars. *Sov. Astron. (Engl. Transl.)* **16,** 331–335 (1972).

C. C. Allen, Rayed craters on the moon and Mercury. *Phys. Earth Planet. Inter.* **15,** 179–188 (1977).

C. C. Allen, Volcano-ice interactions on Mars. *JGR, J. Geophys. Res.* **84,** 8048–8060 (1979).

D. A. Allen, The method of determining infrared diameters. *NASA [Spec. Publ.] SP* **NASA SP-267,** 41–44 (1971).

L. W. Alvarez, W. Alvarez, F. Asaro, and H. V. Michel, Extraterrestrial cause for the Cretaceous-Tertiary extinction. *Science* **208,** 1095–1108 (1980).

F. Alyea, D. Cunnold, and R. G. Prinn, Stratospheric ozone destruction by aircraft-induced nitrogen oxides. *Science* **188,** 117–121 (1975).

E. Anders, Chemical processes in the early solar system, as inferred from meteorites. *Acc. Chem. Res.* **1,** 289–298 (1968).

E. Anders, Elements 112 to 119: Were they present in meteorites? *Science* **164,** 821–823 (1969).

E. Anders, Interrelations among meteorites, asteroids, and comets. *NASA SP* **267,** 429–446 (1971).

E. Anders, On the depletion of moderately volatile elements in ordinary chondrites. *Meteoritics* **10,** 283–286 (1976).

E. Anders and J. W. Larimer, Extinct superheavy elements in meteorites: Attempted characterization. *Science* **175,** 981–983 (1972).

E. Anders and J. W. Larimer, Validity of trace-element cosmothermometer-reply. *Geochim. Cosmochim. Acta* **39,** 1320–1324 (1975).

E. Anders and T. Owen, Mars and Earth: Origin and abundance of volatiles. *Science* **198,** 453–465 (1977).

E. Anders, H. Higuchi, J. Gros, H. Takahashi, and J. W. Morgan, Extinct superheavy elements in the Allende meteorite. *Science* **190,** 1262–1271 (1975).

E. Anders, H. Higuchi, R. Ganapathy, and J. W. Morgan, Chemical fractionation in meteorites. IX. C3 chondrites. *Geochim. Cosmochim. Acta* **40,** 1131–1140 (1976).

A. T. Anderson, Some basaltic and andesitic gases. *Rev. Geophys. Space Phys.* **13,** 37–55 (1975).

C. A. Anderson, K. Keil, and B. Mason, Silicon oxynitride: A meteoritic mineral. *Science* **146,** 256–257 (1964).

D. E. Anderson, Jr., The Mariner 5 ultraviolet photometer experiment: Analysis of hydrogen Lyman alpha data. *JGR, J. Geophys. Res.* **81,** 1213–1216 (1976).

D. E. Anderson, Jr. and C. W. Hord, Mariner 6 and 7 ultraviolet spectrometer experiment: Analysis of hydrogen Lyman alpha data. *JGR, J. Geophys. Res.* **76**, 6666–6673 (1971).

D. L. Anderson, The origin of the Moon. *Nature (London)* **239**, 263 (1972a).

D. L. Anderson, Internal constitution of Mars. *JGR, J. Geophys. Res.* **77**, 789–795 (1972b).

D. L. Anderson and R. L. Kovach, The composition of the terrestrial planets. *Earth Planet. Sci. Lett.* **3**, 19–24 (1967).

D. L. Anderson, R. L. Kovach, G. Latham, F. Press, M. N. Toksöz, and G. Sutton, Seismic investigations: The Viking Mars lander. *Icarus* **16**, 205–216 (1972).

D. L. Anderson, F. K. Duennebier, G. V. Latham, M. N. Toksöz, R. L. Kovach, T. C. D. Knight, A. R. Lazarewicz, W. F. Miller, Y. Nakamura, and G. Sutton, The Viking seismic experiment. *Science* **194**, 1318–1321 (1976).

D. L. Anderson, W. F. Miller, G. V. Latham, Y. Nakamura, M. N. Toksöz, A. M. Dainty, F. K. Duennebier, A. R. Lazarewicz, R. L. Kovach, and T. C. D. Knight, Seismology on Mars. *JGR, J. Geophys. Res.* **82**, 4524–4546 (1977).

D. M. Anderson, K. Biemann, L. E. Orgel, J. Oro, T. Owen, G. P. Shulman, P. Toulmin, III, and H. C. Urey, Mass spectrometric analysis of organic compounds, water, and volatile constituents in the atmosphere and surface of Mars: The Viking Mars lander. *Icarus* **16**, 111–138 (1972).

J. D. Anderson, D. L. Cain, L. Efron, R. M. Goldstein, W. G. Melbourne, D. A. O'Handley, G. E. Pease, and R. C. Tausworthe, The radius of Venus as determined by planetary radar and Mariner 5 radio tracking data. *J. Atmos. Sci.* **25**, 1171–1173 (1968).

R. C. Anderson, J. G. Pipes, A. L. Broadfoot, and L. Wallace, Spectra of Venus and Jupiter from 1800 to 3200Å. *J. Atmos. Sci.* **26**, 874–888 (1969).

B. N. Andreev, V. T. Guslyakov, V. V. Kerzhanovich, Yu M. Kruglov, V. P. Lysov, M. Ya. Marov, L. V. Onishchenko, M. K. Rozhdestvenskii, V. P. Sorokin, and Yu.N. Shnygin, Venera 8 wind-velocity measurements in the atmosphere of Venus. *Cosmic Res. (Engl. Transl.)* **12**, 385–393 (1974).

N. M. Antsibor, R. V. Bakit'ko, A. L. Ginzburg, V. T. Guslyakov, V. V. Kerzhanovich, Y. F. Makarov, M. Y. Marov, E. P. Molotov, V. I. Rogal'skii, M. K. Rozhdestvenskii, V. P. Sopokin, and Y. N. Shnygin, Estimates of wind velocity and turbulence from relayed Doppler measurements of the velocity of instruments dropped from Venera 9 and Venera 10. *Cosmic Res. (Engl. Transl.)* **14**, 625–631 (1975).

D. C. Applebaum, P. Hateck, R. R. Reeves, Jr., and B. A. Thompson, Some comments on the Venus temperature. *JGR, J. Geophys. Res.* **71**, 5541–5544 (1966).

J. Arkani-Hamed, Geophysical implications of the Martian gravity field. *Icarus* **26**, 313–320 (1975).

A. Arking and C. R. Nagaraja Rao, Refractive index of aqueous HCl and the clouds of Venus. *Nature (London)* **229**, 116–117 (1971).

S. J. Arnold, A. Boksenberg, and W. L. W. Sargent, Measurement of the diameter of Pluto by speckle interferometry. *Astrophys. J.* **234**, L159-163 (1979).

J. R. Aronson and A. G. Emslie, Composition of the Martian dust and derived by infrared spectroscopy from Mariner 9. *JGR, J. Geophys. Res.* **80**, 4925–4931 (1975).

R. E. Arvidson, Morphologic classification of Martian craters and some implications. *Icarus* **22**, 264–271 (1974).

R. E. Arvidson, K. A. Goettel, and C. M. Hohenberg, A post-Viking view of Martian geologic evolution. *Rev. Geophys. Space Phys.* **18**, 565–603 (1980).

M. E. Ash, R. P. Ingalls, G. H. Pettengill, I. I. Shapiro, W. B. Smith, M. A. Slade, D. B. Campbell, R. B. Dyce, R. Jurgens, and T. W. Thompson, The case for the radar radius of Venus. *J. Atmos. Sci.* **25**, 560–563 (1968).

F. W. Aston, The rarity of the inert gases on the Earth. *Nature (London)* **114**, 786–788 (1924).

S. K. Atreya, T. M. Donahue, and W. R. Kuhn, The distribution of ammonia and its photochemical products on Jupiter. *Icarus* **31**, 348–355 (1977a).

S. K. Atreya, Y. L. Yung, T. M. Donahue, and E. S. Barker, Search for Jovian auroral hot spots. *Astrophys. J.* **218**, L83–L87 (1977b).

S. K. Atreya, T. M. Donahue, and W. R. Kuhn, Evolution of a nitrogen atmosphere on Titan. *Science* **201**, 611–613 (1978).

H. H. Aumann and G. S. Orton, Jupiter's spectrum between 12 and 24 micrometers. *Science* **194**, 107–109 (1976).

H. H. Aumann and G. S. Orton, The 12- to 20-micron spectrum on Venus: Implications for temperature and cloud structure. *Icarus* **38**, 251–266 (1979).

V. S. Avduevskii, M. Ya. Marov, and M. K. Rozhdestvenskii, Model of the atmosphere of the planet Venus based on results of measurements made by the soviet automatic interplanetary station Venera 4. *J. Atmos. Sci.* **25**, 537–546 (1968).

V. S. Avduevskii, M. Ya. Marov, and M. K. Rozhdestvenskii, Results of measurement of parameters of the atmosphere of Venus by the Soviet probe Venus 4. *Cosmic Res. (Engl. Transl.)* **7**, 209–219 (1969).

V. S. Avduevskii, M. Ya. Marov, and M. K. Rozhdestvenskii, Results of measurements made on Venera 5 and Venera 6 space probes and a model of the Venusian atmosphere. *Cosmic Res. (Engl. Transl.)* **8**, 800–808 (1970a).

V. S. Avduevskii, M. Ya. Marov, and M. K. Rozhdestvenskii, A tentative model of the Venus atmosphere based on the measurements of Venera 5 and 6. *J. Atmos. Sci.* **27**, 561–568 (1970b).

V. S. Avduevskii, M. Ya. Marov, and M. K. Rozhdestvenskii, Preliminary results of measurements by space probes Venera 5 and Venera 6 in the atmosphere of Venus. *Radio Sci.* **5**, 333–338 (1970c).

V. S. Avduevskii, M. Ya. Marov, M. K. Rozhdestvenskii, N. F. Borodin, and V. V. Kerzhanovich, Soft landing of Venera 7 on the Venus surface and preliminary results of investigations of the Venus atmosphere. *J. Atmos. Sci.* **28**, 263–269 (1971).

V. S. Avduevskii, M. Ya. Marov, B. E. Moshkin, and A. P. Ekonomov, Venera 8: Measurements of the solar illumination through the atmosphere of Venus. *J. Atmos. Sci.* **30**, 1215–1218 (1973).

V. S. Avduevskii, E. L. Akim, V. I. Aleshkin, N. F. Borodin, V. V. Kerzhanovich, Ya. V. Malkov, M. Ya. Marov, S. F. Morozov, M. K. Rozhdestvenskii, O. L. Ryabov, M. I. Subbotin, V. M. Suslov, Z. P. Cheremukhina, and V. I. Shkirina, Martian atmosphere in the vicinity of the landing site of the descent vehicle Mars 6 (preliminary results). *Cosmic Res. (Engl. Transl.)* **13**, 18–27 (1975).

V. S. Avduevskii, S. L. Vishnevetskii, I. A. Golov, Y. Y. Karpeiskii, A. D. Lavrov, V. Y. Likhushin, M. Ya. Marov, D. A. Mel'nikov, N. I. Pomogin, N. N. Pronina, K. A. Razin, and V. G. Fogin, Measurement of wind velocity on the surface of Venus during the operation of stations Venera 9 and Venera 10. *Cosmic Res. (Engl. Transl.)* **14**, 622–625 (1976a).

V. S. Avduevskii, Y. M. Golovin, F. S. Zavelevich, V. Y. Likhushin, M. Ya. Marov, D. A. Mel'nikov, Y. I. Merson, B. E. Moshkin, K. A. Razin, L. I. Chernoshchekov, and A. P. Ekonomov, Preliminary results of an investigation of the lighting conditions in the atmosphere and on the surface of Venus. *Cosmic Res. (Engl. Transl.)* **14**, 643–649 (1976b).

L. Axel, Inhomogeneous models of the atmosphere of Jupiter. *Astrophys. J.* **173**, 451–468 (1972).

W. I. Axford, The polar wind and the terrestrial helium budget. *JGR, J. Geophys. Res.* **73**, 6855–6859 (1968).

J. L. Bada and S. L. Miller, Ammonium ion concentration in the primitive ocean. *Science* **159**, 423–425 (1968).

A. K. Baird, P. Toulmin, III, B. C. Clark, H. J. Rose, Jr., K. Keil, R. P. Christian, and J. L. Gooding, Mineralogic and petrologic implications of Viking geochemical results from Mars: Interim report. *Science* **194**, 1288–1293 (1976).

A. K. Baird, A. J. Castro, B. C. Clark, III, P. Toulmin, III, H. Rose, Jr., K. Keil, and J. I. Gooding, The Viking X-ray fluorescence experiment: Sampling strategies and laboratory simulations. *JGR, J. Geophys. Res.* **82**, 4595–4624 (1977).

V. R. Baker and R. C. Kochel, Martian channel morphology: Maja and Kasei Valleys. *JGR, J. Geophys. Res.* **84**, 7961–7984 (1979).

V. R. Baker and D. J. Milton, Erosion by catastrophic floods on Mars and Earth. *Icarus* **23**, 27–41 (1974).

A. S. Balcham and G. R. Cowan, Shock compression of two iron-silicon alloys to 2.7 megabars. *JGR, J. Geophys. Res.* **71**, 3377–3388 (1966).

R. R. Baldwin, R. W. Walker, and S. J. Webster, The carbon monoxide-sensitized decomposition of hydrogen peroxide. *Combustion and Flame* **15**, 167–172 (1970).

A. Banin and J. Narrot, Origin of life: Clues from relations between chemical compositions of living organisms and natural environments. *Science* **189**, 550–551 (1975).

H. P. Banks, Major evolutionary events and the geological record of plants: Introduction and summary. *Biol. Rev. Cambridge Philos. Soc.* **45**, 317–318 (1970a).

H. P. Banks, Major evolutionary events and the geological record of plants. *Biol. Rev. Cambridge Philos. Soc.* **47**, 451–454 (1970b).

P. M. Banks and T. E. Holzer, High-latitude plasma transport: The Polar wind. *JGR, J. Geophys. Res.* **74**, 6317–6323 (1969).

P. M. Banks and G. Kockarts, "Aeronomy." Academic Press, New York, 1973.

P. M. Banks, H. E. Johnson, and W. I. Axford, The atmosphere of Mercury. *Comments Astrophys. Space Phys.* **2**, 214–220 (1970).

F. T. Bareth, A. H. Barrett, I. Copeland, D. C. Jones and A. E. Lilley, Mariner II. Preliminary reports on measurements of Venus: Microwave radiometers. *Science* **139**, 908–909 (1963).

E. S. Barghoorn and J. W. Schopf, Microorganisms three billion years old from the Precambrian of South Africa. *Science* **152**, 758–763 (1966).

E. S. Barker, Detection of molecular oxygen in the Martian atmosphere. *Nature (London)* **238**, 447–448 (1972).

E. S. Barker, Observations of Venus water vapor over the disk of Venus: The 1972–74 data using the H_2O lines at 8197Å and 8176Å. *Icarus* **25**, 268–281 (1975).

E. S. Barker, Detection of SO_2 in the UV spectrum of Venus. *Geophys. Res. Lett.* **6**, 117–119 (1979).

E. S. Barker and L. M. Trafton, Ultraviolet reflectivity and geometrical albedo of Titan. *Icarus* **20**, 444–454 (1973).

E. S. Barker, R. A. Schorn, A. Woszczyk, R. G. Tull, and S. J. Little, Mars: Detection of atmospheric water vapor during the southern hemisphere spring and summer season. *Science* **170**, 1308–1310 (1970).

E. S. Barker, W. D. Cochran, and A. L. Cochran, Spectrophotometry of Pluto from 3500 to 7350Å. *Icarus* **44**, 43–52 (1980).

J. L. Barker, Jr. and E. Anders, Accretion rate of cosmic matter from iridium and osmium contents of deep-sea sediments. *Geochim. Cosmochim. Acta* **32**, 627–645 (1968).

A. Bar-Nun, Thunderstorms on Jupiter. *Icarus* **24**, 86–94 (1975).

A. Bar-Nun, Acetylene formation on Jupiter: Photolysis or thunderstorms? *Icarus* **38**, 180–191 (1979).

A. Bar-Nun, Production of nitrogen and carbon species by thunderstorms on Venus. *Icarus* **42**, 338–342 (1980).

A. Bar-Nun and M. Podolak, The photochemistry of hydrocarbons in Titan's atmosphere. *Icarus* **38**, 115–122 (1979).

A. Bar-Nun and A. Shaviv, Dynamics of the chemical evolution of Earth's primitive atmosphere. *Icarus* **24**, 197–210 (1975).

A. Bar-Nun, N. Bar-Nun, S. H. Bauer, and C. Sagan, Shock synthesis of amino acids in simulated primitive environments. *Science* **168**, 470–473 (1970).

A. H. Barrett, Microwave absorption and emission in the atmosphere of venus. *JGR, J. Geophys. Res.* **65**, 1835–1838 (1960).

S. S. Barshay and J. S. Lewis, Chemistry of solar material. In "The Dusty Universe" (G. B. Field and A. G. W. Cameron, eds.), pp. 33–40. Neale Watson Academic Publ. Inc., New York, 1975.

S. S. Barshay and J. S. Lewis, Chemistry of primitive solar material. *Annu. Rev. Astron. Astrophys.* **14**, 81–90 (1976).

S. S. Barshay and J. S. Lewis, Chemical structure of the deep atmosphere of Jupiter. *Icarus* **33**, 593–607 (1978).

S. S. Barshay and J. S. Lewis, Accretion and the equilibrium condensation model. Preprint (1981).

V. L. Barsukov, V. G. Zolotchin, V. M. Kovtupenko, and R. Z. Sagdeyev, The Venera 11 and Venera 12 Soviet interplanetary automatic Stations. *Astron. Zh.* **5**, 2–3 (1979).

V. L. Barsukov, I. L. Khodakovsky, V. P. Volkov, and K. P. Florensky, Composition of the Venus clouds. *Space Res.* **20**, 197 (1980a).

V. L. Barsukov, V. Volkov, and I. L. Khodakovsky, The mineral composition of Venus surface rocks: A preliminary prediction. *Geochim. Cosmochim. Acta, Suppl.* **14**, 765–773 (1980b).

C. A. Barth, Interpretation of the Mariner 5 Lyman-Alpha measurements. *J. Atmos. Sci.* **25**, 564–567 (1968).

C. A. Barth and M. L. Dick, Ozone and the polar hood of Mars. *Icarus* **22**, 205–211 (1974).

C. A. Barth and C. W. Hord, Mariner ultraviolet spectrometer: Topography and polar cap. *Science* **173**, 197–201 (1971).

C. A. Barth, L. Wallace, and J. B. Pearce, Mariner 5 measurements of Lyman-Alpha radiation near Venus. *JGR, J. Geophys. Res.* **73**, 2541–2545 (1968).

C. A. Barth, W. G. Fastie, C. W. Hord, J. B. Pearce, K. K. Kelly, A. E. Stewart, G. E. Thomas, G. P. Anderson, and O. F. Raper, Mariner 6: Ultraviolet spectrum of Mars' upper atmosphere. *Science* **165**, 1004–1005 (1969).

C. A. Barth, C. W. Hord, A. I. Stewart, A. L. Lane, M. L. Dick, and G. P. Anderson, Mariner 9 ultraviolet spectrometer experiment: Seasonal variation of ozone on Mars. *Science* **179**, 795–796 (1973).

P. Bartholdi and F. Owen, The occultation at Beta Scorpii by Jupiter and Io. II. *Astron. J.* **77**, 60–64 (1972).

W. A. Bassett, The diamond cell and the nature of the earth's mantle. *Annu. Rev. Earth Planet. Sci.* **7**, 357–384 (1978).

R. M. Batson, Cartographic products from the Mariner 9 mission. *JGR, J. Geophys. Res.* **78**, 4424–4434 (1973).

S. J. Bauer and R. E. Hartle, Venus ionosphere: An interpretation of Mariner 10 observations. *Geophys. Res. Lett.* **1**, 7–9 (1974).

S. J. Bauer, R. E. Hartle, and J. R. Herman, Topside ionosphere of Venus and its interaction with the solar wind. *Nature (London)* **225**, 533–534 (1970).

M. E. Baur, Thermodynamics of heterogeneous iron carbon systems: Implications for the terrestrial primitive reducing atmosphere. *Chem. Geol.* **22**, 189–206 (1978).

R. H. Becker and R. N. Clayton, Carbon isotopic evidence for the origin of a banded iron-formation in Western Australia. *Geochim. Cosmochim. Acta* **36**, 577–95 (1972).

R. H. Becker and S. Epstein, Carbon, hydrogen and nitrogen isotopes in solvent-extractable organic matter from carbonaceous chondrites. *Geochim. Cosmochim. Acta* **46**, 97–103 (1982).

J. E. Beckman, The measurement of abundances in planetary atmospheres using an image intensifier and a solar spectrograph. *Planet. Space Sci.* **15**, 1211–1218 (1967).

R. Beer, Detection of carbon monoxide in Jupiter. *Astrophys. J.* **200**, L167–L169 (1975).

R. Beer and F. W. Taylor, The abundance of CH_3D and the D/H ratio in Jupiter. *Astrophys. J.* **179**, 309–327 (1973).

R. Beer, R. H. Norton, and J. V. Martonchik, Absorption by Venus in the 3-4 micron region. *Astrophys. J.* **168**, L121–124 (1971a).

R. Beer, R. H. Norton, and J. V. Martonchik, Astronomical infrared spectroscopy with a Connes type interferometer. II. Mars, 2500–3500 cm^{-1}. *Icarus* **15**, 1–10 (1971b).

R. Beer, C. B. Farmer, R. H. Norton, J. V. Martonchik, and T. G. Barnes, Jupiter: Observation of deuterated methane in the atmosphere. *Science* **175**, 1360–1361 (1972).

F. Begemann, H. W. Weber, and H. Hintenberger, On the primordial abundance of Argon 40. *Astrophys. J.* **203**, L155–L157 (1976).

D. Belcher, J. Veverka, and C. Sagan, Mariner photography of Mars and aerial photography of the Earth: Some analogies. *Icarus* **15**, 241–252 (1971).

T. Belsky and I. R. Kaplan, Light hydrocarbon gases,[13]C, and the origin of organic matter in carbonaceous chondrites. *Geochim. Cosmochim. Acta* **34**, 257–278 (1970).

M. J. S. Belton, Theory of the curve of growth and phase effects in a cloudy atmosphere: Applications to Venus. *J. Atoms. Sci.* **25**, 596–609 (1968).

M. J. S. Belton and S. H. Hayes, An estimate of the temperature and abundance of CH_4 and other molecules in the atmosphere of Uranus. *Icarus* **24**, 348–357 (1975).

M. J. S. Belton and D. M. Hunten, Water vapor in the atmosphere of Venus. *Astrophys. J.* **146**, 307–308 (1966).

M. J. S. Belton and D. M. Hunten, A search for O_2 on Mars and Venus—A possible detection of oxygen in the atmosphere of Mars. *Astrophys. J.* **153**, 963–965 (1968).

M. J. S. Belton and D. M. Hunten, Errata: A search for O_2 on Mars and Venus: A possible detection of oxygen in the atmosphere on Mars. *Astrophys. J.* **156**, 797 (1969).

M. J. S. Belton and D. M. Hunten, The distribution of CO_2 on Mars: A spectroscopic determination of surface topography. *Icarus* **15**, 204–232 (1971).

M. J. S. Belton, D. M. Hunten, and M. B. McElroy, A search for an atmosphere on Mercury. *Astrophys. J.* **150**, 1111–1124 (1967).

M. J. S. Belton, A. L. Broadfoot, and D. M. Hunten, Abundance and temperature of CO_2 on Mars during the 1967 opposition. *JGR, J. Geophys. Res.* **73**, 4795–4805 (1968a).

M. J. S. Belton, A. L. Broadfoot, and D. M. Hunten, Upper limit to O_2 on Venus. *J. Atmos. Sci.* **25**, 582–585 (1968b).

M. J. S. Belton, M. B. McElroy, and M. J. Price, The atmosphere of Uranus. *Astrophys. J.* **164**, 191–209 (1971).

D. C. Benner, U. Fink, and R. H. Cromwell, Image tube spectra of Pluto and Triton from 6800 to 9000Å. *Icarus* **36**, 82–91 (1978).

J. T. Bergstralh, Methane absorption in the atmosphere of Saturn: Rotational temperature and abundance from the $3\nu_3$ band. *Icarus* **18**, 605–611 (1973).

J. T. Bergstralh, Uranian methane abundance, rotational temperature, and effective pressure from the 6800Å band. *Astrophys. J.* **202**, 832–838 (1975).

J. T. Bergstralh and H. J. Smith, A search for a CO_2 Mercury atmosphere. *Astron. J.* **72**, 786 (1967).

J. T. Bergstralh, D. L. Matson, and T. V. Johnson, Sodium D- Line Emission from Io: Synoptic observations from Table Mountain Observatory. *Astrophys. J.* **195**, L131–L135 (1975).

L. V. Berkner and L. C. Marshall, On the origin and rise of oxygen concentration in the Earth's atmosphere. *J. Atmos. Sci.* **22**, 225–261 (1965).

J. D. Bernal, "The Physical Basis of Life." Routledge and Paul, London, 1951.

R. A. Berner, Goethite stability and the origin of red beds. *Geochim. Cosmochim. Acta* **33**, 267–273 (1969).

W. Bernstein, R. W. Fredericks, J. L. Vogle, and W. A. Fowler, The lunar atmosphere and the solar wind. *Icarus* **2**, 233–248 (1963).

A. L. Betz, R. A. McLaren, E. C. Sutton, and M. A. Johnson, Infrared heterodyne spectroscopy of CO_2 in the atmosphere of Mars. *Icarus* **30**, 650–662 (1977).

R. Bibron, R. Chesselet, G. Crozaz, G. Leger, J. P. Mennessier, and E. Picciotto, Extraterrestrial [53]Mn in antarctic ice. *Earth Planet. Sci. Lett.* **21**, 109–116 (1974).

K. Biemann and J. M. Lavoie, Jr., Some final conclusions and supporting experiments related to the search for organic compounds on the surface of Mars. *JGR, J. Geophys. Res* **84**, 8385–8390 (1979).

K. Biemann, J. Oro, P. Toulmin, III, L. E. Orgel, A. O. Nier, D. M. Anderson, P. G. Simmonds, D. Flory, A. V. Diaz, D. R. Rushneck, and J. A. Biller, Search for organic and volatile inorganic compounds in two surface samples from the Chryse Planitia region of Mars. *Science* **194**, 72–76 (1976a).

K. Biemann, T. Owen, D. R. Rushneck, A. L. LaFleur, and D. W. Howarth, The atmosphere of Mars near the surface: Isotope ratios and upper limits on noble gases. *Science* **194**, 76–77 (1976b).

K. Biemann, J. Oro, P. Toulmin, III, L. E. Orgel, A. O. Nier, D. M. Anderson, P. G. Simmonds, D.

Flory, A. V. Diaz, D. R. Rushneck, J. E. Biller, and A. L. Lafleur, The search for organic substances and inorganic volatile compounds in the surface of Mars. *JGR, J. Geophys. Res.* **82**, 4641–4658 (1977).

R. W. Bild and J. T. Wasson, Netschaëvo: A new class of chondritic meteorite. *Science* **197**, 58–62 (1977).

B. G. Bills and A. J. Ferrari, Mars topography harmonics and geophysical implications. *JGR, J. Geophys. Res.* **83**, 3497–3508 (1978).

A. B. Binder, Topography and surface features of Mars. *Icarus* **11**, 24–35 (1969a).

A. B. Binder, Internal structure of Mars. *JGR, J. Geophys. Res.* **74**, 3110–3117 (1969b).

A. B. Binder and J. Colin Jones, Spectrophotometric studies of the photometric function, composition, and distribution of the surface materials of Mars. *JGR, J. Geophys. Res.* **77**, 3005–3019 (1972).

A. B. Binder and D. P. Cruikshank, Comparison of the infrared spectrum of Mars with the spectra of selected terrestrial rocks and minerals. *Comm. Lunar Planet. Lab.* **2**, 193–196 (1964a).

A. B. Binder and D. P. Cruikshank, Evidence for an atmosphere on Io. *Icarus* **3**, 299–305 (1964b).

A. B. Binder and D. P. Cruikshank, Photometric search for atmospheres on Europa and Ganymede. *Icarus* **5**, 7–9 (1966).

A. B. Binder and D. P. Cruikshank, Mercury: New observations of the infrared bands of carbon dioxide. *Science* **155**, 1135 (1967).

A. B. Binder and D. R. Davis, Internal structure of Mars. *Phys. Earth Planet. Inter.* **7**, 477–485 (1973).

A. B. Binder and D. W. McCarthy, Jr., Mars: The lineament systems. *Science* **176**, 279–281 (1972).

A. B. Binder, R. E. Arvidson, G. A. Guiness, K. L. Jones, E. C. Morris, T. A. Mutch, D. C. Pieri, and C. Sagan, The geology of the Viking lander site. *JGR, J. Geophys. Res.* **82**, 4439–4451 (1977).

F. Birch, Elasticity and constitution of the Earth's interior. *JGR, J. Geophys. Res.* **57**, 227–286 (1952).

F. Birch, Density and composition of mantle and core. *JGR, J. Geophys. Res.* **69**, 4377–4388 (1964).

T. Birkelund and R. G. Bromley, eds., "Cretaceous-Tertiary Boundary Events," Vol. 1. University of Copenhagen, 1979.

Yu. L. Biryukov and A. S. Panfilov, Study of optical characteristics of atmosphere of Venus based on measurements of Venera 8. *Cosmic Res. (Engl. Transl.)* **12**, 819–824 (1974).

Yu. L. Biryukov and L. G. Titarchuk, Determination of the characteristics of light-scattering particles in the atmosphere of Venus from photometric measurements. *Cosmic Res. (Engl. Transl.)* **10**, 517–520 (1973).

D. C. Black, On the origins of trapped helium, neon, and argon isotopic variations in meteorites. I. Gas-rich meteorites, lunar soil and breccia. *Geochim. Cosmochim. Acta* **36**, 347–375 (1972a).

D. C. Black, On the origins of trapped helium, neon, and argon isotopic variations in meteorites. II. Carbonaceous meteorites. Geochim. Cosmochim. Acta **36**, 377–394 (1972b).

T. R. Blackburn, H. D. Holland, and G. P. Ceasar, Viking gas exchange reaction: Simulation on UV-irradiated manganese dioxide. *JGR, J. Geophys. Res.* **84**, 8391–8394 (1979).

J. Blamont and B. Ragent, Further results of the pioneer Venus nephelometer experiment. *Science* **205**, 67–70 (1979).

K. R. Blasius, The record of impact cratering on the great volcanic shields of the Tharsis region on Mars. *Icarus* **29**, 343–361 (1976).

K. R. Blasius, J. A. Cutts, J. E. Guest, and H. Masursky, Geology of the Valles Marineris: First analysis of imaging from the Viking 1 orbiter primary mission. *JGR, J. Geophys. Res.* **82**, 4067–4092 (1977).

G. Boato, The isotopic composition of hydrogen and carbon in the carbonaceous chondrites. *Geochim. Cosmochim. Acta* **6**, 209–220 (1954).

P. Bodenheimer, Contraction models for the evolution of Jupiter. *Icarus* **29**, 165–171 (1976).

P. Bodenheimer, A. S. Grossman, W. DeCampli, G. Marcy, and J. B. Pollack, "Calculations of the Evolution of the Giant Planets," Preprint (1980).

R. W. Boese, J. B. Pollack, and P. M. Silvaggio, First results from the large probe infrared radiometer experiment. *Science* **203**, 797–801 (1979).

D. D. Bogard, D. S. Burnett, P. Eberhardt, and G. J. Wasserburg, $^{87}Rb–^{87}Sr$ isochron and $^{40}K–^{40}Ar$ ages of the Norton County achondrite. *Earth Planet. Sci. Lett.* **3**, 179–189 (1967).

B. Bolin, On the exchange of CO_2 between atmosphere and ocean. *Tellus* **12**, 274–281 (1960).

B. Bolin, E. Degens, P. Duvigneaud, and S. Kempe. In "The Global Carbon Cycle," (B. Bolin, ed.), pp. 1–56. Wiley, New York, 1979.

W. J. Borucki, Comparison of Venusian lightning observations. *Icarus* **52**, 354–364 (1982).

K. Boström and K. Frederiksson, Surface conditions of the original meteorite parent body as indicated by mineral associations. *Smithson. Misc. Collect.* **151**, No. 3, 3966 (1966).

M. Bottema, W. Plummer J. Strong, and R. Zander, Composition of the clouds of Venus. *Astrophys. J.* **140**, 1640–1641 (1964).

M. Bottema, W. Plummer, J. Strong, and R. Zander, The composition of the Venus clouds and implications for model atmospheres. *JGR, J. Geophys. Res.* **70**, 4401–4402 (1965a).

M. Bottema, W. Plummer, and J. Strong, A quantitative measurement of water-vapor in the atmosphere of Venus. *Ann. Astrophys.* **28**, 225–230 (1965b).

E. Bowell and B. Zellner, Polarizations of asteroids and satellites. *In* "Planets, Stars and Nebulae Studied by Photo-polarimetry" (T. Gehrels, ed.), pp. 381–404. Univ. of Arizona Press, Tucson, 1974.

C. Boyer, The 4-day rotation of the upper atmosphere of Venus. *Planet. Space Sci.* **21**, 1559–1561 (1973).

C. Boyer and P. Guérin, Etude de la rotation rétrograde, en 4 jours, de le couche extérieure nuageuse de Vénus. *Icarus* **11**, 338–355 (1969).

P. J. Brancazio and A. G. W. Cameron, eds., "The Origin and Evolution of Atmospheres and Oceans." Wiley, New York, 1963.

G. W. Brass, Stability of brines on Mars. *Icarus* **42**, 20–28 (1980).

A. Brecher, P. L. Briggs, and G. Simmons, The low temperature electrical properties of carbonaceous chondrites. *Earth Planet. Sci. Lett.* **28**, 37–45 (1975).

J. D. Bregman, D. F. Lester, and D. M. Rank, Observation of the v_2 band of PH$_3$ in the atmosphere of Saturn. *Astrophys. J.* **202**, L55–56 (1975).

R. Brett, The Earth's core: Speculations on its chemical equilibrium with the mantle. *Geochim. Cosmochim. Acta* **35**, 203–221 (1971).

R. Brett, A. lunar core of Fe-Ni-S. *Geochim. Cosmochim. Acta* **37**, 165–170 (1973a).

R. Brett, The lunar crust: Product of heterogeneous accretion or differentiation of a homogeneous moon? *Geochim. Cosmochim. Acta* **37**, 2697–2701 (1973b).

R. Brett, Sulfur in the core of the earth. *Meteoritics* **10**, 372–373 (1975).

R. Brett, The current status of speculations of the composition of the core of the Earth. *Rev. Geophys. Space Phys.* **14**, 375–383 (1976).

H. S. Bridge, A. J. Lazarus, C. W. Snyder, E. J. Smith, L. Davis, P. J. Coleman, and D. E. Jones, Mariner V: Plasma and magnetic fields observed near Venus. *Science* **158**, 1669–1673 (1967).

H. S. Bridge, A. J. Lazarus, J. D. Scudder, K. W. Ogilvie, R. E. Hartle, J. R. Asbridge, S. J. Bame, W. C. Feldman, and G. L. Siscoe, Observations at Venus encounter by the plasma science experiment on Mariner 10. *Science* **183**, 1293–1296 (1974).

H. S. Bridge and 11 others, Plasma observations near Jupiter: Initial results from Voyager 1. *Science* **204**, 987–991 (1979).

M. H. Briggs, Venus—A summary of present knowledge. *J. Br. Interplanet. Soc.* **19**, 45–52 (1963).

P. L. Briggs, M. S. Thesis, Massachusetts Institute of Technology, Cambridge (1976).

R. T. Brinkman, Dissociation of water vapor and evolution of oxygen in the terrestrial atmosphere. *JGR, J. Geophys. Res.* **74**, 5355–5368 (1969).

R. T. Brinkman, Departures from Jeans' escape rate for H and He in the Earth's atmosphere. *Planet. Space Sci.* **18**, 449–478 (1970).

R. T. Brinkman, More comments on the validity of Jeans' escape rate. *Planet. Space Sci.* **19**, 791–794 (1971a).

R. T. Brinkman, Mars: Has nitrogen escaped? *Science* **174**, 944–945 (1971b).

A. L. Broadfoot, Ultraviolet spectrometry of the inner solar system from Mariner 10. *Rev. Geophys. Space Phys.* **14**, 625–627 (1976).

A. L. Broadfoot, S. Kumar, M. J. S. Belton, and M. B. McElroy, Ultraviolet observations of Venus from Mariner 10: Preliminary results. *Science* **183**, 1315–1318 (1974a).

A. L. Broadfoot, S. Kumar, M. J. S. Belton, and M. B. McElroy, Mercury's Atmosphere from Mariner 10: Preliminary results. *Science* **185**, 166–169 (1974b).

A. L. Broadfoot, D. E. Shemansky, and S. Kumar, Mariner 10: Mercury atmosphere. *Geophys. Res. Lett.* **3**, 577–580 (1976).

A. L. Broadfoot and 16 others, Extreme ultraviolet observations from Voyager 1 encounter with Jupiter. *Science* **204**, 979–982 (1979).

A. L. Broadfoot and 15 others, Extreme ultraviolet observations from Voyager 1 encounter with Saturn. *Science* **212**, 206–211 (1981).

H. P. Broida, O. R. Lundell, H. I. Schiff, and R. D. Ketcheson, Is ozone trapped in the solid carbon dioxide polar cap of Mars? *Science* **170**, 1402 (1970).

V. A. Bronshtén, The nature of Venus. *Sol. Syst. Res. (Engl. Transl.)* **1**, 2–28 (1967).

V. A. Bronshtén and N. B. Ibragimov, Spectrophotometry of the continents, Maria, and polar caps of Mars with the 2-m Shemakha Observatory reflector in 1969. *Sol. Syst. Res. (Engl. Transl.)* **5**, 12–16 (1971).

P. Brosche, Eine Schätzung der Masse und Dichte von Pluto. *Icarus* **7**, 132–138 (1967).

H. Brown, On The compositions and structures of the planets. *Astrophys. J.* **111**, 641–653 (1950).

H. Brown, Rare gases and the formation of the Earth's atmosphere. *In* "The Atmospheres of the Earth and Planets" (G. P. Kuiper, ed.), pp. 258–266. Univ. of Chicago Press, Chicago, Illinois, 1952.

R. A. Brown and F. H. Chaffee, Jr., High-resolution spectra of sodium emission from Io. *Astrophys. J.* **187**, L125–L126 (1974).

R. A. Brown, R. M. Goody, F. J. Murcray, and F. H. Chaffee, Jr., Further studies of line emission from Io. *Astrophys. J.* **200**, L49–L52 (1975).

D. Brunt, The possibility of condensation by descent of air. *Q. J. R. Meteorol. Soc.* **60**, 279–284 (1934).

M. Budyko, "Climate and Life." Hydrological Publishing House, Leningrad, USSR (1971).

O. I. Bugaenko, Zh. M. Dlugach, A. V. Morozhenko, and E. E. Yanovitskii, Optical properties of Saturn's cloud layer in the visible spectral range. *Sol. Syst. Res. (Engl. Transl.)* **9**, 9–15 (1975).

F. Bühler, W. I. Axford, H. J. A. Chivers, and K. Marti, New evidence concerning the origin of atmospheric ^{3}He. *Meteoritics* **8**, 334 (1973).

M. S. T. Bukowinski, The effect of pressure on the physics and chemistry of potassium. *Geophys. Res. Lett.* **3**, 491–496 (1976).

D. Bukvic and J. S. Lewis, Outgassing of chondritic planets. In preparation (1981).

T. E. Bunch and S. Chang, Carbonaceous chondrites. II. Carbonaceous chondrite phyllosilicates and light element geochemistry as indicators of parent body processes and surface conditions. *Geochim. Cosmochim. Acta* **44**, 1543–1578 (1980).

J. A. Burns, ed., "Planetary Satellites. Univ. of Arizona Press, Tucson, 1977.

R. Burns and R. Hardy. "Nitrogen Fixation in Bacteria and Higher Plants." Springer-Verlag, Berlin, 1975.

J. A. Burns and M. Harwit, Towards a more habitable Mars—or the coming Martian spring. *Icarus* **19**, 126–130 (1973).

J. Burt, J. Veverka, and K. Cook, Depth-diameter relation for large Martian craters determined from Mariner 9 UVS altimetry. *Icarus* **29**, 83–90 (1976).

D. M. Butler, M. J. Newman, and R. J. Talbot, Jr., Interstellar cloud material: Contribution to planetary atmospheres. *Science* **201**, 522–525 (1978).

W. A. Butler, P. M. Jeffery, J. H. Reynolds, and G. J. Wasserburg, Isotopic variations in terrestrial xenon. *JGR, J. Geophys. Res.* **68**, 32 83–3291 (1963).

A. Cairns-Smith, Origin of life and the nature of the primitive gene. *J. Theor. Biol.* **10**, 53–88 (1965).

J. Caldwell, Retrograde rotation of the upper atmosphere of Venus. *Icarus* **17**, 608–616 (1972).

J. Caldwell, The atmosphere of Saturn: An infrared perspective. *Icarus* **30**, 493–510 (1977a).

J. Caldwell, Thermal radiation from Titan's atmosphere. *In* "Planetary Satellites" (J. A. Burns, ed.), pp. 438–450. Univ. of Arizona Press, Tucson, 1977b.

G. S. Callendar, On the amount of carbon dioxide in the atmosphere. *Tellus* **10**, 243–248 (1958).

M. Calvin, Chemical evolution. *Proc. R. Soc. London, Ser.* A **288**, 441–466 (1965).

A. G. W. Cameron, The origin of atmospheric xenon. *Icarus* **1**, 314–316 (1963a).

A. G. W. Cameron, The origin of the atmospheres of Venus and the Earth. *Icarus* **2**, 249–260 (1963b).

A. G. W. Cameron, Interpretation of xenon measurements. *In* "The Origin and Evolution of Atmospheres and Oceans" (P. J. Brancazio and A. G. W. Cameron, eds.), pp. 235–248. Wiley, New York, 1963c.

A. G. W. Cameron, The orbital eccentricity of Mercury and the origin of the moon. *Nature (London)* **240**, 299 (1972).

A. G. W. Cameron, Formation of the outer planets. *Space Sci. Rev.* **14**, 383–396 (1973).

A. G. W. Cameron, Physics of the primitive solar accretion disk. *Moon Planets* **18**, 5–40 (1978).

A. G. W. Cameron, "Elementary and Nuclidic Abundances in the Solar System," Preprint (1980).

A. G. W. Cameron and M. R. Pine, Numerical models of the primitive solar nebula. *Icarus* **18**, 377–406 (1973).

D. B. Campbell, R. B. Dyce, R. P. Ingalls, G. H. Pettengill, and I. I. Shapiro, Venus: Topography revealed by radar data. *Science* **175**, 514–515 (1972).

D. B. Campbell, J. F. Chandler, G. H. Pettengill, and I. I. Shapiro, Galilean satellites of Jupiter: 12.6-cm radar observations. *Science* **196**, 650–653 (1977).

C. F. Capen, Jr., Martian albedo feature variations with season: Data of 1971 and 1973. *Icarus* **28**, 213–230 (1976).

L. A. Capone, R. C. Whitten, J. Dubach, S. S. Prasad, and W. T. Huntress, Jr., The lower ionosphere of Titan. *Icarus* **28**, 367–378 (1976).

L. A. Capone, J. Dubach, R. C. Whitten, S. S. Prasad, and K. Santhanam, Cosmic ray synthesis of organic molecules in Titan's atmosphere. *Icarus* **44**, 72–84 (1980).

N. P. Carleton and W. A. Traub, Detection of molecular oxygen on Mars. *Science* **177**, 988–992 (1972).

N. P. Carleton, A. Sharma, R. M. Goody, W. L. Liller, and F. Roesler, Measurement of the abundance of CO_2 in the Martian atmosphere. *Astrophys. J.* **155**, 323–331 (1969).

R. W. Carlson and D. L. Judge, Pioneer 10 ultraviolet photometer observations at Jupiter encounter. *JGR, J. Geophys. Res.* **79**, 3623–3633 (1974).

R. W. Carlson, J. C. Bhattacharyya, B. A. Smith, T. V. Johnson, B. Hidayat, S. A. Smith, G. E. Taylor, B. O'Leary, and R. T. Brinkman, An atmosphere on Ganymede from its occultation of SAO-186800 on 7 June 1972. *Science* **182**, 53–55 (1973).

R. W. Carlson, D. L. Matson, and T. V. Johnson, Electron impact ionization of Io's sodium emission cloud. *Geophys. Res. Lett.* **2**, 469–472 (1975).

M. H. Carr, Volcanism on Mars. *JGR, J. Geophys. Res.* **78**, 4049–4061 (1973).

M. H. Carr, Tectonism and volcanism of the Tharsis region of Mars. *JGR, J. Geophys. Res.* **79**, 3943–3949 (1974a).

M. H. Carr, The role of lava erosion in the formation of lunar rilles and martian channels. *Icarus* **22**, 1–23 (1974b).

M. H. Carr and G. G. Schaber, Martian permafrost features. *JGR, J. Geophys. Res.* **82**, 4039–4054 (1977).

M. H. Carr, W. A. Baum, G. A. Briggs, H. Masursky, D. W. Wise, and D. R. Montgomery, Imagining experiment: The Viking Mars orbiter. *Icarus* **16**, 17–33 (1972).

M. H. Carr, H. Masursky, and R. S. Saunders, A generalized geologic map of Mars. *JGR, J. Geophys. Res.* **78**, 4031–4036 (1973).

M. H. Carr, H. Masursky, W. A. Baum, K. R. Blasius, G. A. Briggs, J. A. Cutts, T. Duxbury, R. Greeley, J. E. Guest, B. A. Smith, L.A. Soderblom, J. Veverka, and J. B. Wellman, Preliminary results from the Viking orbiter imaging experiment. *Science* **193**, 766–776 (1976).

M. H. Carr, R. Greeley, K. R. Blasius, J. E. Guest, and J. B. Murray, Some Martian volcanic features viewed from the Viking orbiters. *JGR, J. Geophys. Res.* **82**, 3985–4015 (1977a).

M. H. Carr, L. S. Crumpler, J. A. Cutts, R. Greeley, J. E. Guest, and H. Masursky, Martian impact craters and emplacement of eject by surface flow. *JGR, J. Geophys. Res.* **82**, 4055–4066 (1977b).

M. H. Carr, H. Masursky, R. G. Strom, and R. J. Terrile, Volcanic features of Io. *Nature (London)* **280**, 729–733 (1979).

P. Cassen, Planetary magnetism and the interiors of the moon and Mercury. *Phys. Earth Planet. Inter.* **15**, 113–120 (1977).

P. Cassen, R. T. Reynolds, and S. J. Peale, Is there liquid water on Europa? *Geophys. Res. Lett.* **6**, 731–734 (1979).

P. Cassen, S. J. Peale, and R. T. Reynolds, On the comparative evolution of Ganymede and Callisto. *Icarus* **41**, 232–239 (1980).

R. D. Cess, The thermal structure within the stratospheres of Venus and Mars. *Icarus* **17**, 560–569 (1972).

M. S. Chadha, J. J. Flores, J. G.Lawless and C. Ponnamperuma, Organic synthesis in a simulated Jovian atmosphere. II. *Icarus* **15**, 39–44 (1971).

P. M. Chalk and D. R. Keeney, Nitrate and ammonium contents of Wisconsin limestones. *Nature (London)* **229**, 42 (1971).

J. W. Chamberlain, The atmosphere of Venus near her cloud tops. *Astrophys. J.* **141**, 1184–1205 (1965).

J. W. Chamberlain and G. P. Kuiper, Rotational temperature and phase variations of the carbon dioxide bands of Venus. *Astrophys. J.* **124**, 399–405 (1956).

J. W. Chamberlain and G. R. Smith, Comments on the rate of evaporation of a non-Maxwellian atmosphere. *Planet. Space Sci.* **19**, 675–684 (1971).

J. W. Chamberlain and G. R. Smith, Spectrum of Venus with a double-cloud model. *Astrophys. J.* **173**, 469–475 (1972).

C. R. Chapman, Cratering on Mars. I. Cratering and obliteration history. *Icarus* **22**, 272–291 (1974a).

C. R. Chapman, Cratering on Mars. II. Implications for future cratering studies from Mariner 4 reanalysis. *Icarus* **22**, 292–300 (1974b).

C. R. Chapman, Asteroid size distribution: Implications for the origin of stony-iron and iron meteorites. *Geophys. Res. Lett.* **1**, 341–343 (1974c).

C. R. Chapman, The nature of asteroids. *Sci. Am.* **232**, 24–33 (1975).

C. R. Chapman, Asteroids as meteorite parent-bodies: The astronomical perspective. *Geochim. Cosmochim. Acta* **40**, 701–719 (1976).

C. R. Chapman, J. B. Pollack, and C. Sagan, An analysis of the Mariner 4 cratering statistics. *Astron. J.* **74**, 1039–1051 (1969).

C. R. Chapman, D. Morrison, and B. Zellner, Surface properties of the asteroids: A synthesis of polarimetry, radiometry and spectrophotometry. *Icarus* **25**, 104–130 (1975).

S. Chapman, A theory of upper-atmospheric ozone. *Mem. Roy. Meteorol. Soc.* **3**, 103–125 (1930).

S. Chapman and T. G. Cowling, "The Mathematical Theory of Nonuniform Gases." Cambridge Univ. Press, London and New York 1952.

J. Charney, W. Quirk, S. Chaio, and J. Kornfield, A comparative study of the effects of albedo change on drought in semi-arid regions. *J. Atmos. Sci.* **34**, 1366–1385 (1977).

S. C. Chase, E. D. Miner, D. Morrison, G. Münch, and G. Neugebauer, Preliminary infrared radiometry of Venus from Mariner 10. *Science* **183**, 1291–1292 (1974).

Z. P. Cheremukhina, S. F. Morozov, and N. F. Borodin, Estimate of temperature of Venus' stratosphere from data on deceleration forces acting on the Venera 8 probe. *Cosmic Res. (Engl. Transl.)* **12**, 238–244 (1974).

S. K. Christensen and T. Birkelund, eds., "Cretaceous-Tertiary Boundary Events," Vol. 2. University of Copenhagen, 1979.

J. W. Christy and R. S. Harrington, The satellite of Pluto. *Astron. J.* **83**, 1005–1008 (1978).

J. W. Christy and R. S. Harrington, The discovery and orbit of Charon. *Icarus* **44**, 38–40 (1980).

S. F. S. Chun, K. D. Pang, J. A. Cutts, and M. Ajello, Photocatalytic oxidation of organic compounds on Mars. *Nature (London)* **274**, 875–877 (1978).

CIAP (1975): see Grobecker *et al.* (1974).

R. J. Cicerone, R. S. Stolarsky, and S. Walters, Stratospheric ozone destruction by man-made chlorofluoromethanes. *Science* **185**, 1165–1167 (1974).

M. J. Cintala, J. W. Head, and T. A. Mutch, Characteristics of fresh Martian craters as a function of diameter: Comparison with the moon and Mercury. *Geophys. Res. Lett.* **3**, 117–120 (1976).

B. C. Clark, Implications of abundant hygroscopic minerals in the Martian regolith. *Icarus* **34**, 645–655 (1978).

B. C. Clark and A. K. Baird, Ultraminiature X-ray flourescence spectrometer for in-situ geological analysis on Mars. *Earth Planet. Sci. Lett.* **19**, 359–368 (1973).

B. C. Clark and A. K. Baird, Volatiles in the Martian regolith. *Geophys. Res. Lett.* **6**, 811–814 (1979a).

B. C. Clark and A. K. Baird, Is the Martian lithosphere sulfur rich? *JGR, J. Geophys. Res.* **84**, 8395–8403 (1979b).

B. C. Clark, P. Toulmin, III, A. K. Baird, K. Keil, and H. J. Rose, Argon content of the Martian atmosphere at the Viking I landing site: Analysis by X-ray fluorescence spectroscopy. *Science* **193**, 804–805 (1976a).

B. C. Clark, A. K. Baird, H. J. Rose, Jr., P. Toulmin, III, K. Keil, A. J. Castro, W. C. Kelliher, C. D. Rowe, and P. H. Evans, Inorganic analyses of Martian surface samples at the Viking landing sites. *Science* **194**, 1283–1288 (1976b).

B. C. Clark III, A. K. Baird, H. J. Rose, Jr., P. Toulmin, III, R. P. Christian, W. C. Kelliher, A. J. Castro, C. D. Rowe, K. Keil, and G. R. Huss, The Viking X-ray fluorescence experiment: Analytical methods and early results. *JGR, J. Geophys. Res.* **82**, 4577–4594 (1977).

B. R. Clark and R. P. Mullin, Martian glaciation and the flow of solid CO_2. *Icarus* **27**, 215–228 (1976).

I. D. Clark, The chemical kinetics of CO_2 atmospheres. *J. Atmos. Sci.* **28**, 847–858 (1971).

J. A. Clark and C. S. Lingle, Future sea level changes due to west Antarctic ice sheet fluctuations. *Nature (London)* **269**, 206–209 (1977).

R. N. Clark, Ganymede, Europa, Callisto and Saturn's rings: Compositional analysis from reflectance spectroscopy. *Icarus* **44**, 388–409 (1980).

D. B. Clarke, G. G. Pe, R. M. Mackay, K. R. Gill, M. J. O'Hara, and J. A. Gard, A new potassium-iron-nickel sulphide from a nodule in kimberlite. *Earth Planet. Sci. Lett.* **35**, 421–428 (1977).

W. B. Clarke, M. A. Beg, and H. Craig, Excess ^3He in the sea: Evidence for terrestrial primordial helium. *Earth Planet. Sci. Lett.* **6**, 213–220 (1969).

D. D. Clayton, Extinct Radioactivities: Trapped Residuals of presolar grains. *Astrophys. J.* **199**, 765–769 (1975).

R. N. Clayton, N. Onuma, and T. K. Mayeda, A classification of meteorites based on oxygen isotopes. *Earth Planet. Sci. Lett.* **30**, 10–18 (1976).

R. N. Clayton, N. Onuma, L. Grossman, and T. K. Mayeda, Distribution of the pre-solar component in Allende and other carbonaceous chondrites. *Earth Planet. Sci. Lett.* **34**, 209–224 (1977).

P. Cloud, Jr., Atmospheric and hydrospheric evolution on the primitive Earth. *Science* **160**, 729–736 (1968).

P. Cloud, Jr., A working model of the primitive Earth. *Am. J. Sci.* **272**, 537–548 (1972).

P. A. Coutier, M. B. McElroy, and F. C. Michel, Modification of the Martian ionosphere by the solar wind. *JGR, J. Geophys. Res.* **74**, 6215–6227 (1969).

D. L. Coffeen, Optical polarization of Venus. *J. Atmos. Sci.* **25**, 643–648 (1968).

D. L. Coffeen, Wavelength dependence of polarization. XVI. Atmosphere of Venus. *Astron. J.* **74**, 446–460 (1969).

L. Colin, Encounter With Venus. *Science* **203**, 743–745 (1979a).

L. Colin, Encounter with Venus: An update. *Science* **205**, 44–46 (1979b).

N. B. Colthup, Identification of aldehyde in Mars vegetation regions. *Science* **134**, 529 (1961).

M. Combes, T. Encrenaz, L. Vapillon, and Y. Zeau, Confirmation of the identification of C_2H_2 and C_2H_6 in the Jovian atmosphere. *Astron. Astrophys.* **34**, 33–35 (1974).

J. Connes, P. Connes, and M. Kaplan, Mars: New absorption bands in the spectrum. *Science* **153**, 739–740 (1966).

P. Connes, J. Connes, W. S. Benedict, and L. D. Kaplan, Traces of HCl and HF in the atmosphere of Venus. *Astrophys. J.* **147**, 1230–1237 (1967).

P. Connes, J. Connes, L. D. Kaplan, and W. S. Benedict, Carbon monoxide in the Venus atmosphere. *Astrophys. J.* **152**, 731–742 (1968).

B. Conrath, R. Curran, R. Hanel, V. Kunde, W. Maguire, J. Pearl, J. Pirraglia, J. Welker, and T. Burke, Atmospheric and surface properties of Mars obtained by infrared spectroscopy on Mariner 9. *JGR, J. Geophys. Res.* **78**, 4267–4278 (1973).

B. J. Conrath, Thermal structure of the Martian atmosphere during the dissipation of the dust storm of 1971. *Icarus* **24**, 36–46 (1975).

G. J. Consolmagno, Sulfur volcanoes on Io. *Science* **205**, 397–398 (1979).

G. J. Consolmagno, Io: Thermal models and chemical evolution. *Icarus* **47**, 36–45 (1981).

G. J. Consolmagno and J. S. Lewis, Preliminary thermal history models of icy satellites. In "Planetary Satellites" (J. A. Burns, ed.), pp. 492–500. Univ. of Arizona Press, Tucson, 1977.

G. J. Consolmagno and J. S. Lewis, Structural and thermal models of icy Galilean satellites. In "Jupiter: The Giant Planet" (T. Gehrels, ed.), pp. 1035–1051. Univ. of Arizona Press, Tucson, 1976.

G. J. Consolmagno and J. S. Lewis, The evolution of icy satellite interiors and surfaces. *Icarus* **34**, 280–93 (1978).

A. F. Cook, II, T. C. Duxbury, and G. E. Hunt, First results on Jovian lightning. *Nature (London)* **280**, 794 (1979).

M. Coradini and E. Flemini, A thermodynamic study of the Martian permafrost. *JGR, J. Geophys. Res.* **84**, 8115–8130 (1979).

B. M. Cordell and R. G. Strom, Global tectonics of Mercury and the moon. *Phys. Earth Planet. Inter.* **15**, 146–155 (1977).

B. M. Cordell, R. E. Lingenfelter, and G. Schubert, South polar and equatorial differences in central peaked Martian craters. *Nature (London)* **234**, 335–337 (1971).

B. M. Cordell, R. E. Lingenfelter, and G. Schubert, Martian cratering and central peak statistics: Mariner 9 results. *Icarus* **21**, 448–456 (1974).

R. J. Corice, Jr. and K. Fox, The hypothetical chemical and spectroscopic activity of germane in the atmosphere of Jupiter. *Icarus* **16**, 388–391 (1972).

C. C. Counselman, III, Observations of Mars from Earth between 1965 and 1969. *Icarus* **18**, 1–7 (1973).

C. C. Counselman, III, S. A. Gourevitch, R. W. King, G. H. Pettingill, R. G. Prinn, I. I. Shapiro, R. B. Miller, J. R. Smith, R. Ramos, and P. Liebrecht, Wind velocities on Venus: Vector determination by radio interferometry. *Science* **203**, 805–806 (1979a).

C. C. Counselman, III, S. A. Gourevitch, R. W. King, G. B. Loriot, and R. G. Prinn, Venus winds are zonal and retrograde below the clouds. *Science* **205**, 85–87 (1979b).

C. C. Counselman III, S. A. Gourevitch, R. W. King, G. B. Loriot, and E. Ginsberg, zonal and meridional circulation of the lower atmosphere of Venus determined by radio interferometry, *JGR, J. Geophys. Res.* **85**, 8026–8030 (1980).

V. E. Courtillot, C. J. Allegre, and M. Mattauer, On the existence of lateral relative motions on Mars. *Earth Planet. Sci. Lett.* **25**, 279–285 (1975).

L. P. Cox and J. S. Lewis, Numerical simulation of the final stages of terrestrial planet accretion. *Icarus* **44**, 706–721 (1980).

L. P. Cox, J. S. Lewis, and M. Lecar, A model for close encounters in the planetary problem. *Icarus* **34**, 415–428 (1978).

H. Craig and W. B. Clarke, Oceanic ^{3}He: Contribution from cosmogonic tritium. *Earth Planet. Sci. Lett.* **9**, 45–48 (1970).

H. Craig and D. Lal, The production rate of natural tritium. *Tellus* **13**, 85–91 (1961).

H. Craig and J. Lupton, Primordial neon, helium and hydrogen in oceanic basalts. *Earth Planet. Sci. Lett.* **31**, 369–385 (1976).

C. A. Cross, The heat balance of the Martian polar caps. *Icarus* **15**, 110–114 (1971).

D. P. Cruikshank, Sulfur compounds in the atmosphere of Venus. II. Upper limit for the abundance of COS and H_2O. *Comm. Lunar Planet. Lab.* **6**, 199–206 (1967).

D. P. Cruikshank and A. B. Binder, Minor constituents of the atmosphere of Jupiter. *Astrophys. Space Sci.* **3**, 347–356 (1969).

D. P. Cruikshank and G. P. Kuiper, Sulfur compounds in the atmosphere of Venus. I. An upper limit for the abundance of SO_2. *Comm. Lunar Planet. Lab.* **6**, 199–200 (1967).

D. P. Cruikshank and J. L. Morgan, Titan: Suspected near-infrared variability. *Astrophys J.* **235**, L53–L54 (1980).

D. P. Cruikshank and D. Morrison, Radii and albedos of asteroids 1, 2, 3, 4, 6, 15, 51, 433 and 511. *Icarus* **20**, 477–481 (1973).

D. P. Cruikshank and R. E. Murphy, The post-eclipse brightening of Io. *Icarus* **20**, 7–17 (1973).

D. P. Cruikshank and P. M. Silvaggio, Triton: A satellite with an atmosphere. *Astrophys. J.* **233**, 1016–1020 (1979).

D. P. Cruikshank and P. M. Silvaggio, The surface and atmosphere of Pluto. *Icarus* **41**, 96–102 (1980).

D. P. Cruikshank and A. B. Thomson, On the occurrence of ferrous chloride in the clouds of Venus. *Icarus* **15**, 497–503 (1971).

D. P. Cruikshank, C. B. Pilcher, and D. Morrison, Pluto: Evidence for methane frost. *Science* **194**, 835–837 (1976).

D. P. Cruikshank, A. Stockton, H. M. Dyck, E. E. Becklin, and W. Macy, Jr., The diameter and reflectance of Triton. *Icarus* **40**, 104–114 (1979).

P. J. Crutzen, Ozone production rates in an oxygen-hydrogen-nitrogen oxide atmosphere. *JGR, J. Geophys. Res.* **76**, 7311–7327 (1971).

P. J. Crutzen, "Artificial Increases of the Stratospheric Nitrogen Oxide Content and Possible Consequences for the Atmospheric Ozone," Rep. AP-15. Int. Meteorol. Inst., Stockholm 1974.

P. J. Crutzen, The possible impact of CSO for the sulfate layer of the stratosphere. *Geophys. Res. Lett.* **3**, 73–75 (1976).

D. Cunnold, F. Alyea, N. Phillips, and R. G. Prinn, A three-dimensional dynamical-chemical model of atmospheric ozone. *J. Atmos. Sci.* **32**, 170–194 (1975).

J. A. Cutts, Wind erosion in the Martian polar regions. *JGR, J. Geophys. Res.* **78**, 4211–4221 (1973a).

J. A. Cutts, Nature and origin of layered deposits of the Martian polar regions. *JGR, J. Geophys. Res.* **78**, 4231–4248 (1973b).

J. A. Cutts, and R. S. U. Smith, Eolian deposits and dunes on Mars. *JGR, J. Geophys. Res.* **78**, 4139–4153 (1973).

J. A. Cutts, L. A. Soderblom, R. P. Sharp, B. A. Smith, and B. C. Murray, The surface of Mars. 3. Light and dark markings. *JGR, J. Geophys. Res.* **76**, 343–356 (1974).

J. A. Cutts, K. R. Blasius, G. A. Briggs, M. H. Carr, R. Greeley, and H. Masursky, North polar region of Mars: Imaging results from Viking 2. *Science* **194**, 1329–1337 (1976).

J. N. Cuzzi and D. O. Muhleman, The microwave spectrum and nature of the subsurface of Mars. *Icarus* **17**, 548–560 (1972).

A. Dalgarno and M. B. McElroy, Mars: Is nitrogen present? *Science* **170**, 167–168 (1970).

G. B. Dalrymple and J. G. Moore, Argon 40: Excess in submarine pillow basalts from Kilauea Volcano, Hawaii. *Science* **161**, 1132–1134 (1968).

P. E. Damon and J. L. Kulp, Inert gases and evolution of the atmosphere. *Geochim. Cosmochim. Acta* **13**, 280–292 (1958).

R. E. Danielson, The structure of the atmosphere of Uranus. *Icarus* **30**, 462–478 (1977).

R. E. Danielson, J. J. Caldwell, and D. R. Larach, An inversion in the atmosphere of Titan. *Icarus* **20**, 437–443 (1973).

A. D. Danilov, Radioastronomical investigations and modern concepts concerning the Venusian atmosphere. *Cosmic Res. (Engl. Transl.)* **2**, 107–117 (1964).

A. D. Danilov and S. P. Yatsenko, The ionospheric interpretation of the results of radar observations of Venus. *Geomagn. Aeron. (Engl. Transl.)* **3**, 475–483 (1963).

A. Dauvillier, Sur la nature des nuages de Vénus. C. R. Hebd. Seances Acad. Sci. 243, 1257–1268 (1956).

A. Dauvillier, Sur l'atmosphère primitive du globe et les atmosphères de Vénus et de Mars. C. R. Hebd. Seances Acad. Sci. Ser. D 267, 697–700 (1968).

D. W. Davidson, Clathrate hydrates. In "Water: A Comprehensive Treatise" (F. Franks, ed.), Vol. 2, pp. 115–234. Plenum, New York, 1973.

M. E. Davies, Photogrammetric measurements of Olympus Mons on Mars. Icarus 21, 230–236 (1974).

B. W. Davis, Some speculations on absoption and desorption of CO_2 in Martian bright areas. Icarus 11, 155–158 (1969).

V. D. Davydov, The nature of the Hellas region on Mars. Sol. Syst. Res. (Engl. Transl.) 5, 196–201 (1971).

K. L. Day, Synthetic phyllosilicates and the matrix material of C1 and C2 chrondites. Icarus 27, 561–568 (1976).

M. O. Dayhoff, R. V. Eck, E. R. Lippincott, and C. Sagan, Venus: Atmosphere evolution. Science 155, 566–568 (1967).

C. de Bergh, Venus: Microwave opacity of the minor atmospheric constituents. Astron. Astrophys. 23, 467–470 (1973).

W. M. DeCampli, Giant gaseous protoplanets. Ph.D. Dissertation, Harvard University, Cambridge, Massachusetts (1978).

W. M. DeCampli and A. G. W. Cameron, Structure and evolution of isolated giant gaseous protoplanets. Icarus 38, 367–391 (1979).

D. Deirmendjian, A water cloud interpretation of Venus' microwave continuum. Icarus 3, 109–120 (1964).

A. H. Delsemme, ed., "Comets, Asteroids and Meteorites: Interrelations, Evolution and Origns." University of Toledo, Toledo, Ohio, 1977.

W. C. DeMarcus, The constitution of Jupiter and Saturn. Astron. J. 63, 2–13 (1958).

K. G. Denbigh, "The Principles of Chemical Equilibrium." Cambridge Univ. Press, London and New York, 1957.

G. De Vaucouleurs, J. Roth, and C. Mulholland, Preliminary albedo map of the South polar region. JGR, J. Geophys. Res. 78, 4436–4439 (1973).

R. E. Dickinson, Circulation and thermal structure of the Venusian thermosphere. J. Atmos. Sci. 28, 885–893 (1971).

R. E. Dickinson, Infrared radiative heating and cooling in the Venusian mesosphere. I. Global mean radiative equilibrium. J. Atmos. Sci. 29, 1531–1555 (1972).

R. E. Dickinson, Infrared radiative heating and cooling in the Venusian mesophere. II. Day-to-night variation. J. Atmos. Sci. 30, 296–301 (1973).

R. E. Dickinson, Venus mesosphere and thermosphere temperature structure. I. Global mean radiative and conductive equilibrium. Icarus 27, 479–493 (1976).

R. E. Dickinson and E. C. Ridley, Numerical solution for the composition of a thermosphere in the presence of a steady subsolar-to-antisolar circulation with application to Venus. J. Atmos. Sci. 29, 1557–1570 (1972).

R. E. Dickinson and E. C. Ridley, Venus mesophere and thermosphere temperature structure. II. Day-night variations. Icarus 30, 163–178 (1977).

K. E. Dierenfeldt, V. Fink, and H. P. Larson, Temperature and pressure determinations in the Venus atmosphere by means of high-resolution spectra from 1 to 2.5 μm Icarus 31, 11–24 (1977).

D. J. Diner, J. A. Westphal, and F. P. Schloerb, Infrared imaging of Venus: 8–14 micrometers. Icarus 27, 191–195 (1976).

M. T. Dmitrieva and V. V. Ilyukhin, Crystal structure of djerfisherite. Dokl. Akad. Nauk SSSR 223, 343–346 (1975).

A. Dobrovolskis and A. P. Ingersoll, Carbon dioxide-water clathrate as a reservoir of CO_2 on Mars. Icarus 26, 353–357 (1975).

J. S. Dohnanyi, Interplanetary objects in review: Statistics of their masses and dynamics. *Icarus* 17, 1–48 (1972).

Sh. Sh. Dolginov, E. G. Eroshenko, and D. N. Zhuzgov, Magnetic field investigation with interplanetary station "Venera-4." *Cosmic Res. (Engl. Transl.)* 6, 469–480 (1968).

Sh. Sh. Dolginov, E. G. Ersoshenko, and L. Davis, Nature of the Magnetic field in the neighborhood of Venus. *Cosmic Res. (Engl. Transl.)* 7, 675–680 (1969).

Sh. Sh. Dolginov, E. G. Eroshenko, and L. N. Zhuzgov, magnetic field in the very close neighborhood of Mars according to data from Mars 2 and Mars 3 spacecraft. *JGR, J. Geophys. Res.* 78, 4779–4785 (1973).

Sh. Sh. Dolginov, E. G. Eroshenko, and L. N. Zhuzgov, Magnetic field of Mars from data of Mars 3 and Mars 5. *Cosmic Res. (Engl. Transl.)* 13, 94–106 (1975).

Sh. Sh. Dolginov, E. G. Eroshenko, and L. N. Zhuzgov, The magnetic field of Mars from the Mars 3 and Mars 5. *JGR, J. Geophys. Res.* 81, 3353–3362 (1976).

A. Dollfus, Polarization studies of planets. *In* "Planets and Satellites" (G. P. Kuiper and B. M. Middlehurst, eds.), pp. 343–399. Univ. of Chicago Press, Chicago, Illinois, 1961.

A. Dollfus, Observation de la vapeur d'eau sur la planète Vénus. *C. R. Hebd. Seances Acad.* 256, 3250–3253 (1963).

A. Dollfus and D. L. Coffeen, Polarization of Venus. I. Disk observations. *Astron. Astrophys.* 8, 251–266 (1970).

A. Dollfus, J. Focas, and E. Bowell, The planet mars: The nature of its surface, the characteristics of its atmosphere from the polarimetry of its light. Part II. The nature of the soil. *Astron. Astrophys.* 2, 105–121 (1969).

T. M. Donahue, The upper atmosphere of Venus: A review. *J. Atmos. Sci.* 25, 568–573 (1968).

T. M. Donahue, Deuterium in the upper atmospheres of Venus and Earth. *JGR, J. Geophys. Res.* 74, 1128–1137 (1969).

T. M. Donahue, Polar ion flow: Wind or breeze? *Rev. Geophys. Space Phys.* 9, 1–9 (1971).

T. M. Donahue, Pioneer Venus results: An overview. *Science* 205, 41–44 (1979).

T. M. Donahue, J. H. Hoffman, and R. R. Hodges, Jr., Krypton and xenon in the atmosphere of Venus. *Geophys. Res. Lett.* 8, 513–516 (1981).

T. M. Donahue, J. H. Hoffman, R. R. Hodges, Jr., and A. J. Watson, Venus was wet: A measurement of the ratio of deuterium to hydrogen. *Science* 216, 630–632 (1982).

F. D. Drake, Improbability of non-thermal radio emission from Venus water clouds. *Astrophys. J.* 149, 459–461 (1967).

A. L. Draper, J. A. Adamcik, and E. K. Gibson, Comparison of the spectra of Mars and a goethite-hematite mixture in the 1 to 2 micron region. *Icarus* 3, 63–65 (1964).

G. Dreibus, B. Spettel, and H. Wänke, Halogens in meteorites and their primordial abundances. *In* "Origin and Distribution of the Elements" (L. H. Ahrens, ed.), pp. 33–38. Pergamon, Oxford, 1979.

A. J. Dressler and P. A. Cloutier, Discussion of letter by P. M. Banks and T. E. Holzer: "The Polar Wind." *JGR, J. Geophys. Res.* 74, 3730–3733 (1969).

E. R. du Fresne and E. Andres, On the chemical evolution of the carbonaceous chondrites. *Geochim. Cosmochim. Acta* 26, 1085–1114 (1962).

R. L. Duncombe and P. K. Seidelmann, A history of the determination of Pluto's mass. *Icarus* 44, 12–18 (1980).

R. L. Duncombe, W. Klepczynski, and P. K. Seidelmann, Note on the mass of Pluto. *Publ. Astron. Soc. Pac.* 82, 916–917 (1970).

J. A. Dunne, Mariner 10 Venus encounter. *Science* 183, 1289–1291 (1974a).

J. A. Dunne, Mariner 10 Mercury encounter. *Science* 185, 141–142 (1974b).

J. A. Dunne, W. D. Stromberg, R. M. Ruiz, S. A. Collins, and T. E. Thrope, Maximum discriminability versions of the near-encounter Mariner pictures. *JGR, J. Geophys. Res.* 76, 438–472 (1971).

J. Dymond and L. Hogan, Noble gas abundance patterns in deep-sea basalts-primordial gases from the mantle. *Earth Planet. Sci. Lett.* 20, 131–139 (1973).

D. Dzurisin and K. R. Blasius, Topography of the polar layered deposits of Mars. *JGR, J. Geophys. Res.* **80**, 3286–3306 (1975).

D. Dzurisin and A. P. Ingersoll. Seasonal buffering of atmospheric pressure on Mars. *Icarus* **26**, 437–440 (1975).

P. Eberhardt, J. Geiss, and N. Grögler, Über die Verteilung der Uredelgase im Meteoriten Khor Temiki. *Tschermaks Mineral. Petrogr. Mitt.* **10**, 535–551 (1965).

I. J. Eberstein, B. N. Khare, and J. B. Pollack, Infrared transmission properties of CO, HCl, SO_2 and their significance for the greenhouse effect on Venus. *Icarus* **11**, 159–170 (1969).

W. E. Egan, Polarimetric and photometric simulation of the Martian surface. *Icarus* **10**, 223–237 (1969).

W. G. Egan, Size classification of Mars simulation samples. *JGR, J. Geophys. Res.* **76**, 6213–6219 (1971).

D. H. Ehhalt, The atmosphere cycle of methane. *Tellus* **26**, 58–70 (1974).

A. El Goresy, N. Grögler, and J. Ottemann, Djerfisherite composition in Bishopville, Peña Blanca Springs, St. Marks and Toluca meteorites. *Chem. Erde* **30**, 77–82 (1971).

A. El Goresy, P. Ramdohr, M. Pavicevič, O. Mendenbach, O. Müller, and W. Gentner, Zinc, lead, chlorine and FeOOH-bearing assemblages in the Apollo 16 sample 66095: Origin by impact of a comet or a carbonaceous chondrite. *Earth Planet. Sci. Lett.* **18**, 411–419 (1973).

W. E. Elston and E. I. Smith, Mars: Evidence for dynamic processes from Mariner 6 and 7. *Icarus* **19**, 130–194 (1973).

T. Encrenaz and T. Owen, The abundance of ammonia on Jupiter, Saturn and Titan. *Astron. Astrophys.* **37**, 49–55 (1974).

T. Encrenaz, M. Combes, Y. Zeau, L. Vapillon, and J. Berezne, A tentative identification of C_2H_4 in the spectrum of Saturn. *Astron. Astrophys.* **42**, 355–356 (1975).

E. Eriksson, The yearly circulation of chloride and sulfur in nature: Meteorological, geochemical and pedological implications. *Tellus* **11**, 375–443 (1959).

V. R. Eshleman, Atmospheres of Mars and Venus: A review of the Mariner 4 and 5 and Venera 4 experiments. *Radio Sci.* **5**, 325–332 (1970).

V. R. Eshleman, G. Fjeldbo, J. D. Anderson, A. Kliore, and R. B. Dyce, Venus: Lower atmosphere not measured. *Science* **162**, 661–665 (1968).

A. Eucken, Physikalisch-chemisch Betrachtungen über die Früheste Entwicklungsgeschichte der Erde. *Nachr. Akad. Wiss. Goettingen* (1944).

J. V. Evans and R. P. Ingalls, Absorption of radar signals by the atmosphere of Venus. *J. Atmos. Sci.* **25**, 555–559 (1968).

A. W. Fairhall, Concerning the source of the excess ^{3}He in the sea. *Earth Planet. Sci. Lett.* **7**, 249–250 (1969).

F. W. Fallon and R. E. Murphy, Absence of post-eclipse brightening of Io and Europa in 1970. *Icarus* **15**, 494–496 (1971).

F. P. Fanale, History of Martian volatiles: Implication for organic synthesis. *Icarus* **15**, 279–303 (1971a).

F. P. Fanale, A case for catastrophic early degassing of the Earth. *Chem. Geol.* **8**, 79–105 (1971b).

F. P. Fanale, Martian volatiles: Their degassing history and geochemical fate. *Icarus* **28**, 179–202 (1976).

F. P. Fanale and W. A. Cannon, Absorption on the Martian regolith. *Nature (London)* **230**, 502–504 (1971a).

F. P. Fanale and W. A. Cannon, Physical adsorption of rare gas on terrigenous sediments. *Earth Planet. Sci Lett.* **11**, 362–368 (1971b).

F. P. Fanale and W. A. Cannon, Origin of planetary primordial rare gas: The possible role of absorption. *Geochim. Cosmochim. Acta* **36**, 319–328 (1972).

F. P. Fanale and W. A. Cannon, Exchange of adsorbed H_2O and CO_2 between the Regolith and atmosphere of Mars caused by changes in surface insolation. *JGR, J. Geophys. Res.* **79**, 3397–3402 (1974).

F. P. Fanale and W. A. Cannon, Mars: The role of the regolith in determining atmospheric pressure and the atmosphere's response to insolation changes. *JGR, J. Geophys. Res.* **83**, 2321–2325 (1978).

F. P. Fanale and W. A. Cannon, Mars: CO_2 adsorption and capillary condensation on clays—significance for volatile storage and atmospheric history. JGR, J. Geophys. Res. **84**, 8404–8414 (1979).

F. P. Fanale, T. V. Johnson, and D. L. Matson, Constraints on the composition of Io's surface. Trans. Am. Geophys. Union **57**, 276 (1976).

F. P. Fanale, T. V. Johnson, and D. L. Matson, Io's surface composition: Observational constraints and theoretical considerations. Geophys. Res. Lett. **4**, 303–306 (1977).

F. P. Fanale, W. A. Cannon, and T. Owen, Mars: Regolith adsorption and the relative concentrations of atmospheric rare gases. Geophys. Res. Lett. **5**, 77–81 (1978).

F. P. Fanale, R. H. Brown, D. P. Cruikshank, and R. N. Clarke, Significance of adsorption features in Io's IR reflectance spectrum. Nature (London) **280**, 761–763 (1979).

C. B. Farmer, Liquid water on Mars. Icarus **28**, 279–296 (1976).

C. B. Farmer, D. W. Davies, and D. D. LaPorte, Viking: Mars atmospheric water vapor mapping experiment—preliminary report of results. Science **193**, 776–780 (1976a).

C. B. Farmer, D. W. Davies, and D. D. LaPorte, Mars: Northern summer ice cap-water vapor observations from Viking 2. Science **194**, 1339–1340 (1976b).

C. B. Farmer, D. W. Davies, A. L. Holland, D. D. LaPorte, and P. E. Doms, Mars: Water vapor observations from the Viking orbiters. JGR, J. Geophys. Res. **82**, 4225–4248 (1977).

B. Fegley, Jr. and J. S. Lewis, Thermodynamics of selected trace elements in the Jovian atmosphere. Icarus **38**, 166–179 (1979).

B. Fegley, Jr. and J. S. Lewis, Volatile element chemistry in the solar nebula: Na, K, F. Cl, Br and P. Icarus **41**, 439–455 (1980).

B. Fegley, Jr. A condensation-accretion model for volatile element retention. Pap. Conf. Planet. Volatiles, (1982).

M. A. Feierberg and M. J. Drake, The meteorite-asteroid connection: The infrared spectra of eucrites, shergottites and vesta. Science **209**, 805–806 (1980).

M. H. Feierberg, H. Larson, U. Fink, and H. Smith, Spectroscopic evidence for at least two achondrite parent bodies. Geochim. Cosmochim. Acta **44**, 513–521 (1980).

E. E. Ferguson, F. C. Fehsenfeld, and A. L. Schmetekopf, A new speculation on terrestrial helium loss. Planet. Space Sci. **13**, 925–928 (1965).

J. P. Ferris and C. T. Chen, Chemical evolution. XXVI. Photochemistry of methane, nitrogen, and water mixtures as a model for the atmosphere of the primitive Earth. J. Am. Chem. Soc. **97**, 11–16 (1975a).

J. P. Ferris and C. T. Chen, Photosynthesis of organic compounds in the atmosphere of Jupiter. Nature (London) **258**, 587–588 (1975b).

J. P. Ferris, E. H. Edelson, N. M. Mount, and A. E. Sullivan, The effect of clays on the oligomerization of HCN. J. Mol. Evol. **13**, 317–330 (1979).

V. G. Fesenkov, Increase in the turbidity of the atmosphere due to the fall of the Tunguska meteorite on June 30, 1908. Meteoritika **6**, 8–12 (1949).

V. G. Fesenkov, Can comets consist of antimatter? J. Brit. Astron. Assoc. **78**, 126–128 (1968).

G. B. Field, Atmosphere of Mercury. Atron. J. **67**, 575–576 (1962).

G. B. Field, The atmosphere of Mercury. In "The Origin and Evolution of Atmospheres and Oceans" (P. J. Brancazio and A. G. W. Cameron, eds.), p. 269. Wiley, New York, 1963.

U. Fink and H. P. Larson, Temperature dependence of the water-ice spectrum between 1 and 4 Microns: Application to Europa, Ganymede, and Saturn's rings. Icarus **24**, 411–420 (1975).

U. Fink and H. P. Larson, Deuterated methane observed on Saturn. Science **201**, 543–544 (1978).

U. Fink and H. P. Larson, The infrared spectra of Uranus, Neptune and Titan from 0.8 to 2.5 microns. Astrophys. J. **233**, 1021–1040 (1979).

U. Fink, H. P. Larson, G. P. Kuiper, and R. F. Poppen, Water vapor in the atmosphere of Venus. Icarus **17**, 617–631 (1972).

U. Fink, N. H. Dekkers, and H. P. Larson, Infrared spectra of the Galilean satellites of Jupiter. Astrophys. J. **179**, 155–159 (1973).

U. Fink, H. P. Larson, and T. N. Gautier, III, New upper limits for the atmospheric constituents on Io. Icarus **27**, 439–446 (1976).

U. Fink, H. P. Larson, and R. R. Treffers, Germane in the atmosphere of Jupiter. *Icarus* **34**, 344–354 (1978).

U. Fink, B. A. Smith, D. C. Bonner, J. R. Johnson, H. J. Reitsema, and J. A. Westphal, Detection of a CH_4 atmosphere on Pluto. *Icarus* **44**, 62–71 (1980).

F. F. Fish, Jr., The stability of goethite on Mars. *JGR, J. Geophys. Res.* **71**, 3063–3068 (1966).

D. E. Fisher, Trapped He and Ar and formation of the atmosphere by degassing. *Nature (London)* **256**, 113–114 (1975).

D. E. Fisher, A search for primordial atmospheric-like argon in an iron meteorite. *Geochim. Cosmochim. Acta* **45**, 245–249 (1981).

J. D. Fix, Comments on the interior of pluto. *Icarus* **16**, 569–570 (1972).

G. Fjeldbo, A. J. Kliore, and V. R. Eshleman, The neutral atmosphere of Venus as studied with the Mariner V radio occultation experiments. *Astron. J.* **76**, 123–140 (1971).

G. Fjeldbo, D. Sweetnam, J. Brenkle, E. Christensen, D. Farless, J. Mehta, S. Seidel, W. H. Michael, Jr., A. Wallio, and M. Grossi, Viking radio occultation measurements of the Martian atmosphere and topography: Primary mission coverage. *JGR, J. Geophys. Res.* **82**, 4317–4324 (1977).

F. M. Flasar and R. M. Goody, Diurnal behaviour of water on Mars. *Planet. Space Sci.* **24**, 161–181 (1976).

C. P. Florenskii, A. T. Bazilevskii, R. O. Kuz'min, and I. M. Chernaya, Geological and morphological analysis of some Mars-4 and 5 photographs. *Cosmic Res. (Engl. Transl.)* **13**, 56–63 (1975).

C. P. Florenskii, V. P. Volkov, and O. V. Nikolayeva, On a geochemical model of the troposhphere of Venus. *Geokhimiya* pp. 1135–1150 (1976).

C. P. Florenskii, A. T. Basilevskii, G. A. Burba, O. V. Nikolaeva, A. A. Pronin, and V. P. Volkov, First panoramas of the Venusian surface. *Geochim. Cosmochim. Acta, Suppl.* **8**, 2655–2664 (1977).

C. P. Florenskii, V. P. Volkov, and O. V. Nikolaeva, A. geochemical model of the Venus troposphere. *Icarus* **33**, 537–553 (1978).

J. H. Focas, Seasonal evolution of the fine structure of the dark areas of Mars. *Planet. Space Sci.* **9**, 371–381 (1962).

F. F. Forbes, Infrared polarization of Venus. *Astrophys. J.* **165**, L21–25 (1971).

P. G. Ford and G. H. Pettengill, Venus: Global surface radio emissivity. *Science* **220**, 1379–1381 (1983).

O. G. Franz and R. L. Millis, A. search for an anomalous brightening of Io after eclipse. *Icarus* **14**, 13–15 (1971).

O. G. Franz and R. L. Millis, A search for posteclipse brightening of Io in 1973. II. *Icarus* **23**, 431–436 (1974).

H. Frey, Surface features on Mars: Ground-based Albedo and radar compared with Mariner 9 topography. *JGR, J. Geophys. Res.* **79**, 3907–3916 (1974).

H. Frey, Post-eclipse brightening and non-brightening of Io. *Icarus* **25**, 439–446 (1975).

H. Frey, B. L. Lowry, and S. A. Chase, Pseudocraters on Mars. *JGR, J. Geophys. Res.* **84**, 8075–8086 (1979).

P. E. Fricker, R. T. Reynolds, A. L. Summers, and P. M. Cassen, Does Mercury have a molten core? *Nature (London)* **259**, 293–294 (1976).

E. Frieden, Chemical elements of life. *Sci. Am.* **227**, 52–61 (1972).

J. P. Friend, R. Leifer, and M. Trichau, On the formation of stratospheric aerosols. *J. Atmos. Sci.* **30** 465–479 (1973).

P. M. Frolov, Kvoprosy sushchestvovaniya glubinnoy vody na planete Mars (The Possible Existence of Deep Water on the Planet Mars). *Mosk. Obshch. Ispyt. Prir. Byull., Otd. Geol.* **41**, 150 (1966).

L. H. Fuchs, Djerfisherite, alkali copper-iron sulfide: A new mineral from enstatite chondrites. *Science* **153**, 166–167 (1966).

N. Fukuta, T. L. Wang, and W. F. Libby, Ice nucleation in a Venus atmosphere. *J. Atmos. Sci.* **26**, 1142–1145 (1969).

N. Fukuta, T. L. Wang, and W. F. Libby, Reply. *J. Atmos. Sci.* **27**, 334–335 (1970).

M. J. Gaffey, Spectral reflectance characteristics of the meteorite classes. *JGR, J. Geophys. Res.* **81**, 905–920 (1976).

M. F. Gaffey and T. B. McCord, Asteroid surface materials: Mineralogical characterizations from reflectance spectra. *Space Sci. Rev.* **21**, 555–628 (1978).

M. J. Gaffey and T. B. McCord, Mineralogical and petrological characterizations of asteroid surface materials. *In* "Asteroids" (T. Gehrels, ed.), pp. 688–723. Univ. of Arizona Press, Tuscon, 1979.

A. R. Gaiduk, Short-wave spectrum of Mars in the region of the NO_2 bands. *Sov. Astron. (Engl. Transl.)* **15**, 454–456 (1971).

W. Gale, M. Liwschitz, and A. C. E. Sinclair, Venus: An isothermal lower atmosphere? *Science* **164**, 1059–1060 (1969).

R. Ganapathy, A major meteorite impact on the earth 65 million years ago: Evidence from the Cretaceous-Tertiary boundary clay. *Science* **209**, 921–923 (1980).

R. Ganapathy and E. Anders, Bulk compositions of the moon and Earth, estimated from meteorites. *Geochim. Cosmochim. Acta, Suppl.* **5**, 2, 1181–1206 (1974).

J. Ganguly and G. C. Kennedy, Solubility of K in Fe-S liquid, silicate-K-(Fe-S) equilibria and their planetary implications. *Earth Planet. Sci. Lett.* **35**, 411–420 (1977).

P. W. Gast, Limitations on the composition of the upper mantle. *JGR, J. Geophys. Res.* **65**, 1287–1297 (1960).

P. W. Gast, Terrestrial ratio of K to Rb and the composition of the Earth's mantle. *Science* **147**, 858–860 (1965).

H. A. Gebbie, L. Delbouille, and G. Roland, The use of Michelson interferometer to obtain infrared sprectra of Venus. *Mon. Not. R. Astron. Soc.* **123**, 497–500 (1962).

T. Gehrels, ed., "Jupiter: the Giant Planet." Univ. of Arizona Press, Tucson, 1976.

T. Gehrels, ed., "Asteroids." Univ. of Arizona Press, Tucson, 1979.

B. G. Gel'man, V. G. Zolotukhin, B. V. Kazakov, N. I. Lamonov, A. N. Lipatov, B. V. Isachuk, L. M. Mukhin, D. F. Neparokov, B. N. Okhotnikhov, A. V. Sipel'mikov, V. A. Rotin, and V. N. Khokhlov, An analysis of the chemical composition of the atmosphere by a gas chromatograph method. *Cosmic Res. (Engl. Transl.)* **17**, 585–589 (1979a).

B. G. Gel'man, V. G. Zolotukhin, N. L. Lamonov, B. Levchuk, A. N. Lipatov, L. M. Mukhin, D. F. Nenarokov, V. A. Rotin, and B. N. Okhotnikov, An analysis of the chemical composition of the atmosphere of Venus on an AMS of the Venera 12 using a gas chromatograph. *Kosm. Issled.* **17**, 708–712 (1979b).

E. Gerling, Helium isotope composition in some rocks. *Geokhimiya* **10**, 1209–1220 (1971).

H. Gerstenkorn, Über Gezeitenreibung beim Zweikorper Problem. *Z. Astrophys.* **36**, 245–274 (1955).

H. Gerstenkorn, The earliest past of the Earth-Moon system. *Icarus* **11**, 189–207 (1969).

N. Getoff, Reduktion der Kohlensäure in Wässeriger Lösung unter Entwirkung von UV-Licht. *Z. Naturforsch B: Anorg. Chem., Org. Chem., Biochem., Biophys., Biol.* **17**, 87–90 (1962).

E. K. Gibson, Jr., Ph.D. Dissertation, Arizona State University, Tempe, 1969.

E. K. Gibson, Jr., A comparison of the spectral reflectivity of Mars with oxidized meteoritic material. *Icarus* **13**, 96–99 (1970).

E. K. Gibson, Jr., Production of simple molecules on the surface of mercury. *Phys. Earth Planet. Inter.* **15**, 303–312 (1977).

E. K. Gibson, Jr. and C. B. Moore, The distribution of total nitrogen in iron meteorites. *Geochim. Cosmochim. Acta* **35**, 877–890 (1971).

E. K. Gibson, Jr., C. B. Moore, and C. F. Lewis, Total nitrogen and carbon abundances in carbonaceous chondrites. *Geochim. Cosmochim. Acta* **35**, 599–604 (1971).

P. J. Gierasch, Meridional circulation and maintenance of Venus' atmospheric rotation. *J. Atmos. Sci.* **2**, 1038–1044 (1975).

P. J. Gierasch and R. Goody, A study of the thermal and dynamical structure of the Martian lower atmosphere. *Planet. Space Sci.* **16**, 615–646 (1968).

P. J. Gierasch and O. B. Toon, Atmospheric pressure variations and the climate of Mars. *J. Atmos. Sci.* **30**, 1502–1508 (1973).

F. C. Gillett, Further observations of the 8-13 micron spectrum of Titan. *Astrophys. J.* **201**, L41–L43 (1975).

F. C. Gillett and W. J. Forrest, The 7.5 to 13.5 micron spectrum of Saturn. *Astrophys. J.* **187**, L37–L39 (1974).

F. C. Gillett and G. S. Orton, Center-to-limb observations of Saturn in the thermal infrared. *Astrophys. J.* **195**, L47–L49 (1975).

F. C. Gillett, F. J. Low, and W. A. Stein, Absolute spectrum of Venus from 2.8 to 14 microns. *J. Atmos. Sci.* **25**, 594–595 (1968).

F. C. Gillett, K. M. Merrill, and W. A. Stein, Albedo and thermal emission of Jovian satellites. I–IV. *Astrophys. Lett.* **6**, 247–249 (1970).

F. C. Gillett, W. J. Forrest, and K. M. Merrill, 8–13 micron observations of Titan. *Astrophys. J.* **184**, L93–L95 (1973).

L. P. Giver, E. C. Y. Inn, J. H. Miller, and R. W. Boese, The Martian CO_2 abundance from measurements in the 1.05 μ band. *Astrophys. J.* **153**, 285–289 (1968).

M. F. Glaessner, Precambrian paleontology. *Earth Sci. Rev.* **1**, 29–50 (1966).

K. A. Goettel, Partitioning of potassium between silicates and sulphide melts: Experiments relevant to the Earth's core. *Phys. Earth Planet. Inter.* **6**, 161–166 (1972).

K. A. Goettel, Models for the origin and composition of the Earth, and the hypothesis of potassium in the Earth's core. *Geophys. Surv.* **2**, 369–397 (1976).

K. A. Goettel, Density of the mantle of Mars. *Geophys. Res. Lett.* **8**, 497–500 (1981).

K. A. Goettel and J. S. Lewis, Comments on a paper by V. H. Oversby and A. E. Ringwood. *Earth Planet. Sci. Lett.* **18**, 148–150 (1973).

K. A. Goettel and J. S. Lewis, Ammonia in the atmosphere of Venus. *J. Atmos. Sci.* **31**, 828–830 (1974).

K. A. Goettel, J. A. Shields, and D. A. Decker, Density constraints on the composition of Venus. *Geochim. Cosmochim. Acta, Suppl.* **16**, 265–266 (1982).

N. R. Goins and A. R. Lazarewicz, Martian seismicity. *Geophys. Res. Lett.* **6**, 368–370 (1979).

T. Gold, Outgassing processes on the moon and Venus. *In* "The Origin and Evolution of Atmospheres and Oceans" (P. J. Brancazio and A. G. W. Cameron, eds.), pp. 249–255. Wiley, New York, 1964.

T. Gold, M. J. Campbell, and B. T. O'Leary, Optical and high-frequency electrical properties of the lunar sample. *Geochim. Cosmochim. Acta, Suppl.* **3**, 2149–2154 (1971).

P. Goldreich, The history of the lunar orbit. *Rev. Geophys.* **4**, 411–439 (1966).

P. Goldreich and W. R. Ward, The formation of planetesimals. *Astrophys. J.* **183**, 1051–1061 (1973).

V. M. Goldschmidt, Der Stoffwechsel der Erde. *Skr. Nor. Vidensk. Akad. [Kl.]I: Mat.-Naturvidensk. Kl.* **11** (1922).

V. M. Goldschmidt, "Geochemistry." Oxford Univ. Press (Clarendon), London and New York, 1954.

R. M. Goldstein an H. C. Rumsey, A radar image of Venus. *Icarus* **17**, 899–703 (1972).

R. M. Goldstein, R. R. Green, and H. C. Rumsey, Venus radar images. *JGR, J. Geophys. Res.* **81**, 4807–4817 (1976).

G. S. Golitsyn, Another look at atmospheric dynamics on Titan and some of its general consequences. *Icarus* **24**, 70–75 (1975).

Yu. M. Golovin, B. E. Moshkin, and A. P. Ekonomov, Aerosol component of the atmosphere of Venus according to the results of spectrophotometric measurements aboard Venera-II and -12 descenders. *Cosmic Res.* **19**, 295–302 (1981).

R. M. Goody, "Atmospheric Radiation." Oxford Univ. Press (Clarendon), London and New York, 1964.

R. M. Goody, The scale height of the Venus haze layer. *Planet. Space Sci.* **15**, 1817–1819 (1967).

G. Gopala Rao, The origin of nitric oxide in the atmosphere. *Bull. Acad. Sci. United Prov. Agra Oudh, India* **1**, 82–87 (1931).

H. C. Graboske, Jr., J. B. Pollack, A. S. Grossman, and R. J. Olness, The structure and evolution of Jupiter: The fluid contraction stage. *Astrophys. J.* **199**, 265–281 (1975).

J. Gradie and E. Tedesco, Compositional structure of the asteroid belt. *Science* **216**, 1405–1407 (1982).

J. Gradie, P. Thomas, and J. Veverka, The surface composition of Amalthea. *Icarus* **44**, 373–387 (1980).

L. D. Gray and R. A. Schorn, High-dispersion spectroscopic studies of Venus. *Icarus* **8**, 409–422 (1968).

L. D. Gray, R. A. Schorn, and E. Barber, High dispersion spectroscopic observation of Venus. IV. The weak carbon dioxide band at 7883 Å. *Appl. Opt.* **8**, 2087–2093 (1969).

K. V. Grechnev, V. G. Istomin, L. N. Ozerov, and V. G. Klimovitskii, The mass spectrometer for Venera 11 and 12. *Cosmic Res. (Engl. Transl.)* **17**, 575–580 (1979).

R. Greeley and P. D. Spudis, Volcanism on Mars. *Rev. Geophys.* **19**, 13–41 (1981).

K. I. Gringauz and T. K. Breus, Comparative characteristics of the ionospheres of the planets of the terrestrial group: Mars, Venus, and the Earth. *Space Sci. Rev.* **10**, 743–769 (1968).

K. I. Gringauz, V. V. Bezrukikh, L. S. Musatov, and T. K. Breus, Plasma measurements in the vicinity of Venus by the space vehicle "Venus-4." *Cosmic Res. (Engl. Transl.)* **6**, 350–355 (1968).

K. I. Gringauz, V. V. Bezrukihh, G. I. Volkov, L. S. Musatov, and T. K. Breus, Interplanetary plasma disturbances near Venus determined from "Venera-4" and "Venera-6" data. *Cosmic Res. (Engl. Transl.)* **8**, 393–397 (1970).

A. Grobecker, S. Coroniti and R. Connor, CIAP report of findings: The effects of stratospheric pollution by aircraft. Nat. Tech. Inf. Serv. Springfield, VI (1974).

J. Gros and E Anders, Gas-rich minerals in the Allende meteorite: Attempted chemical characterization. *Earth Planet. Sci. Lett.* **33**, 401–406 (1977).

S. H. Gross, Evolutionary aspects of the atmosphere of Titan and the Galilean satellites. *Bull. Am. Astron. Soc.* **5**, 305–306 (1973).

S. H. Gross and G. V. Ramanathan, The atmosphere of Io. *Icarus* **29**, 493–507 (1976).

A. S. Grossman, H. Graboske, J. Pollack, R. Reynolds, and A. Summers, An evolutionary calculation of Jupiter. *Phys. Earth Planet. Inter.* **6**, 91–98 (1972).

A. S. Grossman, J. B. Pollack, R. Reynolds, A. L. Summers, and H. C. Graboske, Jr., The effect of dense cores on the structure and evolution of Jupiter and Saturn. *Icarus* **42**, 358–379 (1980).

L. Grossman, Condensation in the primitive solar nebula. *Geochim. Cosmochim. Acta* **36**, 597–619 (1972).

L. Grossman, Refractory trace elements in Ca-Al rich inclusions in the Allende meteorite. *Geochim. Cosmochim. Acta* **37**, 1119–1140 (1973).

L. Grossman and J. W. Larimer, Early chemical history of the solar system. *Rev. Geophys. Space Phys.* **12**, 71–101 (1974).

W. E. Groth and H. von Weyssenhoff, Photochemical formation of organic compounds from mixtures of simple gases. *Planet. Space Sci.* **2**, 79–85 (1960).

K. S. Groves, S. R. Mattingly, and A. F. Tuck, Increased atmospheric carbon dioxide and stratospheric ozone. *Nature (London)* **273**, 711–715 (1978).

D. Gubbins, Speculations on the origin of the magnetic field of Mercury. *Icarus* **30**, 186–191 (1977).

S. Gulkis and R. Poynter, Thermal radio emission from Jupiter and Saturn. *Phys. Earth Planet. Inter.* **6**, 36–43 (1972).

S. Gulkis, M. A. Janssen, and E. T. Olsen, Evidence for the depletion of ammonia in the Uranus atmosphere. *Icarus* **34**, 10–19 (1978).

R. Gull, C. R. O'Dell, and R. A. R. Parker, Water vapor in Venus determined by airborne observations of the 8200 Å band. *Icarus* **21**, 213–218 (1974).

D. A. Gurnett, R. R. Shaw, R. R. Anderson, W. S. Kurth, and F. L. Scarf, Whistlers observed by Voyager 1: Detection of lightning on Jupiter. *Geophys. Res. Lett.* **6**, 511–514 (1979).

S. D. Gutshabash and A. S. Safrai, Near-infrared solar-radiation flux (E) of lower atmosphere of Venus. *Cosmic Res.* **13**, 678–681 (1975).

H. T. Hall and V. Rama Murthy, The early chemical history of the Earth: Some critical elemental fractionations. *Earth Planet. Sci. Lett.* **11**, 239–244 (1971).

R. W. Hall and N. F. B. A. Branson, High resolution radio observations of the Planet Venus at a wavelength of 6 cm. *Mon. Not. R. Astron. Soc.* **151**, 185–196 (1971).

Y. Hamano, Argon degassing models of Mars. *Nature (London)* **266**, 41–42 (1977).

R. A. Hanel, M. Forman, G. Stambach, and T. Meilleur, Preliminary results of Venus observations between 8 and 13 microns. *J. Atmos. Sci.* **25**, 586–593 (1968).

R. A. Hanel, B. J. Conrath, W. A. Hovis, V. G. Kunde, P. D. Lowman, J. C. Pearl, C. Prabhakara, B.

Schlachman, and G. V. Levin, Infrared spectroscopy experiment on the Mariner 9 mission: Preliminary results. *Science* **175**, 305–308 (1972a).

R. A. Hanel, B. J. Conrath, W. A. Hovis, V. G. Kunde, P. D. Lowman, W. Maguire, J. C. Pearle, J. Pirraglia, C. Prabhakara, B. Schlachman, G. V. Levin, P. Straat, and T. Burke, Investigation of the Martian environment by infrared spectroscopy on Mariner 9. *Icarus* **17**, 423–442 (1972b).

R. A. Hanel and 12 others, Infrared observations of the Jovian system from Voyager 1. *Science* **204**, 972–976 (1979).

R. A. Hanel and 15 others, Infrared observations of the Saturnian system from Voyager 1. *Science* **212**, 192–200 (1981).

T. C. Hanks and D. L. Anderson, The early thermal history of the Earth. *Phys. Earth Planet. Inter.* **2**, 19–29 (1969).

J. E. Hansen and A. Arking, Clouds of Venus: Evidence for their nature. *Science* **171**, 669–671 (1971).

J. E. Hansen and H. Cheyney, Comments on the paper by D. G. Rea and B. T. O'Leary 'On the composition of the Venus coulds.' *JGR, J. Geophys. Res.* **73**, 6136–6137 (1968a).

J. E. Hansen and H. Cheyney, Near infrared reflectivity of Venus and ice clouds. *J. Atmos. Sci.* **25**, 629–633 (1968b).

J. E. Hansen and J. W. Hovenier, Interpretation of the polarization of Venus. *J. Atmos. Sci.* **31**, 1137–1160 (1974).

J. E. Hansen, D. Johnson, A. Lacis, S. Lebedeff, P. Lee, D. Rind, and G. Russell, Climate impact of increasing atmospheric carbon dioxide. *Science* **213** 957–966 (1981).

O. L. Hansen, An explication of the radiometric method for size and albedo determination. *Icarus* **31**, 456–482 (1977).

P. L. Hanst, L. L. Spiller, D. M. Watts, J. W. Spence, and M. F. Miller, Infrared measurement of fluorocarbons, carbon tetrachloride, carbonyl sulfide, and other atmospheric trace gases. *J. Air Pollut.* **25**, 1220–1226 (1975).

B. Hapke, Venus clouds: A dirty hydrochloric acid model. *Science* **175**, 748–751 (1972).

B. Hapke, Interpretations of optical observations of Mercury and the Moon. *Phys. Earth Planet. Inter.* **15**, 264–274 (1977).

B. Hapke, Io's surface and environs: A magmatic-volatile model. *Geophys. Res. Lett.* **6**, 799–802 (1979).

B. Hapke and R. Nelson, Evidence for an elemental sulfur component of the clouds from Venus spectrophotometry. *J. Atmos. Sci.* **32**, 1212–1218 (1975).

B. Hapke and E. Wells, Lunar bi-directional reflectance spectroscopy. *Lunar Sci.* **7**, 345–347 (1976).

R. W. F. Hardy and U. D. Havelka, Nitrogen-fixation research—the key to world food. *Science* **188**, 633–643 (1975).

R. B. Hargraves and N. Peterson, Magnetic properties investigation: The Viking Mars lander. *Icarus* **16**, 223–227 (1972).

R. B. Hargraves, D. W. Collinson, and C. R. Spitzer, Viking magnetic properties investigation: Preliminary results. *Science* **194**, 84–86 (1976a).

R. B. Hargraves, D. W. Collinson, R. E. Arvidson, and C. R. Spitzer, Viking magnetic properties investigation: Further results. *Science* **194**, 1303–1309 (1976b).

R. B. Hargraves, D. W. Collinson, R. E. Arvidson, and C. R. Spitzer, The Viking magnetic properties experiment: Primary mission results. *JGR, J. Geophys. Res.* **82**, 4547–4558 (1977).

R. B. Hargraves, D. W. Collinson, R. E. Arvidson, and P. M. Cates, Viking magnetic properties experiment: Extended mission results. *JGR, J. Geophys. Res.* **84**, 8379–8389 (1979).

D. L. Harris, Photometry and Colorimetry of the Planets and Satellites. *In* "The Solar System" (G. P. Kuiper and B. M. Middlehurst, eds.), Vol. 3, pp. 272–342. Chicago, Illinois, 1961.

J. E. Harris, Ratio of HNO_3 to NO_2 Concentrations in daytime stratosphere. *Nature (London)* **274**, 235 (1978).

H. Harrison, D. M. Scattergood, and M. R. Shupe, The condensation and Sublimation of CO_2 with H_2O: Carbonic acid on Mars? *Planet. Space Sci.* **16**, 495–499 (1968).

M. H. Hart, A possible atmosphere for Pluto. *Icarus* **21**, 242–247 (1974).

R. E. Hartle, S. A. Curtis, and G. E. Thomas, Mercury's helium exosphere. *JGR, J. Geophys. Res.* **80,** 3689–3692 (1975a).

R. E. Hartle, K. W. Ogilvie, J. D. Scudder, H. S. Bridge, G. L. Siscoe, A. J. Lazarus, V. M. Vasyliunas, and C. M. Yeates, Preliminary interpretation of plasma electron observations at the third encounter of Mariner 10 with Mercury. *Nature (London)* **255,** 206–208 (1975b).

H. Hartman, Speculations on the origin of metabolism. *J. Mol. Evol.* **4,** 359–370 (1975).

W. K. Hartmann, Martian cratering. II. Asteroid impact history. *Icarus* **15,** 396–409 (1971a).

W. K. Hartmann, Martian cratering. III. Theory of crater obliteration. *Icarus* **15,** 410–428 (1971b).

W. K. Hartmann, Martian Cratering. IV. Mariner 9 initial analysis of cratering chronology. *JGR, J. Geophys. Res.* **78,** 4096–4115 (1973a).

W. K. Hartmann, Martian surface and crust: Review and synthesis. *Icarus* **19,** 550–575 (1973b).

W. K. Hartmann, Martian and terrestrial paleoclimatology: Relevance of solar variability. *Icarus* **22,** 301–311 (1974a).

W. K. Hartmann, Geological observations of Martian arroyos. *JGR, J. Geophys. Res.* **79,** 3951–3957 (1974b).

W. K. Hartmann, Planet formation: Compositional mixing and lunar compositional anomalies. *Icarus* **27,** 553–559 (1976).

W. K. Hartmann, Relative crater production rates on planets. *Icarus* **31,** 260–276 (1977).

C. Hayashi, Stellar evolution in early phases of gravitational contraction. *Publ. Astron. Soc. Jpn.* **13,** 450–452 (1961).

C. Hayashi, K. Nakazawa, and H. Mizumo, Earth's melting due to the blanketing effect of the primordial dense atmosphere. *Earth Planet. Sci. Lett.* **43,** 22–28 (1979).

R. Hayatsu, R. G. Scott, M. H. Studier, R. S. Lewis, and E. Anders, Carbynes in meteorites: Detection, low-temperature origin, and implications for interstellar molecules. *Science* **209,** 1515–1518 (1980).

J. W. Head, M. Settle, and C. A. Wood, Origin of Olympus Mons escarpment by erosion of pre-volcano substrate *Nature (London)* **265,** 667–668 (1976).

A. Henderson-Sellers and A. W. Schwartz, Chemical evolution and ammonia in the early Earth atmosphere. *Nature (London)* **287,** 526–528 (1980).

E. W. Hennecke and O. K. Manuel, Argon, krypton and xenon in iron meteorites. *Earth Planet. Sci. Lett.* **36,** 29–43 (1977).

F. Herbert, M. Wiskerchen, C. P. Sonett, and J. K. Chao, Solar wind induction in Mercury: Constraints on the formation of a magnetosphere. *Icarus* **28,** 489–500 (1976).

J. R. Herman, Helium in the topside Venus ionosphere. *JGR, J. Geophys. Res.* **78,** 4669–4673 (1973).

J. R. Herman, R. E. Hartle, and J. S. Bauer, The dayside ionosphere of Venus. *Planet. Space Sci.* **19,** 443–460 (1971).

J. M. Herndon and M. L. Rudee, The chemical state of carbon in the enstatite chondrites. *Meteoritics* **11,** 297 (1976).

J. M. Herndon, M. W. Rowe, E. E. Larson, and D. E. Watson, Origin of magnetite and pyrrhotite in carbonaceous chondrites. *Nature (London)* **253,** 516–518 (1975).

K. C. Herr and G. C. Pimentel, Infrared absorptions near three microns recorded over the polar cap of Mars. *Science* **166,** 496–499 (1969).

K. C. Herr, D. Horn, J. M. McAfee, and G. C. Pimentel, Martian topography from the Mariner 6 and 7 infrared spectra. *Astron. J.* **75,** 883–894 (1970).

J. Hertogen, J. Vizgirda, and E. Anders, Composition of the parent bodies of eucritic meteorites. *Bull. Am. Astron. Soc.* **9,** 458–459 (1977).

H. G. Hertz, Mass of Vesta. *Science* **160,** 299–300 (1968).

G. Herzberg, Spectroscopic evidence of molecular hydrogen in the atmospheres of Uranus and Neptune. *Astrophys. J.* **115,** 337–340 (1952).

S. L. Hess, The vertical distribution of water vapor in the atmosphere of Mars. *Icarus* **28,** 269–278 (1976).

S. L. Hess, R. M. Henry, C. B. Leovy, J. A. Ryan, J. E. Tillman, T. E. Chamberlain, H. L. Cole, R. G. Dutton, G. C. Greene, 'V. E. Simon, and J. L. Mitchell, Preliminary meteorological results on Mars from the Viking 1 lander. *Science* **193**, 788–791 (1976a).

S. L. Hess, R. M. Henry, C. B. Leovy, J. A. Ryan, J. E. Tillman, T. E. Chamberlain, H. L. Cole, R. G. Dutton, G. C. Greene, W. E. Simon, and J. L. Mitchell, Mars climatology from Viking I after 20 sols. *Science* **194**, 78–80 (1976b).

S. L. Hess, R. M. Henry, C. B. Leovy, J. L. Mitchell, J. A. Ryan, and J. E. Tillman, Early meterological results from the Viking 2 lander. *Science* **194**, 1352–1353 (1976c).

S. L. Hess, R. M. Henry, C. B. Leovy, J. A. Ryan, and J. E. Tillman, Meterological Results from the surface of Mars: Viking 1 and 2. *JGR, J. Geophys. Res.* **82**, 4599–4574 (1977).

D. Heymann, The inert gases: He(2), Ne(10), Ar(18), Kr(36) and Xe(54). *In* "Handbook of Elemental Abundances in Meteorites" (B. Mason, ed.), pp. 29–66. Gordon & Breach, New York, 1971.

R. D. Hill, A survey of lightning energy estimates. *Rev. Geophys. Space Phys.* **17**, 155–164 (1979).

A. R. Hochstim (ed.), Bibliography of chemical kinetics and collision processes. IFI/Plenum, New York (1969).

C. A. Hodges and H. J. Moore, The subglacial birth of Olympus Mons and its aureoles. *JGR, J. Geophys. Res.* **84**, 8061–8074 (1979).

R. R. Hodges, Jr., Model atmospheres for Mercury based on a lunar analogy. *JGR, J. Geophys. Res.* **79**, 2881–2885 (1974).

J. H. Hoffman, R. R. Hodges, Jr., M. B. McElroy, T. M. Donahue, and M. Kolpin, Venus lower atmospheric compsition: Preliminary results from Pioneer Venus. *Science* **203**, 800–802 (1979a).

J. H. Hoffman, R. R. Hodges, Jr., M. B. McElroy, T. M. Donahue, and M. Kolpin, Composition and structure of the Venus atmosphere: Results from Pioneer Venus. *Science* **205**, 49–52 (1979b).

J. H. Hoffman, R. R. Hodges, Jr. T. M. Donahue, and M. B. McElroy, Composition of the Venus lower atmosphere from the Pioneer Venus mass spectrometer. *JGR, J. Geophys. Res.* **85**, 7882–7891 (1980).

J. S. Hogan and R. W. Stewart, Exospheric temperatures on Mars and Venus. *J. Atmos. Sci.* **26**, 332–333 (1969).

H. D. Holland, Model for the evolution of the Earth's atmosphere, *Geol. Soc. Am., Buddington Vol.* pp. 447–477 (1962).

H. D. Holland, On the chemical evolution of the terrestrial and cytherian atmospheres. *In* "The Origin and Evolution of Atmospheres and Oceans" (P. J. Brancazio and A. G. W. Cameron, eds.), pp. 86–101. Wiley, New York, 1963.

H. D. Holland, The geologic history of sea water—an attempt to solve the problem. *Geochim. Cosmochim. Acta* **36**, 637–652 (1972).

H. D. Holland, Systematics of the isotopic composition of sulfur in the oceans during the Phanerozoic and its implications for atmospheric oxygen. *Geochim. Cosmochim. Acta* **37**, 2605–2616 (1973).

H. D. Holland, "The Chemistry of the Atmosphere and Oceans." Wiley (Interscience) New York 1978.

D. Horn, J. M. McAfee, A. M. Winer, K. C. Herr, and G. C. Pimentel, The composition of the Martian atmosphere: Minor constituents. *Icarus* **16**, 543–556 (1972).

N. H. Horowitz and J. S. Hubbard, The origin of life. *Annu. Rev. Genet.* **8**, 393–410 (1974).

N. H. Horowitz, J. S. Hubbard, and G. L. Hobby, The carbon-assimilation experiment: The Viking Mars lander. *Icarus* **16**, 147–152 (1972).

N. H. Horowitz, G. L. Hobby, and J. S. Hubbard, The Viking carbon assimilation experiments: Interim report. *Science* **194**, 1321–1322 (1976).

N. H. Horowitz, G. L. Hobby, and J. S. Hubbard, Viking on Mars: The carbon assimilation experiments. *JGR, J. Geophys. Res.* **82**, 4659–4662 (1977).

J. R. Houck, J. B. Pollack, C. Sagan, D. Schaack, and J. A. Decker, Jr., High altitude infrared spectroscopic evidence for bound water on Mars. *Icarus* **18**, 470–480 (1973).

W. A. Hovis, Jr., Infrared reflectivity of Fe_2O_3. nH_2O: Influence on Martian reflection spectra. *Icarus* **4**, 41–42 (1965).

H. T. Howard and 29 others, Venus: Mass, gravity field, atmosphere, and ionosphere as measured by the Mariner 10 dual-frequency radio system. *Science* **183**, 1297–1301 (1974).

F. Hoyle, "The Frontiers of Astronomy" Harper, New York, 1955.

K. S. Hsü, Terrestrial catastrophe caused by cometary impact at the end of Cretaceous. *Nature (London)* **285**, 201–203 (1980).

J. S. Hubbard, The pyrolytic release experiment: Measurement of carbon-assimilation. *Origins Life* **7**, 281–292 (1976).

J. S. Hubbard, J. P. Hardy, and N. H. Horowitz, Photocatalytic production of organic compounds from CO and H_2O in a simulated Martian atmosphere. *Proc. Natl. Acad. Sci. U.S.A.* **68**, 574–578 (1971).

W. B. Hubbard, Thermal structure of Jupiter. *Astrophys. J.* **152**, 745 (1968).

W. B. Hubbard, Thermal models of Jupiter and Saturn. *Astrophys. J.* **155**, 333–344 (1969).

W. B. Hubbard, Structure of Jupiter: Chemical composition, contraction, and rotation. *Astrophys. J.* **162**, 687–698 (1970).

W. B. Hubbard, Observational constraints on the structure of hydrogen planets. *Astrophys. J.* **182**, L35–L38 (1973).

F. O. Huck, D. J. Jobson, S. K. Park, S. D. Wall, R. E. Arvidson, W. R. Patterson, and W. D. Benton, Spectrophotometric and color estimates of the Viking lander sites. *JGR, J. Geophys. Res.* **82**, 4401–4411 (1977).

D. W. Hughes and P. A. Daniels, The magnitude distribution of comets. *Mon. Not. R. Astron. Soc.* **191**, 511–520 (1980).

R. L. Huguenin, Photostimulated oxidation of magnetite. 1. Kinetics and alteration phase identification. *JGR, J. Geophys. Res.* **78**, 8481–8493 (1973a).

R. L. Huguenin, Photostimulated oxidation of Magnetite. 2. Mechanism. *JGR, J. Geophys. Res.* **78**, 8495–9506 (1973b).

R. L. Huguenin, The formation of goethite and hydrated clay minerals on Mars. *JGR, J. Geophys. Res.* **79**, 3895–3904 (1974).

R. L. Huguenin, Surface oxidation: A major sink for water on Mars. *Science* **192**, 138–139 (1975).

R. L. Huguenin, Mars: Chemical weathering as a massive volatile sink. *Icarus* **28**, 203–212 (1976).

R. L. Huguenin and S. M. Clifford, Martian oases. *Bull. Am. Astron. Soc.* **11**, 580 (1979).

R. L. Huguenin, R. P. Prinn, and M. Maderazzo, Photocatalytic stability of the Martian atmosphere. *Icarus* **32**, 270–298 (1977).

G. Hulme, The determination of the rheological properties and effusion rate of an Olympus Mons lava. *Icarus* **27**, 207–213 (1976).

J. R. Hulston and H. G. Thode, Variations in the ^{33}S, ^{34}S and ^{36}S contents of the meteorites and their relation to chemical and nuclear effects. *JGR, J. Geophys. Res.* **70**, 3475–3484 (1965).

B. G. Hunt, Photochemistry of ozone in a moist atmosphere. *JGR, J. Geophys. Res.* **71**, 1385–1398 (1966).

G. E. Hunt, Interpretation of hydrogen quadruple and methane observations of Jupiter and the radiative properties of the visible clouds. *Mon. Not. R. Astron. Soc.* **161**, 347–363 (1973).

G. E. Hunt, Possible climatic and biological impact of nearby supernovae. *Nature (London)* **271**, 430–431 (1978).

G. E. Hunt and R. A. J. Schorn, Height variations of Venusian clouds. *Nature (London), Phys. Sci.* **233**, 39–40 (1971).

G. R. Hunt, L. M. Logan, and J. W. Salisbury, Mars: Components of infrared spectra and the composition of the dust cloud. *Icarus* **18**, 459–469 (1973).

J. N. Hunt, R. Palmer, and W. Penney, Atmospheric waves caused by large explosions. *Philos. Trans. R. Soc. London, Ser. A* **252**, 275–315 (1960).

D. M. Hunten, The structure of the lower atmosphere of Venus. *JGR, J. Geophys. Res.* **73**, 1093–1094 (1968).

D. M. Hunten, Composition and structure of planetary atmospheres. *Space Sci. Rev.* **12**, 539–599 (1971).

D. M. Hunten, The escape of H_2 from Titan. *J. Atmos. Sci.* **30**, 726–731 (1973a).

D. M. Hunten, The escape of light gases from planetary atmospheres. *J. Atmos. Sci.* **30**, 1481–1493 (1973b).

D. M. Hunten, Aeronomy of the lower atmosphere of Mars. *Rev. Geophys. Space Phys.* **12**, 529–535 (1974a).

D. M. Hunten, The atmosphere of Titan. *Icarus* **22**, 111–116 (1974b).

D. M. Hunten, The atmosphere of Venus: Conference review. *J. Atmos. Sci.* **32**, 1262–1265 (1975).

D. M. Hunten, Thermal and nonthermal escape mechanisms for terrestrial bodies. *Planet. Space Sci.* **30**, 773–783 (1982).

D. M. Hunten and L. Colin, eds. "Venus." Univ. of Arizona Press, Tucson (1982).

D. M. Hunten and R. M. Goody, Venus: The next phase of planetary exploration. *Science* **165**, 1317–1323 (1969).

D. M. Hunten and M. B. McElroy, Production and escape of hydrogen on Mars. *JGR, J. Geophys. Res.* **75**, 5989–6000 (1970).

D. M. Hunten and A. J. Watson, Stability of Pluto's atmosphere. *Icarus* **51**, 665–667 (1982).

D. M. Hunten, M. J. S. Belton, and H. Spinrad, Water vapor on Venus—reply. *Astrophys. J.* **150**, L125–L126 (1967).

P. M. Hurley, Test on the possible chondritic composition of the Earth's mantle and its abundance of uranium, thorium, and potassium. *Geol. Soc. Am. Bull.* **68**, 379–382 (1957).

P. M. Hurley, Correction to: Absolute abundance and distribution of Rb, K and Sr in the Earth. *Geochim. Cosmochim. Acta* **32**, 1025–1030 (1968).

R. Hutchison, The formation of the Earth. *Nature (London)* **250**, 556–558 (1974).

J. I. Inge and W. A. Baum, A comparison of Martian Albedo features with topography. *Icarus* **19**, 323–328 (1973).

A. P. Ingersoll, The runaway greenhouse: A history of water on Venus. *J. Atmos. Sci.* **26**, 1191–1198 (1969).

A. P. Ingersoll, Mars: Occurrence of liquid water. *Science* **168**, 972–973 (1970).

A. P. Ingersoll, Jupiter's Great Red Spot: A free atmospheric vortex. *Science* **182**, 1346–1348 (1973).

A. P. Ingersoll, Mars: The case against permanent CO_2 frost caps. *JGR, J. Geophys. Res.* **79**, 3403–3410 (1974).

A. P. Ingersoll and C. B. Leovy, The atmospheres of Venus and Mars. *Annu. Rev. Astron. Astrophys.* **9**, 147–182 (1971).

E. C. Y. Inn and J. M. Heimerl, Photolysis of carbon dioxide at wavelengths exceeding 1740 Å. *J. Atmos. Sci.* **28**, 838–841 (1971).

D. S. Intriligator and E. J. Smith, Mars in the solar wind. *JGR, J. Geophys. Res.* **84**, 8427–8435 (1979).

D. S. Intriligator, H. R. Collard, J. D. Mihalov, R. C. Whitten, and J. H. Wolfe, Electron observations and ion flows from the Pioneer Venus orbiter plasma analyzer experiment. *Science* **205**, 116–119 (1979).

W. M. Irvine, Monochromatic phase curves and albedos for Venus. *J. Atmos. Sci.* **25**, 610–616 (1968).

V. G. Istomin and K. V. Grechnev, Argon in the Martian atmosphere: Evidence from the Mars 6 descent module. *Icarus* **28**, 155–158 (1976).

V. G. Istomin, K. V. Grechnev, L. N. Ozerov, M. E. Slutskii, V. A. Pavlenko, and V. N. Tsvetkov, Experiment to measure the atmospheric composition of Mars on the Mars 6 entry capsule. *Cosmic Res. (Engl. Transl.)* **13**, 13–17 (1975).

V. G. Istomin, K. V. Grechnev, V. A. Kochnev, and L. N. Ozerov, Composition of Venus lower atmosphere from mass-spectrometer data. *Cosmic Res. (Engl. Transl.)* **17**, 581–584 (1979).

V. G. Istomin, K. V. Grechnev, and V. A. Kochnev, Preliminary results of mass-spectrometric measurements on board the, Venera 13 and Venera 14 probes. *Pis'ma Astron. Zh.* **8**, 391–398 (1982).

M. N. Izakov, Influence of turbulence on the thermal conditions of planetary thermospheres. *Cosmic Res. (Engl. Transl.)* **16**, 324–331 (1978).

M. N. Izakov, Inert gases in the atmosphere of Venus, Earth and Mars and the origin of planetary atmospheres. *Cosmic Res. (Engl. Transl.)* **17**, 493–500 (1979).

M. N. Izakov and S. K. Morozov, Structure and dynamics of the thermosphere of Venus. *Cosmic Res. (Engl. Transl.)* **13**, 359–367 (1975).

W. Jaffe, J. Caldwell, and T. Owen, The brightness temperature of Titan at 6 centimeters from the very large array. *Astrophys. J.* **232**, L75–L76 (1979).

M. A. Janssen, R. E. Hills, D. D. Thornton, and W. J. Welch, Venus: New microwave measurements show no atmospheric water vapor. *Science* **179**, 994–996 (1973).

J. Jeans, "Astronomy and Cosmogony." Cambridge Univ. Press, London and New York, 1928.

H. Jeffreys, The density distributions in the inner planets. *Mon. Not. R. Astron. Soc., Geophys. Suppl.* **4**, 62–91 (1937).

E. B. Jenkins, D. C. Morton, and A. V. Sweigart, Rocket spectra of Venus and Jupiter from 2000 to 3000 Å. *Astrophys. J.* **157**, 913–924 (1969).

W. J. Jenkins, M. A. Beg, W. B. Clarke, P. J. Waugersky, and H. Craig, Excess ^{3}He in the Atlantic Ocean. *Earth Planet. Sci. Lett.* **16**, 122–126 (1972).

W. Jenkins, J. Edmond, and J. Corliss, Excess ^{3}He and ^{4}He in Galapagos submarine hydrothermal waters. *Nature (London)* **272**, 156–158 (1978).

F. S. Johnson, The atmosphere of Venus: A conference review. *J. Atmos. Sci.* **25**, 658–660 (1968).

F. S. Johnson, Origin of planetary atmospheres. *Space Sci. Rev.* **9**, 303–324 (1969).

H. E. Johnson and W. I. Axford, Production and loss of ^{3}He in the Earth's atmosphere. *JGR, J. Geophys. Res.* **74**, 2433–2437 (1969).

H. E. Johnson, Reduction of stratospheric ozone by nitrogen oxide catalysts from supersonic transport exhaust. *Science* **173**, 517–522 (1971).

T. V. Johnson, Albedo and spectral reflectivities of the Galilean satellites of Jupiter. Ph.D. Dissertation, California Institute of Technology, Pasadena (1969).

T. V. Johnson, Galilean satellites: Narrowband photometry 0.30 to 1.10 microns. *Icarus* **14**, 94–111 (1971).

T. V. Johnson, The Galilean satellites of Jupiter: Four Wolds. *Annu. Rev. Earth Planet. Sci.* **6**, 93–125 (1978).

T. V. Johnson and F. P. Fanale, Optical properties of carbonaceous chondrites and their relationship to asteroids. *JGR, J. Geophys. Res.* **78**, 8507–8518 (1973).

T. V. Johnson and T. B. McCord, Galilean satellites. The spectral reflectivity 0.30–1.10 micron. *Icarus* **13**, 37–42 (1970).

T. V. Johnson and T. B. McCord, Spectral geometric albedo of the Galilean satellites 0.3 to 2.5 microns. *Astrophys. J.* **169**, 589–593 (1971).

T. V. Johnson, D. L. Matson, and R. W. Carlson, Io's atmosphere and ionosphere: New limits on surface pressure from plasma models. *Geophys. Res. Lett.* **3**, 293–296 (1976).

T. V. Johnson, A. F. Cook, II, C. Sagan, and L. A. Soderblom, Volcanic resurfacing rates and implications for volatiles on Io. *Nature (London)* **280**, 746–749 (1979).

D. H. Johnston and M. N. Toksöz, Internal structure and properties of Mars. *Icarus* **32**, 73–84 (1977).

D. H. Johnston, T. R. McGetchin, and M. N. Toksöz, The thermal state and internal structure of Mars. *JGR, J. Geophys. Res.* **79**, 3959–3970 (1974).

H. S. Johnston, Global ozone balance in the natural stratosphere. *Rev. Geophys. Space Phys.* **13**, 637–649 (1975).

A. Johnstone, Does Mars have a megnetosphere? *Nature (London)* **272**, 394–395 (1978).

D. E. Jones, The microwave temperature of Venus. *Planet. Space Sci.* **5**, 166–172 (1961).

D. E. Jones, D. M. Wrathall, and B. L. Meredith, Spectral observations of Venus in the frequency interval 18.5–24.0 GHz; 1964 and 1967–8. *Publ. Astron. Soc. Pac.* **84**, 435–442 (1972).

K. L. Jones, Evidence for an episode of crater obliteration intermediate in Martian history. *JGR, J. Geophys. Res.* **79**, 3917–3931 (1974).

J. F. Jordan and J. Lorell, Mariner 9: An instrument of dynamical science. *Icarus* **25**, 146–165 (1975).

C. E. Junge, Sulfur in the atmosphere. *JGR, J. Geophys. Res.* **65**, 227–237 (1960).

C. E. Junge, "Air Chemistry and Radioactivity." Academic Press, New York, 1963.

C. E. Junge and T. G. Ryan, Study of the SO_2 oxidation in solution and its role in atmospheric chemistry. *Quart. J. Roy Meteorol. Soc.* **84**, 64–55 (1958).

R. K. Kakar, J. W. Waters, and W. J. Wilson, Venus: Microwave detection of carbon monoxide. *Science* 191, 379–380 (1976).

R. K. Kakar, J. W. Waters, and W. J. Wilson, Mars: Microwave detection of carbon monoxide. *Science* 196, 1090–1091 (1977).

J. Kane, J. Kasold, M. Suda, P. Metcalf, and S. Caccamo, Alpine glacial features of Mars. *Nature (London)* 244, 20–21 (1973).

I. Kaneoka and N. Takaoka, Excess [129]Xe and high [3]He/[4]He ratios in olivine phenocrysts of Kapuho lava and xenolithic dunites from Hawaii. *Earth Planet. Sci. Lett.* 39, 382–386 (1978).

I. R. Kaplan, Hydrogen (1). *In* "Handbook of Elemental Abundances in Meteorites" (B. Mason, ed.), pp. 21–27. Gordon & Breach, New York, 1971.

I. R. Kaplan, Stable isotopes as a guide to biochemical processes. *Proc. R. Soc. London. Ser. B* 189, 183–211 (1975).

L. D. Kaplan, A new interpretation of the structure and CO_2 content of the Venus atmosphere. *Planet. Space Sci.* 8, 23–29 (1961).

L. D. Kaplan, Spectroscopic investigations of Venus. *J. Quant. Spectrosc. Radiat. Transfer* 3, 537–542 (1963).

L. D. Kaplan, G. Münch, and H. Spinrad, An analysis of the spectrum of Mars. *Astrophys. J.* 139, 1–15 (1964).

L. D. Kaplan, J. Connes, and P. Connes, Carbon monoxide in the Martian atmosphere. *Astrophys. J.* 157, L187–L192 (1969).

G. W. Kattawar, G. N. Plass, and C. N. Adams, Flux and polarization calculations of the radiation reflected from the clouds of Venus. *Astrophys. J.* 170, 371–386 (1971).

G. N. Katterfeld and P. Hédervári, Ring-shaped and linear structures on Mars. *Sov. Astron. (Engl. Transl.)* 12, 863–870 (1968).

W. M. Kaula, Dynamical aspects of lunar origin. *Rev. Geophys.* 9, 217–238 (1971).

W. M. Kaula, Comments on the origin of Mercury. *Icarus* 28, 429–433 (1976).

W. M. Kaula, The moment of inertia of Mars. *Geophys. Res. Lett.* 6, 194–196 (1979).

W. M. Kaula and P. E. Bigeleisen, Early scattering by Jupiter and its collision effects in the terrestrial zone. *Icarus* 25, 18–33 (1975).

G. M. Keating, R. H. Tolson, and E. W. Hinson, Venus thermosphere and exosphere: First satellite drag measurements on an extraterrestrial atmosphere. *Science* 203, 772–774 (1979a).

G. M. Keating, F. W. Taylor, J. Y. Nicholson, and E. W. Hinson, Short-term cyclic variations and diurnal variatons of the Venus upper atmosphere. *Science* 205, 62–64 (1979b).

C. S. L. Keay, F. J. Low, G. H. Rieke, and R. B. Minton, High-resolution maps of Jupiter at 5 microns. *Astrophys. J.* 183, 1063–1073 (1973).

K. Keil and K. G. Snetsinger, Niningerite: A new meteoritic sulfide. *Science* 155, 451–453 (1967).

M. V. Keldysh, Venus exploration with Venera 9 and Venera 10 spacecraft. *Icarus* 30, 605–625 (1977).

J. F. Kerridge, J. D. MacDougall, and K. Marti, Clues to the origin of sulfide minerals in CI chondrites. *Earth Planet. Sci. Lett.* 43, 359–367 (1979).

V. V. Kerzhanovich, Wind velocity and turbulence in the Venusian atmosphere, as obtained from the data of doppler measurements of the velocity of the automatic interplanetary stations Venera-4, Venera-5, and Venera-6. *Cosmic Res. (Engl. Transl.)* 10, 232–242 (1972).

V. V. Kerzhanovich, Mars 6: Improved analysis of the descent module measurements. *Icarus* 30, 1–25 (1977).

V. V. Kerzhanovich and M. Ya. Marov, On the wind-velocity measurements from Venera spacecraft data. *Icarus* 30, 320–325 (1977).

V. V. Kerzhanovich, M. Ya. Marov and M. K. Rozhdestvensky, Data on dynamics of the subcloud Venus atmosphere from Venera spaceprobe measurements. *Icarus* 17, 659–674 (1972).

V. V. Kerzhanovich, F. I. Kozlov, A. S. Selivanov, Yu. S. Tyuflin, and V. I. Khizhnichenko, Structure of Venusian cloud layer according to Venera 9 televised pictures. *Cosmic Res. (Engl. Transl.)* 17, 57–66 (1979).

B. N. Khare and C. Sagan, Red clouds in reducing atmospheres. *Icarus* 20, 311–321 (1973).

B. N. Khare and C. Sagan, Cyclic octatomic sulfur: A possible infrared and visible chromophore in the clouds of Jupiter. *Science* 189, 722–723 (1975).

B. N. Khare and C. Sagan, On the temperature dependence of possible S_8 infrared bands in planetary atmospheres. *Icarus* **30**, 231–233 (1977).

B. N. Khare, C. Sagan, E. Bandurski, and B. Nagy, Ultraviolet- photoproduced organic solids synthesized under simulated Jovian conditions: Molecular analysis. *Science* **199**, 1199–1201 (1978).

I. L. Khodakovskii, V. P. Volkov, Yu. I. Sidorov, M. V. Borisov, and M. V. Lomonosov, Venus: Preliminary prediction of the mineral composition of surface rocks. *Icarus* **39**, 352–363 (1979a).

I. L. Khodakovskii, V. P. Volkov, Yu. I. Sidorov, V. A. Dorofeeva, M. V. Borisov, and V. L. Barsukov, Geochemical model of the crust and troposphere of the planet Venus from new data. *Geokhimiya* pp. 1747–1758 (1979b).

H. Kieffer, Spectal reflectance of CO_2–H_2O frosts. *JGR, J. Geophys. Res.* **75**, 501–509 (1970a).

H. Kieffer, Interpretation of the Martian polar cap spectra. *JGR, J. Geophys. Res.* **75**, 510–514 (1970b).

H. H. Kieffer, Soil and surface temperatures at the Viking landing sites. *Science* **194**, 1344–1346 (1976).

H. H. Kieffer, Mars south polar spring and summer temperatures: A residual CO_2 frost. *JGR, J. Geophys. Res.* **84**, 8263–8288 (1979).

H. H. Kieffer and W. D. Smythe, Frost spectra: Comparison with Jupiter's satellites. *Icarus* **21**, 506–512 (1974).

H. H. Kieffer, S. C. Chase, Jr., E. Miner, G. Münch, and G. Neugebauer, Preliminary report on infrared radiometric measurements from the Mariner 9 spacecraft. *JGR, J. Geophys. Res.* **78**, 4291–4312 (1973).

H. H. Kieffer, S. C. Chase, Jr., E. D. Miner, F. D. Pallucconi, G. Münch, G. Neugebauer, and T. Z. Martin, Infrared thermal mapping of the Martian surface and atmosphere: First results. *Science* **193**, 780–785 (1976a).

H. H. Kieffer, S. C. Chase, Jr., T. Z. Martin, E. D. Miner, and F. D. Pallucconi, Martian North Pole summer temperatures: Dirty water ice. *Science* **194**, 1341–1343 (1976b).

H. H. Kieffer, P. R. Christensen, T. Z. Martin, E. D. Miner, and F. D. Pallucconi, Temperatures of the Martian surface and atmosphere: Viking observation of diurnal and geometric variations. *Science* **194**, 1346–1351 (1976c).

H. H. Kieffer, T. Z. Martin, A. R. Peterfreund, B. M. Jakosky, E. D. Miner, and F. D. Pallucconi, Thermal and Albeda mapping of Mars during the Viking primary mission. *JGR, J. Geophys. Res.* **82**, 4249–4292 (1977).

C. C. Kiess, K. Karber, and H. K. Kiess, Oxides of nitrogen in the Martian atmosphere. *Publ. Astron. Soc. Pac.* **75**, 50–60 (1963).

D. A. King and T. J. Ahrens, Shock compression of iron sulfide and the possible sulphur content of the Earth's core. *Nature (London), Phys. Sci.* **243**, 82–84 (1973).

J. S. King and J. R. Riehle, A proposed origin of the Olympus Mons escarpment. *Icarus* **23**, 300–317 (1974).

H. P. Klein, The Viking biological investigation: General aspects. *JGR, J. Geophys. Res.* **82**, 4677–4680 (1977).

H. P. Klein, J. Lederberg, and A. Rich, Biological experiments: The Viking Mars lander. *Icarus* **16**, 139–146 (1972).

H. P. Klein, N. H. Horowitz, G. V. Levin, V. I. Oyama, J. Lederber, A. Rich, J. S. Hubbard, G. L. Hobby, P. A. Straat, B. J. Berdahl, G. C. Carle, F. S. Brown and R. D. Johnson, The Viking biological investigations: Preliminary results. *Science* **194**, 99–105 (1976).

A. J. Kliore and D. L. Cain, Mariner 5 and the radius of Venus. *J. Atmos. Sci.* **25**, 549–554 (1968).

A. J. Kliore, G. L. Levy, D. L. Cain, G. Fjeldbo, and S. I. Rasool, Atmosphere and ionosphere of Venus from Mariner V S-band radio occultation measurements. *Science* **158**, 1683–1688 (1967).

A. Kliore, D. L. Cain, G. Fjeldbo, G. Levy, and S. I. Rasool, Atmosphere of Venus as observed by the Mariner V S-band radio occultation experiment. *Astron. J.* **73**, 521–526 (1968).

A. Kliore, G. Fjeldbo, B. L. Seidel, and S. I. Rasool, Mariners 6 and 7: Radio occultation measurements of the atmosphere of Mars. *Science* **166**, 1393–1397 (1969).

A. Kliore, G. Fjeldbo, and B. L. Seidel, First results of the Mariner-6 radio occultation measurement of the lower atmosphere of Mars. *Radio Sci.* **5**, 373–380 (1970).

A. J. Kliore, D. L. Cain, G. Fjeldbo, B. L. Seidel, and S. I. Rasool, Mariner 9 S-Band Martian occultation experiment: Initial results on the atmosphere and topography of Mars. *Science* 175, 313–317 (1972).

A. J. Kliore, G. Fjeldbo, B. L. Seidel, M. J. Sykes and P. M. Woiceshyn, S band radio occultation measurements of the atmosphere and topography of Mars with Mariner 9: Extended mission coverage of polar and intermediate latitudes. *JGR, J. Geophys. Res.* 78, 4331–4350 (1973).

A. J. Kliore, G. Fjeldbo, B. L. Seidel, D. N. Sweetnam, T. T. Sesplaukis, P. M. Woiceshyn, and S. I. Rasool, The atmosphere of Io from Pioneer 10 radio occultation measurements. *Icarus* 24, 407–410 (1975).

R. G. Knollenberg and D. M. Hunten, Clouds of Venus: Particle size distribution measurements. *Science* 203, 792–795 (1979a).

R. G. Knollenberg and D. M. Hunten, Clouds of Venus: A preliminary assessment of microstructure. *Science* 205, 70–74 (1979b).

R. Knollenberg and D. Hunten, The microphysics of the clouds of Venus: Results of the Pioneer Venus particle size spectrometer experiment. *JGR, J. Geophys. Res.* 85, 8039–8058 (1980).

L. Knopoff and M. S. T. Bukowinski, Physics and chemistry of iron and potassium at lower mantle and core pressures. *In* "High Pressure Research" (M. H. Manghmani and S. Akimoto, eds.), pp. 367–388. Academic Press, New York, 1977.

W. C. Knudsen and A. D. Anderson, Estimate of radiogenic ^{4}He and ^{40}Ar concentration in the Cytherean atmosphere. *JGR, J. Geophys. Res.* 74, 5629–5632 (1969).

W. C. Knudsen, K. Spenner, R. C. Whitten, J. R. Spreiter, K. L. Miller, and V. Novak, Thermal structures and energy influx to the day- and nightside Venus ionosphere. *Science* 205, 105–107 (1979).

M. A. Kolosov, O. I. Yakovlev, G. D. Yakoleva, A. I. Efimof, B. P. Trusov, T. S. Timofeeva, Yu. M. Kruglov, and V. P. Oreshkin, Results of investigations of the atmosphere of Mars by the method of radio transillumination by means of the automatic interplanetary stations "Mars 2," "Mars 4," and "Mars 6." *Cosmic Res. (Engl. Transl.)* 13, 46–50 (1975).

R. L. Kovach and D. L. Anderson, Interiors of the terrestrial planets. *JGR, J. Geophys. Res.* 70, 2873–2883 (1965).

I. K. Koval', Atmosphere and surface of Mars. *Sol. Syst. Res. (Engl. Transl.)* 5, 107–118 (1971).

I. K. Koval' and E. G. Yanovitskii, Optical parameters of the Martian surface and atmosphere. *Sov. Astron. (Engl. Transl.)* 13, 499–505 (1969).

Y. Kozai, Secular perturbations of asteroids with high inclination and eccentricity. *Astron. J.* 67, 591–598 (1962).

S. V. Kozlovskaya, On the internal constitution and chemical composition of Mercury. *Astrophys. Lett.* 4, 1–3 (1969).

N. A. Kozyrev, The atmosphere of Mercury. *Sky Telescope* 27, 339–341 (1964).

V. A. Krasnopol'skii, Lightning on Venus according to information obtained by the satellites Venera 9 and 10. *Cosmic Res. (Engl. Transl.)* 18, 325–330 (1980).

V. Krasnopol'skii and V. Parshev, "Initial Data for Calculation of Venus' Atmospheric Photochemistry at Heights Down to 50 km," Publ. 590. Space Res. Inst., USSR Acad. Sci., 1980a.

V. A. Krasnopolskii and V. Parshev, "Photochemistry of the Venus Atmosphere Down to 50 km: Results of Calculations," Publ. 591. Space Res. Inst. USSR Acad. Sci., 1980b.

K. B. Krauskopf, "Introduction to Geochemistry," 2nd ed. McGraw-Hill, New York, 1979.

L. V. Ksanfomaliti, Daytime minima of the infrared brightness temperature of Venus. *Cosmic Res. (Engl. Transl.)* 17, 349–357 (1979).

L. V. Ksanfomaliti, Discovery of frequent lightning discharges in clouds on Venus. *Nature (London)* 284, 244–246 (1980).

L. V. Ksanfomaliti and V. I. Moroz, Infrared radiometry on board Mars-5. *Cosmic Res. (Engl. Transl.)* 13, 65–67 (1975).

L. V. Ksanfomaliti, B. S. Kunashev, and V. I. Moroz, Preliminary results of pressure and surface heights from Mars 5 measurements of the CO_2 band intensity. *Cosmic Res. (Engl. Transl.)* 13, 72–74 (1975a).

L. V. Ksanfomaliti, V. I. Moroz, B. S. Kunashev, and V. S. Zhegulev, Pressures and altitudes according to CO_2 altimetry with the Mars-3 automatic interplanetary station. *Cosmic Res. (Engl. Transl.)* **13**, 501–519 (1975b).

L. V. Ksanfomaliti, E. V. Dedova, L. F. Obukhova, N. V. Temnaya, and G. F. Filippov, Infrared radiation of the clouds of Venus. *Cosmic Res. (Engl. Transl.)* **14**, 670–677 (1976a).

L. V. Ksanfomaliti, E. V. Dedova, L. F. Obukhova, V. M. Pokras, N. V. Temnaya, and G. F. Filippova, Thermal asymmetry of Venus. *Sov. Astron.* **20**, 476–480 (1976b).

L. V. Ksanfomaliti, N. M. Vasilchikov, O. F. Ganpantzerova, E. V. Petrova, A. P. Suvarov, G. F. Filippova, O. V. Yablouskaya, and L. V. Yabrova, Electrical discharges in the Venus atmosphere. *Pisma Astron. Zh.* **5**, 229–236 (1979).

K. V. Kuhi, Mass-loss from T-Tauri stars. *Astrophys. J.* **140**, 1409–1433 (1964).

W. Kuhn and A. Rittmann, Über den Zustand des Erdinnern und seine Enstehung aus einem homgenen Urzustand. *Geol. Rundsch.* **32**, 215–256 (1941).

G. P. Kuiper, Titan: A satellite with an atmosphere. *Astrophys. J.* **100**, 378–383 (1944).

G. P. Kuiper, The diameter of Pluto. *Publ. Astron. Soc. Pac.* **62**, 133–137 (1950).

G. P. Kuiper, ed., "The Atmospheres of the Earth and Planets." Univ. of Chicago Press, Chicago, Illinois, 1952.

G. P. Kuiper, Identification of the Venus cloud layers. *Comm. Lunar Planet. Lab.* **6**, 229–250 (1969).

G. P. Kuiper and G. T. Sill, Identification of the Venus cloud layers. *Bull. Am. Astron. Soc.* **1**, 351 (1969).

S. Kumar, Mercury's atmosphere: A perspective after Mariner 10. *Icarus* **28**, 579–591 (1976).

S. Kumar, The stability of an SO_2 atmosphere on Io. *Nature (London)* **280**, 758–760.

S. Kumar and A. L. Broadfoot, He 584 Å Airglow emission from Venus: Mariner 10 observations. *Geophys. Res. Lett.* **2**, 357–360 (1975).

S. Kumar and D. M. Hunten, Venus: An ionospheric model with an exospheric temperature of 350°K. *JGR, J. Geophys. Res.* **79**, 2529–2532, 1974.

S. Kumar and D. M. Hunten, The atmospheres of Io and other satellites. *In* "The Satellites of Jupiter" (T. Gehrels, ed.), pp. 782–806. Univ. of Arizona Press, Tuscon, 1982.

S. Kumar, D. Hunten, and A. Broadfoot, Non-thermal H in the Venus exosphere: The ionispheric source and the H budget. *Planet. Space Sci.* **26**, 1063–1075 (1978).

S. Kumar, D. Hunten, and H. Taylor, H_2 Abundance in the atmosphere of Venus. *Geophys. Res. Lett.* **8**, 237–240 (1981).

C. C. Kung and R. N. Clayton, Nitrogen abundance and isotopic compositions in stony meteorites. *Earth Planet. Sci. Lett.* **38**, 421–435 (1978).

I. Kupo, Y. Mekler, and A. Eviatar, Detection of ionized sulfur in the Jovian magnetosphere. *Astrophys. J.* **205**, L51–L53 (1976).

P. K. Kuroda, Temperature of the sun in the early history of the solar system. *Nature (London), Phys. Sci.* **230**, 40–42 (1971).

P. K. Kuroda and W. A. Crouch, Jr., On the chronology of the formation of the solar system. 2. Iodine in terrestrial rocks and the xenon 129/136 formation interval of the Earth. *JGR, J. Geophys. Res.* **67**, 4863–4866 (1962).

P. K. Kuroda and O. K. Manuel, On the chronology of the formation of the solar system. 1. Radiogenic xenon 129 in the Earth's atmosphere. *JGR, J. Geophys. Res.* **67**, 4859–4862 (1962).

V. G. Kurt and V. S. Zhegulev, Preliminary results of research performed on the Venera 11 and Venera 12 spacecraft. *Cosmic Res. (Engl. Transl.)* **17**, 529–531 (1979).

T. Kusaka, T. Nakano, and C. Hayashi, Growth of solid particles in the primordial solar nebula. *Prog. Theor. Phys.* **44**, 1580–1592 (1970).

A. D. Kuz'min, The atmosphere of the planet Venus. *Radio Sci.* **5**, 339–346 (1978).

A. D. Kuz'min, B. Ya. Losovsky, and N. Vetukhnovskaya, Measurements of Mars radio emissions at 8.22 mm and evaluation of thermal and electrical properties of its surface. *Icarus* **14**, 192–195 (1971).

A. D. Kuz'min, A. P. Naumov, and T. V. Smirnova, Estimate of ammonia concentration in subcloud atmosphere of Saturn from radio-astronomy measurements. *Sol. Syst. Res. (Engl. Transl.)* **6**, 10–13 (1972).

F. T. Kyte, Z. Zhou, and J. T. Wasson, Siderophile-enriched sediments from the Cretaceous-Tertiary boundary. *Nature (London)* **288**, 651–656 (1980).

A. A. Lacis, Cloud structure and heating rates in the atmosphere of Venus. *J. Atmos. Sci.* **32**, 1107–1124 (1975).

A. A. Lacis and J. E. Hansen, Atmosphere of Venus: Implications of Venera 8 sunlight measurements. *Science* **184**, 979–982 (1974).

L. Lahav and D. H. White, A possible role of fluctuating clay-water systems in the production of ordered prebiotic oligomers. *J. Mol. Evol.* **16**, 11–22 (1980).

N. Lahav, D. White, and S. Chang, Peptide formation in the prebiotic era: Thermal condensation of glycine in fluctuating clay environments. *Science* **201**, 67–69 (1978).

D. L. Lamar, Optical ellipticity and internal structure of Mars. *Icarus* **1**, 258–265 (1962).

M. S. Lancet and E. Anders, Solubilities of noble gases in magnetite: Implications for planetary gases in meteorites. *Geochim. Cosmochim. Acta* **37**, 1371–1388 (1973).

A. L. Lane, C. A. Barth, C. W. Hord, and A. I. Stewart, Mariner 9 ultraviolet spectrometer experiment: Observations of ozone on Mars. *Icarus* **18**, 102–108 (1973).

J. W. Larimer, Chemical fractionation in meteorites. I. Condensation of the elements. *Geochim. Cosmochim. Acta* **31**, 1215–1238 (1967).

J. W. Larimer, Composition of the Earth: Chondritic or achondritic? *Geochim. Cosmochim. Acta* **35**, 769–786 (1971).

J. W. Larimer and E. Anders, Chemical fractionation in meteorites. II. Abundance patterns and their interpretation. *Geochim. Cosmochim. Acta* **31**, 1239–1270 (1967).

J. W. Larimer and E. Anders, Chemical fractionation in meteorites. III. Major element fractionation in chondrites. *Geochim. Cosmochim. Acta* **34**, 367–387 (1970).

H. P. Larson and U. Fink, Identification of carbon dioxide frost on the Martian polar caps. *Astrophys. J.* **171**, L91–L95 (1972).

H. P. Larson and U. Fink, Infrared and spectral observations of asteroid 4 Vesta. *Icarus* **26**, 420–427 (1975).

H. P. Larson, R. R. Treffers, and U. Fink, Phosphine in Jupiter's atmosphere: The evidence from high-altitude observations at 5 micrometers. *Astrophys. J.* **211**, 977–979 (1977).

H. P. Larson, U. Fink, H. A. Smith, and D. S. Davis, The middle infrared spectrum of Saturn: Evidence for phosphine and upper limits to other trace atmospheric constituents. *Astrophys J.* **240**, 327–337 (1980).

R. B. Larson, The evolution of protostars—theory. *Fundam. Cosmic Phys.* **1**, 1–70 (1974).

A. C. Lasaga, H. D. Holland, and M. J. Dwyer, Primordial oil slicks. *Science* **174**, 53–55 (1971).

W. M. Latimer, Astrochemical problems in the formation of the Earth. *Science* **112**, 101–104 (1950).

J. G. Lawless and N. Levi, The role of metal ions in chemical evolution: Polymerization of alanine and glycine. *J. Mol. Evol.* **13**, 281–286 (1979).

L. A. Lebofsky, Stability of frosts in the solar system. *Icarus* **25**, 205–217 (1975).

L. A. Lebofsky, Identification of water frost on Callisto. *Nature (London)* **269**, 785–787 (1977).

L. A. Lebofsky, Asteroid 1 Ceres: Evidence for water of hydration. *Mon. Not. R. Astron. Soc.* **182**, 17b–21b (1978).

L. A. Lebofsky and M. B. Fegley, Jr., Laboratory reflection spectra for the determination of chemical composition of icy bodies. *Icarus* **28**, 379–387 (1976).

L. A. Lebofsky, G. H. Rieke, and M. J. Lebofsky, Surface composition of Pluto. *Icarus* **37**, 554–558 (1979).

T. Lee, New Isotopic clues to solar system formation. *Rev. Geophys. Space Phys.* **17**, 1591–1611 (1979).

T. Lee and D. Papanastassiou, Mg isotopic anomalies in the Allende meteorite and correlation with O and Sr effects. *Geophys. Res. Lett.* **1**, 225–228 (1974).

T. Lee, D. A. Papanastassiou, and G. J. Wasserburg, Correction: Demonstration of ^{26}Mg excess in Allende and evidence for ^{26}Al. *Geophys. Res. Lett.* **3**, 109–112 (1976).

W. H. K. Lee, Effects of selective fusion on the thermal history of the Earth's mantle. *Earth Planet. Sci. Lett.* **4**, 270–276 (1968).

R. B. Leighton and B. C. Murray, Behavior of carbon dioxide and other volatiles on Mars. *Science* **153**, 136–144 (1966).

R. B. Leighton and B. C. Murray, One year's processing and interpretation—an overview. *JGR, J. Geophys. Res.* **76**, 293–296 (1971).

R. B. Leighton, N. H. Horowitz, B. C. Murray, R. P. Sharp, A. G. Herriman, A. T. Young, B. A. Smith, M. E. Davies, and C. B. Leovy, Mariner VI television pictures: First report. *Science* **165**, 684–690 (1969a).

R. B. Leighton, N. H. Horowitz, B. C. Murray, R. P. Sharp, A. G. Herriman, A. T. Young, B. A. Smith, M. E. Davies, and C. B. Leovy, Mariner VII television pictures: First report. *Science* **165**, 788–795 (1969b).

R. B. Leighton, N. H. Horowitz, B. C. Murray, R. P. Sharp, A. G. Herriman, A. T. Young, B. A. Smith, M. E. Davies, and C. B. Leovy, Mariner 6 and 7 television pictures: Preliminary analysis. *Science* **166**, 49–67 (1969c).

J. LeMaire and M. Scherer, Kinetic models of the solar and polar winds. *Rev. Geophys.* **11**, 427–468 (1973).

C. B. Leovy, Exchange of water vapor between the atmosphere and the surface of Mars. *Icarus* **18**, 120–125 (1973).

C. B. Leovy and J. B. Pollack, A first look at atmospheric dynamics and temperature variations on Titan. *Icarus* **19**, 195–201 (1973).

G. V. Levin, Detection of metabolically produced labeled gas: The Viking Mars lander. *Icarus* **16**, 153–166 (1972).

G. V. Levin and P. A. Straat, Labeled release–an experiment in radiorespirometry. *Origins Life* **7**, 293–311 (1976a).

G. V. Levin and P. A. Straat, Viking labeled release biology experiment: Interim results. *Science* **194**, 1322–1328 (1976b).

G. V. Levin and P. A. Straat, Recent results from the Viking labeled release experiments on Mars. *JGR, J. Geophys. Res.* **82**, 4663–4668 (1977).

J. S. Levine, Argon in the Martian atmosphere. *Geophys. Res. Lett.* **1**, 285–287 (1974).

E. C. Levinthal, W. Green, K. L. Jones, and R. Tucker, Processing the Viking lander camera data. *JGR, J. Geophys. Res.* **82**, 4412–4420 (1977a).

E. C. Levinthal, K. L. Jones, P. Fox, and C. Sagan, Lander imaging as a detector of life on Mars. *JGR, J. Geophys. Res.* **82**, 4468–4478 (1977b).

H. Levy, II, Normal atmosphere: Large radical and formaldehyde concentrations predicted. *Science* **173**, 141–143 (1971).

J. S. Lewis, A possible origin for sulfates and sulfur in meteorites. *Earth Planet. Sci. Lett.* **2**, 29–34 (1967).

J. S. Lewis, An estimate of the surface conditions of Venus. *Icarus* **8**, 434–456 (1968a).

J. S. Lewis, Composition and structure of the clouds of Venus. *Astrophys. J.* **152**, L79–L81 (1968b).

J. S. Lewis, The clouds of Jupiter and the NH_3–H_2O and NH_3–H_2S systems. *Icarus* **10**, 365–378 (1969a).

J. S. Lewis, Observability of spectroscopically active compounds in the atmosphere of Jupiter. *Icarus* **10**, 393–407 (1969b).

J. S. Lewis, Geochemistry of the volatile elements on Venus. *Icarus* **11**, 367–385 (1969c).

J. S. Lewis, Geochemistry of Venus and the interpretation of the radar data. *Radio Sci.* **5**, 363–368 (1970a).

J. S. Lewis, Ice clouds on Venus? *J. Atmos. Sci.* **27**, 333–335 (1970b).

J. S. Lewis, Venus: Atmospheric and lithospheric composition. *Earth Planet. Sci. Lett.* **10**, 73–80 (1970c).

J. S. Lewis, Consequences of the presence of sulfur in the core of the Earth. *Earth Planet. Sci. Lett.* **11**, 130–135 (1971a).

J. S. Lewis, Satellites of the outer planets: Thermal models. *Science* **172**, 1127–1128 (1971b).

J. S. Lewis, Refractive index of aqueous HCl solutions and the composition of the Venus clouds. *Nature (London)* **230**, 295–296 (1971c).

J. S. Lewis, The atmosphere, clouds and surface of Venus. *Am. Sci.* **59**, 557–566 (1971d).

J. S. Lewis, Satellites of the outer planets: Their physical and chemical nature. *Icarus* **15**, 174–185 (1971e).

J. S. Lewis, Low temperature condensation from the solar nebula. *Icarus* **16**, 241–252 (1972a).

J. S. Lewis, Metal/silicate fractionation in the solar system. *Earth Planet, Sci. Lett.* **15**, 286–290 (1972b).

J. S. Lewis, Composition of the Venus cloud tops in light of recent spectroscopic observation. *Astrophys. J.* **171**, L75–L78 (1972c).

J. S. Lewis, Chemistry of the outer solar system. *Space Sci. Rev.* **14**, 401–410 (1973a).

J. S. Lewis, Origin and composition of the terrestrial planets and satellites of the outer planets. *In* "The Origin of the Solar System," H. Reeves, ed. pp. 202–205. CNRS, Paris, 1973b.

J. S. Lewis, The chemistry of the solar system. *Sci. Amer.* **230**, No. 3, 50–65 (1974a).

J. S. Lewis, Volatile element influx on Venus from cometary impacts. *Earth Planet. Sci. Lett.* **22**, 239–244 (1974b).

J. S. Lewis, The temperature gradient in the solar nebula. *Science* **186**, 440–442 (1974c).

J. S. Lewis, Equilibrium and disequilibrium chemistry of adiabatic solar composition planetary atmospheres. *In* "Chemical Evolution of the Giant Planets" (C. Ponnamperuma, ed.), pp. 13–25. Academic Press, New York, 1976.

J. S. Lewis, Electrical discharge synthesis of organic compounds on Jupiter. *Icarus* **43**, 85–100 (1980a).

J. S. Lewis, Lightning on Jupiter: Rate, energetics and effects. *Science* **210**, 1351–1352 (1980b).

J. S. Lewis, Io: Geochemistry of sulfur. *Icarus* **50**, 103–114 (1982).

J. S. Lewis and B. Fegley, Jr., Hot-atom synthesis of organic compounds on Jupiter. *Astrophys. J.* **232**, L135–L138 (1979).

J. S. Lewis and B. Fegley, Jr., Venus: Halide cloud condensation and volatile element inventories. *Science* **216**, 1223–1225 (1982).

J. S. Lewis and F. A. Kreimendahl, Oxidation state of the atmosphere and crust of Venus from pioneer Venus results. *Icarus* **42**, 330–337 (1980).

J. S. Lewis and H. R. Krouse, Isotopic composition of sulfur and sulfate produced by oxidation of FeS. *Earth Planet. Sci. Lett.* **5**, 425–434 (1969).

J. S. Lewis and E. P. Ney, Iron and the formation of astrophysical dust grains. *Astrophys. J.* **234**, 154–157 (1979).

J. S. Lewis and R. G. Prinn, Jupiter's clouds: Structure and composition. *Science* **169**, 472–473 (1970).

J. S. Lewis and R. G. Prinn. Chemistry and photochemistry of the atmosphere of Jupiter. *In* "Theory and Experiment in Exobiology" (A. W. Schwartz, ed.), pp. 123–142. Wolters-Nordhoff, Gröningen, 1971.

J. S. Lewis and R. G. Prinn, Titan revisited. *Comments Astrophys. Space Phys.* **5**, 1–7 (1973).

J. S. Lewis and R. G. Prinn, Kinetic inhibition of CO and N_2 reduction in the solar nebula. *Astrophys. J.* **238**, 357–364 (1980).

J. S. Lewis, S. S. Barshay, and B. Noyes, Primordial retention of carbon by the terrestrial planets. *Icarus* **37**, 190–206. (1979).

J. S. Lewis, G. H. Watkins, H. Hartman, and R. G. Prinn, Chemical consequences of major impact events on earth. *In* "Geological Implications of Impacts of Large Asteroids and Comets on the Earth" L. T. Silver and P. H. Schultz, eds.) *Geol. Soc. Am. Special Paper* **190**, 215–222 (1982).

W. F. Libby, Ice caps on Venus? *Science* **159**, 1097 (1968a).

W. F. Libby, Venus: Ice sheets. *Science* **160**, 1474 (1968b).

S. Liebes, Jr. and A. A. Schwartz, Viking 1975 Mars lander interactive computerized video sterophotogrammetry. *JGR, J. Geophys. Res.* **82**, 4421–4429 (1977).

S. C. Lin, Cometary impact and the origin of tektites. *JGR, J. Geophys. Res.* **71**, 2427–2437 (1966).

R. S. Lindzen, Internal gravity-waves. *Bull. Am. Meteorol. Soc.* **51**, 387 (1970).

E. R. Lippincott, R. V. Eck, M. D. Dayhoff, and C. Sagan, Thermodynamic equilibria in planetary atmospheres. *Astrophys. J.* **147**, 753–763 (1967).

S. C. Liu and T. M. Donahue, Realistic model of hydrogen constituents in the lower atmosphere and escape flux from the upper atmospheres. *J. Atmos. Sci.* **31**, 2238–2242 (1974).

S. C. Liu and T. M. Donahue, The aeronomy of the upper atmosphere of Venus. *Icarus* **24**, 148–156 (1975).

S. C. Liu and T. M. Donahue, The regulation of hydrogen and oxygen escape from Mars. *Icarus* **28**, 231–246 (1976).

S. C. Liu, R. J. Cicerone, T. M. Donahue and W. L. Chameides, Limitation of fertilizer induced ozone reduction by the long lifetime of the reservoir of fixed nitrogen. *Geophys. Res. Lett.* **3**, 157–160 (1976).

H. C. Lord, III, Molecular equilibria and condensation in a solar nebula and cool stellar atmospheres. *Icarus* **4**, 279–288 (1965).

J. Lorell, G. H. Born, E. J. Christensen, J. F. Jordan, P. A. Laing, W. L. Martin, W. L. Sjogren, I. I. Shapiro, R. D. Reasenberg, and G. L. Slater, Mariner 9 celestial mechanics experiment: Gravity field and pole direction of Mars. *Science* **175**, 317–320 (1972).

J. Lorell, G. H. Born, E. J. Christensen, P. B. Esposito, J. F. Jordan, P. A. Laing, W. L. Sjogren, S. K. Wong, R. D. Reasenberg, I. Shapiro, and G. L. Slater, Gravity field of Mars from Mariner 9 tracking data. *Icarus* **18**, 304–316 (1973).

V. M. Loscutov, Interpretation of polarimetric observations of the planets. *Sov. Astron. (Engl. Transl.)* **15**, 129–133 (1971).

J. E. Lovelock and L. Margulis, Atmospheric homeostasis by and for the biosphere: The Gaia hypothesis. *Tellus* **26**, 2–10 (1974).

F. J. Low and G. H. Rieke, Infrared photometry of Titan. *Astrophys. J.* **190**, L143–145 (1974).

B. K. Lucchitta, Landslides in Valles Marineris, Mars. *JGR, J. Geophys. Res.* **84**, 8097–8114 (1979).

B. K. Lucchitta, Grooved terrain on Ganymede. *Icarus* **44**, 481–501 (1980).

N. L. Lukashevich, M. Ya. Marov, and E. M. Feigel'son, Interpretation of illumination measurements in Venus' atmosphere. *Cosmic Res. (Engl. Transl.)* **12**, 246–52 (1974).

G. Lunde and A. Bjorseth, Polycyclic aromatic hydrocarbons in long-range transported aerosols. *Nature (London)* **268**, 518–519 (1977).

M. J. Lupo, Mass-radius relationships in icy satellites after Voyager. *Icarus* **52**, 40–53 (1982).

M. J. Lupo and J. S. Lewis, Mass-radius relationships in icy satellites. *Icarus* **40**, 157–170 (1979).

M. J. Lupo and J. S. Lewis, Mass-radius relationships and constraints on the composition of Pluto. *Icarus* **42**, 29–34 (1980a).

M. J. Lupo and J. S. Lewis, Mass-radius relationships and constraints on the composition of Pluto. II. *Icarus* **44**, 41–42 (1980b).

M. J. Lupo and J. S. Lewis, "Gaseous Transport of Iron and Nickel in Meteorite Parent Bodies," Preprint (1981).

J. Lupton, R. Weiss, and H. Craig, Mantle helium in Red Sea brines. *Nature (London)* **266**, 244–245 (1977).

B. L. Lutz and T. Owen, The search for HD in the spectrum of Uranus: An upper limit to [D/H]. *Astrophys. J.* **190**, 731–734 (1974).

D. Lynden-Bell and J. E. Pringle, The evolution of viscous discs and the origin of the nebular variables. *Mon. Not. R. Astron. Soc.* **168**, 603–621.

B. Lyot, Recherches sur la polarisation de la lumière des planètes et de quelques substances terrestres. *Ann. Obs. Paris, Meudon* **8**, 1–161 (1929).

R. A. Lyttleton, On the internal structure of the planet Mars. *Mon. Not. R. Astron. Soc.* **129**, 21–39 (1965).

R. A. Lyttleton, On the theory of the structure of Mars. *Mon. Not. R. Astron. Soc.* **141**, 251–253 (1968).

D. C. McAdoo and J. A. Burns, The Coprates trough assemblage: More evidence for Martian polar wander. *Earth Planet. Sci. Lett.* **25**, 347–354 (1975).

J. F. McCarthy, J. B. Pollack, J. R. Houck, and W. J. Forrest, 16–30 micron spectroscopy of Titan. *Astrophys. J.* **236**, 701–705 (1980).

J. F. McCauley, Mariner 9 evidence for wind erosion in the equatorial and mid-latitude regions of Mars. *JGR, J. Geophys. Res.* **78**, 4123–4137 (1973).

J. F. McCauley, M. H. Carr, J. A. Cutts, W. K. Hartmann, H. Masursky, D. J. Milton, R. P. Sharp, and D. E. Wilhelms, Preliminary Mariner 9 report on the geology of Mars. *Icarus* **17**, 289–327 (1972).

J. F. McCauley, B. A. Smith, and L. A. Soderblom, Erosional scarps on Io. *Nature (London)* **280**, 736–738 (1979).

J. C. McConnell, Atmospheric ammonia. *JGR, J. Geophys. Res.* **78**, 7812–7821 (1973a).

J. C. McConnell, The atmosphere of Mars. *In* "Physics and Chemistry of Upper Atmospheres" (B. M. McCormack, ed.), pp. 309–334. Reidel Publ., Dordrecht, Netherlands, 1973b.

J. C. McConnell and H. I. Schiff, Methyl chloroform: Impact on stratospheric ozone. *Science* **199**, 174–176 (1978).

J. C. McConnell, M. B. McElroy, and S. C. Wofsy, Natural sources of atmospheric CO. *Nature (London)* **233**, 187–188 (1971).

T. B. McCord, Comparison of the reflectivity and color of bright and dark regions on the surface of Mars. *Astrophys. J.* **156**, 79–86 (1969).

T. B. McCord and J. B. Adams, Mercury: Interpretation of optical observations. *Icarus* **17**, 585–588 (1972).

T. B. McCord and R. N. Clark, The Mercury soil: Presence of Fe^{2+}. *Lunar Planet. Sci.* **10**, 789–791 (1979a).

T. B. McCord and R. N. Clark, The Mercury soil: Presence of Fe^{2+}. *JGR, J. Geophys. Res.* **84**, 7664–7668 (1979b).

T. B. McCord and J. A. Westphal, Mars: Narrow-band photometry, from 0.3 to 2.5 microns, of surface regions during the 1969 apparition. *Astrophys. J.* **168**, 141–153 (1971).

T. B. McCord, J. B. Adams, and T. V. Johnson, Asteroid Vesta: Spectral reflectivity and compositional implications. *Science* **168**, 1445–1447 (1970).

T. B. McCord, J. H. Elias, and J. A. Westphal, Mars: The spectral albedo ($0.3–2.5\mu$) of small bright and dark regions. *Icarus* **14**, 245–251 (1971).

T. B. McCord, R. L. Huguenin, D. Mink, and C. Pieters, Spectral reflectance of Martian areas during the 1973 opposition: Photoelectric filter photometry $0.33–1.10$ μm. *Icarus* **31**, 25–39 (1977).

W. H. McCrea, Densities of the terrestrial planets. *Nature (London)* **224**, 28–29 (1969).

G. J. F. MacDonald, Calculations on the thermal history of the Earth. *JGR, J. Geophys Res.* **64**, 1967–2000 (1959).

G. J. F. MacDonald, On the internal constitution of the inner planets. *JGR, J. Geophys Res.* **67**, 2945–2974 (1962).

G. J. F. MacDonald, The internal constitution of the inner planets and the moon. *Space Sci. Rev.* **2**, 473–557 (1963a).

G. J. F. MacDonald, The escape of helium from the Earth's atmosphere. *In* "The Origin and Evolution of Atmospheres and Oceans" (P. J. Brancazio and A. G. W. Cameron, eds.), pp. 127–182. Wiley, New York, 1963b.

G. J. F. MacDonald and L. Knopoff, On the chemical composition of the outer core. *Geophys. J.* **1**, 284–297 (1958).

T. R. McDonough and N. M. Brice, New kind of ring around Saturn? *Nature (London)* **242**, 513 (1973a).

T. R. McDonough and N. M. Brice, A Saturnian gas ring and the recycling of Titan's atmosphere. *Icarus* **20**, 136–145 (1973b).

M. B. McElroy, The upper atmosphere of Venus. *JGR, J. Geophys. Res.* **73**, 1513–1521 (1968a).

M. B. McElroy, The upper atmosphere of Venus in light of the Mariner 5 measurements. *J. Atmos. Sci.* **25**, 574–577 (1968b).

M. B. McElroy, Structure of the Venus and Mars atmospheres. *JGR, J. Geophys. Res.* **74**, 29–41 (1969a).

M. B. McElroy, Atmospheric composition of the Jovian planets. *J. Atmos. Sci.* **26**, 798–811 (1969b).

M. B. McElroy, Mars: An evolving atmosphere. *Science* **175**, 443–445 (1972).

M. B. McElroy and T. M. Donahue, Stability of the Martian atmosphere. *Science* **177**, 986–988 (1972).

M. B. McElroy and D. M. Hunten, The ratio of deuterium to hydrogen in the Venus atmosphere. *JGR, J. Geophys. Res.* **74**, 1720–1737 (1969).

M. B. McElroy and D. M. Hunten, Photochemistry of CO_2 in the atmosphere of Mars. *JGR, J. Geophys. Res.* **75**, 1188–1202 (1970).

M. B. McElroy and J. C. McConnell, Dissociation of CO_2 in the Martian atmosphere. *J. Atmos. Sci.* **28**, 879–884 (1971).

M. B. McElroy and M. J. Prather, Noble gases in the terrestrial planets: Clues to evolution.*Nature* **293**, 535–539 (1981).

M. B. McElroy and D. F. Strobel, Models for the nighttime Venus ionosphere. *JGR, J. Geophys. Res.* **74**, 1118–1127 (1969).

M. B. McElroy and Y. L. Yung, The atmosphere and ionosphere of Io. *Astrophys. J.* **196**, 227–249 (1975).

M. B. McElroy, N. D. Sze, and Y. L. Yung, Photochemistry of the Venus atmosphere. *J. Atmos. Sci.* **30**, 1437–1446 (1973).

M. B. McElroy, Y. L. Yung, and R. A. Brown, Sodium emission from Io: Implications. *Astrophys. J.* **187**, L127–L130 (1974).

M. B. McElroy, Y. L. Yung, and A. O. Nier, Isotopic composition of Nitrogen: Implications for the past history of Mars' atmosphere. *Science* **194**, 70–72 (1976a).

M. B. McElroy, J. W. Elkins, S. C. Wofsy, and Y. L. Yung, Sources and sinks for atmospheric N_2O. *Rev. Geophys.* **14**, 143–150 (1976b).

M. B. McElroy, M. Prather, and J. Rodriguez, Escape of hydrogen from Venus. *Science* **215**, 1614–1615 (1982).

M. McEwan and L. Phillips, "Chemistry of the Atmosphere." Wiley, New York, 1975.

L. A. McFadden, J. F. Bell, and T. B. McCord, Visible spectral reflectance measurements (0.33–1.1 μm) of the Galilean satellites at many orbial phase angles. *Icarus* **44**, 410–430 (1980).

T. R. McGetchin and J. R. Smyth, Mantle of Mars: Some possible geological implications of its high density. *Icarus* **34**, 512–536 (1978).

G. E. McGill, Venus tectonics: Another Earth or another Mars? *Geophys. Res. Lett.* **6**, 739–742 (1979).

G. E. McGill and D. U. Wise, Regional variations in degradation and density of Martian craters. *JGR, J. Geophys. Res.* **77**, 2433–2441 (1972).

W. E. McGovern, The primitive Earth: Thermal models of the upper atmosphere for a methane-dominated environment. *J. Atmos. Sci.* **26**, 623–635 (1969).

C. P. McKay and G. E. Thomas, Consequences of a past encounter of the Earth with an interstellar cloud. *Geophys. Res. Lett.* **5**, 215–218 (1978).

W. B. McKinnon and H. J. Melosh, Evolution of planetary lithospheres: Evidence from multiringed structures on Ganymede and Callisto. *Icarus* **44**, 454–471 (1980).

D. J. McLaren, Time, life and boundaries. *J. Paleontol.* **44**, 801–815 (1970).

D. M. McLean, Terminal Mesozoic "greenhouse": Lessons from the past. *Science* **201**, 401–406 (1978).

W. Macy, Jr., An analysis of Saturn's methane $3v_3$ band profiles in terms of an inhomogeneous atmosphere. *Icarus* **29**, 49–56 (1976).

W. M. Macy, Jr. and W. Sinton, Detection of methane and ethane emission on Neptune but not on Uranus. *Astrophys. J.* **218**, L79–81 (1977).

W. M. Macy, Jr. and L. M. Trafton, A model for Io's atmosphere and sodium cloud. *Astrophys. J.* **200**, 510–519 (1975a).

W. M. Macy, Jr. and L. M. Trafton, Io's sodium emission cloud. *Icarus* **25**, 432–438 (1975b).

W. M. Macy, Jr. and L. M. Trafton, Neptune's atmosphere: The source of the thermal inversion. *Icarus* **26**, 428–436 (1975c).

S. V. Majeva, The thermal history of the terrestrial planets. *Astrophys. Lett.* **4**, 11–16 (1969).

M. C. Malin, Salt weathering on Mars. *JGR, J. Geophys. Res.* 79, 3888–3893 (1974).

M. C. Malin, Age of Martian channels, *JGR, J. Geophys. Res.* 81, 4825–4844 (1976).

M. C. Malin and R. S. Saunders, Surface of Venus: Evidence of diverse landforms from radar observations. *Science* 196, 987–990 (1977).

P. Mange, Diffusion in the thermosphere. *Ann. Geophys.* 17, 277–291 (1961).

R. H. Manka and F. C. Michel, Detection of lunar atmosphere in lunar samples. *Trans. Am. Geophys. Union* 52, 266 (1971).

P. G. Manning, Is Pluto an iron-rich planet? *Nature (London)* 230, 234–235 (1971).

S. Manabe and R. T. Wetherald, Thermal equilibrium of the atmosphere with a given distribution of relative humidity. *J. Atmos. Sci.* 24, 241–259 (1967).

O. K. Manuel, E. W. Hennecke, and D. D. Sabu, Xenon in carbonaceous chondrites. *Nature (London), Phys. Sci.* 240, 99–100 (1972).

H. K. Mao, A discussion of the iron oxides at high pressure with implications for the chemical and thermal evolution of the Earth. *Year Book—Carnegie Inst. Washington* pp. 510–518 (1974).

H. K. Mao and P. M. Bell, High-pressure physics: Sustained static generation of 1.36 to 1.72 megabars. *Science* 200, 1145–1147 (1978).

A. H. Marcus, Martian craters: Number density. *Science* 160, 1333–1335 (1968).

J. S. Margolis, R. A. J. Schorn, and L. D. G. Young, High dispersion spectroscopic studies of Mars. V. A. search for oxygen in the atmosphere of Mars. *Icarus* 15, 197–203 (1971).

L. Margulis, "Origin of Eukaryotic Cells." Yale Univ. Press, New Haven, Connecticut (1970).

L. Margulis and J. E. Lovelock, Biological modulation of the Earth's atmosphere. *Icarus* 21, 471–489 (1974).

Mariner Stanford Group, Venus: Ionosphere and atmosphere as measured by dual-frequency radio occultation of Mariner V. *Science* 158, 1678–1683 (1967).

M. Ya. Marov, Venus: A perspective at the beginning of planetary exploration. *Icarus* 16, 415–461 (1972).

M. Ya. Marov, Results of Venus missions. *Annu. Rev. Astron. Astrophys.* 16, 141–169 (1978).

M. Ya. Marov and G. I. Petrov, Investigations of Mars from the Soviet automatic stations Mars 2 and 3. *Icarus* 19, 163–179 (1973).

M. Ya. Marov, V. S. Avduevskii, M. K. Rozhdestvenskii, N. F. Borodin, and V. V. Kerzhanovich, Preliminary results of the investigation of Venus' atmosphere with the help of the Venera 7 automatic interplanetary spacecraft. *Cosmic Res. (Engl. Transl.)* 9, 521–529 (1971).

M. Ya. Marov, V. S. Avduevsky, N. F. Borodin, A. P. Ekonomov, V. V. Kerzhanovich, V. P. Lysov, B. Ye. Moshkin, M. R. Rozhdestvensky, and O. L. Ryabov, Preliminary results on the Venus atmosphere from the Venera 8 descent module. *Icarus* 20, 407–421 (1973a).

M. Ya. Marov, V. S. Avduevsky, V. V. Kerzhanovich, M. K. Rozhdestvensky, N. F. Borodin, and O. L. Ryabov, Venera 8: Measurements of temperature, pressure and wind velocity on the illuminated side of Venus. *J. Atmos. Sci.* 30, 1210–1214 (1973b).

M. Ya. Marov, B. V. Byvshev, K. N. Manuilov, Y. P. Baranov, I. S. Kuznetsov, V. N. Lebedev, V. E. Lystsev, A. V. Maksimov, G. K. Popandopulo, V. A. Razdolin, V. A. Sandimirov, and A. M. Frolov, Nephelometric measurements by the Venera 9 and Venera 10 spacecraft. *Cosmic Res. (Engl. Transl.)* 14, 637–642 (1976).

K. Marti, Trapped rare gases in meteorites. *Science* 156, 540–541 (1967).

J. V. Martonchik, Sulfuric acid cloud interpretation of the infrared spectrum of Venus. *Astrophys. J.* 193, 495–500 (1974).

B. Mason, "Meteorites." Wiley, New York, 1962.

B. Mason, Composition of the Earth. *Nature (London)* 211, 616–618, (1966).

B. Mason, ed. "Handbook of Elemental Abundances in Meteorites." Gordon & Breach, New York, 1971.

B. J. Mason, "Physics of Clouds." Academic Press, New York, 1957.

P. Masson, Structure pattern analysis of the Noctis Labyrinthus-Valles Marineris regions of Mars. *Icarus* 30, 49–62 (1977).

H. Masursky, An overview of geological results from Mariner 9. *JGR, J. Geophys. Res.* **78**, 4009–4029 (1973).

H. Masursky, Martian channels—a classification. *Meteoritics* **9**, 379–381 (1974).

H. Masursky and 30 others, Mariner 9 television reconaissance of Mars and its satellites: Preliminary results. *Science* **175**, 294–330 (1972).

H. Masursky, J. M. Boyce, A. L. Dial, G. G. Schaber, and M. E. Strobell, Classification and time of formation of Martian channels based on Viking data. *JGR, J. Geophys. Res.* **82**, 4016–4038 (1977).

H. Masursky, G. G. Schafer, L. A. Soderblom, and R. G. Strom, Preliminary geological mapping of Io. *Nature (London)* **280**, 725–729 (1979).

D. L. Matson, Infrared observations of asteroids. *NASA SP* **267**, 45–50 (1971).

D. L. Matson, T. V. Johnson, and F. P. Fanale, Sodium d-line emission from Io: Sputtering and resonant scattering hypotheses. *Astrophys. J.* **192**, L43–L46 (1974).

D. L. Matson, T. V. Johnson, and G. J. Veeder, Asteroid infrared reflectances and compositional implications. *In* "Comets, Asteroids, Meteorites: Interrelations, Evolution and Origins," (A. H. Delsemme, ed.) U. of Toledo Press, Toledo, Ohio (1977).

D. L. Matson, B. A. Goldberg, T. V. Johnson, and R. W. Carlson, Images of Io's sodium cloud. *Science* **199**, 541–533 (1978).

D. L. Matson, G. A. Ransford, and T. V. Johnson, Heat flow from Io (JI). *JGR, J. Geophys. Res.* **86**, 1664–1672 (1981).

K. Mauersberger and U. von Zahn, Upper limits on argon isotope abundances in the Venus thermosphere. *Geophys. Res. Lett.* **6**, 671–674 (1979).

N. U. Mayall and S. B. Nicholson, Positions, orbit and mass of Pluto. *Astrophys. J.* **73**, 1–12 (1931).

C. H. Mayer, T. P. Cullough, and R. M. Sloanaker, Observations of Venus at 3.15 cm wavelength. *Astrophys. J.* **127**, 1–10 (1958).

A. J. Meadows, The origin and evolution of the atmospheres of the terrestrial planets. *Planet. Space Sci.* **21**, 1467–1474 (1973).

G. H. Megrue, Distribution of gases within Apollo 15 samples: Implications for the incorporation of gases within solid bodies of the solar system. *JGR, J. Geophys. Res.* **78**, 4875–4883 (1973).

A. B. Meinel and D. T. Hoxie, On the spectrum of lightning in the Venus atmosphere. *Publ. Astron. Soc. Pac.* **74**, 329–330 (1962).

Yu. Mekler and A. Eviatar, Spectroscopic observations of Io. *Astrophys. J.* **193**, L151–L152 (1974).

D. H. Menzel, Water-cell transmissions and planetary temperatures. *Astrophys. J.* **58**, 65–69 (1923).

D. H. Menzel and F. L. Whipple, The case of H_2O clouds on Venus. *Publ. Astron. Soc. Pac.* **67**, 161–168 (1955).

G. A. Meshcheryakov, Model of Mars' internal structure constructed from its gravityfield parameters according to Mariner-9 data. *Sov. Astron. (Engl. Transl.)* **17**, 699–700 (1973).

G. A. Meshcheryakov, The figure of Mars. *Sov. Astron. (Engl. Transl.)* **19**, 229–231 (1975).

F. C. Michel, Solar wind interaction with planetary atmospheres. *Rev. Geophys.* **8**, 427–440 (1971).

V. V. Mikhnevich and V. A. Sokolov, A model atmosphere of Venus based on the results of direct temperature and density measurements. *Cosmic Res. (Engl. Transl.)* **7**, 197–207 (1969).

V. V. Mikhnevich, A. I. Livshits, and B. G. Gel'man, Certain characteristics of the Venusian cloud layer determined from ionization densimeter measurements aboard the Venera 4 automatic interplanetary station. *Cosmic Res. (Engl. Transl.)* **14**, 248–252 (1976).

S. Miki, The Gaseous Flow around a Protoplanet in the Primordial Solar Nebula, *Progr. Theor. Phys.* **67**, 1053–1063 (1983).

S. L. Miller, Production of some organic compounds under possible primitive Earth conditions. *J. Am. Chem. Soc.* **77**, 2351–2361 (1955).

S. L. Miller and W. D. Smythe, Carbon dioxide clathrate in the Martian ice cap. *Science* **170**, 531–532 (1970).

S. L. Miller and H. C. Urey, Organic compound synthesis on the primitive Earth. *Science* **130**, 245–251 (1959).

R. L. Millis, D. T. Thompson, B. J. Harris, P. Birch, and R. Sefton, A search for posteclipse brightening of Io in 1973. I. *Icarus* **23**, 425–430 (1974).

D. J. Milton, Water and processes of degradation in the Martian landscape. *JGR, J. Geophys. Res.* **78**, 4037–4046 (1973).

D. J. Milton, Carbon dioxide hydrate and the floods on Mars. *Science* **183**, 654–655 (1974).

Y. Mintz, Temperature and circulation of the Venus atmosphere. *Planet. Space Sci.* **5**, 141–152 (1961).

A. Miyashiro, Effect of the high-luminosity stage of the protosun on the composition of planets and meteorites. *Chem. Erde* **27**, 252–259 (1968).

H. Mizumo, K. Nakazawa, and C. Hayashi, Instability of a gaseous envelope surrounding a planetary core and formation of giant planets. *Prog. Theor. Phys.* **60**, 699–710 (1978).

M. Molina and F. Rowland, Stratospheric sink for chlorofluoromethanes: Chlorine-catalysed destruction of ozone. *Nature (London)* **249**, 810–812 (1974).

F. Möller, On the influence of changes in the CO_2 concentration in air on the radiation balance of the Earth's surface and on the climate. *JGR. J. Geophys. Res.* **68**, 3877–3886 (1963).

P. M. Molton and C. Ponnamperuma, Organic synthesis in a simulated Jovian atmosphere. III. Synthesis of aminonitriles. *Icarus* **21**, 166–174 (1974).

C. B. Moore, Nitrogen (7). *In* "Handbook of Elemental Abundances in Meteorites" (B. Mason, ed.), pp. 93–98. Gordon & Breach, New York, 1971a.

C. B. Moore, Phosphorus (15). *In* "Handbook of Elemental Abundances in Meteorites" (B. Mason, ed.), pp. 131–136. Gordon & Breach, New York, 1971b.

C. B. Moore, Sulfur (16). *In* "Handbook of Elemental Abundances in Meteorites" (B. Mason, ed.), pp. 137–142. Gordon & Breach, New York, 1971c.

C. B. Moore and E. K. Gibson, Nitrogen abundances in chondritic meteorites. *Science* **163**, 174–175 (1969).

C. B. Moore and C. F. Lewis, Carbon abundances in chondritic meteorites. *Science* **149**, 317–318 (1965).

C. B. Moore and C. F. Lewis, The distribution of total carbon content in enstatite chondrites. *Earth Planet. Sci. Lett.* **1**, 376–378 (1966).

C. B. Moore and C. F. Lewis, Carbon in ordinary chondrites, *JGR, J. Geophys. Res.* **72**, 6289–6301 (1967).

C. B. Moore, E. K. Gibson, and K. Keil, Nitrogen abundances in enstatite chondrites. *Earth Planet. Sci. Lett.* **6**, 457–460 (1969).

H. J. Moore, R. E. Hutton, R. F. Scott, C. R. Spitzer, and R. W. Shorthill, Surface materials of the Viking landing sites. *JGR, J. Geophys. Res.* **82**, 4497–4523 (1977).

L. A. Morabito, S. P. Synnott, P. N. Kupferman, and S. A. Collins, Discovery of currently active extraterrestrial volcanism. *Science* **204**, 972 (1979).

J. W. Morgan and E. Anders, Chemical composition of the Earth, Venus, and Mercury, *Proc. Natl. Acad. Sci. USA* **77**, 6973–6977 (1980).

J. W. Morgan, U. Krahenbuhl, R. Ganapathy, and E. Anders, Luna-20 soil: Abundance of 17 trace elements. *Geochim. Cosmochim. Acta* **37**, 953–961 (1973).

J. W. Morgan, J. Hertogen, and E. Anders, The moon: Composition determined by nebular processes. *Moon Planets* **18**, 465–478 (1978).

W. J. Morgan, Convection plumes in the lower mantle. *Nature (London)* **230**, 42–43 (1971).

H. Morowitz and C. Sagan, Life in the clouds of Venus? *Nature (London)* **215**, 1259–1260 (1967).

V. I. Moroz, Infrared spectrum of Mercury ($\lambda = 1.0 - 3.9$ μ). *Sov. Astron. (Engl. Transl.)* **8**, 882–889 (1965).

V. I. Moroz, IR spectrophotometry of the moon and Galilean satellites of Jupiter. *Sov. Astron. (Engl. Transl.)* **9**, 999–1006 (1966).

V. I. Moroz, Polar tropopause of Venus. *Sov. Astron. (Engl. Transl.)* **15**, 448–453 (1971).

V. I. Moroz, Preliminary results of studies conducted on the Soviet automatic stations Mars 4, Mars 5, Mars 6, and Mars 7. *Cosmic Res. (Engl. Transl.)* **13**, 1–6 (1975).

V. I. Moroz, Argon in the Martian atmosphere: Do the results of Mars 6 agree with the optical and radio occulatation measurements? *Icarus* **28**, 159–163 (1976a).

V. I. Moroz, Structure of the Martian surface from optical and infrared observations. *Cosmic Res. (Engl. Transl.)* **14**, 76–84 (1976c).

V. I. Moroz, Panorama of the surface of Venus, some conclusions regarding the structure of the boundary layer of the atmosphere. *Cosmic Res. (Engl. Transl.)* **14**, 607–608 (1976d).

V. I. Moroz, The atmosphere of Venus. *Space Sci. Rev.* **29**, 3–127 (1981).

V. I. Moroz and D. P. Cruikshank, Spectroscopic determination of the Martian atmospheric pressure from the CO_2 bands. *Sov. Astron. (Engl. Transl.)* **15**, 822–827 (1971).

V. I. Moroz and L. V. Ksanfomaliti, Preliminary results of astrophysical observations of Mars from Mars-3. *Icarus* **17**, 408–422 (1972).

V. I. Moroz and V. G. Kurt, The atmosphere of Venus (a comparison of the results of astronomical observations and direct experiment). *Cosmic. Res. (Engl. Transl.)* **6**, 481–487 (1968).

V. I. Moroz and A. E. Nadzhip, Preliminary results of water vapor measurement in the Martian atmosphere made by the Mars-5 interplanetary probe. *Cosmic Res. (Engl. Transl.)* **13**, 28–30 (1975a).

V. I. Moroz and A. E. Nadzhip, Water vapor in the atmosphere of Mars based on measurements on board Mars-3. *Cosmic Res. (Engl. Transl.)* **13**, 658–669 (1975b).

V. I. Moroz, N. A. Parfent'ev, D. P. Cruikshank, and L. V. Gromova, Experimental determination of height differences on Mars from the intensity of the 2.16 μ CO_2 bands. *Sov. Astron. (Engl. Transl.)* **15**, 624–627 (1971).

V. I. Moroz, L. V. Ksanfomaliti, G. N. Krasovskii, V. D. Davydov, N. A. Parfent'ev, V. S. Zhegulev, and G. F. Folippov, Infrared temperatures and thermal properties of the Martian surface measured by the Mars 3 orbiter. *Cosmic Res. (Engl. Transl.)* **13**, 346–357 (1975).

V. I. Moroz, B. Moshkin, A. Ekonomov, N. Sanko, N. Parfentev, and Y. Golovin, "Spectrophotometric Experiment on Board the Venera 11, 12 Descenders: Some Results of the Analysis of the Venus Day-Side Spectrum," Publ. 117 Space Res. Inst. Acad. Sci. USSR, 1979.

V. I. Moroz, B. E. Moshkin, A. P. Ekonomov, Yu. M. Golovin, V. I. Gnedykh, and A. V. Grigor'ev, Spectrophotometrical experiment on Venera 13 and Venera 14. *Pis'ma Astron. Zh.* **8**, 404–410 (1982).

R. V. Morris and H. V. Lauer, Jr., The case against photostimulated oxidation of magnetite. *Geophys. Res. Lett.* **7**, 605–608 (1980).

D. Morrison, Thermophysics of the planet Mercury. *Space Sci. Rev.* **11**, 271–302 (1970).

D. Morrison, The densities and bulk compositions of Ceres and Vesta. *Geophys. Res. Lett.* **3**, 701–704 (1976).

D. Morrison, and C. R. Chapman, Radiometric diameters for an additional 22 asteroids. *Astrophys. J.* **204**, 934–939 (1976).

D. Morrison and D. P. Cruikshank, Thermal Properties of the Galilean satellites. *Icarus* **18**, 224–236 (1973).

D. Morrison and C. M. Tedesco, Io: Observational constraints on internal energy and thermophysics of the surface. *Icarus* **44**, 226–233 (1980).

D. Morrison, C. Sagan, and J. B. Pollack, Martian temperatures and thermal properties. *Icarus* **11**, 36–45 (1969).

D. Morrison, D. P. Cruikshank, and R. E. Murphy, Temperatures of Titan and the Galilean satellites at 20 μ. *Astrophys. J.* **173**, 143–146 (1972).

D. Morrison, D. P. Cruikshank, and J. A. Burns, Introducing the satellites. *In* "Planetary Satellites" (J. A. Burns, ed.) Univ. of Arizona, Tucson, 3–17 (1977).

D. Morrison, D. P. Cruikshank, C. B. Pilcher, and G. H. Rieke, Surface compositions of satellites of Saturn from infrared photometry. *Astrophys. J.* **207**, 1213–1216 (1976).

P. Morrison and J. Pine, Radiogenic origin of the helium isotopes in rock. *Ann. N.Y. Acad. Sci.* **62**, 69–74 (1955).

P. J. Mouginis-Mark, Morphology of Martian rampart craters. *Nature (London)* **272**, 691–694 (1978)

P. Mouginis-Mark, Martian fluidized crater morphologies: Variations with crater size, latitude and target. *JGR, J. Geophys. Res.* **84**, 8011–8022 (1979).

A. Muan and E. F. Osborne, Phase equilibria at liquidus temperatures in the system $MgO\text{-}FeO\text{-}Fe_2O_3\text{-}SiO_2$. *J. Am. Ceram. Soc.* **39**, 121–140 (1956).

R. F. Mueller, Chemistry and petrology of Venus: Preliminary deductions. *Science* **141**, 1046–1047 (1963).

R. F. Mueller, Stability of hydrogen in compounds on Venus. *Nature (London)* **203**, 625–626 (1964a).

R. F. Mueller, A chemical model for the lower atmosphere of Venus. *Icarus* **3**, 285–298 (1964b).

R. F. Mueller, Stability of sulfur compounds on Venus. *Icarus* **4**, 506–512 (1965).

R. F. Mueller, Effect of temperature on the strength and composition of the upper lithosphere of Venus. *Nature (London)* **224**, 354–356 (1969).

R. F. Mueller and S. J. Kridelbaugh, Kinetics of CO_2 production on Venus. *Icarus* **19**, 531–541 (1973).

H. R. Mukherjee and G. L. Siscoe, Possible sources of water on the moon. *JGR, J. Geophys. Res.* **78**, 1741–1752 (1973).

L. M. Mukhin, B. G. Gel'man, N. I. Lamonov, V. V. Mel'nikov, D. F. Nenarokov, B. P. Okhotnikov, V. A. Rotin, and V. N. Khokhlov, Gas-Chromatographical analysis of chemical composition of the atmosphere of Venus done by Venera 13 and Venera 14 probes. *Pis'ma Astron. Zh.* **8**, 399–403 (1982).

G. Münch, J. T. Trauger, and P. L. Roesler, A search for the H_2 (3,0) S1 line in the spectrum of Titan. *Astrophys. J.* **216**, 963–966 (1977).

R. E. Murphy and L. M. Trafton, Evidence for an internal heat source in Neptune. *Astrophys. J.* **193**, 253–255 (1974).

B. C. Murray, Mercury. *Sci. Am.* **233** (3) 59–68 (1975a).

B. C. Murray, The Mariner 10 pictures of Mercury: An overview. *JGR, J. Geophys. Res.* **80**, 2342–2344 (1975b).

B. C. Murray, R. L. Wildey, and F. A. Westphal, Infrared photometric mapping of Venus through the 8-14- micron atmospheric window. *JGR, J. Geophys. Res.* **68**, 4813–4818 (1963).

B. C. Murray, R. L. Wildey, and F. A. Westphal, Observations of Jupiter and the Galilean satellites at 10 microns. *Astrophys. J.* **139**, 986–993 (1964).

B. C. Murray, L. A. Soderblom, R. P. Sharp, and J. A. Cutts, The surface of Mars 1. Cratered terrains. *JGR, J. Geophys. Res.* **76**, 313–329 (1971).

B. C. Murray, L. A. Soderblom, J. A. Cutts, R. P. Sharp, D. J. Milton, and R. B. Leighton, Geological framework of the south polar region of Mars. *Icarus* **17**, 328–345 (1972).

B. C. Murray, M. J. S. Belton, G. E. Danielson, M. E. Davies, D. Gault, B. Hapke, B. O'Leary, R. G. Strom, V. Suomi, and N. Trask, Venus: Atmospheric motion and structure from Mariner 10 pictures. *Science* **183**, 1307–1315 (1974a).

B. C. Murray, M. J. S. Belton, G. E. Danielson, M. E. Davies, D. Gault, B. Hapke, B. O'Leary, R. G. Strom, V. Suomi, and N. Trask, Mariner 10 pictures of Mercury: First results. *Science* **184**, 459–461 (1974b).

B. C. Murray, M. J. S. Belton, G. E. Danielson, M. E. Davies, D. E. Gault, B. Hapke, B. O'Leary, R. G. Strom, V. Suomi, and N. Trask, Mercury's surface: Preliminary description and interpretation from Mariner 10 pictures. *Science* **185**, 169–178 (1974c).

B. C. Murray, R. G. Strom, N. J. Trask, and D. E. Gault, Surface history of Mercury: Implications for terrestrial planets. *JGR, J. Geophys. Res.* **80**, 2508–2514 (1975).

V. R. Murthy and H. T. Hall, The chemical composition of the Earth's core: Possibility of sulphur in the core. *Phys. Earth Planet. Inter.* **2**, 276–282 (1970).

V. R. Murthy and H. T. Hall, The origin and chemical composition of the Earth's core. *Phys. Earth Planet. Inter.* **6**, 123–130 (1972).

P. Mutch and A. Woronow, Martian rampart and pedestal craters' ejecta-emplacement: Coprates quadrangle. *Icarus* **41**, 259–268 (1980).

T. A. Mutch, The Viking lander imaging investigation. *JGR, J. Geophys. Res.* **82**, 4389–4390 (1977).

T. A. Mutch and R. S. Saunders, The geological development of Mars: A review. *Space Sci. Rev.* **19**, 3–57 (1976).

T. A. Mutch, A. B. Binder, F. O. Huck, E. C. Levinthal, E. C. Morris, C. Sagan, and A. T. Young, Imaging experiment: The Viking lander. *Icarus* **16**, 92–110 (1972).

T. A. Mutch, A. B. Binder, F. O. Huck, E. C. Levinthal, S. Liebes, Jr., E. C. Morris, W. R. Patterson, J. B. Pollack, C. Sagan, and G. R. Taylor, The surface of Mars: The view from the Viking one lander. *Science* **193**, 791–800 (1976a).

T. A. Mutch and 24 others, The surface of Mars: The view from the Viking 2 lander. *Science* **194**, 1277–1283 (1976b).

T. A. Mutch, R. E. Arvidson, A. B. Binder, E. A. Guiness, and E. C. Morris, The geology of the Viking lander 2 site. *JGR, J. Geophys. Res.* **82**, 4452–4467 (1977).

T. Nagata, Paleomagnetic data in connection with the evolution of the Earth's core. *Phys. Earth Planet. Inter.* **2**, 311–317 (1970).

B. Nagy, W. G. Meinschein, and D. J. Henessy, Aqueous, low-temperature environment of the Orgueil meteorite parent body. *Ann N.Y. Acad. Sci.* **108**, 534–552 (1963).

R. F. Naill, D. L. Meadows, and M. J. Stanley, Transition to coal. *Technol. Rev.* 78(1), 18–29 (1975).

M. P. Nakoda and J. D. Mihalov, Accretion of solar wind to form a lunar atmosphere. *JGR. J. Geophys. Res.* **67**, 1670–1671 (1962).

W. Napier and S. V. M. Clube, A theory of terrestrial catastrophism. *Nature (London)* **282**, 455–459 (1979).

D. B. Nash, The relative age of the escarpments in the Martian polar laminated terrain based on morphology. *Icarus* **22**, 385–396 (1974).

D. B. Nash and F. P. Fanale, Io's surface composition based on reflectance spectra of sulfur/salt mixtures and proton irradiation experiments. *Icarus* **31**, 40–80 (1977).

D. B. Nash and R. M. Nelson, Spectral evidence for sublimates and adsorbates on Io. *Nature (London)* **280**, 763–766 (1979).

D. B. Nash, D. L. Matson, T. V. Johnson, and F. P. Fanale, Na-D line emission from rock specimens by proton bombardment: Implications for emissions from Jupiter's satellite Io. *JGR, J. Geophys. Res.* **80**, 1975–1979 (1975).

National Academy of Sciences, "Environmental Impact of Stratospheric Flight." NAS, Washington, D.C., 1975.

A. P. Naumov and G. M. Strelkov, Theoretical submillimeter emission spectrum of Venus. *Sol. Syst. Res. (Engl. Transl.)* **4**, 187–189 (1970).

G. M. Nedyalkova and I. E. Turchinovich, Dissipation of atmosphere from Venus. *Sov. Astron. (Engl. Transl.)* **17**, 424–425 (1973).

G. M. Nedyalkova and I. E. Turchinovich, Escape of ions from the atmosphere of Venus. *Sov. Astron. (Engl. Transl.)* **19**, 81–84 (1975).

V. B. Neiman, A hyposometric analysis of the relief of Mars based on Mariner 9 data. *Sol. Syst. Res. (Engl. Transl.)* **9**, 201–206 (1975).

N. F. Ness, K. W. Behannon, R. P. Lepping, Y. C. Whang, and K. H. Schatten, Magnetic field observations near Venus: Preliminary results from Mariner 10. *Science* **183**, 1301–1307 (1974a).

N. F. Ness, K. W. Behannon, R. P. Lepping, Y. C. Whang, and K. H. Schatten, Magnetic field observations near Mercury: Preliminary results from Mariner 10. *Science* **185**, 151–159 (1974b).

N. F. Ness, K. W. Behannon, R. D. Lepping, and Y. C. Whang, Magnetic field of Mercury confirmed. *Nature (London)* **255**, 204–205. (1975a).

N. F. Ness, K. W. Behannon, R. P. Lepping, and Y. C. Whang, The magnetic field of Mercury. 1. *JGR, J. Geophys. Res.* **80**, 2708–2715 (1975b).

N. F. Ness, K. W. Behannon, R. P. Lepping, and Y. C. Whang, Observations of Mercury's magnetic field. *Icarus* **28**, 479–488 (1976).

G. Neugebauer, G. Münch, S. C. Chase, Jr., H. Hatzenbeler, E. Miner, and D. Schofield, Mariner 1969: Preliminary results of the infrared radiometer experiment. *Science* **166**, 98–99 (1969).

G. Neugebauer, G. Münch, H. Kieffer, S. C. Chase, Jr., and E. Miner, Mariner 1969 infrared radiometer results: Temperatures and thermal properties of the Martian surface. *Astron. J.* **76**, 719–727 (1971).

G. Neukum and D. U. Wise, Mars: A standard crater curve and possible new time scale. *Science* **194**, 1381–1386 (1976).

R. L. Newburn, Jr. and S. Gulkis, A survey of the outer planets Jupiter, Saturn, Uranus, Neptune, Pluto and their satellites. *Space Sci. Rev.* **14**, 179–271 (1973).

R. E. Newell, Venus—a contribution to the greenhouse-ionosphere debate. *Icarus* **7**, 114–131 (1967).

M. J. Newman and R. T. Rodd, Implications of solar evolution for the Earth's early atmosphere. *Science* **198**, 6035–1036 (1975).

M. Nicolet, Stratospheric ozone: An introduction to its study. *Rev. Geophys.* **13**, 593–636 (1975).

H. B. Niemann, R. E. Hartle, W. T. Kasprzak, N. W. Spencer, D. M. Hunten, and G. R. Carignan, Venus upper atmosphere neutral composition: Preliminary results from the pioneer Venus orbiter. *Science* **203**, 770–772 (1979a).

H. B. Niemann, R. E. Hartle, A. E. Hedin, W. T. Kasprzak, N. W. Spencer, D. M. Hunten, and G. R. Carignan, Venus upper atmosphere neutral gas composition: First observations of the diurnal variations. *Science* **205**, 54–56 (1979b).

A. O. Nier and M. B. McElroy, Composition and structure of Mars' upper Atmosphere: Results from the neutral mass spectrometers on Viking 1 and 2. *JGR, J. Geophys. Res.* **82**, 4341–4349 (1977).

A. O. Nier, W. B. Hanson, A. Seiff, M. B. McElroy, N. W. Spencer, R. J. Ducket, T. C. D. Knight, and W. S. Cook, Composition and structure of the Martian atmosphere: Preliminary results from Viking 1. *Science* **193**, 786–788 (1976a).

A. O. Nier, M. B. McElroy, and Y. L. Yung, Isotopic composition of the Martian atmosphere. *Science* **194**, 68–70 (1976b).

J. F. Noxon, W. A. Traub, N. P. Carleton, and P. Connes, Detection of O_2 dayglow emission from Mars and the Martian ozone abundance. *Astrophys. J.* **207**, 1025–1035 (1976).

N. Noy, A. Bar-Nun, and M. Podolak, Acetylene photopolymers in Jupiter's stratosphere. *Icarus* **40**, 199–204 (1979).

S. Nozette and J. S. Lewis, Venus: Chemical weathering of igneous rocks and buffering of atmospheric composition. *Science* **216**, 181–183 (1982).

M. D. Nussinov, Y. B. Chernyak, and J. L. Ettinger, Model of the fine-grain component of Martian soil based on Viking lander data. *Nature (London)* **274**, 859–861 (1978).

J. T. O'Connor, "Fossil" Martian weathering. *Icarus* **8**, 513–517 (1968a).

J. T. O'Connor, Mineral stability at the Martian surface. *JGR, J. Geophys. Res.* **73**, 5301–5311 (1968b).

M. J. O'Hara and G. M. Biggar, A point of phase-equilibria interpretation in connection with lavas from the Apollo 12 site. *Earth Planet. Sci. Lett.* **16**, 388–390 (1972).

G. Ohring and A. Lacser, The ammonia profile in the atmosphere of Saturn from inversion of its microwave emission spectrum. *Astrophys. J.* **206**, 622–626 (1976).

E. A. Okal and D. L. Anderson, Theoretical models for Mars and their seismic properties. *Icarus* **33**, 514–528 (1978).

B. O'Leary, Venus halo: Photometric evidence for ice in the Venus clouds. *Icarus* **13**, 292–298 (1970).

B. O'Leary and E. Miner, Another possible post eclipse brightening of Io. *Icarus* **20**, 18–20 (1973).

B. T. O'Leary and J. B. Pollack, A critique of the paper by Walter G. Egan, "Polarimetric and photometric simulation of the Martian surface." *Icarus* **10**, 238–240 (1969).

B. T. O'Leary and D. G. Rea, On the polarimetric evidence for an atmosphere on Mercury. *Astrophys. J.* **148**, 249–253 (1967).

B. T. O'Leary and D. G. Rea, The opposition effect of Mars and its implications. *Icarus* **9**, 405–428 (1968a).

B. T. O'Leary and D. G. Rea, Mars: Visible and near-infrared studies and the composition of the surface. *Astron. J.* **73**, 529–530 (1968b).

B. T. O'Leary and J. Veverka, On the anomalous brightening of Io after Eclipse. *Icarus* **14**, 265–268 (1971).

R. K. O'Nions, N. M. Evensen, and P. J. Hamilton, Geochemical modeling of mantle differentiation and crustal growth. *JGR, J. Geophys. Res.* **84**, 6091–6101 (1979).

E. Öpik, "Physics of Meteor Flight in the Atmosphere." Wiley (Interscience), New York, 1958.

E. J. Öpik, The aeolosphere and atmosphere of Venus. *JGR, J. Geophys. Res.* **66**, 2807–2820 (1961).

E. J. Öpik, The moon's surface. *Annu. Rev. Astron. Astrophys.* **7**, 473–526 (1969).

G. S. Orton and H. H. Aumann, The abundance of C_2H_2 in Jupiter. *Bull. Am. Astron. Soc.* **9**, 478 (1977).

G. S. Orton and A. P. Ingersoll, Pioneer 10 and 11 and ground-based infrared data on Jupiter: The thermal structure and He/H_2 ratio. *In* "Jupiter: The Giant Planet" (T. Gehrels, ed.), pp. 206–215, Univ. of Arizona Press, Tucson, 1976.

R. G. Ostic, Physical conditions in gaseous spheres. *Mon. Not. R. Astron. Soc.* **131**, 191–197 (1965).

V. M. Oversby and A. E. Ringwood, Potassium distribution between metal and silicate and its bearing on the occurrence of potassium in the Earth's core. *Earth Planet. Sci. Lett.* **14**, 345–347 (1972).

T. Owen, Water vapor on Venus—a dissent and a clarification. *Astrophys. J.* **150**, L121–L123 (1967).

T. Owen, A search for minor constituents in the atmosphere of Venus. *J. Atmos. Sci.* **25**, 583–585 (1968).

T. Owen, The atmosphere of Jupiter. *Science* **167**, 1675–1681 (1970).

T. Owen and K. Biemann, Composition of the atmosphere at the surface of Mars: Detection of argon-36 and preliminary analysis. *Science* **193**, 801–803 (1976).

T. Owen and H. P. Mason, Mars: Water vapor in its atmosphere. *Science* **165**, 893–895 (1969).

T. Owen and C. Sagan, Minor constituents in planetary atmospheres: Ultraviolet spectroscopy from the orbiting astronomical observatory. *Icarus* **16**, 557–568 (1972).

T. Owen, T. Scattergood, and J. H. Woodman, On the abundance of NO_2 in the Martian atmosphere. *Icarus* **24**, 193–196 (1975).

T. Owen, K. Biemann, D. R. Rushneck, J. E. Biller, D. W. Howarth, and A. L. LaFleur, The atmosphere of Mars: Detection of krypton and xenon. *Science* **194**, 1293–1295 (1976).

T. Owen, A. R. W. McKellar, T. Encrenaz, J. Lecacheux, C. De Bergh, and J. P. Maillard, A study of the 1.56μ NH_3 band on Jupiter and Saturn. *Astron. Astrophys.* **54**, 291–295 (1977a).

T. Owen, K. Biemann, D. R. Rushneck, J. E. Biller, D. Howarth, and A. L. Lafleur, The composition of the atmosphere at the surface of Mars. *JGR, J. Geophys. Res.* **82**, 4635–4640 (1977b).

T. Owen, R. D. Cess, and V. Ramanathan, Enhanced CO_2 greenhouse to compensate for reduced solar luminosity on early Earth. *Nature (London)* **277**, 640–642 (1979).

V. I. Oyama, The gas exchange experiment for life detection: The Viking Mars lander. *Icarus* **16**, 167–184 (1972).

V. I. Oyama and B. J. Berdahl, The Viking gas exchange experiment results from Chryse and utopia surface samples. *JGR, J. Geophys. Res.* **82**, 4669–4676 (1977).

V. I. Oyama, B. J. Berdahl, G. C. Carle, M. E. Lehwalt, and H. S. Ginoza, The search for life on Mars: Viking 1976 gas changes as indicators of biological activity. *Origins Life* **7**, 313–333 (1976).

V. I. Oyama, B. J. Berdahl, and G. C. Carle, Preliminary findings of the Viking gas exchange experiment and a model for Martian surface chemistry. *Nature (London)* **265**, 110–113 (1977).

V. I. Oyama, G. C. Carle, F. Woeller, and J. B. Pollack, Venus lower atmospheric composition: Analysis by gas chromatography. *Science* **203**, 802–805 (1979a).

V. I. Oyama, G. C. Carle, F. Woeller, and J. B. Pollack, Laboratory corroboration of the pioneer Venus gas chronatograph analyses. *Science* **205**, 52–54 (1979b).

V. I. Oyama, G. C. Carle, and F. Woeller, Corrections on the Pioneer Venus sounder probe gas chromatographic analysis of the lower Venus atmosphere. *Science* **208**, 399–400 (1980a).

V. I. Oyama, G. Carle, F. Woeller, J. Pollack, R. Reynolds, and R. Craig, Pioneer Venus gas chromatography of the lower atmosphere. *JGR, J. Geophys. Res.* **85**, 7891–7902 (1980b).

M. Ozima, Was the evolution of the atmosphere continuous or catastrophic? *Nature (London)* **246**, 41–42 (1973).

M. Ozima and E. Alexander, Rare gas fractionation patterns in terrestrial samples and the Earth atmosphere evolution model. *Rev. Geophys. Space Phys.* **14**, 385–390 (1976).

M. Ozima and K. Kudo, Excess argon in submarine basalts and an earth-atmosphere evolution model. *Nature (London), Phys. Sci.* **239**, 23 (1972).

M. Ozima and K. Nakazawa, The origin of rare gases in the Earth atmosphere. *Lunar Planet. Symp. 12th*, pp. 111–112 (1979).

M. Paecht-Horowitz, J. Berger, and A. Katchalsky, Prehistoric synthesis of polypetides by heterogeneous polycondensation of amino-acid adenylates. *Nature (London)* **228**, 636–639 (1970).

A. Palm, The evolution of Venus' atmosphere. *Planet. Space Sci.* **17**, 1021–1028 (1969).

K. Pang and C. W. Hord, Mariner 7 ultraviolet spectrometer experiment: Photometric function and roughness of Mars' polar cap surface. *Icarus* **15**, 443–453 (1971).

C. Park, Nitic oxide production by Tunguska meteor. *Acta Astronaut.* **5**, 523–542 (1978).

T. D. Parkinson and D. M. Hunten, Spectroscopy and aeronomy of O_2 on Mars. *J. Atmos. Sci.* **29**, 1380–1390 (1972).

E. M. Parmentier and J. W. Head, Internal Processes affecting surfaces of low-density satellites: Ganymede and Callisto. *JGR, J. Geophys. Res.* **84**, 6263–6271 (1979).

W. R. Patterson, III, F. O. Huck, S. D. Wall, and M. R. Wolf, Calibration and performance of the Viking landers cameras. *JGR, J. Geophys. Res.* **82**, 4391–4400 (1977).

K. Pavlovski, Two-color search for the post eclipse brightening of Io. *Icarus* **29**, 509–512 (1976).

S. J. Peale, Does mercury have a molten core? *Nature (London)* **262**, 765–767 (1976).

S. J. Peale, P. Cassen, and R. T. Reynolds, Melting of Io by tidal dissipation. *Science* **204**, 892–894 (1979).

J. Pearl, R. Hanel, V. Kunde, W. Maquire, K. Fox, S. Gupta, C. Ponnamperuma, and F. Raulin, Identification of gaseous SO_2 and new upper limits for other gases on Io. *Nature (London)* **280**, 755–758 (1979).

P. J. E. Peebles, The structure and composition of Jupiter and Saturn. *Astrophys. J.* **140**, 328–340 (1964).

B. M. Peek, "The Planet Jupiter." Faber & Faber, London, 1958.

T. A. Perls, Carbon suboxide on Mars: A working hypothesis. *Icarus* **14**, 252–264 (1971).

T. A. Perls, Carbon suboxide on Mars: Evidence against formation. *Icarus* **20**, 511–512 (1973).

F. Perri and A. G. W. Cameron, Hydrodynamic instability of the solar nebula in the presence of a planetary core. *Icarus* **22**, 416–425 (1974).

E. C. Perry Jr., J. Monster, and T. Reimer, Sulfur isotopes in Swaziland system barites and the evolution of the Earth's atmosphere. *Science* **171**, 1015–1016 (1971).

C. Peterson, A source mechanism for meteorites controlled by the Yarkovsky effect. *Icarus* **29**, 91–111 (1976).

H. F. Petersons, Shape of Mars. *Nature (London)* **253**, 103–104 (1975).

G. H. Pettengill, C. C. Counselman, L. P. Rainville, and I. I. Shapiro, Radar measurements of Martian topography. *Astron. J.* **74**, 461–482 (1969).

G. H. Pettengill, A. E. E. Rogers, and I. I. Shapiro, Martian craters and a scarp as seen by radar. *Science* **174**, 1321–1324 (1971).

G. H. Pettengill, P. G. Ford, W. E. Brown, W. M. Kaula, C. H. Keller, H. Masursky, and G. E. McGill, Pioneer Venus radar mapper experiment. *Science* **203**, 806–808 (1979a).

G. H. Pettengill, P. G. Ford, W. E. Brown, W. M. Kaula, H. Masursky, E. Eliason, and G. E. McGill, Venus: Topographic and surface imaging results from the Pioneer orbiter. *Science* **205**, 90–93 (1979b).

R. J. Phillips and R. S. Saunders, The isostatic state of Martian Topography. *JGR, J. Geophys. Res.* **80**, 2893–2897 (1975).

R. J. Phillips, R. S. Saunders, and J. E. Conel, Mars: Crustal structure inferred from Bouguer gravity anomalies. *JGR, J. Geophys. Res.* **78**, 4815–4820 (1973).

D. Phinney, ^{36}Ar, Kr and Xe in terrestrial materials. *Earth Planet. Sci. Lett.* **16**, 413–420 (1972).

D. Phinney, J. Tennyson, and V. Frick, Xenon in CO_2 well gas revisited. *JGR, J. Geophys. Res.* **83**, 2313–2319 (1978).

D. Pieri, Distribution of small channels on the Martian surface. *Icarus* **27**, 25–50 (1976).

C. B. Pilcher, S. T. Ridgway, and T. B. McCord, Galilean satellites: Identification of water frost. *Science* **178**, 1087 (1972).

G. C. Pimental, P. B. Forney, and K. C. Herr, Evidence about hydrate and solid water in the Martian surface from the 1969 Mariner infrared spectrometer. *JGR, J. Geophys. Res.* **79**, 1623–1634 (1974).

F. Pisani, Etude chimique et analyse de l'aérolithe d'Orgueil. *C. R. Hebd. Séances Acad. Sci.* **59**, 132–135 (1864).

W. T. Plummer, Venus clouds: Test for hydrocarbons. *Science* **163**, 1191–1192 (1969a).

W. T. Plummer, Infrared reflectivity of frost and the Venus clouds. *JGR, J. Geophys. Res.* **74**, 3331–3336 (1969b).

W. T. Plummer, The Venus spectrum: New evidence for ice. *Icarus* **12**, 233–237 (1970).

W. T. Plummer and R. K. Carson, Mars: Is the surface colored by carbon suboxide? *Science* **166**, 1141–1142 (1969).

W. T. Plummer and R. K. Carson, Venus clouds: Test for carbon suboxide. *Astrophys. J.* **159**, 159–165 (1970).

M. Podolak, Methane rich models of Uranus. *Icarus* **27**, 473–477 (1976).

M. Podolak, The abundance of water and rock in Jupiter as derived from interior models. *Icarus* **30**, 155–162 (1977).

M. Podolak and A. Bar-Nun, A constraint on the distribution of Titan's atmospheric aerosol. *Icarus* **39**, 272–276 (1979).

M. Podolak and A. G. W. Cameron, Models of the giant planets. *Icarus* **22**, 123–148 (1974).

M. Podolak and A. G. W. Cameron, Further investigations of Jupiter models. *Icarus* **25**, 627–634 (1975).

M. Podolak, N. Noy, and A. Bar-Nun, Photochemical aerosols in Titan's atmosphere. *Icarus* **40**, 193–198 (1979).

F. A. Podosek, Isotopic structures in solar system materials. *Annu. Rev. Astron. Astrophys.* **16**, 293–334 (1978).

F. A. Podosek, M. Honda, and M. Ozima, Sedimentary noble gases. *Geochim. Cosmochim. Acta* **44**, 1875–1889 (1980).

J. B. Pollack, Temperature structure of Oangrey planetary atmospheres. *Icarus* **10**, 301–316 (1969).

J. B. Pollack, A nongrey calculation of the runaway greenhouse: Implications for Venus' past and present. *Icarus* **14**, 295–306 (1971).

J. B. Pollack, Greenhouse models of the atmosphere of Titan. *Icarus* **19**, 43–58 (1973).

J. B. Pollack and D. C. Black, Implications of the gas compositional measurements of Pioneer Venus for the origin of planetary atmospheres. *Science* **205**, 56–59 (1979).

J. B. Pollack and D. Morrison, Venus: Determination of atmospheric parameters from the microwave spectrum. *Icarus* **12**, 376–390 (1970).

J. B. Pollack and R. T. Reynolds, Implications of Jupiter's early contraction history for the composition of the Galilean satellites. *Icarus* **21**, 248–253 (1974).

J. B. Pollack and C. Sagan, An analysis of the Mariner 2 microwave observations of Venus. *Astrophys. J.* **150**, 327–344 (1967a).

J. B. Pollack and C. Sagan, A critical test of the electrical discharge model of the Venus microwave emission. *Astrophys. J.* **150**, 699–706 (1967b).

J. B. Pollack and C. Sagan, The case for ice clouds on Venus. *JGR, J. Geophys. Res.* **73**, 5943–5948 (1968).

J. B. Pollack and C. Sagan, An analysis of Martian photometry and polarimetry. *Space Sci. Rev.* **9**, 243–299 (1969).

J. B. Pollack and F. C. Witteborn, Evolution of Io's volatile inventory. *Icarus* **44**, 249–257 (1980).

J. B. Pollack and A. T. Wood, Jr., Venus: Implications from microwave spectroscopy of the atmospheric content of water vapor. *Science* **161**, 1125–1127 (1968).

J. B. Pollack, D. Pitman, B. N. Khare, and C. Sagan, Goethite on Mars: A laboratory study of physically and chemically bound water in Ferric oxides. *JGR, J. Geophys. Res.* **75**, 7480–7490 (1970a).

J. B. Pollack, R. N. Wilson, and G. G. Goles, A re-examination of the stability of Goethite on Mars. *JGR, J. Geophys. Res.* **75**, 7491–7499 (1970b).

J. B. Pollack, E. F. Erickson, F. C. Witteborn, C. Chackerian, Jr., A. L. Summers, W. van Camp, B. J. Baldwin, G. C. Augason, and L. J. Caroff, Aircraft observations of Venus' near-infrared reflection spectrum: Implications for cloud composition. *Icarus* **23**, 8–26 (1974).

J. B. Pollack, B. Ragent, R. Boese, M. G. Tomasko, J. Blamont, R. G. Knollenberg, L. W. Esposito, A. L. Stewart, and L. Travis, Nature of the absorber in the Venus clouds: Inferences based on Pioneer Venus data. *Science* **205**, 76–79 (1979).

C. Ponnamperuma, The organic chemistry and biology of the atmosphere of the planet Jupiter. *Icarus* **29**, 321–328 (1976).

R. F. Poppen, U. Fink, and H. P. Larson, A new upper limit for an atmosphere of CO_2 on Mercury. *Bull. Am. Astron. Soc.* **5**, 302 (1973).

J. F. Potter, Effect of cloud scattering on line formation in the atmosphere of Venus. *J. Atmos. Sci.* **26**, 511–517 (1969).

J. F. Potter, On mercury clouds in the atmosphere of Venus. *Icarus* **17**, 79–87 (1972).

S. S. Prasad and L. A. Capone, The photochemistry of ammonia in the Jovian atmosphere. *JGR, J. Geophys. Res.* **81**, 5596–5599 (1976).

M. J. Prather, J. A. Logan, and M. B. McElroy, Carbon monoxide in Jupiter's upper atmosphere. *Astrophys. J.* **223**, 1072–1081 (1978).

F. Press and R. Sievers, "Earth." Freeman, San Francisco, California, 1974.

R. G. Prinn, UV radiative transfer and photolysis in Jupiter's atmosphere. *Icarus* **13**, 424–436 (1970).

R. G. Prinn, Photochemistry of HCl and other minor constituents in the atmosphere of Venus. *J. Atmos. Sci.* **28**, 1058–1067 (1971).

R. G. Prinn, Venus atmosphere: Structure and stability of the ClOOO radical. *J. Atmos. Sci.* **29**, 1004–1007 (1972).

R. G. Prinn, Venus: Composition and structure of the visible clouds. *Science* **182**, 1132–1134 (1973a).

R. G. Prinn, The upper atmosphere of Venus: A review. *In* "Physics and Chemistry of the Upper Atmosphere" (B. M. McCormae, ed.). D. Reidel, Dordrecht, pp. 334–335 (1973b).

R. G. Prinn, Venus: Vertical transport rates in the visible atmosphere. *J. Atmos. Sci.* **31**, 1691–1697 (1974).

R. G. Prinn, Chemical and dynamical processes in the stratosphere and mesosphere. *J. Atmos. Sci.* **32**, 1237–1247 (1975).

R. G. Prinn, Atmospheric chemistry of the planet Venus. *IUPAC Int. Bull.* **1**, 20–25 (1978a).

R. G. Prinn, Venus: Chemistry of the lower atmosphere prior to the Pioneer Venus mission. *Geophys. Res. Lett.* **5**, 973–976 (1978b).

R. G. Prinn, On the possible roles of gaseous sulfur and sulfanes in the atmosphere of Venus. *Geophys. Res. Lett.* **6**, 807–810 (1979).

R. G. Prinn and S. S. Barshay, Carbon monoxide on Jupiter and implications for atmospheric convection. *Science* **198**, 1031–1034 (1977).

R. G. Prinn and M. B. Fegley, Jr., Kinetic inhibition of CO and N_2 reduction in circumplanetary nebulae: Implications for satellite composition. *Astrophys J.* **249**, 308–317 (1981).

R. G. Prinn and J. S. Lewis, Uranus atmosphere: Structure and composition. *Astrophys. J.* **179**, 333–342 (1973).

R. G. Prinn and J. S. Lewis, Phosphine on Jupiter and implications for the Great Red Spot. *Science* **190**, 294–296 (1975).

R. G. Prinn and T. Owen, Chemistry and spectroscopy of the Jovian atmosphere. *In* "Jupiter: The Giant Planet" (T. Gehrels, ed.), pp. 319–372. Univ. of Arizona Press, Tucson, 1976.

R. G. Prinn, F. N. Alyea, and D. M. Cunnold, Stratospheric distributions of odd nitrogen and odd hydrogen in a two-dimensional model. *JGR, J. Geophys. Res.* **80**, 4997–5004 (1975).

R. G. Prinn, F. N. Alyea, and D. M. Cunnold, Photochemistry and dynamics of the ozone layer. *Ann. Rev. Earth Planet. Sci.* **6**, 43–74 (1978).

V. K. Prokofyev and N. N. Petrova, On the presence of oxygen in the atmosphere of Venus. *Mem. Soc. R. Sci. Liege* **7**, 311–321 (1963).

B. Ragent and J. Blamont, Preliminary results of the Pioneer Venus nephelometer experiment. *Science* **203**, 790–792 (1979).

K. Rages and J. B. Pollack, Titan aerosols: Optical properties and vertical distribution. *Icarus* **41**, 119–130 (1980).

V. Ramanathan, The role of ocean-atmosphere interactions in the CO_2 climate problem. *J. Atmos. Sci.* **38**, 918–930 (1981).

K. Rankama and T. G. Sahama, "Geochemistry." Univ. of Chicago Press, Chicago, Illinois, 1950.

S. I. Rasool, Loss of water from Venus. *J. Atmos. Sci.* **25**, 663–664 (1968).

S. I. Rasool, The structure of Venus clouds. *Radio Sci.* **5**, 367–368 (1970).

S. I. Rasool and C. de Bergh, The runaway greenhouse and the accumulation of CO_2 in the Venus atmosphere. *Nature (London)* **226**, 1037–1039 (1970).

S. I. Rasool and L. Le Sergeant, Implications of the Viking results for volatile outgassing from Earth and Mars. *Nature (London)* **266**, 822–823 (1977).

S. I. Rasool and S. H. Schneider, Atmospheric carbon dioxide and aerosols: Effects of large increases on global climate. *Science* **173**, 138–141 (1971).

S. I. Rasool, S. H. Gross, and W. E. McGovern, The atmosphere of Mercury. *Sci. Rev.* **5**, 565–584 (1966).

F. Raulin, A. Bossard, D. Toupance, and C. Ponnamperuma, Abundance of organic compounds photochemically produced in the atmospheres of the outer planets. *Icarus* **38**, 358–366 (1979).

D. G. Rea, Composition of the upper clouds of Venus. *Rev. Geophys. Space Phys.* **10**, 369–378 (1972).

D. G. Rea and B. T. O'Leary, Visible polarization data on Mars. *Nature (London)* **206**, 1138–1140 (1965).

D. G. Rea and B. T. O'Leary, On the composition of the Venus clouds. *Astron. J.* **72**, 317 (1967).

D. G. Rea and B. T. O'Leary, On the composition of the Venus clouds. *JGR, J. Geophys. Res.* **73**, 665–675 (1968).

D. G. Rea, T. Belsky, and M. Calvin, Interpretation of the 3-to-4 micron infrared spectrum of Mars. *Science* **141**, 923–927 (1963a).

D. G. Rea, T. Belsky, and M. Calvin, Reflection spectra of bio-organic materials in the 2.5-4 μ region and the interpretation of the infrared spectrum of Mars. *Life Sci. Space Res.* **2**, 86–100 (1963b).

D. G. Rea, B. T. O'Leary, and W. M. Sinton, Mars: The origin of the 3.58 - and 3.69 micron minima in the infrared spectra. *Science* **147**, 1286–1288 (1965).

R. D. Reasenberg, The moment of inertia and isostasy of Mars. *JGR, J. Geophys. Res.* **82**, 369–375 (1977).

R. D. Reasenberg, I. I. Shapiro, and R. D. White, The gravity field of Mars. *Geophys. Res. Lett.* **2**, 89–92 (1975).

R. J. Reed and K. E. German, A contribution to the problem of stratospheric diffusion by large-scale mixing. *Monthly Weather Rev.* **93**, 313–321 (1965).

C. E. Rees, The sulfur isotope balance of the ocean: An improved model. *Earth Planet. Sci. Lett.* **7**, 366–370 (1970).

D. E. Reese and P. R. Swan, Venera 4 probes atmosphere of Venus. *Science* **159**, 1228–1230 (1967).

R. R. Reeves, P. Harteck, B. A. Thompson, and R. W. Waldron, Photochemical equilibrium studies of CO_2 and their significance for the Venus atmosphere. *J. Phys. Chem.* **70**, 1637–1640 (1966).

R. Revelle and H. E. Suess, Carbon dioxide exchange between atmosphere and ocean and the question of an increase of atmospheric CO_2 during the past decades. *Tellus* **9**, 18–27 (1957).

J. H. Reynolds, Isotopic composition of primordial xenon. *Phys. Rev. Lett.* **4**, 351–354 (1960).

J. H. Reynolds and G. Turner, Rare gases in the chondrite Renazzo. *J. Geophys. Res.* **69**, 3263–3281 (1964).

R. T. Reynolds and P. M. Cassen, On the internal structure of the major satellites of the outer planets. *Geophys. Res. Lett.* **6**, 121–124 (1979).

R. T. Reynolds, S. J. Peale, and P. Cassen, Io: Energy constraints and plume volcanism. *Icarus* **44**, 234–239 (1980).

E. H. Richardson, The spectrum of Venus. *Astron. J.* **65**, 56–59 (1960).

S. M. Richardson, Vein formation in the Cl carbonaceous chondrites. *Meteoritics* **13**, 141–159 (1978).

S. M. Richardson and H. Y. McSween, Jr., The matrix composition of carbonaceous chondrites. *Meteoritics* **11**, 355–356 (1976).

S. T. Ridgway, Detection of phosphine on Jupiter. *Bull. Am. Astron. Soc.* **6**, 376 (1974a).

S. T. Ridgway, Jupiter: Identification of ethane and acetylene. *Astrophys. J.* **187**, L41–L43 (1974b).

S. T. Ridgway, L. Wallace, and G. R. Smith, The 800–1200 inverse centimeter absorption spectrum of Jupiter. *Astrophys. J.* **207**, 1002–1006 (1976).

A. E. Ringwood, On the chemical evolution and densities of the planets. *Geochim. Cosmochim. Acta* **15**, 257–283 (1959).

A. E. Ringwood, Chemical evolution of the terrestrial planets. *Geochim. Cosmochim. Acta* **30**, 41–104 (1966).

A. E. Ringwood, Origin of the moon: The precipitation hypothesis. *Earth Planet. Sci. Lett.* **8**, 131–140 (1970).

A. E. Ringwood, Core-mantle equilibrium: Comments on a paper by R. Brett. *Geochim. Cosmochim. Acta* **35**, 223–230 (1971).

A. E. Ringwood, Some comparative aspects of lunar origin. *Phys. Earth Planet. Inter.* **6**, 366–376 (1972).

A. E. Ringwood, Heterogeneous accretion and the lunar crust. *Geochim. Cosmochim. Acta* **38**, 983–984 (1974).

A. E. Ringwood, "Composition and petrology of the Earth's Mantle." McGraw-Hill, New York, 1975.

A. E. Ringwood, Limits on the bulk composition of the moon. *Icarus* **28**, 325–349 (1976).

A. E. Ringwood, Composition of the core and implications for the origin of the Earth. *Geochem. J.* **11**, 111–135 (1977).

A. E. Ringwood, "Origin of the Earth and Moon." Springer-Verlag, Berlin and New York, 1979.

A. E. Ringwood and D. L. Anderson, Earth and Venus—comparative study. *Icarus* **30**, 243–253 (1977).

A. E. Ringwood and S. P. Clark, Internal constitution of Mars. *Nature (London)* **234**, 89–92 (1971).

J. A. Ripmeester and D. W. Davidson, Some new clathrate hydrates. *Mol. Cryst.* **43**, 189–195 (1977).

R. C. Robbins, The reaction products of solar hydrogen and components of the high atmosphere of Venus—a possible source of the Venusian cloud. *Planet. Space Sci.* **12**, 1143–1146 (1964).

F. Robert and S. Epstein, The concentration and isotopic composition of hydrogen, carbon and nitrogen in carbonaceous meteorites. *Geochim. Cosmochim. Acta* **46**, 81–95 (1982).

J. C. Robinson, Mars: Correlation of optical and radar observations. *Science* **164**, 176–177 (1969).

E. Robinson and R. C. Robbins, Gaseous sulfur pollutants from urban and natural sources. *J. Air Poll. Control Assoc.* **20**, 233–235 (1970).

J. C. Roddick and E. Farrar, High initial argon ratios in hornblendes. *Earth Planet. Sci. Lett.* **12**, 208–214 (1971).

A. E. E. Rogers and R. P. Ingalls, Venus: Mapping the surface reflectivity by radar interferometry. *Science* **165**, 797–799 (1969).

A. E. E. Rogers and R. P. Ingalls, Radar mapping of Venus with interferometric resolution of the range-Doppler ambiguity. *Radio Sci.* **5**, 425–434 (1970).

A. E. E. Rogers, R. P. Ingalls, and L. P. Rainville, The topography of a swath around the equator of the planet Venus from the wavelength dependence of the radar cross section. *Astron. J.* **77**, 100–103 (1972).

A. E. E. Rogers, R. P. Ingalls, and G. H. Pettengill, Radar map of Venus at 3.8 cm wavelength. *Icarus* **21**, 237–241 (1974).

A. B. Ronov and A. A. Yaroshevskii, Chemical structure of the Earth's crust. *Geokhimiya* **11**, 1285–1309 (1967).

W. Rossow, S. Fels, and P. Stone, Comments on "A three-dimensional model of dynamical processes in the Venus atmosphere." *J. Atmos. Sci.* **37**, 250–252 (1980).

G. J. Rottman and H. W. Moos, The ultraviolet (1200–1900 angstrom) spectrum of Venus. *JGR, J. Geophys. Res.* **78**, 8033–8048 (1973).

P. W. Rowe and P. K. Kuroda, Fissiogenic xenon from the Pasamonte meteorite. *JGR, J. Geophys. Res.* **70**, 709–711 (1965).

W. W. Rubey, Geologic history of sea water: An attempt to state the problem. *Geol. Soc. Am. Bull.* **62**, 1111–1147 (1951).

H. C. Rumsey, G. A. Morris, R. R. Green, and R. M. Goldstein, A radar brightness and altitude image of a portion of Venus. *Icarus* **23**, 1–7 (1974).

C. T. Russell, The magnetic moment of Venus: Venera-4 measurements reinterpreted. *Geophys. Res. Lett.* **3**, 125–128 (1976a).

C. T. Russell, Venera-9 magnetic field measurements in the Venus wake: Evidence for an Earth-like interaction. *Geophys. Res. Lett.* **3**, 413–416 (1976b).

C. T. Russell, The magnetosphere of Venus: Evidence for a boundary layer and a magnetotail. *Geophys. Res. Lett.* **3**, 589–592 (1976c).

C. T. Russell, The magnetic field of Mars: Mars 3 evidence reexamined. *Geophys. Res. Lett.* **5**, 81–84 (1978a).

C. T. Russell, The magnetic field of Mars: Mars 5 evidence reexamined. *Geophys. Res. Lett.* **5**, 85–88 (1978b).

C. T. Russell, R. C. Elphic, and J. A. Slavin, Initial Pioneer Venus magnetic field results: Dayside observations. *Science* **203**, 745–748 (1979a).

C. T. Russell, R. C. Elphic, and J. A. Slavin, Initial Pioneer Venus magnetic field results: Nightside observations. *Science* **205**, 114–116 (1979b).

D. A. Russell, The biotic crisis at the end of the Cretaceous period. *Syllogeus* **12**, 11–23 (1977).

D. A. Russell, The enigma of the extinction of the dinosaurs. *Annu. Rev. Earth Planet. Sci.* **7**, 163–182 (1979).

H. N. Russell, R. S. Dugan, and J. Q. Stewart, "Astronomy." Ginn, Boston, Massachusetts, 1945.

R. D. Russell and M. Ozima, The potassium/rubidium ratio of the Earth. *Geochim. Cosmochim. Acta* **35**, 679–685 (1971).

M. G. Rutten, Geological data on atmospheric history. *Paleogeogr., Paleoclimatol., Paleoecol.* **2**, 47–57 (1966).

D. D. Sabu and O. K. Manuel, Xenon record of the early solar system. *Nature (London)* **262**, 28–32 (1976).

D. D. Sabu, E. W. Hennecke, and O. K. Manuel, Trapped xenon in meteorites. *Nature (London)* **251**, 21–24 (1974).

V. S. Safronov, "Evolution of the Protoplanetary Cloud and Formation of the Earth and Planets." Israel Program for Scientific Translations, Tel Aviv, 1972.

C. Sagan, The planet Venus. *Science* **133**, 849–859 (1961).

C. Sagan, Structure of the lower atmosphere of Venus. *Icarus* **1**, 151–169 (1962).

C. Sagan, Mariner IV observations and the possibility of iron oxides on the Martian surface. *Icarus* **5**, 102–103 (1966).

C. Sagan, The long winter model of Martian biology: A speculation. *Icarus* **15**, 511–514 (1971).

C. Sagan, Sandstorms and eolian erosion on Mars. *JGR, J. Geophys. Res.* **78**, 4155–4161 (1973a).

C. Sagan, Liquid carbon dioxide and the Martian polar laminas. *JGR, J. Geophys. Res.* **78**, 4250–4251 (1973b).

C. Sagan, The greenhouse of Titan. *Icarus* **18**, 649–656 (1973c).

C. Sagan, Planetary engineering on Mars. *Icarus* **20**, 513–514 (1973d).

C. Sagan, Windblown dust on Venus. *J. Atmos. Sci.* **32**, 1079–1083 (1975).

C. Sagan, Erosion and the rocks of Venus. *Nature (London)* **261**, 31 (1976).

C. Sagan and P. Fox, The canals of Mars: An assessment after Mariner 9. *Icarus* **25**, 602–612 (1975).

C. Sagan and B. N. Khare, Long-wavelength ultraviolet photoproduction of amino acids on the primitive Earth. *Science* **173**, 417–420 (1971a).

C. Sagan and B. N. Khare, Experimental Jovian photochemistry: Initial results. *Astrophys. J.* **168**, 563–569 (1971b).

C. Sagan and J. Lederberg, The prospects for life on Mars: A pre-Viking assessment. *Icarus* **28**, 291–300 (1976).

C. Sagan and G. Mullen, Earth and Mars: Evolution of atmospheres and surface temperatures. *Science* **177**, 52–55 (1972).

C. Sagan and J. B. Pollack, Elevation differences on Mars. *JGR, J. Geophys. Res.* **73**, 1373–1386 (1968).

C. Sagan and J. B. Pollack, On the structure of the Venus atmosphere. *Icarus* **10**, 274–289 (1969).

C. Sagan and J. B. Pollack, Differential transmission of sunlight on Mars. Biological implications. *Icarus* **21**, 490–495 (1974).

C. Sagan and J. Veverka, The microwave spectrum of Mars: An analysis. *Icarus* **14**, 222–234 (1971).

C. Sagan, J. P. Phaneuf, and M. Ihnat, Total reflection spectrophotometry and thermoparametric analyses of simulated Martian surface materials. *Icarus* **4**, 43–61 (1965a).

C. Sagan, P. L. Hanst, and A. T. Young, Nitrogen oxides on Mars. *Planet. Space Sci.* **13**, 73–88 (1965b).

C. Sagan, E. R. Lippincott, M. O. Dayhoff, and R. V. Eck, Organic molecules and the coloration of Jupiter. *Nature (London)* **213**, 273–274 (1967).

C. Sagan, O. Toon, and P. Gierasch, Climatic change on Mars. *Science* **181**, 1045–1049 (1973).

R. Z. Sagdeev and V. I. Moroz, Venera 13 and Venera 14. *Pis'ma Astron. Zh.* **8**, 387–390 (1982).

C. E. St. John and W. S. Adams, An attempt to detect water-vapor and oxygen lines in the spectrum of Mars. *Astrophys. J.* **63**, 133–137 (1926).

C. E. St. John and S. B. Nicholson, Absence of oxygen and water-vapor lines from the spectrum of Venus. *Astrophys. J.* **56**, 380–319 (1922).

F. B. Salisbury, Martian biology. *Science* **136**, 17–26 (1962).

J. W. Salisbury and G. R. Hunt, Compositional implications of the spectral behavior of the Martian surface. *Nature (London)* **222**, 132–136 (1969).

W. Sandner, "The Planet Mercury." Faber & Faber, London, 1963.

N. San'ko, Gaseous sulfur in the atmosphere of Venus. (InCrus.) *Kos. Issled.* **18**, 600–608 (1980).

C. N. Satterfield, R. Reid, and D. Briggs, Rate of oxidation of hydrogen sulfide by hydrogen peroxide. *J. Am. Chem. Soc.* **76**, 3922–3923 (1954).

F. L. Scarf, Plasma instability and the microwave radiation from Venus. *JGR, J. Geophys. Res.* **68**, 141–146 (1963).

F. L. Scarf, D. A. Gurnett, and W. S. Kurth, Jupiter plasma wave observations: An initial Voyager 1 overview. *Science* **204**, 991–995 (1979).

F. L. Scarf, W. Taylor, C. Russell, and L. Brace, Lightning on Venus: Orbiter detection of whistler signals. *JGR, J. Geophys. Res.* **85**, 8158–8166 (1980).

F. L. Scarf, D. A. Gurnett, W. S. Kurth, R. R. Anderson, and R. R. Shaw, An upper bound to the lightning flash rate in Jupiter's atmosphere. *Science* **213**, 684–685 (1981).

M. W. Schaefer, Low-temperature photostimulated oxidation of magnetite. In preparation (1981).

J. Schilling, M. Bender, and C. Unni, Origin of Cl and Br in the oceans. *Nature (London)* **273**, 631–632 (1978).

R. A. Schmitt, G. G. Goles, R. H. Smith, and T. W. Osborn, Elemental abundances in stone meteorites. *Meteoritics* **7**, 131–214 (1972).

S. H. Schneider and R. E. Dickinson, Climate modelling. *Revs. Geophys. Space. Phys.* **12**, 447–493 (1974).

J. W. Schopf, Precambrian microorganisms and evolutionary events prior to the origin of vascular plants. *Biol. Rev. Cambridge Philo. Soc.* **45**, 319–352 (1970).

T. J. M. Schopf, Cretaceous endings. *Science* **211**, 571–572 (1981).

R. A. Schorn, *Planet. Atmos. Proc. IAU Symp.* **40**, 223–236 (1971).

R. A. Schorn and L. G. Young, Comments on "The Venus spectrum: New evidence for ice." *Icarus* **15**, 103–109 (1971).

R. A. Schorn, H. Spinrad, R. C. Moore, H. J. Smith, and L. P. Giver, High-dispersion spectroscopic observations of Mars 2. The water-vapor variations. *Astrophys. J.* **147** 743–752 (1967).

R. A. Schorn, E. S. Barker, L. D. Gray, and R. C. Moore, High-dispersion spectroscopic studies of Venus. II. The water vapor variations. *Icarus* **10**, 98–104 (1969a).

R. A. Schorn, L. D. Gray, and E. S. Barker, High-dispersion spectroscopic observations of Venus. III. The carbon dioxide band at 7820 Å. *Icarus* **10**, 241–257 (1969b).

R. A. Schorn, C. B. Farmer, and S. J. Little, High-dispersion spectroscopic studies of Mars. III. Preliminary results of 1968–69 water-vapor studies. *Icarus* **11**, 283–288 (1969c).

R. A. Schorn, L.G. Young, and E. S. Barker, High-dispersion spectroscopic observations of Venus. V. The carbon dioxide band at 10364 Å. *Icarus* **12**, 391–410 (1970).

R. A. Schorn, L. D. G. Young, and E. S. Barker, High-dispersion spectroscopic observations of Venus. VIII. The carbon dioxide band at 10,627 Å. *Icarus* **14**, 21–35 (1971).

R. A. Schorn, A. Woszczyk, and L. D. G. Young, High-dispersion spectroscopic observations of Venus during 1968 and 1969. II. The carbon-dioxide band at 8689 Å. *Icarus* **25**, 64–88 (1975).

D. N. Schramm, F. Tera, and G. J. Wasserburg, The isotopic abundance of ^{26}Mg and limits of ^{26}Al in the early solar system. *Earth Planet. Sci. Lett.* **10**, 44–59 (1970).

J. S. Schubart, Asteroid masses and densities. NASA SP 267, 33–38 (1971).

J. S. Schubart, The mass of Pallas. *Astron. Astrophys.* **39**, 147–148 (1975).

G. Schubert and R. E. Young, The 4-Day Venus circulation driven by periodic thermal forcing. *J. Atmos. Sci.* **27**, 523–528 (1970).

G. Schubert, R. E. Young, and J. Hinch, Prograde and retrograde motion in a fluid layer: Consequences for thermal diffusion in the Venus atmosphere. *JGR, J. Geophys. Res.* **76**, 2126–2129 (1971).

G. Schubert, C. Covey, A. Del Genio, L. S. Elson, G. Keating, A. Seiff, R. Young, J. Apt, C. Counselman, A. Kliore, S. Limaye, H. Revercomb, L. Sromovsky, V. Suomi, F. Taylor, R. Woo, and U. von Zahn, Structure and circulation of the Venus atmosphere. *JGR, J. Geophys. Res.* **85**, 8007–8025 (1980).

R. D. Schuiling, The Earth as a cosmochemical body. *Meteoritics* **10**, 482–486 (1975).

P. H. Schultz, Endogenic modification of impact craters on Mercury. *Phys. Earth Planet. Inter.* **15**, 202–219 (1977).

P. H. Schultz and H. Glicken, Impact crater and basin control of igneous processes on Mars. *JGR, J. Geophys. Res.* **84**, 8033–8047 (1979).

P. H. Schultz and F. E. Ingersol, Martian lineaments from Mariner 6 and 7 images. *JGR, J. Geophys. Res.* **78**, 8415–8428 (1973).

S. A. Schum, Structural origin of large Martian channels. *Icarus* **22**, 371–384 (1974).

D. W. Schwartzman, Argon degassing models of the Earth. *Nature* (*London*) **245**, 20–21 (1973).

A. A. Scott and E. J. Reese, Venus: Atmospheric rotation. *Icarus* **17**, 589–601 (1972).

P. K. Seidelmann, W. J. Klepczynski, R. L. Duncombe, and E. S. Jackson, Determination of the mass of Pluto. *Astron. J.* **76**, 488–492 (1971).

A. Seiff and D. B. Kirk, Structure of Mars' atmosphere up to 100 km from the entry measurements of Viking 2. *Science* **194**, 1300–1302 (1976).

A. Seiff and D. B. Kirk, Structure of the atmosphere of Mars in summer at mid-latitudes. *JGR, J. Geophys. Res.* **82**, 4364–4378 (1977).

A. Seiff, D. B. Kirk, S. C. Sommer, R. E. Young, R. C. Blanchard, D. W. Juergens, J. E. Lepetich, P. F. Intrievi, J. T. Findlay, and J. S. Derr, Structure of the atmosphere of Venus up to 110 kilometers: Preliminary results from the four Pioneer Venus entry probes. *Science* **203**, 787–790 (1979a).

A. Seiff, D. B. Kirk, R. E. Young, S. C. Sommer, R. C. Blanchard, J. T. Findlay, and G. M. Kelly, Thermal contrast in the atmosphere of Venus: Initial appraisal from Pioneer Venus probe data. *Science* **205**, 46–49 (1979b).

W. Seiler, The cycle of atmospheric CO. *Tellus* **26**, 117–135 (1974).

A. S. Selivanov, M. K. Naraeva, B. A. Suvorov, I. F. Sinel'nikova, and M. I. Bokhomov, Instrumentation and some photographic results obtained with Mars 4 and 5. *Cosmic Res.* (*Engl. Transl.*) **13**, 51–55 (1975).

W. D. Sellers, A new global climatic model. *J. Appl. Meteorol.* **12**, 241–254 (1973).

V. P. Shari, Fluxes of thermal radiation in the lower atmosphere of Venus. *Cosmic Res.* (*Engl. Transl.*) **14**, 86–97 (1976).

R. P. Sharp, Surface processes modifying the Martian craters. *Icarus* **8**, 472–480 (1968).

R. P. Sharp, Mars: Troughed terrain. *JGR, J. Geophys. Res.* **78**, 4063–4072 (1973a).

R. P. Sharp, Mars: Fretted and chaotic terrains. *JGR, J. Geophys. Res.* **78**, 4073–4083 (1973b).

R. P. Sharp, Mars: South polar pits and etched terrain. *JGR, J. Geophys. Res.* **78**, 4222–4230 (1973c).

R. P. Sharp, L. A. Soderblom, B. C. Murray, and J. A. Cutts, The surface of Mars. 2. Uncratered terrains. *JGR, J. Geophys. Res.* **76**, 331–341 (1971a).

R. P. Sharp, B. C. Murray, R. B. Leighton, L. A. Soderblom, and J. A. Cutts, The surface of Mars. 4. South polar cap. *JGR, J. Geophys. Res.* **76**, 357–368 (1971b).

H. N. Sharpe and D. W. Strangway, The magnetic field of Mercury and models of thermal evolution. *Geophys. Res. Lett.* **3**, 285–288 (1976).

Y. Shimazu and T. Urabe, An energetic study of the evolution of the terrestrial and cytherean atmospheres. *Icarus* **9**, 498–506 (1968).

M. Shimizu, Instability of a highly reducing atmosphere on the primitive Earth. *Precambian Res.* **3**, 463–472 (1976).

M. Shimizu, Implications of ^{36}Ar excess on Venus. *Moon Planets* **10**, 317–319 (1979).

J. S. Shirk, W. A. Haseltine, and G. C. Pimentel, Sinton bands: Evidence for deuterated water on Mars. *Science* **147**, 48–49 (1965).

E. M. Shoemaker, Astronomically observable crater-forming projectiles. *In* "Impact and Explosion Cratering" (D. J. Roddy, R. Pepin, and R. B. Merrill, eds.), pp. 617–628. Pergamon, Oxford, 1977.

E. M. Shoemaker and C. J. Lowery, Airwaves associated with large fireballs and the frequency distribution of energy of large meteoroids. *Meteoritics* **3**, 123–124 (1967).

R. W. Shorthill, R. E. Hutton, H. J. Moore, II, and R. F. Scott, Martian physical properties experiments: The Viking Mars lander. *Icarus* **16**, 217–222 (1972).

R. W. Shorthill, R. E. Hutton, H. J. Moore, II, R. F. Scott, and C. R. Spitzer, Physical properties of the Martian surface from the Viking I lander: Preliminary results. *Science* **193**, 805–809 (1976a).

R. W. Shorthill, H. J. Moore, II, R. F. Scott, R. E. Hutton, S. Liebes, Jr., and C. R. Spitzer, The "soil" of Mars (Viking 1). *Science* **194**, 91–97 (1976b).

R. W. Shorthill, H. J. Moore, II, R. E. Hutton, R. F. Scott, and C. R. Spitzer, The environs of Viking 2 lander. *Science* **194**, 1309–1318 (1976c).

U. Siegenthaler and H. Oeschger, Predicting future atmospheric carbon dioxide levels. *Science* **199**, 388–395 (1978).

R. W. Siegfried, II and S. C. Solomon, Mercury: Internal structure and thermal evolution. *Icarus* **23**, 192–205 (1974).

B. Siffert, Genesis and synthesis of clays and clay minerals: Recent developments and future prospects. *Dev. Sedimentol.* **27**, 337–347 (1979) (Int. Clay Conf., Proc., 6th, 1978).

G. T. Sill, Sulfuric acid in the Venus clouds. *Comm. Lunar Planet. Lab.* **9**, 191–197 (1972).

G. T. Sill, The composition of the ultraviolet dark markings on Venus. *J. Atmos. Sci.* **32**, 1201–1204 (1975).

G. T. Sill and L. L. Wilkening, Ice clathrate as a possible source of the atmospheres of the terrestrial planets. *Icarus* **33**, 13–22 (1978).

L. T. Silver and P. H. Schultz, eds., "Geological Implications of Impacts of Large Asteroids and Comets on the Earth," Geol. Soc. Am., Spec. Pap. 190. GSA, Boulder, Colorado, 1982.

A. C. E. Sinclair, J. P. Basart, D. Buhl, and W. A. Gale, Precision interferometric observations of Venus at 11.1 centimeter wavelength. *Astrophys. J.* **175**, 555–572 (1972).

R. B. Singer, T. B. McCord, R. N. Clark, J. B. Adams, and R. L. Huguenin, Mars surface composition from reflectance spectroscopy: A summary. *JGR, J. Geophys. Res.* **84**, 8415–8426 (1979).

W. M. Sinton, Spectroscopic evidence for vegetation on Mars. *Astrophys. J.* **126**, 231–239 (1957).

W. M. Sinton, Further evidence of vegetation on Mars. *Science* **130**, 1234–1237 (1959).

W. M. Sinton, Identification of aldehyde in Mars vegetation regions. *Science* **134**, 529 (1961a).

W. M. Sinton, An upper limit to the concentration of NO_2 and N_2O_4 in the Martian atmosphere. *Publ. Astron. Soc. Pac.* **73**, 125–128 (1961b).

W. M. Sinton, Limb and polar brightening of Uranus at 8870 Å. *Astrophys. J.* **176**, L131–L133 (1972).

W. M. Sinton, Does Io have an ammonia atmosphere? *Icarus* **20**, 284–296 (1973).

W. M. Sinton, Io's 5 micron variability. *Astrophys. J.* **235** L49–L51 (1980).

W. M. Sinton, The thermal emission spectrum of Io and a determination of the heat flux from its hot spots. *JGR, J. Geophys. Res.* **86**, 3122–3128 (1981).

W. M. Sinton, A. T. Tokunaga, E. E. Becklin, I. Gatley, T. J. Lee, and C. J. Lonsdale, Io: Ground-based observations of hot spots. *Science* 210, 1015–1017 (1980).

G. Siscoe and L. Christopher, Variations in the solar wind stand-off distance at Mercury. *Geophys. Res. Lett.* 2, 158–160 (1975).

M. A. Slade and I. I. Shapiro, Interpretation of radar and radio observations of Venus. *JGR, J. Geophys. Res.* 75, 3301–3318 (1970).

W. L. Slattery, Protoplanetary core formation by rain-out of iron drops. *Moon Planets* 19, 443–457 (1978).

V. I. Slysh, Identification of the CO molecule in the radiation spectrum of the Venusian nighttime sky. *Cosmic Res. (Engl. Transl.)* 14, 693–695 (1976).

T. V. Smirnova and A. D. Kus'min, Water-vapor and ammonia abundance in the lower atmosphere of Venus estimated from radar measurements. *Sov. Astron. (Engl. Transl.)* 18, 357–358 (1974).

J. S. Smit and J. Hertogen, An extraterrestrial event at the Cretaceous-Tertiary boundary. *Nature (London)* 285, 198–200 (1980).

B. A. Smith and J. C. Robinson, Mars Mariner IV: Identification of some Martian surface features. *Icarus* 9, 466–473 (1968).

B. A. Smith, E. M. Shoemaker, S. W. Kieffer, and A. F. Cook, II, The role of SO_2 in volcanism on Io. *Nature (London)* 280, 738–743 (1979).

B. A. Smith and 26 others, Encounter with Saturn: Voyager I imaging science results. *Science* 212, 163–191 (1981).

E. I. Smith, Comparison of the crater morphology-size relationship for Mars, Moon, and Mercury. *Icarus* 28, 543–550 (1976).

L. L. Smith and S. H. Gross, The evolution of water vapor in the atmosphere of Venus. *J. Atmos. Sci.* 29, 173–178 (1972).

P. J. Smith, Continental drift changes climate. *Nature (London)* 266, 592–594 (1977).

W. B. Smith, Venus radius controversy. *Science* 169, 1001–1002 (1970).

R. Smoluchowski, Mars: Retention of ice. *Science* 159, 1348–1350 (1967a).

R. Smoluchowski, Internal structure and energy emission of Jupiter. *Nature (London)* 215, 691–695 (1967b).

W. D. Smythe, R. M. Nelson, and D. B. Nash, Spectral evidence for SO_2 frost or adsorbate on Io's surface. *Nature (London)* 280, 766 (1979).

C. W. Snyder, Mariner V flight past Venus. *Science* 158, 1665–1669 (1967).

C. W. Snyder, The missions of the Viking orbiters. *JGR, J. Geophys. Res.* 82, 3971–3984 (1977).

C. W. Snyder, The extended mission of Viking. *JGR, J. Geophys. Res.* 84, 7917–7933 (1979a).

C. W. Snyder, The planet Mars as seen at the end of the Viking mission. *JGR, J. Geophys. Res.* 84, 8487–8519 (1979b).

V. V. Sobolev, An investigation of the atomosphere (sic) of Venus II. *Sov. Astron. (Engl. Transl.)* 12, 135–140 (1968).

L. A. Soderblom, Viking orbital colorimetric images of Mars: Preliminary results. *Science* 194, 97–99 (1976).

L. A. Soderblom, T. J. Kreidler, and H. Masursky, Latitudinal distribution of a debris mantle on the Martian surface. *JGR, J. Geophys. Res.* 78, 4117–4122 (1973a).

L. A. Soderblom, M. C. Malin, J. A. Cutts, and B. C. Murray, Mariner 9 observations of the surface of Mars in the north polar region. *JGR, J. Geophys. Res.* 78, 4197–4209 (1973b).

L. A. Soderblom, C. D. Condit, R. A. West, B. M. Herman, and T. J. Kreidler, Martian planetwide crater distributions: Implications for geologic history and surface processes. *Icarus* 22, 239–263 (1974).

G. A. Soffen, Status of the Viking missions. *Science* 194, 57–58 (1976a).

G. A. Soffen, Scientific results of the Viking missions. *Science* 194, 1274–1276 (1976b).

G. A. Soffen, The Viking Project. *JGR, J. Geophys. Res.* 82, 3959–3970 (1977).

G. A. Soffen and G. W. Snyder, The first Viking mission to Mars. *Science* 193, 759–765 (1976).

G. A. Soffen and A. T. Young, The Viking missions to Mars. *Icarus* 16, 1–16 (1972).

B. T. Soifer, G. Neugebauer, and K. Matthews, The 1.5–2.5 μm spectrum of Pluto. *Astron. J.* **85**, 166–167 (1980).

M. N. Sokolova, M. G. Dobrovol'skaya, N. I. Organova and A. L. Dmitrik, A sulfide of iron and potassium, the new mineral rasvumite. *Zapiski Vses. Mineral. Obshch.* **99**, 712–714 (1970).

S. C. Solomon, Same aspects of core formation in Mercury. *Icarus* **28**, 509–521 (1976).

C. P. Sonett, D. S. Colburn, and K. Schwartz, Electrical heating of meteorite parent bodies and planets by dynamo induction from a pre-Main Sequence T-Tauri "solar wind." *Nature* **219**, 924–926 (1968).

C. P. Sonett, D. S. Colburn, K. Schwartz, and K. Keil, The melting of asteroidal-sized bodies by unipolar dynamo induction from a primordial T-Tauri Sun. *Astrophys. Space Sci.* **7**, 446–488 (1970).

H. Spinrad, Spectroscopic temperature and pressure measurements in the Venus atmosphere. *Publ. Astron. Soc. Pacific* **74**, 187–201 (1962a).

H. Spinrad, A search for water vapor and trace constituents in the Venus atmosphere. *Icarus* **1**, 266–270 (1962b).

H. Spinrad, The nitrogen dioxide content of the Martian atmosphere. *Publ. Astron. Soc. Pacific* **75**, 190–191 (1963).

H. Spinrad, Lack of a noticeable methane atmosphere on Triton. *Publ. Astron. Soc. Pacific* **81**, 895–896 (1969).

H. Spinrad and P. W. Hodge, An explanation of Kozyrev's hydrogen emission lines in the spectrum of Mercury. *Icarus* **4**, 105–108 (1965).

H. Spinrad and E. H. Richardson, High dispersion spectra of the outer planets. II. A new upper limit for the water vapor Content of the Martian atmosphere. *Icarus* **2**, 49–53 (1963).

H. Spinrad, G. Field, and P. W. Hodge, Spectroscopic observations of Mercury. *Astrophys. J.* **141**, 1155–1160 (1965).

H. Spinrad and E. H. Richardson, An upper limit to the molecular oxygen content of the Venus atmosphere. *Astrophys. J.* **141**, 282–286 (1964).

H. Spinrad and S. J. Shawl, Water vapor on Venus: A confirmation. *Astrophys. J.* **146**, 328–329 (1966).

H. Spinrad, G. Münch, and L. D. Kaplan, The detection of water vapor on Mars. *Astrophys. J.* **137**, 1319–1321 (1963).

L. Spitzer Jr., The terrestrial atmosphere above 300 km. In "The Atmospheres of the Earth and Planets" (G. P. Kuiper, ed.), pp. 211–247. Univ. Chicago Press, Chicago, 1952.

S. W. Squyres, Topographic domes on Ganymede: Ice vulcanism or isostatic upwarping. *Icarus* **44**, 472–480 (1980).

B. Šrinivasan, R. S. Lewis, and E. Anders, Noble gases in the Allende and Abee meteorites and a gas-rich mineral fraction: Investigation by stepwise heating. *Geochim. Cosmochim. Acta* **42**, 183–195 (1978).

D. O. Staley, The adiabatic lapse rate in the Venus atmosphere. *J. Atmos. Sci.* **27**, 219–223 (1970).

H. Stauffer, Primordial argon and neon in carbonaceous chondrites and ureilites. *Geochim. Cosmochim. Acta* **24**, 70–82 (1961).

R. H. Steinbacher and S. Z. Gunter, The Mariner Mars 1971 experiments: Introduction. *Icarus* **12**, 3–9 (1970).

R. H. Steinbacher and N. R. Haynes, Mariner 9 mission profile and project history. *Icarus* **18**, 64–74 (1973).

R. H. Steinbacher, A. Kliore, J. Lorell, H. Hipsher, C. A. Barth, H. Masursky, G. Münch, J. Pearl, and B. Smith, Mariner 9 science experiments: Preliminary results. *Science* **175**, 293–294 (1972).

A. Stephenson, Crustal remanence and the magnetic moment of Mercury. *Earth Planet. Sci. Lett.* **28**, 454–458 (1976).

F. J. Stevenson, Chemical state of the nitrogen in rocks. *Geochim. Cosmochim. Acta* **26**, 797–809 (1962).

A. I. Stewart, Mariner 6 and Mariner 7. Ultraviolet spectrometer experiment: Implications of CO_2^+, CO and airglow. *JGR, J. Geophys. Res.* **77**, 54–68 (1972).

A. I. Stewart, D. Anderson, L. Esposito, and C. Barth, Ultraviolet spectroscopy of Venus: Initial results from the Pioneer Venus orbiter. *Science* **203**, 777–779 (1979).

A. I. Stewart, J. Gerard, D. Rusch, and S. Bougher, Morphology of the Venus ultraviolet airglow. *JGR, J. Geophys. Res.* **85**, 7861–7870 (1980).

R. M. Stewart, Composition and temperature of the outer core. *JGR, J. Geophys. Res.* **78**, 2586–2597 (1973).

R. W. Stewart, Interpretations of Mariner 5 and Venera 4 data on the upper atmosphere of Venus. *J. Atmos. Sci.* **25**, 578–581 (1968).

R. S. Stolarki and R. J. Cicerone, Stratospheric chlorine: A possible sink for ozone. *Can. J. Chem.* **52**, 1610–1615 (1974).

E. C. Stone and E. D. Miner, Voyager 1 encounter with the Saturnian system. *Science* **212**, 159–162 (1981).

P. H. Stone, Stability of rotating atmospheres. *J. Atmos. Sci.* **29**, 405–418 (1972).

P. H. Stone, Effects of large-scale eddies on climate change. *J. Atmos. Sci.* **30**, 521–529 (1973).

P. H. Stone, The dynamics of the atmosphere of Venus. *J. Atmos. Sci.* **32**, 1005–1016 (1975).

P. H. Stone, The meteorology of the Jovian atmosphere. *In* "Jupiter: The Giant Planet" (T. Gehrels, ed.), pp. 586–618. Univ. of Arizona Press, Tucson, 1976.

G. M. Strelkov, The radiation of Venus near 21.35 cm. *Sol. Syst. Res. (Engl. Transl.)* **2**, 185–193 (1968).

D. F. Strobel, The photochemistry of methane in the Jovian atmosphere. *J. Atmos. Sci.* **26**, 906–911 (1969).

D. F. Strobel, The photochemistry of hydrocarbons in the Jovian atmosphere. *J. Atmos. Sci.* **30**, 489–497 (1973a).

D. F. Strobel, The photochemistry of NH_3 in the Jovian atmosphere. *J. Atmos. Sci.* **30**, 1205–1209 (1973b).

D. F. Strobel, Hydrocarbon abundances in the Jovian atmosphere. *Astrophys. J.* **192**, L47–L99 (1974a).

D. F. Strobel, Photochemistry of hydrocarbons in the atmosphere of Titan. *Icarus* **21**, 466–470 (1974b).

D. F. Strobel, NH_3 and PH_3 photochemistry in the Jovian atmosphere. *Astrophys. J.* **214**, L97–L99 (1977).

D. F. Strobel, Chemistry and evolution of Titan's atmosphere. *Planet. Space Sci.* **30**, 839–848 (1982).

D. F. Strobel and Y. L. Yung, The Galilean satellites as a source of CO in the Jovian upper atmosphere. *Icarus* **37**, 256–263 (1979).

R. G. Strom, Mercury: A post-Mariner 10 assessment. *Space Sci. Rev.* **24**, 3–70 (1979).

R. G. Strom, B. C. Murray, M. J. S. Belton, G. E. Danielson, M. E. Davies, D. E. Gault, B. Hapke, B. O'Leary, N. Trask, J. E. Guest, J. Anderson, and K. Klaasen, Preliminary imaging results from the second Mercury encounter. *JGR, J. Geophys. Res.* **80**, 2345–2356 (1975a).

R. G. Strom, N. J. Trask, and J. E. Guest, Tectonism and volcanism on Mercury. *JGR, J. Geophys. Res.* **80**, 2478–2507 (1975b).

M. H. Studier, R. Hayatsu, and E. Anders, Origin of organic matter in early solar system. I. Hydrocarbons. *Geochim. Cosmochim. Acta* **18**, 151–173 (1968).

H. E. Suess, Die Häufigkeit der Edelgase auf Erde und im Kosmos. *J. Geol.* **57**, 600–607 (1949).

H. E. Suess, Remarks concerning the chemical composition of the atmosphere of Venus. *Z. Naturforsch., A.* **19A**, 84 (1964).

H. E. Suess, H. Wänke, and F. Wlotyka, On the origin of gas-rich meteorites. *Geochim. Comsochim. Acta* **28**, 595–607 (1964).

S. S. Sun and R. W. Nesbitt, Chemical heterogeneity of the archaean mantle, composition of the Earth and mantle evolution. *Earth Planet. Sci. Lett.* **35**, 429–448 (1977).

Yu. A. Surkov, Geochemical studies of Venus by Venera.9 and 10 automatic interplanetary stations. *Geochim. Cosmochim. Acta, Suppl.* **8**, 2665–2689 (1977).

Yu. A. Surkov, and G. A. Fedoseev, Argon 40 in the Martian atmosphere. *Cosmic Res. (Engl. Transl.)* **14**, 522–526 (1976).

Yu. A. Surkov, F. F. Kirnozov, O. P. Sobornov, G. A. Fedoseev, L. N. Myasnikova, B. N. Kononov,

S. S. Kurochkin, and D. E. Fertman, Instrumentation, experimental procedure and main results of the Gamma-radiation investigation of Venus' surface by the Venera-8 spacecraft. *Cosmic Res. (Engl. Transl.)* **11**, 699–706 (1973a).

Yu. A. Surkov, B. M. Andreychikov, and O. M. Kalinkina, Ammonia content of the Venus atmosphere from Venera 9 planetary probe data. *Dokl. Akad. Nauk. SSSR.* **213**, 296–299 (1973b).

Yu. A. Surkov, F. F. Kirnozov, V. N. Glazov, A. G. Dunchenko, and L. P. Tatsii, The content of natural radioactive elements in Venusian rock as determined by Venera 9 and Venera 10, *Cosmic Res. (Engl. Transl.)* **14**, 618–622 (1976).

Yu. A. Surkov, F. Kirnozov, V. Guryanov, V. Glazov, A. Dunchenko, S. Kurochkin, V. Rasputniy, E. Kharitonova, L. Tatziy, and V. Ghimadov, Venus cloud aerosol investigation on the Venera 12 space probe. *Geokhimiya* No. 1, p. 3 (1981).

Yu. A. Surkov, V. F. Ivanova, A. N. Pudov, V. A. Pavlenko, N. A. Davydov, and D. M. Shejnin, The measurement of water vapor concentration in the Venus atmosphere by Venera 13 and Venera 14. *Pis'ma Astron. Zh.* **8**, 411–413 (1982).

N. D. Sze and M. B. McElroy, Some problems in Venus' aeronomy. *Planet. Space Sci.* **23**, 763–786 (1975).

H. Takahashi, H. Higuchi, J. Gros, J. W. Morgan, and E. Anders, Allende meteorite: Isotopically anomalous xenon is accompanied by normal osmium. *Proc. Natl. Acad. Sci. U.S.A.* **73**, 4253–4256 (1976).

T. Takahashi, Carbon dioxide in the atmosphere and in the Atlantic Ocean water. *JGR, J. Geophys. Res.* **66**, 477–494 (1961).

S. N. Tandon and J. T. Wasson, Gallium, germanium, indium and iridium variations in a suite of L-group chondrites. *Geochim. Cosmochim. Acta* **32**, 1087–1109 (1968).

M. Tatsumoto, U-Th-Pb measurements of luna-20 soil. *Geochim. Cosmochim. Acta* **37**, 1079–1086 (1973).

M. E. Tauber and D. B. Kirk, Impact craters on Venus. *Icarus* **28**, 351–357 (1976).

S. R. Taylor, Abundance of chemical elements in the continental crust—a new table. *Geochim. Cosmochim. Acta* **28**, 1273–1285 (1964a).

S. R. Taylor, Chondritic Earth model. *Nature (London)* **202**, 281–282 (1964b).

S. R. Taylor, Trace element abundances and the chondritic Earth model. *Geochim. Cosmochim. Acta* **28**, 1989–1998 (1964c).

S. R. Taylor, Chemical evidence for lunar melting and differentiation. *Nature (London)* **245**, 203–205 (1973).

S. R. Taylor, "Lunar Science: A Post-Apollo View." Pergamon, Oxford, 1975.

V. G. Teifel, Giant planets. *Astron. Vestn.* **4**, 81–95 (1970).

V. G. Teifel and G. A. Kharitonova, The molecular rotational temperature of Uranus and an upper limit on the pressure in its outer atmosphere. *Sov. Astron. (Engl. Transl.)* **13**, 865–873 (1970).

E. Theilig and R. Greeley, Plains and channels in the Lunae Planium-Chryse Planitia region of Mars. *JGR, J. Geophys. Res.* **84**, 7994–8010 (1979).

G. E. Thomas, Mercury: Does its atmosphere contain water? *Science* **183**, 1197–1198 (1974).

C. H. Thorman and G. G. Goles, The relative age of the transtion zone between Hellas and the Martian cratered terrain. *Earth Planet. Sci. Lett.* **15**, 45–52 (1972).

T. E. Thorpe, Viking Orbiter photometric observations of the Mars phase function July through November 1976. *JGR, J. Geophys. Res.* **82**, 4161–4166 (1977).

C. H. Thurber, A. T. Hsui, and M. N. Toksöz, Ganymede and Callisto: Thermal evolution models and constraints from Voyager data. *Geochim. Cosmochim. Acta, Suppl.* **14**, 1149–1151 (1980).

B. R. Tittmann, Brief note for consideration of active seismic exploration of Mars. *JGR, J. Geophys. Res.* **84**, 7940–7942 (1979).

A. Tokunaga, R. F. Knacke, and T. Owen, The detection of ethane on Saturn. *Astrophys. J.* **197**, L77–L78 (1975).

A Tokunaga, R. F. Kancke, and T. Owen, Ethane and acetylene abundances in the Jovian atmosphere. *Astrophys. J.* **209**, 294–301 (1976).

C. W. Tolbert and A. W. Straiton, A consideration of microwave radiation associated with particles in the atmosphere of Venus. JGR, J. Geophys. Res. 67, 1741–1744 (1962).

H. G. Tolland, Formation of the Earth's core. Nature (London) 243, 141–142 (1973).

H. G. Tolland, Thermal regime in the Earth's core and lower mantle. Phys. Earth Planet. Inter. 8, 282–286 (1974).

R. H. Tolson, T. C. Duxbury, G. H. Born, E. J. Christensen, R. E. Diehl, D. Farless, C. E. Hildebrand, R. T. Mitchell, P. M. Molko, and L. A. Morabito, Viking first encounter of Phobos: Preliminary results. Science 199, 61–64 (1978).

M. G. Tomasko, L. R. Doose, J. Palmer, A. Holmes, W. Wolfe, N. D. Castillo, and P. H. Smith, Preliminary results of the solar flux radiometer experiment aboard the Pioneer Venus multiprobe mission. Science 203, 795–797 (1979a).

M. G. Tomasko, L. R. Doose, and P. H. Smith, Absorption of sunlight in the atmosphere of Venus. Science 205, 80–82 (1979b).

O. B. Toon, J. B. Pollack, and C. Sagan, Physical properties of the particles composing the Martian dust storm of 1971–1972. Icarus 30, 663–696 (1977).

P. Toulmin, III, A. K. Baird, B. C. Clark, K. Keil, and H. J. Rose, Jr., Inorganic chemical investigation by X-ray fluorescence analysis: The Viking Mars lander. Icarus 20, 153–178 (1973).

P. Toulmin, III, B. C. Clark, A. K. Baird, K. Keil, and H. J. Rose, Jr., Preliminary results from the Viking X-ray fluorescence experiment: The first sample from Chryse Planitia, Mars. Science 194, 81–83 (1976).

P. Toulmin, III, A. K. Baird, B. C. Clark, K. Keil, H. J. Rose, Jr., B. P. Christian, P. H. Evans, and W. C. Kelliher, Geochemical and mineralogical interpretation of the Viking inorganic chemical results. JGR, J. Geophys. Res. 82, 4625–4634 (1977).

K. M. Towe, Early Precambrian oxygen: A case against photosynthesis. Nature (London) 274, 657–661 (1978).

L. M. Trafton, Model atmospheres of the major planets. Astrophys. J. 147, 765–781 (1967).

L. M. Trafton, On the possible detection of H_2 in Titan's atmosphere. Astrophys. J. 175, 285–294 (1972a).

L. Trafton, The bulk composition of Titan's atmosphere. Astrophys. J. 175, 295–306 (1972b).

L. Trafton, Titan: Unidentified strong absorptions in the photometric infrared. Icarus 21, 175–187 (1974).

L. Trafton, The morphology of Titan's methane bands. I. Comparison with a reflecting layer model. Astrophys. J. 195, 805–814 (1975a).

L. Trafton, High resolution spectra of Io's sodium emission. Astrophys. J. 202, L107–L111 (1975b).

L. Trafton, Detection of a potassium cloud near Io. Nature (London) 258, 690–692 (1975c).

L. Trafton, The aerosol distribution in Uranus' atmosphere. Interpretation of the hydrogen spectrum. Astrophys. J. 207, 1007–1024 (1976a).

L. Trafton, A search for emission features in Io's extended cloud. Icarus 27, 429–437 (1976b).

L. Trafton, Does Pluto have a substantial atmosphere? Icarus 44, 53–61 (1980).

L. Trafton and W. Macy, Jr., An oscillating asymmetry to Io's sodium emission cloud. Astrophys. J. 202, L155–L158 (1975).

L. M. Trafton and P. H. Stone, Radiative-dynamical equilibrium states for Jupiter. Astrophys. J. 188, 649–655 (1974).

L. Trafton, T. Parkinson, and W. Macy, Jr., The spatial extent of sodium emission around Io. Astrophys. J. 190, L85–L90 (1974).

N. J. Trask and J. E. Guest, Preliminary geologic terrain map of Mercury. JGR, J. Geophys. Res. 80, 2461–2476 (1975).

N. J. Trask and R. G. Strom, Additional evidence of Martian volcanism. Icarus 28, 559–563 (1976).

W. A. Traub and N. P. Carleton, Observations of O_2, H_2O and HD in planetary atmospheres. In "Exploration of the Planetary System" (A. Woszczyk and C. Iwaniszewska, eds.), p. 223. Reidel Publ., Dordrecht, Netherlands, 1974.

W. A. Traub and N. P. Carleton, Spectroscopic observations of winds on Venus. *J. Atmos. Sci.* **32**, 1045–1059 (1975).

J. T. Trauger, F. L. Roesler, N. P. Carleton, and W. A. Traub, Observation of HD on Jupiter and the D/H ratio. *Astrophys. J.* **184**, L137–L140 (1973).

L. D. Travis, D. L. Coffeen, J. E. Hansen, K. Kawabata, A. A. Lacis, W. A. Lane, S. S. Limaye, and P. H. Stone, Orbiter cloud photopolarimeter investigation. *Science* **203**, 781–785 (1979).

R. R. Treffers, H. P. Larson, U. Fink, and T. N. Gautier, Upper limits to trace constituents in Jupiter's atmosphere from an analysis of its 5-μm spectrum. *Icarus* **34**, 331–343 (1978).

S. S. Tseng and S. Chang, Photo-induced free radicals on a simulated Martian surface. *Nature (London)* **248**, 575–577 (1974).

R. G. Tull, High-dispersion spectroscopic observations of Mars. IV. The latitude distribution of atmospheric water vapor. *Icarus* **13**, 43–57 (1970).

V. S. Tuman, Speculation on the origin and non-symmetry of two thin layers D and F embracing the liquid core inside the Earth. *Nature (London)* **204**, 1146–1148 (1964).

R. P. Turco, O. B. Toon, C. Park, R. C. Whitten, J. B. Pollack, and P. Noerdlinger, Tunguska meteor fall of 1908: Effects on stratospheric ozone. *Science* **214**, 19–23 (1981).

R. P. Turco, O. B. Toon, R. C. Whitten, R. G. Keesee, and D. Hollenbach, "Noctilucent Clouds: Simulation Studies of their Genesis, Properties, and Global Influences," Preprint (1982).

K. K. Turekian, Degassing of argon and helium from the Earth. *In* "The Origin and Evolution of Atmospheres and Oceans" (P. J. Brancazio and A. G. W. Cameron, eds.), pp. 74–82. New York, 1963.

K. K. Turekian and S. P. Clark, Jr., Inhomogeneous accumulation of the Earth from the primitive solar nebula. *Earth Planet. Sci. Lett.* **6**, 346–348 (1969).

G. L. Tyler, V. R. Eshleman, J. D. Anderson, G. S. Levy, G. F. Lindal, G. E. Wood, and T. A. Croft, Radio science investigations of the Saturn system with Voyager I: Preliminary results. *Science* **212**, 201–206 (1981).

R. K. Ulich, An infall model of the T-Tauri phenomenon. *Astrophys. J.* **210**, 337–391 (1976).

H. C. Urey, Structure and chemical composition of Mars. *Phys. Rev.* **80**, 295 (1950).

H. C. Urey, The origin and development of the Earth and other terrestrial planets. *Geochim. Cosmochim. Acta* **1**, 209–277 (1951).

H. C. Urey, The origin and development of the Earth and other terrestrial planets: A correction. *Geochim. Cosmochim. Acta* **2**, 263–268 (1952a).

H. C. Urey, Chemical fractionation in the meteorites and the abundance of the elements. *Geochim. Cosmochim. Acta* **2**, 269–282 (1952b).

H. C. Urey, "The Planets: Their Origin and Development." Yale Univ. Press, New Haven, Connecticut, 1952c.

H. C. Urey, The cosmic abundances of potassium, uranium and thorium and the heat balances of the Earth, the Moon and Mars. *Proc. Nat. Acad. Sci. USA* **41**, 127–144 (1955).

H. C. Urey, Primitive planetary atmospheres and the origin of life. *In* "The Origin of Life on the Earth," pp. 1–16. Macmillan, New York, 1959.

H. C. Urey, Evidence regarding the origin of the Earth. *Geochim. Cosmochim. Acta* **26**, 1–14 (1962).

H. C. Urey, Some general problems relative to the origin of life on Earth or elsewhere. *Am. Nat.* **100**, 285–288 (1966).

H. C. Urey, Was the moon originally cold? *Science* **172**, 403–404 (1971).

H. C. Urey, Cometary collisions and geological periods. *Nature (London)* **242**, 32–33 (1973).

H. C. Urey and H. Craig, The composition of the stone meteorites and the origin of meteorites. *Geochim. Cosmochim. Acta* **4**, 36–82 (1953).

E. A. Ustinov, V. S. Zhegulev, L. V. Zasova, and V. I. Moroz, Model of spectral and altitude distribution of fluxes of scattered solar radiation in the atmosphere of Venus. *Cosmic Res. (Engl. Transl.)* **17**, 67–76 (1979).

V. M. Vakhnin, A review of the Venera 4 flight and its scientific program. *J. Atmos. Sci.* **25**, 533–594 (1968).

V. M. Vakhnin and A. I. Lebedinskii, The nature of radio emission from the surface of Venus. *Cosmic Res.* (*Engl. Transl.*) **3**, 758 (1965).

J. A. Van Allen, S. M. Krimigis, L. A. Frank, and T. P. Armstrong, Venus: An upper limit on intrinsic magnetic dipole moment based on absence of a radiation belt. *Science* **158**, 1673–1675 (1967).

W. R. Van Schmus and J. A. Wood, A chemical petrologic classification for the chondritic meteorites. *Geochim. Cosmochim. Acta* **31**, 747–766 (1967).

R. A. Van Tassel and J. W. Salisbury, The composition of the Martian surface. *Icarus* **3**, 264–269 (1964).

L. Van Valen, The history and stability of atmospheric oxygen. *Science* **171**, 439–443 (1971).

G. P. Vdovykin and C. B. Moore, Carbon (6). *In* "Handbook of Elemental Abundances in Meteorites" (B. Mason, ed.), pp. 81–91. Gordon Breach, New York, 1971.

I. Velikovsky, "Worlds in Collision." Doubleday, New York, 1950.

H. G. Vera Ruiz and F. S. Rowland, Possible scavenging reactions of C_2H_2 and C_2H_4 for phosphorus-containing radicals in the Jovian atmosphere. *Geophys. Res. Lett.* **5**, 407–410 (1978).

J. P. Verdet, Y. Zéau, J. Gay, T. Encrenaz, and F. Sérre, The spectrum of Mars between 8 and 13 microns. *Astron. Astrophys.* **19**, 159–163 (1972).

J. Verhoogen, Thermal regime of the Earth's core. *Phys. Earth Planet. Inter.* **7**, 47–58 (1973).

Yu. N. Vetukhnovskaya and A. D. Kuz'min, Two-layer model of the atmosphere of Venus (interpretation of radar observations). *Sol. Syst. Res.* (*Engl. Transl.*) **1**, 69–71 (1967a).

Yu. N. Vetukhnovskaya and A. D. Kuz'min, The planet Mercury. *Sol. Syst. Res.* (*Engl. Transl.*) **1**, 152–163 (1967b).

Yu. N. Vetukhnovskaya and A. D. Kuz'min, The planet Venus. *Sol. Syst. Res.* (Engl. Transl.) **4**, 5–14 (1970).

Yu. N. Vetukhnovskaya, A. D. Kuz'min, A. P. Naumov, and T. V. Smirnova, Determination of the parameters of the Venusian atmosphere at mean surface level from radioastronomical and radar measurements. *Sov. Astron.* (*Engl. Transl.*) **15**, 113–120 (1971).

J. Veverka, A polarimetric search for a Venus halo during the 1969 inferior conjunction. *Icarus* **14**, 282–283 (1971).

J. Veverka, Titan: Polarimetric evidence for an optically thick atmosphere? *Icarus* **18**, 657–660 (1973a).

J. Veverka, Polarimetric observations of 9 Metis, 15 Eunomia, 89 Julia, and other asteroids. *Icarus* **19**, 114–117 (1973b).

J. Veverka and M. Noland, Asteroidal reflectivities from polarization curves: Calibration of the "slope-albedo" relationship. *Icarus* **19**, 230–239 (1973).

J. Veverka, J. Goguen, S. Yang, and J. E. Elliott, On matching the spectrum of Io: Variations in the photometric properties of sulfur-containing mixtures. *Icarus* **37**, 249–255 (1979).

P. Vidal and L. Dosso, Core formation: Catastrophic or continuous? Sr and Pb isotope geochemistry constraints. *Geophys. Res. Lett.* **5**, 169–172 (1978).

F. Vilas and T. B. McCord, Mercury: Spectral reflectance measurements (0.33–1.06 μm) 1974/1975. *Icarus* **28**, 593–599 (1976).

A. P. Vinogradov, Yu. A. Surkov, K. P. Florensky, and B. M. Andreichikov, Determination of the chemical composition of the Venus atmosphere by the Venera IV space probe. *Dokl. Adad. Nauk SSSR* **179**, 37 (1968a).

A. P. Vinogradov, Yu. A. Surkov, and C. P. Florensky, The chemical composition of the Venus atmosphere based on the data of the interplanetary station Venera 4. *J. Atmos. Sci.* **25**, 535–536 (1968b).

A. P. Vinogradov, Yu. A. Surkov, and B. M. Andreichikov, Research on the composition of the atmosphere of Venus aboard automatic interplanetary stations Venera V and Venera VI. *Dokl. Akad. Nauk SSSR* **190**, 552 (1970a).

A. P. Vinogradov, Y. A. Surkov, B. M. Andreichikov, O. M. Kalinkina, and I. M. Grechishcheva, Chemical composition of the atmosphere of Venus. *Cosmic Res.* (*Engl. Transl.*) **8**, 533–540 (1970b).

A. P. Vinogradov, Yu. A. Surkov, and F. F. Kirmazov, The content of uranium, thorium and potassium in the rocks of Venus as measured by Venera 8. *Icarus* **20**, 253–259 (1973).

G. Visconti, Exospheric temperature of a Primitive terrestrial atmosphere with evolving oxygen content. *J. Atmos. Sci.* **32**, 1631–1637 (1975).

G. Visconti, Hydrogen escape in a terrestrial atmosphere at low oxygen levels. *J. Atmos. Sci.* **34**, 193–204 (1977).

W. V. Vishniac and G. A. Welty, Light scattering experiment: The Viking Mars lander. *Icarus* **16**, 185–195 (1972).

V. P. Volkov, J. L. Khodakovskii, V. A. Dorofeeva, and V. L. Barsikov, Main physico-chemical factors controlling the chemical composition of the clouds of the planet Venus. *Geokhimiya* pp. 1759–1766 (1979).

C. F. von Weizsäcker, Über die Möglichkeit eines dualen β-Zerfalls von Kalium. *Phys. Z.* **38**, 623–624 (1937).

U. von Zahn, D. Krankowsky, K. Mauersberger, A. O. Nier, and D. M. Hunten, Venus thermosphere: *In situ* composition measurements the temperature profile, and the homopause altitude. *Science* **203**, 768–770 (1979a).

U. von Zahn, K. H. Fricke, H. J. Hoffman, and K. Pelka, Venus: Eddy coefficients in the thermosphere and the inferred helium content of the lower atmosphere. *Geophys. Res. Lett.* **6**, 337–341 (1979b).

U. von Zahn, S. Kumar, H. Niemann, and R. Prinn, Composition of the Venus atmosphere. *In* "Venus" (D. M. Hunten and L. Colin, eds.), pp. 299–430. Univ. of Arizona Press, Tuscon, 1982.

F. A. Wade and J. N. de Wys, Permafrost features on the Martian surface. *Icarus* **9**, 175–185 (1968).

J. C. G. Walker, Evolution of the atmosphere of Venus. *J. Atmos. Sci.* **32**, 1248–1256 (1975).

J. C. G. Walker, "Evolution of the Atmosphere." Macmillan, New York, 1977.

J. C. G. Walker, K. K. Turekian, and D. M. Hunten, An estimate of the present-day deep mantle degassing rate from data on the atmosphere of Venus. *JGR, J. Geophys. Res.* **75**, 3558–3561 (1970).

L. Wallace, Analysis of the Lyman- α observations of Venus made from Mariner 5. *JGR, J. Geophys. Res.* **74**, 115–131 (1969).

L. Wallace, On the thermal structure of Uranus. *Icarus* **25**, 538–544 (1975).

L. Wallace and D. M. Hunten, The Jovian spectrum in the region 0.4–1.1 μm: The C/H ratio. *Rev. Geophys. Space Phys.* **16**, 289–319 (1978).

L. Wallace, F. E. Stuart, R. H. Nagel, and M. D. Larson, A search for deuterium on Venus. *Astrophys. J.* **168**, L29–L31 (1971).

L. S. Walter, Petrology of Venus—further deductions. *Science* **143**, 1161–1163 (1964).

W. Wamsteker, R. L. Kroes, and J. A. Fountain, On the surface composition of Io. *Icarus* **23**, 417–424 (1974).

H. Wänke and G. Dreibus, The Earth-Moon system: Chemistry and origin. *In* "Origin and Distribution of the Elements" (L. H. Ahrens, ed.), pp. 99–109. Pergamon, Oxford, 1979.

D. Ward and B. O'Leary, Search for a Venus halo effect during 1970. *Icarus* **16**, 314–317 (1972).

W. R. Ward, Climatic variations on Mars. I. Astronomical theory of insolation. *JGR, J. Geophys. Res.* **79**, 3375–3386 (1974).

W. R. Ward, B. C. Murray, and M. C. Malin, Climatic variations on Mars. II. Evolution of carbon dioxide atmosphere and polar caps. *JGR, J. Geophys. Res.* **79**, 3387–3395 (1974).

P. Warneck, Cosmic radiation as a source of odd nitrogen in the stratosphere. *JGR, J. Geophys. Res.* **77**, 6589–6594 (1972).

P. Warneck and F. F. Marmo, NO_2 in the Martian atmosphere. *J. Atmos. Sci.* **20**, 236–240 (1963).

P. Warneck, F. F. Marmo, and J. O. Sullivan, Ultraviolet absorption of SO_2-dissociation energies of SO_2 and SO. *J. Chem. Phys.* **40**, 1132–1140 (1964).

H. S. Washington, The chemical composition of the Earth. *Am. J. Sci.* **9**, 351 (1925).

G. J. Wasserburg, Comments on the outgassing of the Earth. *In* "The Origin and Evolution of Atmospheres and Oceans" (P. J. Brancazio and A. G. W. Cameron, eds.), pp. 83–84. Wiley, New York, 1963.

G. J. Wasserburg, G. J. F. MacDonald, F. Hoyle, and W. A. Fowler, Relative contributions of uranium, thorium and potassium to heat production in the Earth. *Science* **143**, 465–467 (1964).

L. H. Wasserman and 17 others, The diameter of Pallas from its occultation of SAO 85009. *Astron. J.* **84**, 259–268 (1979).

J. T. Wasson, Primordial rare gases in the atmosphere of the Earth. *Nature (London)* **223**, 163–165 (1969).

J. T. Wasson, Volatile elements on the Earth and the moon. *Earth Planet. Sci. Lett.* **11**, 219–225 (1971).

J. T. Wasson, "Meteorites." Springer-Verlag, Berlin and New York (1974).

J. T. Wasson and C. E. Junge, Terrestrial accretion from the solar wind. *Nature (London)* **94**, 41–42 (1962).

A. J. Watson, T. M. Donahue, D. H. Stedman, R. G. Knollenberg, B. Ragent, and J. Blamont, Oxides of nitrogen and the clouds of Venus. *Geophys. Res. Lett.* **6**, 743–746 (1979).

A. J. Watson, T. Donahue, and J. Walker, The dynamics of a rapidly escaping atmosphere: Applications to the evolution of Earth and Venus. *Icarus* **48**, 150–166 (1981).

K. Watson, B. C. Murray, and H. Brown, On the possible presence of ice on the moon. *JGR, J. Geophys. Res.* **66**, 1598–1600 (1961a).

K. Watson, B. C. Murray, and H. Brown, Stability of volatiles in the solar system. *JGR, J. Geophys. Res.* **66**, 3033–3046 (1961b).

J. Weertman, Venus: Ice sheets. *Science* **160**, 1473–1474 (1968).

P. A. Wehinger, S. Wyckoff, and A. Frohlich, Mapping of the sodium emission associated with Io and Jupiter. *Icarus* **27**, 425–428 (1976).

S. J. Weidenschilling, A model for accretion of the terrestrial planets. *Icarus* **22**, 426–435 (1974).

S. J. Weidenschilling, Mass loss from the region of Mars and the asteroid belt. *Icarus* **26**, 361–366 (1975).

S. J. Weidenschilling, Accretion of the terrestrial planets. II. *Icarus* **27**, 161–170 (1976).

S. J. Weidenschilling, The distribution of mass in the planetary system and solar nebula. *Astrophys. Space Sci.* **51**, 153–158 (1977).

S. J. Weidenschilling, Iron/silicate fractionation and the origin of Mercury. *Icarus* **35**, 99–111 (1978).

S. J. Weidenschilling and J. S. Lewis, Atmospheric and cloud structures of the Jovian planets. *Icarus* **20**, 465–476 (1973).

W. J. Welch and D. G. Rea, Upper limits on liquid water in the Venus atmosphere. *Astrophys. J.* **148**, L151–L153 (1967).

R. A. Wells, Martian topography: Large-scale variations. *Science* **166**, 862–865 (1969).

R. A. Wells, Martian topography from radar observations and the Mariner 6 and 7 ground based CO_2 measurements. *Sov. Astron. (Engl. Transl.)* **16**, 490–495 (1972).

M. West, Martian volcanism: Additional observations and evidence for pyroclastic activity. *Icarus* **21**, 1–11 (1974).

J. A. Westphal, K. Matthews, and R. J. Terrile, Five-micron pictures of Jupiter. *Astrophys. J.* **188**, L111–L112 (1974).

R. T. Wetherald and S. Manabe, Influence of seasonal variation upon the sensitivity of a model climate. *JGR, J. Geophys. Res.* **86**, 1194–1204 (1981).

G. W. Wetherill, Relationships between orbits and sources of chondritic meteorites. *In* "Meteorite Research" (P. M. Millman, ed.), pp. 573–589. Reidel Publ., Dordrecht, Netherlands, 1969.

G. W. Wetherill, Late heavy bombardment of the moon and terrestrial planets. *Geochim. Cosmochim. Acta, Suppl.* **6**, 1539–1561 (1975).

G. W. Wetherill, Where do meteorites come from? A re-evaluation of the Earth-crossing Apollo-objects as sources of chondritic meteorites. *Geochim. Cosmochim. Acta* **40**, 1297–1317 (1976).

G. W. Wetherill, Could the solar wind have been the source of the high concentration of ^{36}Ar in the atmosphere of Venus? *Geochim. Cosmochim. Acta, Suppl.* **14**, 1239–1241 (1980).

G. W. Wetherill, Solar wind origin of ^{36}Ar on Venus. *Icarus* **46**, 70–80 (1981).

F. J. W. Whipple, The great Siberian meteor and the waves, seismic and aerial, which it produced. *Q. J. R. Meteorol. Soc.* **56**, 287–304 (1930).

F. L. Whipple, The meteoritic environment of the moon. *Proc. R. Soc. London* **296**, 3–15 (1967).

F. L. Whipple, A speculation about comets and the Earth. *Mem. Soc. R. Sci. Liege, Collect.* 6° **9**, 101–111 (1976).

A. B. Whitehead, The elevation of Olympus Mons from limb photography. *Icarus* **22**, 189–196 (1974).

A. G. Whittaker, E. J. Watts, R. S. Lewis, and E. Anders, Carbynes: Carriers of primordial noble gases in meteorites. *Science* **209**, 1512–1514 (1980).

R. C. Whitten, Thermal structure of the ionsosphere of Venus. *JGR, J. Geophys. Res.* **74**, 5623–5628 (1969).

R. C. Whitten, The daytime upper inosphere of Venus. *JGR, J. Geophys. Res.* **75**, 3707–3714 (1970).

R. C. Whitten, R. T. Reynolds, and P. F. Michelson, The ionosphere and atmosphere of Io. *Geophys. Res. Lett.* **2**, 49–51 (1975).

R. C. Whitten, L. A. Capone, L. McCulley, and P. F. Michelson, The upper ionosphere of Titan. *Icarus* **31**, 89–96 (1977).

M. A. Whyte, Turning point in phanerozoic history. *Nature (London)* **267**, 679–682 (1977).

H. B. Wiik, The chemical composition of some stony meteorites. *Geochim. Cosmochim. Acta* **9**, 279–289 (1956).

R. Wildt, Ozon and Sauerstoff in den Planetenatmosphären. *Nachr. Ges. Wiss. Goettingen* **1**, 1–9 (1934a).

R. Wildt, The atmospheres of the giant Planets. *Nature (London)* **134**, 418 (1934b).

R. Wildt, Über den innern Aufbau der Grossen Planeten. *Nachr. Ges. Wiss. Goettingen* **1**, 67–78 (1934c).

R. Wildt, Photochemistry of planetary atmospheres. *Astrophys. J.* **86**, 321–336 (1937).

R. Wildt, On the state of matter in the interior of the Planets. *Astrophys. J.* **87**, 508–516 (1938).

R. Wildt, Note on the surface temperature of Venus. *Astrophys. J.* **91**, 266–268 (1940a).

R. Wildt, On the possible existence of formaldehyde in the atmosphere of Venus. *Astrophys. J.* **92**, 247–255 (1940b).

R. Wildt, On the chemistry of the atmosphere of Venus. *Astrophys. J.* **96**, 312–314 (1942).

D. E. Wilhelms, Comparison of Martian and lunar multiringed circular basins. *JGR, J. Geophys. Res.* **78**, 4084–4094 (1973).

D. E. Wilhelms, Comparison of Martian and lunar geologic provinces. *JGR, J. Geophys. Res.* **79**, 3933–3941 (1974).

D. E. Wilhelms, Mercurian volcanism questioned. *Icarus* **28**, 551–558 (1976).

E. M. Wilkens, Seasonal variations in atmospheric carbon dioxide concentration. *JGR, J. Geophys. Res.* **66**, 1314–1315 (1961).

H. C. Willett and F. Sanders, "Descriptive Meteorology." Academic Press, New York, 1959.

I. P. Williams, Atmosphere of Mercury. *Nature (London)* **249**, 234 (1974).

J. G. Williams, Secular perturbations in the solar system. Ph.D. Dissertation, University of California at Los Angeles, 1969.

J. G. Williams, Meteorites from the asteroid belt? *Bull. Am. Geophys. Union* **54**, 233 (1973).

J. Winick and A. Stewart, Photochemistry of SO_2 in Venus' upper atmosphere cloud layer. *JGR, J. Geophys. Res.* **85**, 7849–7860 (1980).

D. U. Wise, An origin of the moon by rotational fission during formation of the Earth's core. *JGR, J. Geophys. Res.* **68**, 1547–1554 (1963).

D. U. Wise, M. P. Golombek, and G. E. McGill, Tectonic evolution of Mars. *JGR, J. Geophys. Res.* **84**, 7934–7939 (1979).

F. Woeller and C. Ponnamperuma, Organic synthesis in a simulated Jovian atmosphere. *Icarus* **10**, 386–392 (1969).

S. C. Wofsy and N. D. Sze, Venus cloud models. *In* "Atmospheres of Earth and the Planets" (B. M. McCormac, ed.), pp. 369–389. Reidel, Dordrecht, The Netherlands, 1975.

S. C. Wofsy, J. C. McConnell, and M. B. McElroy, Atmospheric CH_4, CO and CO_2. *JGR, J. Geophys. Res.* **77**, 4477–4484 (1972).

S. C. Wofsy, M. B. McElroy and N. D. Sze, Freon consumption: implications for atmospheric ozone, *Science* **187**, 535–537 (1975).

S. C. Wofsy, M. B. McElroy, and Y. L. Yung, The chemistry of atmospheric bromine. *Geophys. Res. Lett.* **2**, 215–218 (1975).

P. M. Woiceshyn, Global seasonal atmospheric fluctuations on Mars. *Icarus* **22**, 325–344 (1974).

R. Wolf and E. Anders, Moon and Earth: Compositional differences inferred from siderophiles, volatiles and alkalis in basalts. *Geochim. Cosmochim. Acta* **44**, 2111–2124 (1980).

J. Wolfe, D. S. Intriligator, J. Mihalov, H. Collard, D. McKibbin, R. Whitten, and A. Barnes, Initial observations of the Pioneer Venus orbiter solar wind plasma experiment. *Science* **203**, 750–752 (1979).

R. Woo and A. Ishimaru, Eddy diffusion coefficient for the atmosphere of Venus estimated from radio scintillation measurements. *Nature (London)* **289**, 383–384 (1981).

A. T. Wood, Jr., R. B. Wattson, and J. B. Pollack, Venus: Estimates of the surface temperature and pressure from radio and radar measurements. *Science* **162**, 114–116 (1968).

P. R. Woodward, Theoretical models of star formation. *Annu. Rev. Astron. Astrophys.* **16**, 555–584 (1978).

E. L. Wright, Recalibration of the far- infrared brightness temperatures of the planets. *Astrophys. J.* **210**, 250–253 (1976).

S. S. C. Wu, Photogrammetric portrayal of Mars topography. *JGR, J. Geophys. Res.* **84**, 7955–7960 (1979).

S. S. C. Wu, F. J. Schafer, G. M. Nakata, R. Jordan, and K. R. Blasius, Photometric evaluation of Mariner 9 photography. *JGR, J. Geophys. Res.* **78**, 4405–4410 (1973).

S. Yanagita and M. Imamura, Excess ^{15}N in the Martian atmosphere and cosmic rays in the early solar system. *Nature (London)* **274**, 234–235 (1978).

A. T. Young, Are the clouds of Venus sulfuric acid? *Icarus* **18**, 564–582 (1973).

A. T. Young, Is the four-day "rotation" of Venus illusory? *Icarus* **24**, 1–10 (1975).

A. T. Young, An improved Venus cloud model. *Icarus* **24**, 1–10 (1977).

A. T. Young, Chemistry and thermodynamics of sulfur on Venus. *Geophys. Res. Lett.* **6**, 49–50 (1979a).

A. T. Young, CS_2: Candidate for the 3150 Å Venus band. *Icarus* **37**, 297–300 (1979b).

L. D. G. Young, Interpretation of high-resolution spectra of Venus. I. The $2\nu_3$ band of $^{13}C^{16}O^{18}O$ at 2.21 microns. *Icarus* **11**, 66–75 (1969a).

L. D. G. Young, Interpretation of high resolution spectra of Mars I. CO_2 abundance and surface pressure derived from the curve of growth. *Icarus* **11**, 386–389 (1969b).

L. D. G. Young, Interpretation of high-resolution spectra of Venus. II. The (102–000) band of $^{12}C^{16}O^{18}O$ at 1.71 microns. *Icarus* **13**, 270–275 (1970).

L. D. G. Young, High resolution spectra of Venus—a review. *Icarus* **17**, 632–658 (1972).

L. D. G. Young and A. T. Young, Comments on "The composition of the Venus cloud tops in light of recent spectroscopic data." *Astrophys. J.* **179**, L39–L42 (1973).

L. D. G. Young and A. T. Young, Interpretation of high-resolution spectra of Mars. IV. New calculations of the CO abundance. *Icarus* **30**, 75–79 (1977).

L. D. G. Young, R. A. J. Schorn, and E. S. Barker, High-dispersion spectroscopic observations of Venus. VII. The carbon dioxide band at 10,488 Å. *Icarus* **13**, 58–73 (1970a).

L. D. G. Young, R. A. J. Schorn, and H. J. Smith, High-dispersion spectroscopic observations of Venus. IX. The carbon dioxide bands at 12030 and 12177 Å. *Icarus* **13**, 74–81 (1970b).

L. D. G. Young, A. T. Young, W. J. Young, and J. T. Bergstralh, The planet Venus: A new periodic spectrum variable. *Astrophys. J.* **181**, L5–L8 (1973).

L. D. G. Young, A. T. Young, and A. Woszczyk, High dispersion observations of Venus during 1972: The CO_2 band at 7820 Å. *Icarus* **25**, 239–267 (1975).

L. D. G. Young, R. A. J. Schorn, and A. T. Young, High dispersion spectroscopic observations of Venus near superior conjunction. I. The carbon dioxide band at 7820 Å. *Icarus* **30**, 559–565 (1977).

R. Young and J. Pollack, A three-dimensional model of dynamical processes in the Venus atmosphere. *J. Atmos. Sci.* **34**, 1315–1351 (1977).

R. Young and J. Pollack, Reply. *J. Atmos. Sci.* **37**, 253–255 (1980).

R. E. Young and G. Schubert, Dynamical aspects of the Venus 4-day circulation. *Planet. Space Sci.* **21**, 1563–1580 (1973).

R. E. Young and G. Schubert, Temperatures inside Mars: Is the core liquid or solid? *Geophys. Res. Lett.* **1**, 157–160 (1974).

Y. L. Yung and M. B. McElroy, Stability of an oxygen atmosphere on ganymede. *Icarus* **30**, 97–103 (1977).

Y. L. Yung and J. P. Pinto, Primitive atmosphere and implications for the formation of channels on Mars. *Nature (London)* **273**, 730–732 (1978).

A. Zachs and A. K. Fung, Radar observations of Mars. *Space Sci. Rev.* **10**, 442–454 (1969).

A. Zaikowski and O. A. Schaeffer, Incorporation on noble gases during synthesis of serpentine and implications for meteoritical noble gas abundances. *Meteoritics* **11**, 394–395 (1976).

H. S. Zapolsky and E. E. Salpeter, The mass-radius relationship for cold spheres of low mass. *Astrophys. J.* **158**, 807–813 (1969).

M. Zeilik and A. Dalgarno, Ultraviolet argon dayglow lines in the atmosphere of Mercury. *Planet. Space Sci.* **21**, 383–389 (1973).

B. Zellner, The polarization of Titan. *Icarus* **18**, 661–664 (1973a).

B. Zellner, Minor planets and related objects. VIII. Deimos. *Astron. J.* **77**, 183–185 (1973b).

B. Zellner, Asteroid taxonomy and the distribution of the compositional types. *In* "Asteroids" (T. Gehrels, ed.), pp. 783–806. Univ. of Arizona Press, Tucson, 1979.

B. Zellner and E. Bowell, Asteroid compositional types and their distributions. *In* "Comets, Asteroids, Meteorites" (A. H. Delsemme, ed.), pp. 185–197. Univ. of Toledo Press, Toledo, Ohio, 1977.

B. Zellner, T. Gehrels, and J. Gradie, Minor planets and related objects. XVI. Polarimetric diameters. *Astron. J.* **79**, 1100–1110 (1974).

B. Zellner, T. Lebertre, and K. Day, The asteroid albedo scale. II. Laboratory polarimetry of dark carbon-bearing silicates. *Geochim. Cosmochim. Acta, Suppl.* **8**, 1111–1117 (1977).

V. N. Zharkov and V. P. Trubitsyn, Internal constitution and the figures of the giant planets. *Phys. Earth Planet. Inter.* **8**, 105–107 (1974).

J. R. Zimbelman and H. H. Kieffer, Thermal mapping of the northern equatorial and temperate latitudes of Mars. *JGR, J. Geophys. Res.* **84**, 8239–8251 (1979).

P. D. Zimmerman and G. W. Wetherill, Asteroidal source of meteorites. *Science* **182**, 51–53 (1973).

S. Zisk and P. J. Mouginis-Mark, Anomalous region on Mars: Implications for near-surface liquid water. *Nature (London)* **288**, 126–129 (1980).

S. Zohar and R. M. Goldstein, Venus map: A detailed look at the feature β. *Nature (London)* **219**, 357–358 (1968).

Index